PRINCIPLES OF
TUMORS

PRINCIPLES OF TUMORS

A Translational Approach to Foundations

Leon P. Bignold

ELSEVIER

AMSTERDAM • BOSTON • HEIDELBERG • LONDON
NEW YORK • OXFORD • PARIS • SAN DIEGO
SAN FRANCISCO • SINGAPORE • SYDNEY • TOKYO

Academic Press is an imprint of Elsevier

Academic Press is an imprint of Elsevier
125, London Wall, EC2Y 5AS
525 B Street, Suite 1800, San Diego, CA 92101-4495, USA
225 Wyman Street, Waltham, MA 02451, USA
The Boulevard, Langford Lane, Kidlington, Oxford OX5 1GB, UK

Notices
Knowledge and best practice in this field are constantly changing. As new research and experience broaden
our understanding, changes in research methods, professional practices, or medical treatment may become
necessary.

Practitioners and researchers may always rely on their own experience and knowledge in evaluating and
using any information, methods, compounds, or experiments described herein. In using such information or
methods they should be mindful of their own safety and the safety of others, including parties for whom they
have a professional responsibility.

To the fullest extent of the law, neither the Publisher nor the authors, contributors, or editors, assume any
liability for any injury and/or damage to persons or property as a matter of products liability, negligence
or otherwise, or from any use or operation of any methods, products, instructions, or ideas contained in the
material herein.

Library of Congress Cataloging-in-Publication Data
A catalog record for this book is available from the Library of Congress

British Library Cataloguing-in-Publication Data
A catalogue record for this book is available from the British Library

ISBN: 978-0-12-801565-0

For information on all publications
visit our website at http://store.elsevier.com

Working together
to grow libraries in
developing countries

www.elsevier.com • www.bookaid.org

Publisher: Mica Haley
Acquisition Editor: Catherine Van Der Laan
Editorial Project Manager: Lisa Eppich
Production Project Manager: Melissa Read
Designer: Mark Rogers

Contents

Preface and Acknowledgments

Tumors comprise a large number of complex diseases. Understanding their natures and developing adequate therapies for them have long been recognized as among the greatest challenges to medical science.

In recent decades, there has been enormous growth in our knowledge of all aspects of tumors. These include, particularly, the basic biological abnormalities in tumors, the genomic disturbances in tumor cells, as well as the ways in which tumors may be diagnosed, treated, and prevented.

With this progress almost every area of research and practice relating to tumors has acquired new concepts, which in many instances are supported by their own extensive literature.

It has become difficult for students intending on a career in one or other of the disciplines relating to cancer research or practice—as well as established practitioners and research workers involved with tumors—to approach easily many of the basic issues in areas outside their primary area of activity.

This book has been written as a guide to the principles of the issues in the major areas of cancer research and practice. The emphases have been threefold:

i. To explain the principles, terminology, and concepts in the major areas of tumor research and practice. The overall readability of the overall work has been enhanced by arranging the topics in the traditional order of history, definition, incidence, classifications, etiology and pathogenesis, morphology, therapy, and prevention.

ii. To give accounts of interfaces between different areas where translational approaches may be most valuable. A major example of these is the relationships between the complex morphologies of tumors and their genomic disturbances. Another is the application of appropriate statistical methods to the many different areas of tumor research and practice, especially epidemiology and classification, etiology, therapy, and prevention.

iii. To provide up-to-date references for reviews and further reading in the areas discussed. This has been possible only through the databases accessible through the World Wide Web, especially PubMed, provided by the National Library of Medicine, Washington, DC.

The editors, Cassie van der Laan and Lisa Eppich, are thanked for their patience with what has been a long process of preparing the book.

Familiarity with the works of the early German pathologists and tumor theorists is owed to Emeritus Professor Brian Coghlan and Associate Professor Hubertus Jersmann. Previous collaborations with them have produced translations of works of the nineteenth century pathologists, especially those of David Paul von Hansemann (1858–1920) and Rudolf Virchow (1821–1901).

The University of Adelaide has provided access to the Barr-Smith Library, and the staff there has been of great assistance in retrieving older and recent material on tumors from its collection. The staff of the Library of the Royal Adelaide Hospital and SA Pathology is also thanked for the prompt retrieval of many less-common references.

Many of the photographs have been prepared by the Department of Photography, SA Pathology.

Others who have provided advice and assistance include James Bignold, Monica Bignold, Maria Collis, Peter Dent, Peter Gillam, Soula Grech, Douglas Handley, Wolfgang von Hansemann, Damien Harkin, Sarah Moore, Mary Peterson, Warwick Ruse, Mark Stevens and Peter Sutton-Smith.

Introduction

L.P. Bignold: Principles of Tumors.
DOI: http://dx.doi.org/10.1016/B978-0-12-801565-0.00001-9

Tumors are autonomously growing accumulations of cells which occur in types characterized by distinct, variable combinations of variable differences from their respective kinds of cell of origin. The natures and causes of most of the types of tumors are not well understood. The types which invade adjacent tissues and metastasize to other parts of the body are liable to cause death if untreated.

This chapter summarizes the fundamental aspects of tumors as a basis for the discussions of topics in the following chapters. Much of the material will be at familiar to medical students, as well as to many bioscience students in disciplines such as molecular biology, cell biology, and genetics.

The general principles of normal histology which are relevant to tumors are outlined in Appendix 1. The general aspects of the human genome are described in Appendix 2.

1.1 ESSENTIAL ASPECTS OF THE NATURE, TYPES, AND RATES OF INCIDENCE OF TUMORS

1.1.1 "Tumors" Used in the Sense of "True" Tumors

In its most general sense, a "tumor" is any swelling in the body. The term can be applied to a wide variety of conditions including physiological swellings such as a normal pregnant uterus, as well as to pathological swellings such as inflammations (see Section 11.1). In this book, as in most medical situations, the term "tumor" is used as a synonym for "neoplasm" or "true tumor."

All tumor cells are irreversibly modified normal cells. As a result, tumor cells have two sorts of hereditary characteristics [1]:

i. Those which are particular features of their parent cell and are retained to greater or lesser extents.

ii. Those which are effects of the genomic event associated with tumor.

Tumor cells usually have the following characteristics:

i. The cells have morphological and behavioral abnormalities in comparison with the parent kind of normal cell. They also have different and variable life spans compared to their parent kinds of cells.
ii. While tumor cells individually are not immortal, the cell population of each case of tumor is immortal, in the sense that the population almost invariably continues to grow without any tendency to regress or heal.
iii. In many cases, with time, tumor cell populations show progressive increases in morphological and other abnormalities. The populations virtually never spontaneously revert to less malignant populations.

The degrees of retention of parental features and the degrees of intensity of the acquired abnormalities are highly variable between different tumors. The variability is seen at three levels: between different types of tumors, between different cases of the same kind of tumor, and between different foci in individual cases of tumor.

While the distinction between tumors and nontumorous lesions is usually easy, pathologists recognize many clinicopathological conditions as "tumor-like." Examples include benign hyperplasias of the breast and prostate gland.

1.1.2 Classifications and Terminology of the Tumor Types

Tumors comprise a thousand or so different types. How these types are recognized and the differences between them are described in multivolume works such as the Armed Forces Institute of Pathology's *Atlas of Tumor Pathology* [2] and the *World Health Organization Classification of Tumours* [3]. The diversity of the tumor types also applies to their clinical behavior and responses to therapies. This is the basis of the complexities which are apparent in clinical oncology [4–7].

The different types of tumors are classified according to three main criteria:

i. The first criterion is the broad category of parent cell of origin as listed in Appendix 1 (Section A1.2.1). The categories are mainly epithelial cells, hematopoietic and lymphoid cells, nervous system cells, melanocytes, "soft tissue" cells, "hard tissue" cells, etc.
ii. The second criterion for classification is the exact kind of "parent" cell within the organ system from which the tumor arose. All of the broad categories of cells (Section A1.2.1) comprise more than one kind of cell. Thus the epithelial cells of the epidermis, mucosa of the intestine, kidney tubules, and thyroid follicles have significant differences, one from another. The cells in almost all types of tumors resemble only a single particular kind of normal cell which is located at the site of the tumor.
iii. The third criterion is according to whether the tumor is "benign" or "malignant" (Figure 1.1). Clinically, this is the most significant division of tumors, as has been understood for several centuries [8].

"Benign" types of tumors:

i. Grow slowly.
ii. Have well-circumscribed margins and bulge or displace, but do not invade, local vessels or adjacent organs. No metastases occur.
iii. May compress adjacent structures, but otherwise do little anatomical damage.

Thus benign tumors usually do not harm the individual and are relatively easily removed by surgery.

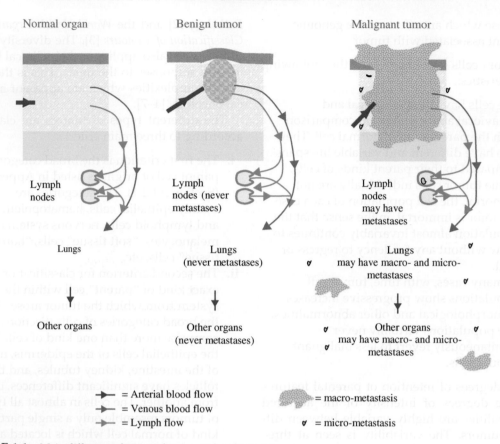

Normal organ   Benign tumor   Malignant tumor

Lymph nodes

Lymph nodes (never metastases)

Lymph nodes may have metastases

Lungs

Lungs (never metastases)

Lungs may have macro- and micro-metastases

Other organs

Other organs (never metastases)

Other organs may have macro- and micro-metastases

➤ = Arterial blood flow
➤ = Venous blood flow
➤ = Lymph flow

= macro-metastasis

ᵥ = micro-metastasis

FIGURE 1.1   Main differences between benign and malignant tumors.

In contrast, "malignant" types of tumors are extremely serious because they commonly grow rapidly. They tend to have irregular margins, and they frequently invade and destroy adjacent structures. These structures often include nearby blood vessels and lymphatic vessels. Via the lumina of these vessels and occasionally other avenues, tumor cells often then spread to distant sites as micro-metastases, which, with time, may grow into macro-metastases.

Many cases of malignant tumors are not removable by surgery, because at the time of diagnosis, they have already invaded a structure in the body which cannot be cut. The tumor is said to be "unresectable."

Because of all these features, malignant tumors are liable to cause the death of the patient.

These criteria of classification are the basis of the general terminology of tumors. The solid tumor types (i.e., excluding hematopoietic tumors) are named primarily by their site, their parent cell, and then as "benign" or "malignant." By convention, benign tumors of epithelial cells are usually called "adenomas," and malignant tumors "carcinomas." Benign tumors of "soft" and "hard" tissue cells are referred to by the kind of cell of origin with the suffix "-oma." Malignant tumors of such cells are indicated by the suffix "-sarcoma."

Thus a tumor might:

 i.   occur in the uterus ("uterine");
 ii. arise from a smooth muscle cell ("leiomy-");
 iii. behave in a benign fashion ("-oma").

Such a tumor is called a "uterine leiomyoma." Another tumor might:

i. occur in the stomach ("gastric");
ii. arise from a glandular epithelial parent cell ("adeno-");
iii. behave in a malignant fashion ("-carcinoma").

Such a tumor thus is named a "gastric adenocarcinoma."

By way of exceptions, leukemias are generally named according to kind of leukocyte involved and the stage of specialization which the majority of leukemic cells appear to have achieved [9]. This is discussed further in Section 8.6.

Within the group of tumors arising from lymphocytes in lymph nodes, the major division is between Hodgkin's and non-Hodgkin's lymphomas. The further classification of these tumors is complex and has been revised frequently in the last few decades (see Section 3.4 and Refs. [9–11]).

Tumors of cells of the nervous system and melanocytes have complex terminologies and classifications, which can be found in relevant special texts [12,13].

1.1.3 Incidences Especially of Malignant Tumor Types

(a) General

If all benign and malignant tumors of all organs including naevi and other benign skin tumors are taken into account, then the majority of people in the world will develop one or other kind of tumor in their lifetime. However, only data on malignant tumors excluding the epidermal malignancies are collected by national agencies such as the Centers for Disease Control [14], and internationally, by the World Health Organization (WHO) [15] in association with the International Agency for Research on Cancer (IARC). Major differences in incidences of tumors occur according to geographical region (Figure 1.2) and by type of tumor (Figure 1.3).

Worldwide, malignant tumors account for approximately 12% of all deaths [15,16], while in the United States, the proportion of deaths due to malignant tumors is approximately 23% [17].

(b) Influence of Ageing Populations

In the last century, tumors of almost all types have become more common in most populations in the world. This is mainly explained by declines in deaths due to infectious, cardiovascular, and other nontumorous diseases. These declines have resulted in longer average life spans of individuals in these populations. Hence more people are living to the ages at which tumors become most frequent (30–70 years) and so the incidence of tumors has risen. The association of the increased incidence of tumors with "modern living conditions" can be explained on this basis of increasing longevity. No evidence of any general age-adjusted increase for most types of tumors has been established [18].

(c) Incidences of Tumor Types According to Kind of Parent Cell

An obvious, but puzzling, point about the tumor types is that their incidences differ enormously according to their kind of cell of origin. For example, skeletal muscle fibers—which contain a large proportion of all the cell nuclei in the entire body, and which are capable of mitosis in healing or reactive nontumorous pathological conditions—very rarely give rise to benign tumors (rhabdomyomas) and only slightly more commonly to malignant tumors (rhabdomyosarcomas). However, the adipocytes and fibrocytes—which are located between the muscle fibers and presumably exposed to carcinogens in similar amounts—much more frequently give rise to both benign and malignant tumors of various types, although their nuclei are much fewer in number than are those of the skeletal muscle fibers.

Another slightly more complicated example is that tumors of smooth muscle cells are much more common in the uterus than in the

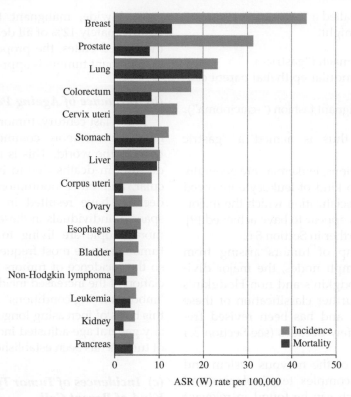

FIGURE 1.2 Incidences and mortality of malignant tumors according to geographical region. The data show significant differences in reported incidence of all cancers combined in different parts of the world. The differences in reported mortality are much less marked than are the differences in incidences. ASR (W) = Age-Standardized Rate (World). *Reproduced with permission from the IARC.*

intestine. There is much more smooth muscle in the latter organ than the former. Both kinds of smooth muscle are subject to the progestogen hormones, but the intestinal smooth muscle is insensitive to them physiologically. Hence the difference between the two organs may be associated with sensitivities to these hormones. Nevertheless, pregnancy is not the cause of uterine leiomyomas; in fact child-bearing protects against these tumors [19].

(d) Use in Medical Practice

Knowledge of the incidences of tumor types is important in everyday medical practice. The different types of tumors occur at markedly different rates not only according to the

kind of parent cell (Section 1.1.3), but also by age and gender of the patient. For example, certain tumors such as retinoblastoma occur only in children. Others such as osteogenic sarcoma are commonest in young adults. This particular type of tumor, however, may complicate Paget's disease of bone, such that a second "peak" of its incidence occurs in old age. Carcinomas of the thyroid gland are more common in females than in males, while the reverse applies to carcinomas of the esophagus.

These facts assist the process of medical diagnosis because the practitioner can have a good idea of what the type of tumor is likely to be simply on the basis of these features of site, age, and gender.

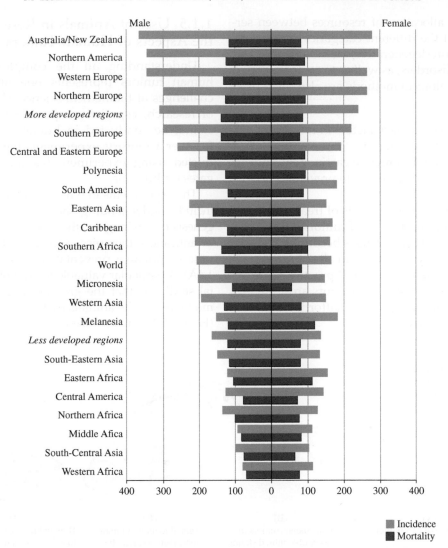

FIGURE 1.3 World incidence and mortality rates of malignant tumors by type. The data show significant differences in the reported incidence of the different types of cancer world-wide. The differences in reported mortality are much less marked than are the differences in incidences. *Reproduced with permission from the IARC.*

1.1.4 Epidemiology of Tumors

Epidemiology is the scientific discipline concerned with collecting and analyzing data—usually involving statistical methods—of the incidences and related issues of the different types of diseases, including tumors.

The incidences of tumors in most Western countries are assessed through reports of diagnoses of malignant tumors to national- or state-based cancer registries (see Chapter 3). The data collected are essential for the planning of provision of health services. This is because government and private agencies must make decisions concerning allocation of resources between services for all the different categories of diseases, such as neurodegenerative, cardiovascular, and infectious disorders, as well as to tumors.

Epidemiological methods are also often used to assess:

i. possible hereditary factors in tumor causation (Chapter 7);
ii. possible environmental carcinogens and hence lead to appropriate preventative measures (Chapter 14);
iii. the efficacies of methods of treatments of the different types of tumors (Chapters 11–13). This is becoming increasingly important in all countries, because the costs of "best practice" care for cases of malignant tumors are threatening to become unaffordable in many countries.

1.1.5 Uses of Animals in Research into the Aspects of Human Tumors

Understanding all the complex issues in human tumors represents one of the major challenges of biomedical science. Certain kinds of research, especially experiments, obviously cannot be conducted in humans. As a result, for over a century, experiments have been conducted using experimental animals as surrogates for humans.

The experiments in which animals can be used include understanding the etiology and pathogenesis of carcinogens, the efficacies or otherwise of anticancer therapies, and the value or otherwise of putative preventative measures (Figure 1.4A). A particular valuable feature of experimental studies is that the growth of tumors can be monitored in the animals by the same techniques (Figure 1.4B) as are used in clinical practice for

Putative carcinogen may be administered orally, cutaneously, or subcutaneously

(A)	(B)	(C)	(D)
Testing putative carcinogens*	Spontaneous tumors in susceptible inbred strains may be used for the purposes described in A, but also for studies of the biology of invasion and metastasis.	Animal to animal tumor transplantation may be used for studies of the biology of invasion and metastasis as well as experimental therapies.	Human tumor transplantation to animals (requires immune-deficient, "nude," animals) may be used for studies of the biology of invasion and metastasis of particular tumor types.
Testing modifying agents for induced tumors in animals:			
May be used in experimental studies of therapies and prevention, including chemoprevention.	Note: Mice genetically engineered for high spontaneous incidence may be used for purposes in B and also for specific genomic analyses.		Experimental therapies with these models may be more relevant to therapies in humans, especially for personalized therapy, than other models.
*certain genetically engineered mice may be more useful than outbred animals.			

FIGURE 1.4   (A) General uses of animals in research into human tumors. (B-D) Techniques which allow also imaging of growth of a tumor in an experimental animal.

cancer patients. This also adds the dimension that tumors in animals may be used to test the qualities of imaging technologies in ways which would otherwise not be possible.

1.2 BASIC ASPECTS OF ETIOPATHOGENESIS

Causation of a disease involves two parts: etiology and pathogenesis (Figure 1.5).

i. Etiology is concerned with the causative factors
ii. Pathogenesis is how the etiological factors cause the lesions in the body.

Considerations of these two parts frequently become intertwined so that the term "etiopathogenesis" is used.

1.2.1 Etiological Factors

This aspect of tumors receives enormous attention in the public arena as well as in

scientific circles. Answers to the question "what things cause tumors?" are of great general interest, as can be seen almost daily in the newspapers and other mass media platforms. The scale of the problem is enormous because in this relative vacuum of reliable information—which determines the advice a clinician might give concerning preventative action—there are many suggestions for causes of these diseases. New findings regarding possible causes of these latter kinds of tumors are reported not just in the biomedical literature but in the lay media as well. Almost every aspect of the general environment, diets, occupations, socioeconomic levels, and lifestyles of humans has at one time or another been reported as associated with—and hence identified as a possible "risk factor" for—causation of tumors.

As one indication of the information given to the public, a search in a leading New York newspaper in the 2013 calendar year for references to causes of cancer retrieved 2,330 results [20] including suggestions that using cell phones may cause brain cancer; breast

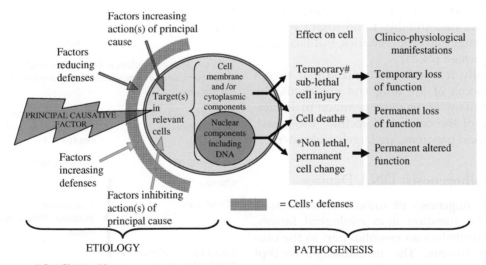

See Chapter 10.

*Tumors are considered the result of a non lethal permanent change in a cell. See Chapters 4–7.

FIGURE 1.5   General scheme of etiology and pathogenesis applicable to all diseases.

implants and estrogen therapy may contribute to breast cancer; particular bacteria may cause colon cancer; formaldehyde may cause myeloid leukemia; and as well, that ethanol, hormones, tungsten particles, sugar, and many other diverse substances or activities may cause cancers of various types.

Various etiological agents for tumors are established, but most of them are relevant to only one or a few types of tumors [21–23] (Table 1.1). For example, ultraviolet B (UVB) radiations cause carcinomas of the epidermis and melanomas (UVA is noncarcinogenic). Some chemicals in tobacco cause carcinoma of the bronchi. A few particular viruses contribute to a small number of types of tumors in humans. For example, the Epstein–Barr virus is associated with Burkitt's lymphoma, human herpes virus 8 with Kaposi's sarcoma in the acquired immunodeficiency syndrome (AIDS), and some strains of human papilloma virus (HPV) with carcinoma of the uterine cervix.

Certain hereditary predispositions to specific types of tumors, such as to neurofibromatosis and retinoblastoma, are known. Inheritance of predispositions to all types of tumors—i.e., a generalized susceptibility to all tumor types—has been suggested for centuries, but has never been confirmed [24,25].

Generally, it must be recognized that, except for carcinoma of the lung and epidermis, the causes of many of the other common malignancies, e.g., of the colon and rectum, the breast, prostate, and pancreas, are largely unknown.

1.2.2 Pathogenesis: DNA Damage

The pathogenesis of tumors attracts much less public attention than etiological factors, but is nevertheless an essential part in the causation of tumors. The fundamental concept in pathogenesis is that if an agent causes a change in a cell, then first, the agent must affect some chemical structure in the cell such that

TABLE 1.1 Common Etiopathogenetic Factors for Common Tumors

1. Agents with associations indicating high risks and recognized pathogenetic mechanisms

UVB radiations	Epidermal carcinomas, malignant melanomas of skin
Tobacco usage	Smoked: lung carcinoma of all histopathological types
	Chewed: squamous cell carcinomas of the mouth
HPV (certain strains only)	Squamous cell carcinoma of the uterine cervix
HIV	Malignant lymphomas; Kaposi's sarcoma
Hepatitis B and C viruses	Carcinoma of liver

2. Factors with associations indicating increased risk of specific malignancies and without definitely understood pathogenic mechanisms

Heredity (see Chapter 7)	
Mutations in the *BRCA* gene[a]	Carcinomas of breast
Mutations in "DNA mismatch repair"	
Genes[a]	Carcinoma of colon and rectum, ovary.

3. Factors with associations indicating increased risk of ranges of malignancies and without definitely understood pathogenetic mechanisms

Excessive ethanol consumption	Carcinoma of liver[a], mouth, esophagus, and other organs
Increasing age	All malignancies[a]
Obesity	
"Poor" diet	Carcinomas of the colon, pancreas, breast, and other organs
Lack of physical exercise	

[a]Substantially increased risk.

the function of the structure is changed. Then, second, the change in function of the "target" structure must then result in the observable changes in the cell.

For most pathological processes, the change caused by the etiological factor is usually one of reduction or loss of one or other metabolic capacity. However, in tumors, the change is characteristically a gain in the function of cell multiplication.

In relation to carcinogens and tumors, the most widely accepted idea of pathogenesis is that carcinogens react with the DNA in the normal cells (Figure 1.6). The major kinds of damage to DNA are shown in Figure 1.7. Damage to individual sites on nucleobases is discussed in Section 5.1. The damage to DNA is thought to cause alterations in the nucleotide sequences during repairs of existing DNA or synthesis of new DNA (see Sections 5.1 and A2.1.4). These altered sequences then cause the phenotypic changes which characterize tumor cells.

There are many kinds of genomic events (see Chapter 5) which can result in errors in nucleotide sequence. First there are substitutions, insertions, and deletions affecting one or two nucleotides. Of note here is that an insertion or deletion of any number of nucleotides, which is not a multiple of three, results in a "frameshift" event, which affects up to thousands of consecutive nucleotides in one section of DNA. Second, there are deletions, insertions, amplifications, transpositions, and inversions of parts of chromosomes from tens or a few hundreds of nucleotides to microscopically visible parts of chromosomes (chromosomal aberrations).

1.2.3 "Cancer Genes" in Pathogenesis

The next step—exactly what part of the genome, when altered, causes tumors—has

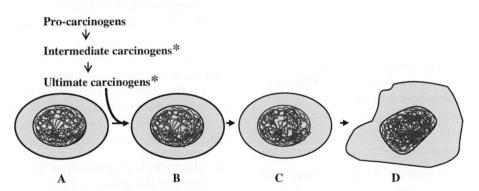

Pro-carcinogens
↓
Intermediate carcinogens*
↓
Ultimate carcinogens*

A. B. C. D.

A. Unspecified normal cell.
B. Nucleobases are randomly damaged throughout the genome.
C. Most of the damage is repaired. However, in the exons of "critical genes," damaged nucleobases which are mis-repaired or cause mis-synthesis of new DNA, result in phenotype-changing genomic events in the cells or their daughter cells (●). Unrepaired damage in inactive parts of the genome has no effect.
D. The events in the exons of the critical genes instigate the "neoplastic" process in cell. It is thought that the same genomic events might apply to all types of tumors; differences between tumor types perhaps due to cytoplasmic factors peculiar to the cell type of origin)

*For discussion of activation of carcinogens, see Section 4. 4. 1.

FIGURE 1.6 Overview of the major theory of tumor formation by carcinogens.

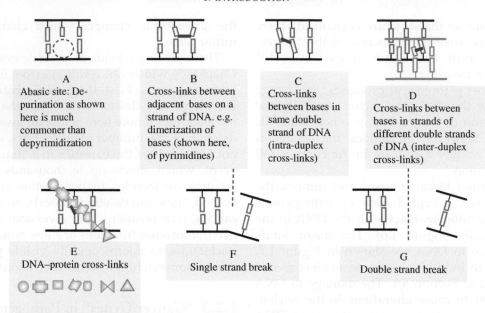

A
Abasic site: De-
purination as shown
here is much
commoner than
depyrimidization

B
Cross-links between
adjacent bases on a
strand of DNA. e.g.
dimerization of
bases (shown here,
of pyrimidines)

C
Cross-links
between bases in
same double
strand of DNA
(intra-duplex
cross-links)

D
Cross-links between
bases in strands of
different double strands
of DNA (inter-duplex
cross-links)

E
DNA–protein cross-links

F
Single strand break

G
Double strand break

FIGURE 1.7   Major categories of damage to DNA. For intercalations see Figure 4.8, and for nucleobase changes and modifications see Figures 5.2 and 5.3.

been pursued mainly in terms of genomic events in exons of genes. Broadly, these occur in two groups (Figure 1.8).

(a) When the Genomic Events Occur Entirely in the Somatic Cell (i.e., Without Germ-Line Genomic Alteration)

The genes in this category are often referred to collectively as "oncogenes" [26]. They include:

  i. Insertion of viral DNA into the genome of a somatic cell. These viral oncogenes have been identified in extracts of the human tumors, e.g., the Epstein–Bar virus in Burkitt's lymphoma [27].
 ii. Induced hyperactivation of one or more endogenous genes. An example of this is the Rous sarcoma virus [28].

All of these genes have been identified mainly on the basis that, when transfected into cells cultured *in vitro*, they induce a change known as "malignant transformation" (Section 14.2.3).

(b) Requiring Germ-Line Alteration as Well as a Somatic Alteration in the Genome

These are less common than the oncogenes and number approximately 20. These are usually referred to as "tumor suppressor genes" (Section 6.5) on the basis that they are genes which normally keep the proliferation of cells under control in physiological and nontumorous pathological conditions [29]. They include the genes in which:

  i. The germ-line alteration is in one allele (copy of the gene), and the somatic event is necessarily on the other allele as in retinoblastoma [30]. These genes have been identified by studies of the genomes of patients and of the genomes of the tumors which occur in them.
 ii. Germ-line alterations are required in both alleles of the same gene—i.e., inheritance is Mendelian recessive. In these patients, one (if not more) additional somatic event, usually in other genes, is required for the tumor [31]. An example of this is

Growth factor/oncogenes

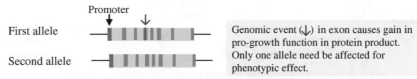

First allele

Second allele

Genomic event (↓) in exon causes gain in pro-growth function in protein product. Only one allele need be affected for phenotypic effect.

All genomic events are somatic; virtually no germ-line genomic events in growth–factor/oncogenes are known.

Tumor suppressor genes (Knudson model, see Figure 6. 6)

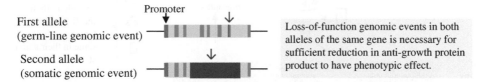

First allele
(germ-line genomic event)

Second allele
(somatic genomic event)

Loss-of-function genomic events in both alleles of the same gene is necessary for sufficient reduction in anti-growth protein product to have phenotypic effect.

The somatic genomic event is not necessarily in the same exon. Any loss-of-function genomic event, such as loss of promoter, loss of whole exon, loss of whole gene, or loss of part of the chromosome, would be sufficient.

Note: The protein products of growth factors are thought to function mainly to be part of the cells signaling cascade/pathway proteins (Figure 1.6), while the functions of the protein products of the tumor suppressor genes are largely unclear but may include regulators of transcription of the cascade/pathway proteins.

FIGURE 1.8 In-principle differences between growth factor–oncogenes and tumor suppressor genes.

xeroderma pigmentosum [32] in which biallelic germ-line genomic events in the particular XP gene result in reduced capacity to repair UVB damage to DNA. The skin tumors which are common in the condition are due to the additional genomic events in the epidermal cells resulting from the damage to DNA.

Genes which require germ-line events in both alleles also include many low-penetrance genomic events which may have their phenotypic effect in adult life [33].

(c) Cell Signaling in Normal Growth Control

The next step in the pathogenesis of tumors concerns the functional roles in cells of the protein products of "cancer genes." Most studies have been directed at whether or not they are part of, or affect, biochemical "signaling pathways" which are involved in proliferation of cells [34–36].

The principle of signaling (Figure 1.9) is that an extracellular factor binds to a specific surface receptor site on a protein. This protein extends through the membrane and has an enzymatic site on its intracellular end. Binding of the factor to the surface receptor activates the intracellular site. The active enzyme site is usually a kinase (an enzyme which adds phosphates to proteins). The signaling usually involves sequential activation of additional phosphorylating enzymes. Ultimately,

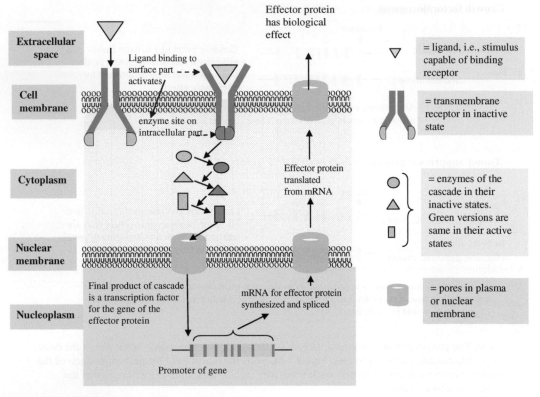

FIGURE 1.9   Cascade/pathway model of intracellular signaling.

transcription factors are produced which lead to the production of "final effector" proteins which induce the relevant effect in the cell. For growth-related signaling pathways, the final effector proteins are cyclins and others involved in cell division (see Section A1.3).

The protein products of oncogenes are mainly one or other component in these cascades, i.e.,

i.   membrane proteins;
ii. intermediary kinases;
iii. transcription factors;
iv. apparent regulators of the cell cycle or cell death;
v.   the final effector protein of the cascade.

Signaling pathways are generally named by the oncogene/membrane receptor protein, and/or the principle kinase(s) involved.

The functions of products of tumor suppressor genes in normal biology and nontumor pathology are assumed to be prevention of excessive growth of cells (Section A1.3). An early idea was that their primary function was to inhibit the activation of oncogenes (hence the term "anti-oncogenes" [37]). At least some products of suppressor genes may act as inhibitors of transcription factors for the components of the signaling cascades (see Section 6.3.3). However currently, the functions of the protein products of many tumor suppressor genes are unknown [38].

Activations of certain pathways tend to be particularly associated with particular types of tumors. However, activations of only a few pathways are individually specific for a single tumor types, and few activations are identifiable in every case of any tumor type (see Section 9.4).

1.3 ABNORMALITIES IN MORPHOLOGY AND "MOLECULAR PATHOLOGY"

1.3.1 Background

The morphological abnormalities of tumors are the bases on which pathologists make most of their diagnoses of tumor types [1,2]. The cells in the different tumor types show different degrees of abnormalities, and no individual abnormality, except accumulation of cells, is shown by all tumor types. No abnormality is seen exclusively in "true" tumors and not in any other nontumorous condition.

The morphological abnormalities include:

i. Alterations in specialized activity of tumor cells.
ii. Changes in the internal structure of the cells. This is seen especially in nuclear size and in the relative proportions of nucleus to cytoplasm seen in the parent cell.
iii. Abnormal micro-anatomical spatial arrangements. This is seen both in the tumor cells to one another and in the tumor cells to the supporting/connective tissue cells of the tumor mass. Invasion

of connective tissue by tumor cells is an extreme example of the latter and is a major indicator of malignancy.

iv. More numerous mitotic figures are seen in the parent cells.
v. Abnormalities in mitotic figures, including unequal distributions of chromosomes during ana(sub)phase (see Section A1.3.4).

1.3.2 Grading of Tumors

(a) General

For most tumor types, the greater degrees of morphological abnormalities of a case are associated with greater degrees of malignancy ("aggressiveness") of the particular case (Figure 1.10). The severity of particular abnormal histological features can be "graded." These grades are of value for assessing prognosis in the individual case and also can be used to guide treatment. High-grade tumors may be frequently treated differently to low-grade tumors of the same type.

There are different grading systems for different tumor types. For most tumor types, new grading systems have been drawn up from time to time.

 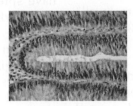 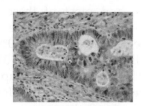

A	B	C
Normal cells show no cyto structural instability.	Adenomas show mild to moderate cytostructural instability.	Carcinomas show marked cytostructural instability.

Note: This principle applies to the tumor types arising from most, but not all, kinds of cells.

FIGURE 1.10   Relationship between degree of cytostructural instability and cell behavior in colonic epithelial tumors.

(b) Grading of Carcinomas of the Lung

There are four major types of lung carcinoma, of which adenocarcinomas and squamous cell carcinomas may be assessed according to their degree of specialization ("differentiation"—see Section 8.1.2). No other grading scheme is widely used.

(c) Grading of Carcinomas of the Breast

Grading has considerable significance for prognosis for carcinoma of the breast and is almost universally used. The main protocol is the Elston–Ellis modification of the Scarff–Bloom–Richardson grading system [39]. Tumor cells are assessed according to three criteria: (i) glandular/tubule formation, (ii) nuclear pleomorphism, and (iii) numbers of mitotic figures. Adding the individual counts gives a range of 3–9, with the higher numbers indicating worse prognosis.

This grading of tumors is taken together with the results of molecular studies (especially of estrogen and progesterone receptors, and HER-2 gene expression), to plan particular components of treatment of the individual case [40].

(d) Grading of Carcinomas of the Large Bowel (Colon and Rectum)

More than 90% of these tumors are adenocarcinomas, and most of these are "well" (Grade 1), "moderately" (Grade 2) "differentiated"/specialized. Only 5–10% of cases are poorly differentiated (grade 3). Cases of Grades 1 and 2 are commonly grouped together as "low grade." "Poorly differentiated"/high-grade/Grade 3 tumors are diagnosed when less than 50% of the cells form tubules. "Undifferentiated" carcinoma shows no glandular, mucus, neuroendocrine, squamous, or sarcomatoid specialization [41].

(e) Grading of Carcinomas of the Prostate

The main grading system for this tumor was devised by Gleason in 1975 and is based entirely on architecture of the tumor masses. Initially, the assessment was intended to include nuclear pleomorphism and stage of the case of tumor. However, nuclear abnormalities were found not to affect outcomes, and assessments of stages were too complicated to include [42].

Most systems of grading for other types of tumors include assessments of the abnormalities listed above, together with abnormalities in antigen expressions and genomic alterations where appropriate (see Chapter 9).

It should be noted that grading systems have only been validated by empirical observations. The biological basis for the association of cytostructural abnormalities—such as loss of specialized activity, loss of spatial regularity, pleomorphism of nuclei, and abnormal mitotic figures—with tendency to invade and metastasize is unknown.

1.3.3 Staging of Cases of Tumor

(a) General

The stage of the tumor is the anatomical extent to which the tumor has already spread in the patient's body at the time of diagnosis. The rationale for documenting the stage of a case is that a tumor which has already spread to, for example, local lymph nodes is more likely to have spread more widely in the body than a tumor which has not already spread to lymph nodes.

As in grading systems, there are various staging systems described for the different types of malignant tumors. In the widely accepted "TNM" scheme, stage is assessed in three components:

i. The degree of local spread of tumor—the "T" number.
ii. The presence or absence of metastases in lymph nodes—the "N" number.
iii. The presence or absence of metastases to other organs—the "M" number. Usually M0 is for no metastases, and M1 indicates distant metastases are present.

For many purposes, a single number for each letter is sufficient. However, there has been a trend to greater and more complex subdivision of each stage. The subdivisions are usually indicated as "A," "B," etc.

The criteria for these subdivisions of the "T" and "N" numbers differ according to the type of malignant tumor. The details are given in publications of the International Union Against Cancer (UICC), the American Joint Committee on Cancer and Cancer Research UK [43–45], as well as other organizations. A summary is as follows.

(b) Staging of Carcinoma of the Lung

In stage 1, the tumor is confined to the lung and is divided into A and B according to size up to 5 cm across. Stage 2 is divided into A and B, referring to sizes 7 cm across or growing into certain thoracic structures (five situations) such as bronchial or hilar lymph nodes. Stage 3 is divided into A and B, referring to a number of possible combinations of (i) size greater than 7 cm, (ii) spread directly to another structure such as the mediastinum, and (iii) metastasis to the other lung, the spine, or lymph nodes of the neck. A tumor is stage 4 if it has already spread to other parts of the body. Further details are given in Ref. [46].

(c) Staging of Carcinoma of the Breast

The system in general use is complicated because it gives emphasis to the size of the primary tumor (especially the measurement of 2 cm) at diagnosis and relatively less weight to a few cells in local lymph nodes (which would be within an actual or potential simple mastectomy specimen). Stages 1–3 are divided into two or more substages (A, B ± C), with up to three alternative situations within each subcategory. Stage 4 indicates spread to distant organs. Details of the stages are given in Refs. [47,48].

(d) Staging of Carcinoma of the Large Bowel (Colon and Rectum)

Stage 0 is *in situ* tumor and stage 1 indicates that slight invasion of the main muscle layer of the bowel. Stage 2 is divided into A, B, and C depending on degree of local invasion. Stage 3 is divided into A, B, and C, with each having two or more alternatives. These are determined by combinations of the depth of invasion by primary tumor and the number of local lymph nodes containing tumor. Stage 4 is when distant metastases are identified. This stage is divided into A and B according to the number of distant organs involved [49].

(e) Staging of Carcinoma of the Prostate

Current schemes for staging these tumors include not only the actual spread of tumor cells in the body, but also the criteria of the histological grade (Gleason score, see Section 1.3.2), as well as the serum prostate-specific antigen (PSA) level. Stage 1 is a tumor which is too small to be detected by clinical examination or imaging. The tumor has a low Gleason grade, and the PSA is low. Stage 2 refers to tumors which have not extended beyond the capsule of the gland. It is divided into "A" (five alternative combinations of features) and "B" (three alternative combinations of features). Stage 3 tumors have spread through the capsule of the gland and may have invaded the seminal vesicles. Stage 4 refers to cases in which greater spread of tumor has occurred: three alternative combinations of features [50].

1.3.4 Importance of Micro-Metastases

There is a considerable literature on the prognostic significance of small collections of tumor cells which are identified adjacent to the main mass in a resection specimen [51]. However, it is important to remember that the techniques used for imaging in the patient (see Section 1.4.2), as well as for microscopic examination of cells in the pathological specimen, do not detect all the malignant cells which have spread in the body before diagnosis and therapy.

These undetectable micro-metastases are the basis of at least some of the secondary tumors which become apparent only during or post-therapy. If these micro-metastases are present just outside the surgical margin—and thus cannot be seen in microscopy sections—a local regrowth of tumor ("recurrence") is possible. If the micro-metastases are present in a distant organ, then new metastases will develop. These situations are sometimes called "relapses."

1.3.5 "Molecular Pathology" and Other Nonmorphological Features of Tumor Cells

The term "molecular pathology" [3,52], when it is applied to tumors, refers to the expressions of antigens, as well as changes in DNA, RNA, and other molecular features in tumor cell populations. In tumors, these aspects are often qualitatively and quantitatively abnormal. Many of these studies are mainly related to oncogenes and proteins of signaling pathways (see Section 1.2.2).

Other nonmorphological abnormalities are mainly the cell turnovers ("cytodynamics"/"cell kinetics") [53] of tumor cell populations. Although the field is little studied at the present time, it is important because the tumor cell populations are significantly different to those of the parent kinds of cells. In tumors, cells may be produced at variable rates and independently die at variable rates. Further, and unlike normal cells, tumor cells may enter a vegetative state in which they remain partly specialized, but do not divide. There are many mechanisms involved in these rate changes of tumor cell life spans (see Sections 6.1, 11.4.2, and A1.3.3).

1.4 FUNDAMENTAL ASPECTS OF CLINICAL FEATURES AND TREATMENTS

1.4.1 Principles of Symptoms and Signs of Tumors

The clinical features of tumors taken together may cause almost every symptom and sign which is known in clinical medicine [4–6]. This is because they can arise in—and in the case of malignant tumors, metastasize to—almost any part of the body.

Both benign and malignant tumors can cause the symptoms and signs through their presence alone, or via their complications (Figure 1.11). In principle, the complications arise because the tumor may:

i. compress or
ii. obstruct an organ or adjacent structure in the body.
    Malignant tumors in addition can:
iii. destroy adjacent structures,
iv. bleed and/or
v. create holes in the part of the organ in which they form.

Malignant tumors can cause the same phenomena in any adjacent organ or tissue which they invade or in which they grow as a metastasis. Also, metastases can develop in more symptom-critical part of the body than the site where the primary tumor arises. Moreover, in some cases, the metastases seem to grow faster than the primary tumor.

The result is that the signs and symptoms of the secondary tumors may appear before those arising from the primary tumor. For example, lung carcinomas can metastasize to the central nervous system and produce symptoms and signs such as epileptic seizures, headaches, and palsies, before any respiratory-tract symptom has appeared.

As another example, carcinomas of the large bowel can metastasize to the liver causing enlargement and pain in that organ, before the primary tumor in the bowel has caused any symptom of a bowel lesion.

Thus, in principle, complications are the bases of most of the clinical features of tumors of noncutaneous (i.e., internal) organs.

The main tumor types and their symptoms and features in each organ system are as follows and are summarized in Table 1.2. Some

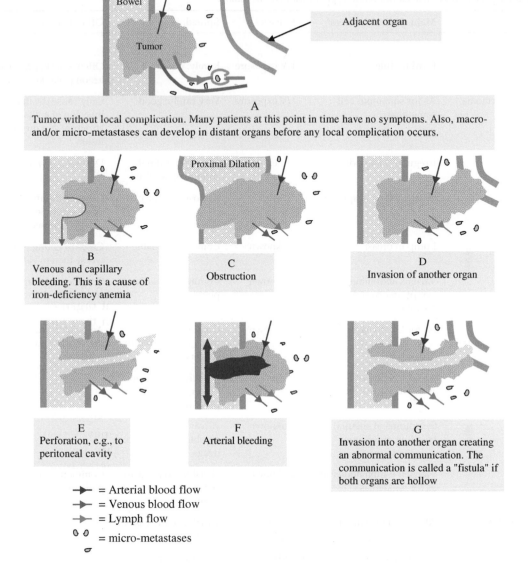

FIGURE 1.11 Complications of malignant solid tumors in approximate order of frequency.

general references for this section are given in Refs. [4–6].

(a) Skin

The most important tumors of this organ are epidermal and melanocytic. Epidermal tumors include many benign types. Of the malignancies, squamous cell carcinomas are mainly low grade, and basal cell carcinomas invade but hardly ever metastasize.

Benign melanocytic tumors comprise naevi of various kinds and are common. Malignant

TABLE 1.2   Features of the Commonest Types of Malignant Tumors

	Main clinical features[a]	Causes	Spread; prognosis	Other features
Epidermis				
Squamous cell carcinoma	Hard nodule	UV exposure	Rarely; good	Often arise in precursor lesion (solar keratosis)
Basal cell carcinoma	As for squamous cell carcinoma	UV exposure	Very rarely; good	Only cause death by invasion of a vital structure. No precursor lesion
Melanocytes	Irregular dark nodule	UV exposure	Often; poor if not completely excised	Dormancy common
Bronchial cells	Cough, blood in sputum	Tobacco smoking	Almost always; poor	Cases of the anaplastic Small cell types are usually unresectable
Female breast	Lump in breast	Unknown[c]	Common; variable	Cases in young women worse than in elderly
Stomach	Various abdominal symptoms; iron-deficiency anemia	Unknown[c]	Common; usually poor	Commoner in Japan— reason(s) unknown
Large bowel (colon and Rectum)	As for Ca stomach	Unknown[c]	Common; many cases cured by surgery	Associated with low-fiber diet, reason unknown
Pancreas	As for Ca stomach Also obstructive jaundice	Unknown	Usually; rarely cured by surgery poor prognosis	
Leukemias	Symptoms of anemia, easy bleeding; infections	Unknown[b]	Always; some cases curable by anticancer drugs	
Lymphomas	Swollen lymph nodes; also as for leukemias	Unknown	Variable; some cases curable by anticancer drugs	Commoner in AIDS patients
Uterine cervix	Abnormal bleeding from vagina	HPV virus implicated	Variable; early cases curable by surgery	
Endometrium (adenocarcinoma)	As for Ca cervix	Unknown[c]	As for Ca cervix	
Placental tissue (choriocarcinoma)	As for Ca cervix abdominal symptoms	Unknown	Some cases can be cured by drugs	Commoner in Chinese
Ovary	Various abdominal symptoms	Unknown[c]	As for Ca cervix	

(continued)

TABLE 1.2 (Continued)

	Main clinical features[a]	Causes	Spread; prognosis	Other features
Testes	Swelling ± pain in testis	Unknown	Some cases can be cured by surgery, radiation, and drugs	
Prostate	Urethral obstruction	Unknown	Common; controversial[d]	More aggressive in men of African descent
Brain (gliomas)	Neurological ± mental symptoms	Unknown	No; generally poor	Life expectancy is closely related histological type
Kidney	Abdominal pain	Unknown	Often; variable	
Bladder	Blood in urine	Unknown[b]	Most cases not spread at diagnosis; usually good	Tobacco smoking may be a factor
Bones	Pain, swelling, sometimes fracture of affected bone	Unknown[b]	Usually; poor	Chondrosarcomas less aggressive than osteosarcomas

Ca, carcinoma
[a]*The first clinical manifestation of some cases of malignant tumor may be via metastases.*
[b]*Particular causes are known for a few cases.*
[c]*Hereditary predispositions in some cases.*
[d]*See Chapter 3.*

melanomas are less common. The outcome of the particular case of melanoma depends mainly on the size of the primary tumor.

Most types of tumors of epidermal cells or melanocytes produce an abnormality of the surface—e.g., a projection, roughness, discoloration, ulceration, and even bleeding. Cases of tumor types arising from cells in the dermis sometimes break the surface, but otherwise appear as thickenings of the skin.

(b) Respiratory Tract

Benign tumors of these organs are uncommon. Most are of connective tissue cells. All carcinomas of the lung/bronchi have poor prognosis. For small cell anaplastic carcinoma, nonsurgical therapies are the main mode of therapy. Nonsmall cell carcinomas—squamous, adeno- and large cell anaplastic types—are of more or less equal malignancy and susceptibility to therapies.

Tumors of the lungs may cause coughing in general, and coughing blood in the sputum in particular. Also, shortness of breath and chest pain (usually on one side) are common. Tumors of the pleura (mainly mesothelioma) almost always cause pain on the side of the tumor and later shortness of breath.

(c) Female Breast

A benign condition known as "mammary dysplasia" or "fibrocystic disease" is very common. It does not tether to skin or other structure, but produces firm areas in the breast tissue which sometimes cannot be distinguished from carcinomas without pathological examination.

Apart from this, the main benign tumor type is the fibroadenoma. This usually occurs in young women. It does not "tether" to skin or any other structure.

Carcinoma of the female breast is common and highly variable, both in age of onset and aggressiveness. Carcinomas may be found at screening, if available, or by the discovery of a mass. Tethering to the skin may occur with superficially placed tumors. Tethering to the underlying skeletal muscles may occur with deeply placed tumors. Some tumors are not palpable because of the amount of surrounding breast tissue, especially if that tissue is fibroadenomatous. These tumors may only manifest through metastases, which are common in local lymph nodes, pleura, and bone as well as liver, lungs, and other organs.

Carcinomas in young women tend to be more aggressive than those in older women. Pathological examination for grade and staging gives valuable prognostic information.

(d) Alimentary System

In the esophagus, benign tumors are rare. Malignancies include squamous cell carcinomas and adenocarcinomas near the esophagogastric junction. These generally have a poor prognosis because they invade deeply before they cause any symptoms.

In the stomach, benign polyps occur but are less common than in the colon. Adenocarcinomas of the stomach have a generally poor prognosis because of their usual high invasiveness.

Tumors of the small intestine are uncommon.

In the colon and rectum, several kinds of benign epithelial tumors ("polyps") occur quite frequently. Adenocarcinomas are common and are of intermediate malignancy (approximately 30% 10-year survival rate).

In the pancreas benign tumors are uncommon. Adenocarcinomas are of more or less uniform malignancy and have a poor prognosis.

Benign and malignant tumors anywhere in the alimentary tract commonly cause obstruction. Carcinomas also cause anemia through bleeding from surface ulceration and hence loss of blood in the feces. Another common symptom is severe abdominal pain through perforating the wall into the peritoneal cavity.

Pancreatic tumors are often silent and cause symptoms and signs only through their metastases. If the tumor is in the head of the pancreas, it can obstruct the bile ducts causing obstructive jaundice.

(e) Hematopoietic and Lymphoid Systems

A proportion of cases of leukemias, and a few types of lymphoma, can be cured. Those cases of acute leukemia which do not respond to therapy can have a life expectancy of only months or a few years. Cases of chronic leukemias can survive for longer periods of time.

Leukemias usually manifest through tiredness in association with anemia, easy bruising, and sometimes severe infections. Lymphomas beginning in the lymph nodes immediately under the skin usually present as lumps. Lymphomas beginning in deeper parts of the body may manifest through interference with the function of an adjacent anatomical structure.

(f) Female Genital

Benign and malignant tumors of the vulva and vagina are relatively uncommon.

The ectocervix gives rise to few benign tumors and more often to squamous cell carcinomas. In situ lesions occur commonly and are screened for and treated in most Western countries.

Benign polyps of the endometrium are very common. Adenocarcinomas are relatively common but usually of low-grade malignancy. These lesions often present with abnormal bleeding per vaginam.

Benign cysts of the ovaries are very common. The commonest malignancies are adenocarcinomas of several different histological types. These are usually of low to medium malignancy. Ovarian tumors are often "silent" for long periods of time and manifest only through torsion of the affected ovary or through abdominal swelling.

(g) Male Genital

In the testis, benign tumors of any kind are uncommon. Malignancies are almost all derived from germ cells. The seminoma type can be cured. Nonseminomatous malignancies have poorer prognoses; being less susceptible to treatment, testicular tumors usually become recognized through swelling or pain of the testis.

Idiopathic hyperplasia of the prostate is very common. No other kind of benign process is common in this organ. Malignancy, being adenocarcinomas of various degrees of malignancy, are common. Any prostatic swelling tends to cause urinary obstruction. Prostatic carcinomas may arise away from the urethra, and so manifest first through metastases, especially to spine and other bones.

(h) Nervous System

There are three major groups of tumors. Meningeal tumors are almost always benign. Glial tumors, although not metastasizing, are malignant because of their tendency to invade, rather than simply compress, adjacent brain tissue. Neuronal tumors are uncommon and seen mainly in infants.

Tumors in the brain manifest mainly through (i) a functional disturbance in the part of the brain in which they begin, (ii) epileptic seizures, or (iii) headaches and other effects of general compression of the whole brain itself in association with raised intracranial pressure.

(i) Urinary System

Tumors of the kidney derive from renal tubular cells. A substantial proportion of cases are malignant.

In the urinary tract (renal pelvis, ureters, and bladder), tumors are of the urothelium. These are not divisible simply into "benign" and "malignant." This is because they form a continuous spectrum in grading.

Tumors of the kidneys may manifest with blood in the urine and/or pain in the flank, or by symptoms produced by metastases. Bladder tumors usually present with blood in the urine or difficulty urinating.

(j) Skeletal System

Various benign osseous and cartilaginous types of tumors are recognized. Of the malignancies, osteogenic sarcomas occur mainly in young adults and have a moderate to poor prognosis. Chondrosarcomas, however, occur in older persons and have a better prognosis.

Bone tumors usually present with swelling and/or pain at the site of the tumor.

(k) Other Systems: Including Endocrine and Special Senses

The clinical manifestations of these organs are too complex to be discussed in terms of general principles. They are strictly related to the particular organ. For example, tumors of the eye usually manifest pain or disturbed vision.

(l) "Progression"

The course of a patient's illness may be affected by "progression" of the tumor. This word is used for two situations:

i. In a benign tumor, the development of a focus of malignant tumor. Part of the benign tumor cell population is said to have "progressed" to malignancy.

ii. In a low-grade malignant tumor, the development of a more malignantly behaving population of tumor cells. The tumor is said to have "progressed" in its degree of malignancy.

(m) Miscellaneous Clinical Features of Tumors

There are some additional clinical features of malignant tumors to be noted.

Cachexia (extreme loss of weight) is a striking feature in some cases of cancer. It may be due to mental depression in patients who are aware of their condition. However, cachexia is often the presenting complaint (before the diagnosis is made). Thus particular metabolism-influencing factors, such tumor necrosis factor beta [54], may well play roles in this clinical manifestation of cancer.

Noninfective thrombi forming on the endocardium ("marantic" thrombosis) is a complication of some kinds of adenocarcinoma. The mechanisms are unknown.

Other uncommon manifestations of cancer include hyperpigmentation of the skin ("pseudoacanthosis nigricans"), osteomalacia [55], and other conditions, all of which have been attributed to "ectopic" hormone production by particular tumor types [56,57].

1.4.2 Diagnosing Suspected Cases of Tumor

This is a large part of medical practice, and any difficulties in diagnosis of cases are often due to the site of the tumor. An abnormal mass of cells is usually obvious if it is located in the skin or subcutaneous tissues. Internal tumors are usually found through:

i. physical examination, e.g., of accessible sites such as the rectum;
ii. visualization, e.g., with endoscopes of the stomach, colon, bronchi, etc.;
iii. imaging, which includes plain X rays, computerized axial tomographic X rays, magnetic resonance images, as well as radioisotope scans and other techniques.

The next step in diagnosis is usually a biopsy of the mass. Pathological examination then allows the particular type of tumor to be established. If malignant and if the biopsy piece is large enough, the grade of the tumor (Section 1.3.2) can be estimated.

If the tumor is malignant, it is necessary to determine the extent of spread which has occurred before diagnosis (i.e., the stage of the tumor, Section 1.3.3). Identification of metastatic masses of tumor involves the same imaging techniques as are used in achieving diagnosis of the primary tumor.

1.4.3 Surgery

Surgical operations are undertaken in for a variety of reasons in the treatment of tumors. These reasons are described in Section 11.2.

This section is concerned only with resection of the primary tumor (Figure 1.12).

After diagnosis and staging of a particular tumor, surgical resection is usually undertaken if there is a reasonable zone of normal tissue at the edge of the tumor.

Pathologists then, as part of their examination of the specimen, often assess the surgical margin for the presence of any tumor cells in this site. If tumor cells are present in the margin, local recurrence of the tumor is likely. However, absence of tumor from the margin of the surgically removed specimen does not guarantee that the tumor will not recur at the site of the resection. This is because the undetectable metastatic micro-foci of tumor (the "micro-metastases," see above) may be present just outside the surgical margin.

The details of the operation depend on the type of tumor [58]. This is because the types of spread of malignant tumors are variable according to type.

Sarcomas as a group spread via the blood vessels and rarely spread via the lymphatics. Hence resecting lymph nodes in the region of a sarcoma is not commonly carried out.

Carcinomas as a group tend to spread by both blood and lymphatic of vessels. Thus many standard operations for carcinomas include

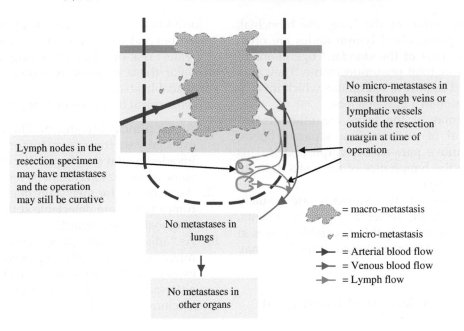

No micro-metastases in transit through veins or lymphatic vessels outside the resection margin at time of operation

Lymph nodes in the resection specimen may have metastases and the operation may still be curative

No metastases in lungs

No metastases in other organs

= macro-metastasis

= micro-metastasis

= Arterial blood flow

= Venous blood flow

= Lymph flow

In principle, all the tumor cells in the primary mass, macro-metastases, and micro-metastases must be included in the resected specimen(s). The resection specimen includes primary tumor and the largest possible margin of normal tissue, including lymph nodes, which might contain micro-metastases.

FIGURE 1.12   Circumstances in which surgery is curative of a tumor.

removal of the draining lymph nodes. For over half a century, the standard operation for carcinoma of the breast included removal of whole breast, underlying muscle, and the axillary, and sometimes additional, nodes all in continuity (Halsted's operation). In recent years, this operation has been almost completely abandoned because it appears to confer little benefit in terms of increased life of the patients over lumpectomy or simple mastectomy. One reason for the failure of lymph node removal to cure all cases of carcinomas is that malignant tumors spread through the blood vessels as well as the lymphatics, so that excision of the lymph nodes does not affect tumor cells which have—in the period of time before diagnosis—invaded local blood vessels and been deposited in distant organs as micro-metastases.

For carcinomas of the colon and rectum, local lymph nodes (i.e., those adjacent to the arteries leading to the tumor) are almost always removed. This improves prognosis because it prevents local recurrence from tumor micro-metastases or larger deposits in the nodes. However, removing additional nodes, e.g., the para-aortic lymph nodes, does not seem to confer any particular benefit.

For carcinomas of the kidney, usually the whole organ is removed with any macroscopically enlarged nodes. If no enlarged nodes are found, the para-aortic region of the body is not resected.

For carcinomas of the lung, the bronchial, hilar, and paratracheal lymph nodes are often resected as part of the standard operations of lobectomy or total pneumonectomy. They may be sampled as part of operations which involve only resection of a part of a lobe.

Carcinomas of the testis are often treated by removal of testis with a separate operation to remove para-aortic lymph nodes. This is because the lymphatics of the testis drain to these nodes first.

Carcinomas of the stomach are usually removed with the lesser and greater omenta, together with any palpable resectable masses in lymph nodes. Any mass found in the liver may also be removed.

1.4.4 General Aspects of Nonsurgical Treatments

Many chemicals and physical agents kill cells. When applied to tissues directly, they have their effects immediately ("acute cell killing"). Examples include:

i. Heat, radiations in high doses, acids, and alkalis. All these agents denature proteins.
ii. Detergents. These dissolve lipids especially of membranes.
iii. Aldehydes. These denature proteins and have the special chemical effect of cross-linking of proteins.
iv. Alcohols. These denature proteins by displacing water from their tertiary structure. Alcohols also dissolve lipids.
v. Specific poisons such as cyanide, which inhibits enzymes of respiration.

Agents which damage normal cells just as much as they damage tumor cells are of little therapeutic value. The only agents which damage tumor cells more than normal cells are therapeutically useful. This effect has been found with radiations given in low doses and with some chemicals. Some further aspects of these therapies are given in the next sections, as well as in Chapter 11.

Individuals whose malignant tumor has spread beyond the field of surgical excision are at greatest risk of ultimately dying of their disease. For these patients, nonsurgical therapies are available [4–6].

Nonsurgical therapies may be given:

i. As the principal treatment. This applies especially when primary tumor has invaded a vital structure in the body, or for some other reason cannot be resected.
ii. As treatment for metastatic masses of tumor which are detectable at the time of diagnosis or afterward.
iii. As "adjuvant" therapy against areas where micro-metastases may be present. An example of this is radiation given to the anterior chest wall after mastectomy for carcinoma of the breast.

1.4.5 Radiation Therapies

Radiations are only useful for those types of tumors whose cell populations are less resistant to radiation damage than normal cells. Radiations only affect the cells which are exposed to them. Disseminated micro-metastases (see above) are not included in the irradiated field in most radiation therapies regimes.

The principles which are the basis of radiotherapy for tumors were established in the early twentieth century. These are mainly:

i. Different sensitivities of different types of tumors.
ii. Greater efficacy when given radiation is administered in split doses ("fractionation").
iii. Different frequencies of relapses for different types of tumors.
iv. Absence of any single target, or cancer-specific target of radiations in tumor cells. Because of this, effective radiotherapy depends on the tumor cells being more sensitive to the treatment than normal cells.

v. Cumulative therapeutic limit for normal tissues. Patients cannot tolerate radiotherapy above particular total lifetime doses.

vi. Radiations have fairly predictable side effects.

For details of radiotherapy of particular tumor types, see Refs. [4–6,59].

1.4.6 Anticancer Drug Therapies

Chemicals which affect tumor cells more than normal cells, and can be given systemically (so that they circulate in the bloodstream), have particular potential for treating disseminated deposits, including micro-metastases in humans.

The principles for anticancer drug therapy were established in the middle of the twentieth century. They are mainly:

i. Diversity in chemical structures of anticancer drugs.

ii. Unpredictability of efficacies against tumors among chemical analogues of known active drugs.

iii. Different sensitivities of different tumor types.

iv. Higher efficacy of administration in split doses ("cycles").

v. Different frequencies of relapses for different types of tumors.

vi. Absence of any cancer-specific target for drugs has been found. All useful drugs are more toxic to tumor cells than to normal cells through the same intracellular targets.

vii. "Treatment memory" of normal tissues. Patients cannot tolerate doses many drugs above particular total doses in their lifetime.

viii. The effects on the tumors and side effects of anticancer drugs are much less predictable than those of radiations.

For details of drug therapies of particular tumor types, see Refs. [4–6].

1.4.7 Relapses

For a few tumor types, nonsurgical treatments cure proportions of these cases (Figure 1.13). However, for the remainder, nonsurgical therapies result in only partial reductions in tumor load and often for only limited periods of time. "Relapses" of tumor are often not sensitive to the agents given earlier in the course of the disease.

Radiotherapies deliver their effect uniformly through tumor masses, without any dependence on blood supply. For these therapies, the likeliest cause of relapses are that tumor cells are present outside the tissue which is irradiated and simply continue to grow after therapy.

The reasons for the partial remissions after drug treatment are unclear and may include the factor that the defenses of some of the cells in the tumor cell population against the drugs are too great (i.e., greater than those of normal cells). Another factor is that vascularization inside the tumor masses may be inadequate for therapeutic doses to be achieved for all cells in a tumor mass.

Until recently, almost all drugs—e.g., derivatives (introduced in the 1940s) of "mustard gas," and anticancer antibiotics (most of which were discovered in the 1950s)—have been assessed simply in terms of their cytotoxic effects. Their exact biochemical "target(s)" have been extensively studied but are still not fully understood.

In recent years, there have been attempts to develop drugs which are selective for particular signaling mechanisms associated with known oncogenes or tumor suppressor genes (see above). These drugs, e.g., BRAF protein inhibitors, are under intensive study at the time of writing.

Other nonsurgical modalities of therapy include immuno-, hormonal, and bio-therapies.

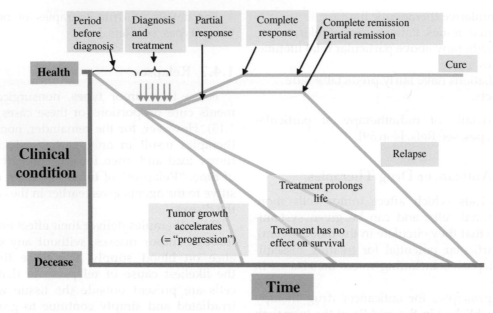

FIGURE 1.13   Overview of potential courses of tumor in relation to nonsurgical anticancer therapies.

1.4.8 Common Side Effects of Nonsurgical Anticancer Therapies

Radiations and almost all drugs have side effects. Almost all of them induce malaise, nausea, and vomiting which in extreme severity can be life-threatening through dehydration and electrolyte imbalance.

Another common, and also life-threatening, side effect is reduction in production of leukocytes in the bone marrow ("bone marrow suppression"). In severe cases, this can be followed by infections by microorganisms which are normally not pathogenic. These organisms include Gram negative bacteria, which are normal saprophytes in the bowel, and fungi, which are normal saprophytes in the mouth. The common infections are of the lungs, the urinary tract, and the skin, all of which can be followed by septicemia and death. The duration and doses of many nonsurgical therapies are limited to the amounts from which the hematopoietic system can recover. A particular

manifestation of neutropenia is ulceration of the oral cavity. This is because the epithelial lining of the mouth is normally protected from bacterial flora by neutrophil leukocytes, as well as mucus. Reductions in the numbers of neutrophil leukocytes can cause overgrowth of these bacteria, followed by loss of epithelial cells and hence ulceration of mucosa.

Other side effects can include skin rashes, renal failure, pulmonary fibrosis, toxicity to the myocardium, and many other disorders.

1.4.9 "Lifetime Limits" of Doses of Certain Anticancer Therapies

In cases of disseminated malignant tumor, the tumor deposits may not be eradicated by nonsurgical therapies. At this point in time, other kinds of nonsurgical treatment can be given. However, ultimately, for many kinds of therapy, there is a limit to the total dose which each patient can accept without the side effects of the treatment becoming intolerable.

This is most clearly seen with radiotherapy, in which only certain maximal doses can be given to any part of the body, or to the body as a whole. The limitations are permanent, and exceeding these limits may cause ulcers which do not heal, and can be followed by infection which may be fatal for the patient. The possible biological bases of these phenomena are discussed in Section 11.5.2.

1.5 PREVENTION

This is the most important dividend which medical research can deliver. As far as tumors are concerned, prevention can be achieved in several ways.

The first is limitation of exposure to carcinogens. If any carcinogen is known to cause a particular kind of tumor, public health measures, as well as alterations in occupational and lifestyle factors can be implemented. General programs of this nature which are already used include:

i. promoting the use of sunscreen lotions for the reduction in skin cancers;
ii. reduction in tobacco smoking in the whole population;
iii. the use of immunization against HPVs in the prevention of cervical cancer;
iv. the reduction of transmission of the human immunodeficiency virus (HIV) in the prevention of Kaposi's sarcoma and other HIV-related malignancies.

The mechanism of prevention in these programs is removing the access of the causative agent to the relevant somatic cells.

A different kind of prevention involves early detection of cases of cancer through screening. These efforts are not aimed at reducing the incidence of new cases of cancer. Rather they are aimed at preventing cancer deaths by detecting and treating the tumors before they have spread in the body (see above). Screening methods are especially used for cancers of the cervix, breast, and large bowel.

Another kind of prevention involves the detection of germ-line mutations which predispose to members of certain families to developing tumors of certain types. This can be done before the individual has actually developed any tumor. These are means of identifying increased risk of particular tumor types. These patients can be followed with more frequent testing so that tumors can be treated early, and hence lives may be saved. Alternatively, identification of genetically at risk individuals may be followed by prophylactic removal of the relevant organ, as in bilateral mastectomy to prevent breast cancer.

Further details of prevention of particular types are given in Chapter 14.

References

[1] Berenblum I. The nature of tumor growth. In: Florey H, editor. General pathology (3rd ed.). London: Lloyd-Luke; 1958. p. 442–5.
[2] (Various authors). Atlas of tumor pathology, Series 1–4. Washington, DC: Armed Forces Institute of Pathology; 1950s to present.
[3] (Various authors). World Health Organization classification of tumours. Several series. Lyon: International Agency for Research on Cancer; 1960s to present.
[4] DeVita Jr VT, Lawrence TS, Rosenberg SA, editors. DeVita, Hellman, and Rosenberg's cancer: principles & practice of oncology (9th ed.). Philadelphia, PA: LWW/Wolters Kluwer; 2011.
[5] Hong WK, Bast RC, Hait WN, editors. Holland–Frei cancer medicine (8th ed.). Shelton, CT: PMPH; 2010.
[6] Alison MR, editor. The cancer handbook, 2 vols. (2nd ed.) Chichester, UK: Wiley; 2007.
[7] Niederhuber JE, Armitage JO, Doroshow JH, editors. Abeloff's clinical oncology (5th ed.). Philadelphia, PA: Elsevier; 2013.
[8] Shimkin MB. Contrary to nature. Washington, DC: U.S. Department of Health and Welfare. DHEW Publication No 79-720, NIH; 1979. p. 67.
[9] Swerdlow SH, Campo E, Harris NL, editors. WHO classification of tumours of haematopoietic and lymphoid tissues (4th ed.). Lyon: IARC; 2008.
[10] Pileri SA, Agostinelli C, Sabattini E, et al. Lymphoma classification: the quiet after the storm. Semin Diagn Pathol 2011;28(2):113–23.

[11] Taylor CR, Hartsock RJ. Classifications of lymphoma; reflections of time and technology. Virchows Arch 2011;458(6):637–48.

[12] Burger P.C., Scheithauer B.W. Tumors of the central nervous system. In: [1], AFIP Atlas Ser 4, Fascicle 7, 2007.

[13] Elder D.E., Murphy G.F. Melanocytic tumors of the skin. In: [1] AFIP Atlas Ser 4, Fascicle 12, 2010.

[14] Centers for Disease Control and Prevention. Cancer data and statistics. Available at: http://www.cdc.gov/cancer/dcpc/data/

[15] World Health Organization. World health statistics 2014. Available at: http://www.who.int/gho/publications/world_health_statistics/2014/en/

[16] American Cancer Society. Global cancer facts and figures. Available at: http://www.cancer.org/acs/groups/content/@epidemiologysurveilance/documents/document/acspc-027766.pdf, Table 4.

[17] Centers for Disease Control and Prevention. Deaths: Final data for 2010. National Vital Statistics Reports. Available at: http://www.cdc.gov/nchs/data/nvsr/nvsr61/nvsr61_04.pdf, Table B.

[18] Parkin DM, Ferlay J, Curado MP, et al. Fifty years of cancer incidence: CI5 I-IX. Int J Cancer 2010;127(12):2918–27.

[19] Walker CL, Cesen-Cummings K, Houle C, et al. Protective effect of pregnancy for development of uterine leiomyoma. Carcinogenesis 2001;22(12):2049–52.

[20] The New York Times. [Internet search - "Causes of cancer"]. Available at: http://query.nytimes.com/search/sitesearch/?action=click&region=Masthead&pgtype=Homepage&module=SearchSubmit&contentCollection=Homepage&t=qry94#/Causes+of+cancer/from20130101to20131231/

[21] IARC monographs on the evaluation of carcinogenic risks to humans. Available at: http://www.iarc.fr/en/publications/list/handbooks/index.php

[22] National Cancer Institute. Causes and prevention. Available at: http://www.cancer.gov/cancertopics/causes

[23] American Cancer Society. Known and probable human carcinogens. Available at: http://www.cancer.org/cancer/cancercauses/othercarcinogens/generalinformation about carcinogens/known-and-probable-human-carcinogens.

[24] Pérez-Losada J, Castellanos-Martín A, Mao JH. Cancer evolution and individual susceptibility. Integr Biol (Camb) 2011;3(4):316–28.

[25] Krepischi AC, Pearson PL, Rosenberg C. Germline copy number variations and cancer predisposition. Future Oncol 2012;8(4):441–50.

[26] Croce CM. Oncogenes and Cancer. N Engl J Med 2008;358:502–11.

[27] Bergonzini V, Salata C, Calistri A, et al. View and review on viral oncology research. Infectious Agents Cancer 2010;5:11.

[28] Weiss RA, Vogt PK. 100 years of Rous sarcoma virus. J Exp Med 2011;208(12):2351–5.

[29] Cosgrove D., Park B.H., Vogelstein B. Tumor suppressor genes. In: [5]. p. 86–101.

[30] Knudson AG. Mutation and cancer: statistical study of retinoblastoma. Proc Natl Acad of Sci 1971;68(4):820–3.

[31] Rahman N, Scott RH. Cancer genes associated with phenotypes in monoallelic and biallelic mutation carriers: new lessons from old players. Hum Mol Genet 2007;16(Spec. No. 1):R60–6.

[32] Cleaver JE. Cancer in xeroderma pigmentosum and related disorders of DNA repair. Nat Rev Cancer 2005;5(7):564–73.

[33] Futreal PA, Coin L, Marshall M, et al. A census of human cancer genes. Nat Rev Cancer 2004;4:177–83.

[34] Harvey A, editor. Cancer cell signalling. New York, NY: Wiley Blackwell; 2013.

[35] Weinberg RA. The biology of cancer. New York, NY: Garland Science; 2013. pp. 159–209.

[36] Reedijk M, McGlade CJ. Cellular signalling pathways. In: Tannock F, Hill RP, Bristow RG, editors. The basic science of oncology (5th ed.). New York, NY: McGraw-Hill; 2013. p. 173–92.

[37] Knudson AG. Antioncogenes and human cancer. Proc Natl Acad Sci USA 1993;90(23):10914–21.

[38] Berger AH, Knudson AG, Pandolfi PP. A continuum model for tumour suppression. Nature 2011;476(7359):163–9.

[39] Ellis IO, Elston CW. Histological grade. In: O'Malley FP, Pinder SE, editors. Breast pathology. Philadelphia, PA: Churchill-Livingstone/Elsevier; 2006. p. 225–33.

[40] Koka R, Ioffe OB. Breast carcinoma: Is molecular evaluation a necessary part of current pathological analysis? Semin Diagn Pathol 2013;30(4):321–8.

[41] Bosman FT, Carniero F, Hruban RH, editors. WHO classification of tumors of the digestive system. Lyon: IARC; 2010.

[42] Bostwick DG, Cheng L. Urologic surgical pathology (2nd ed.). Maryland Heights, MO, Elsevier/Mosby; 2008. pp. 484–6.

[43] Union for International Cancer Control. List of publications. Available at: http://www.uicc.org/resources/tnm/publications

[44] American Joint Committee on Cancer. Desk references. Available at: https://cancerstaging.org/references-tools/deskreferences/Pages/default.aspx

[45] Cancer Research UK. Stages of cancer. Available at: http://www.cancerresearchuk.org/cancer-help/about-cancer/what-is-cancer/the-stages-of-a-cancer

[46] Cancer Research UK. Treating lung cancer. Available at: http://www.cancerresearchuk.org/cancer-help/type/lung-cancer/treatment

[47] American Cancer Society. How is breast cancer staged? Available at: http://www.cancer.org/cancer/breastcancer/detailedguide/breast-cancer-staging

[48] Cancer Research UK. TNM breast cancer staging. Available at: http://www.cancerresearchuk.org/cancer-help/type/breast-cancer/treatment/tnm-breast-cancer-staging

[49] American Cancer Society. Colon cancer treatment. Available at: http://www.cancer.gov/cancertopics/pdq/treatment/colon/Patient/page2

[50] American Cancer Society. Prostate cancer treatment. Available at: http://www.cancer.gov/cancertopics/pdq/treatment/prostate/Patient/page2

[51] Mina LA, Sledge Jr. GW. Rethinking the metastatic cascade as a therapeutic target. Nat Rev Clin Oncol 2011;8(6):325–32.

[52] Harris TJ, McCormick F. The molecular pathology of cancer. Nat Revs Clin Oncol 2010;7:251–65.

[53] Wilson GD. Cell kinetics. Clin Oncol (R Coll Radiol) 2007;19(6):370–84.

[54] Fearon KC, Glass DJ, Guttridge DC. Cancer cachexia: mediators, signalling, and metabolic pathways. Cell Metab 2012;16(2):153–66.

[55] Farrow EG, White KE. Tumor-induced osteomalacia. Expert Rev Endocrinol Metab 2009;4(5):435–42.

[56] DeLellis RA, Xia L. Paraneoplastic endocrine syndromes: a review. Endocr Pathol 2003;14(4):303–17.

[57] Lippitz BE. Cytokine patterns in patients with cancer: a systematic review. Lancet Oncol 2013;14(6):e218–28.

[58] Silberman H, Silberman AW, editors. Principles and practice of surgical oncology. Philadelphia, PA: LWW/Wolters Kluwer; 2010.

[59] Halperin EC, Wazer DE, Perez CA, editors. Perez and Brady's principles and practice of radiation oncology (6th ed.). Philadelphia, PA: LWW/Wolter Kluwer; 2013.

Theories and Definitions of Tumors

L.P. Bignold: Principles of Tumors.
DOI: http://dx.doi.org/10.1016/B978-0-12-801565-0.00002-0

The natures of the different kinds of swellings in the body have been recognized problems in medicine since Ancient times. With the advent of cell theory (in 1839, [1]) and cellular pathology (in 1858, [2]), it was recognized that accumulations of cells occur in many kinds of swellings in the body. Since that time, much research has been directed to (i) distinguishing "true" tumors from other cellular swellings and (ii) identifying the natures and causes of the abnormalities in the cells of the "true" tumors.

The two issues are related because defining "true" tumors (see Section 1.1.1) is an essential aspect of diagnosis. Malignant tumors especially must be precisely diagnosed so that the numbers of each type of tumor, and hence the "cancer burden" on the community, can be accurately determined (Chapter 3). Precise definition of the disordered biology being investigated—in both human lesions and experimental models—is important for satisfactory research into all aspects of tumors: causes, mechanisms of formation, as well as optimal treatments.

This chapter discusses the major theories and definitions for all tumors, including recent descriptive characterizations of cancers.

2.1 DISTINGUISHING TUMORS FROM OTHER SWELLINGS

At the time when cell theory and cellular pathology (see above and Figure 2.1) were first proposed, techniques in microscopy could allow cells and nuclei to be seen, but were not adequate for visualizing smaller structures, such as chromosomes, or small abnormalities in cells. For this reason, accurate assessments of all cases of the different kinds of swellings in the body were not possible.

However in the late 1870s–1880s, several major aspects of microscopic technology, including apoachromatic lenses and substage "condensers" of the light source, were improved almost to their current standards [4]. These inventions enabled the different kinds of swellings to be distinguished from each other with reasonable reliability. The nontumorous swellings in the body were then identified as mainly:

i. All physiological swellings, e.g., of the uterus, breasts, and thyroid gland during pregnancy.

ii. Accumulations of fluid in tissues, as in edema.

iii. Cysts not associated with any tumor cell population (e.g., formed by the obstructions of ducts of secretory glands, as in sebaceous cysts of the skin).

iv. Swellings associated with inflammations (see Chapter 10.8). The cells of inflammatory swellings are almost always morphologically normal or show changes of degeneration or necrosis (Chapter 10). Inflammations have a characteristic general tendency to resolve or heal, which is unlike virtually all cases of "true" tumor.

v. Swellings associated with repairs to tissues (e.g., the callus of bone fracture sites). These, like inflammations, disappear when the broken bone ends are reunited and fully healed with new bone.

vi. Hyperplasias and hypertrophies. Strictly, these terms refer to responses of tissues to abnormal degrees of stimulation. According to this strict definition, they regress when the stimuli to their formation are removed. An example is hyperplasia of the epidermis in lichen simplex chronicus, which is caused by chronic scratching. It disappears when the scratching stops.

However, "hyperplasia" and "hypertrophy" are also applied to enlargement of organs where no stimulus is known. An example is the enlargement of the prostate because of dilatation of glands and proliferation of epithelium. This condition is called "benign hyperplasia" of the prostate, although no stimulus to the enlargement of the gland is known. For this reason, it is sometimes classified as a "tumor-like lesion" (see Section 3.4).

vii. Congenital lesions of organs characterized by abnormal tissue composition.

According to strict sense, hamartomas are developmental (i.e., congenital) malformations which comprise normal tissues of the organ in abnormal proportions and spatial relationships. They grow at the same rate as normal parts of the body in early life and childhood and cease growing in adulthood in coordination with normal tissues. An example is the congenital "birth mark"/"port-wine stain." However, the original [5,6] and many subsequent authors including Willis [7] have used "hamartoma" for lesions having this kind of composition but which appear only in childhood or adult life, and grow slowly and continuously. These lesions are not developmental abnormalities. They are essentially very benign true tumors..

The term "choristoma" is used for lesions which have the characteristics of hamartomas in the strict sense, except that they comprise tissues which are not normal to the organ. Lesions with these features are usually referred to as "ectopias."

viii. Other enlarging "tumor-like" lesions which are not included in the above categories.

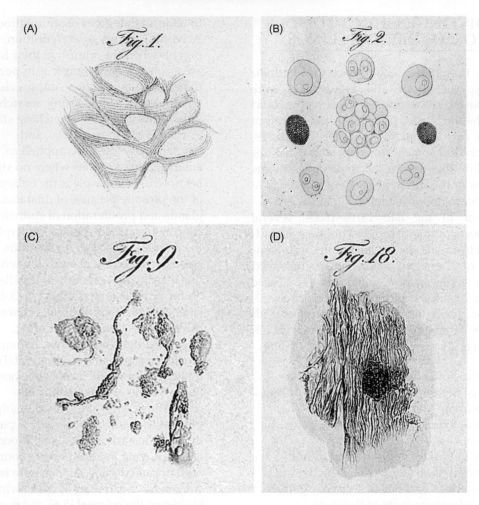

FIGURE 2.1 Early published drawings of cytological abnormalities in the cells of cancers of the breast [3].
The images have been taken from Drawings sheet 2 in the original work (1838). (*Courtesy of The Wellcome Library, London*).
The following legends are from the English translation of the book (1840) [3].

A. "Fig 1". Meshes formed by the bundles of fibres of carcinoma reticulare of the breast as they appear after the globules are removed.
B. "Fig 2". Globules from the reticulum of carcinoma reticulare. Within the globules are germinal cells with their nuclei, and on either side of the figure is a granular opaque corpuscle.
C. "Fig 9". Very irregular caudate bodies from a soft fungus of the female breast, the precise nature of which was never accurately ascertained.
D. "Fig 18". Bundles of fibres from a fibrous tumor of the mamma, in Professor Pockel's museum.

Note absence of detail in cytoplasm and nuclei of cells.
The other figures in the book depicted a variety of tumors, including in stomach, soft tissue, brain and bone.

2.2 EARLY DEFINITIONS OF TUMORS EMPHASIZED UNCONTROLLED GROWTH

In the late nineteenth century, when true tumors could be reliably identified, microscopists looked for a morphological feature, or combination of features, which is present in all true tumors and absent from all normal circumstances and nontumorous conditions. No such feature or combinations of features was ever found. The combinations exhibited across the whole range of tumor types (Section 8.2) are highly variable. All the morphological features of tumor cells, such as altered specialization and changes in size of nuclei, are also seen in various normal or nontumorous pathological conditions. Even abnormalities in mitoses (see Section 8.1.4) are found in nontumorous reactive conditions. In 1890, Hansemann [8] summarized the situation as:

> The absence of any principle is the principle of cancer cells.

Since then, almost all authorities have agreed that the only common feature of tumors is inappropriate and unceasing "autonomous" accumulation of cells [9–12]. As well as to tumorous swellings, the definition also applies to most cases of "*in situ*" tumors (Sections 3.4 and 8.2.9). In these lesions, the whole tissue may not be markedly increased in volume, but tumor cells may fill normally cell-free spaces such as the lumina of ducts.

2.3 THEORIES INVOLVING DISORDERED BIOLOGICAL PROCESSES

In the eighteenth and nineteenth centuries, anatomists and pathologists were studying all aspects of tissues and cells in normal biological processes (e.g. embryonic development), as well as in the nontumorous pathological processes, such as inflammation and repair

(Section 10.7). They were also studying tumors, on the theoretical assumption that there would be one process to fit all tumor types. Two different fundamental ideas about tumors became established:

i. Tumors arise through a deviation in one or other normal biological or nontumor pathological processes.
ii. Tumors arise through the induction or activation of a unique pathological process.

The main examples of these were as follows.

2.3.1 Degeneration

Degeneration as a mechanism of change in populations of living things was suggested in the second half of the eighteenth century [13]. In the same period, the source of new tissues was thought to be the formation of solid particles in "lymph" (at that time, "lymph" meant essentially any translucent fluid in the tissues) [14]. Consistent with this general idea, Le Dran in 1768 [14] suggested that tumors derive from degenerations of "lymph." In 1802, Bichat described the parts of the adult body in terms of individually identifiable "textures" (Fr: *tissu*), hence in English "tissue" (see Section A1.1.3). Bichat [14] suggested that tumors are due to "degeneration" in these tissues. In the next hundred years, a few authors such as Rindfleish [15] suggested that tumors are degenerate cellular growths, and the term "malignant degeneration," particularly of benign tumors (see Section 6.4), persists in medical terminology.

2.3.2 Hypothetical Growth Processes: "Plasias" Including Neoplasia

In the late 1820s, Lobstein [8] described the different kinds of normal and pathological tissues in terms of plasias. The underlying concept was that different morphologies are the results of the actions of different hypothetical "plastic"—meaning "form-making"—influences. Hence he called tissues undergoing normal growth "euplastic," excessively grown normal tissue "hyperplastic," mildly abnormal excessively grown tissues "homeoplastic," and markedly abnormal excessively grown tissues were called "heteroplastic." The suffix "-plasia" is universally used in medical terminology.

The term "neoplasia" was used by Virchow in 1854 [16] to indicate the idea that "true" tumors occur through a particular hypothetical new kind of tissue-forming influence. This process was considered unrelated to any normal biological process, or other pathological process. Initially, the idea was not popular but the term and concept were later was taken up, e.g., by Lancereax (1888) [17] and have been widely used since the early twentieth century [2,4,18].

"Anaplasia" was introduced by Hansemann as a hypothetical common process in all malignant tumors and has been used ever since—although not according to its original meaning—to describe highly malignant tumors [8].

2.3.3 Deviations in Actual Biological and Nontumorous Pathological Processes

These theories are based on the idea that tumor cells are simply normal cells in which there is activation, or excessive activity, in one or more normal biological phenomena (Figure 2.2) such as are found in one or other of the following:

i. Any earlier phase in the life of the particular kind of parent cell.
ii. The repertoire of reactive processes of the particular kind of parent cell.
iii. Phases in the life of any other kind of cell.

(a) Early Ideas of Embryonic Development-Related Phenomena

Possibly the first example of this kind of theory was put forward by Royer-Collard in the 1820s [14]. The idea was that tumors are made up of embryonic-like cells and was based on the observation that embryonic cells grow more rapidly than adult cells. In the later nineteenth century, microscopic studies showed that embryonic cells also occasionally invade adjacent tissues and metastasize (see Section A1.5). The idea of embryonic activation in tumors was then developed in different versions by Cohnheim in 1867 [19], Boll in 1876 [14], Ewing [10], and others including Foulds [18].

In the late twentieth century, there has been renewed interest in possible embryonic phenomena in tumor formation, as is discussed in Section 2.9.1.

(b) Reduced Specialization

Another group of theories relating to normal biological processes invokes aspects of "differentiation" in the sense of specialization of cells (see Section 6.4.5). Hansemann was first to use the term "differentiation" in relation to tumors (see Ref. [8]). However, others have used "differentiation" with different meanings (see Ref. [8]). Among these related concepts, a common suggestion has been that "failed senescence"/"failed physiological cell death—apoptosis" (see Section 10.5) is the major cell biological phenomenon of tumor formation. Essentially, this theory suggests that the normal cessation of cell division associated with full specialization does not occur because of loss of the genes for senescence/physiological cell death, or alternatively, loss of abnormal epigenetic regulation of these genes [20].

This is in contrast to the view that the abnormality of tumors lies in excessive rates of proliferation at the level of the local tissue stem cell (see also Sections 2.9.1 and A1.3.2).

(c) Excessive Angiogenesis

This theory describes excessive growth of blood vessels as a feature of malignant tumors [21–24]. The theory relates to the mechanisms of tumor spread, rather than to the tumorous conversion of normal cells.

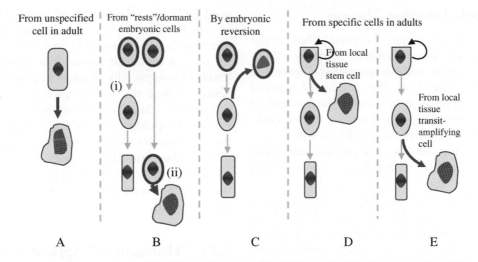

A. In any cell in the body capable of proliferation, direct conversion occurs through "degeneration" or "neoplastic process". All these processes may occur by ± inflammatory, hormonal, or other influences.

B. Normally (i) all embryonic cells develop to adult cells.
In this theory, a few embryonic cells are postulated to "rest" (become dormant) and will be present in adult tissues (ii), where they give rise to form tumors when they are stimulated in some way.

C. The tumor-forming event is in partly specialized cells in adults. The event causes them to revert to embryonic behavior, especially in growth and little specialization. The theory does not explain the morphological variabilities in tumors (Sections 8.2 and 8.3).

D. Local tissue stem cells in adult tissues incur the tumorous event. Immortality of the tumor cells is inherited from stem cell. See Sections A1.3.3.

E. Transit-amplifying cells in adults incur the tumorous event. The result is abnormal specialization. At the same time, the cells must incur an event with confers immortality on the population of descendant cells. This is because all normal descendants of transit-amplifying cells mature to full specialization and die (see Sections 10.3 and A1.3.3).

FIGURE 2.2   The major theories of tumors involving biological processes.

(d) Abnormalities in Inflammatory Responses

In this theory, tumors derive from excessive and/or deviant inflammatory, reactive or regenerative responses to "chronic irritation," or possibly other factors. This idea was put forward by Broussais in1832 [14], and strongly endorsed by Virchow [16], Nicholson [25], and Haddow [26]. At the present time, several authors are exploring further aspects of this possible cause of tumors [27–31].

(e) Abnormal Directions of Specialization

These theories involve altered specialization pathways, as occurs as a nontumorous pathological process. Thus a malignant tumor might develop because its parent cell began to express the features of other kinds of cells, especially leukocytes (see Sections 6.6–6.8). The idea was originally suggested by Virchow in terms of "metaplasia" [32] and is discussed in Ref. [8]. The concept is often described in terms of "reprogramming" of gene activations [33].

2.3.4 Early Infection Theories

At this point, it may be noted that at the beginning of the twentieth century, theories of infection and/or parasitism were the dominant concept of the origin of tumors [14]. The evidence was mainly the finding of microorganism-like structures in tumor cells. One of these discoveries—by Fibiger in 1913—was that a parasite in cockroaches can cause stomach cancer in mice. Fibiger was awarded a Nobel Prize for Medicine on the basis of this discovery, but more recently, his explanations of the phenomenon have been derided [8,34]. Nevertheless, the phenomenon is an example of how a particular hereditary predisposition may require a particular carcinogenic agent for the tumor to occur (see also Section 7.1).

2.3.5 "Blasts" in Tumor Terminology

The use of "blast" in embryology and histology is described in Section A1.2.7. The same word is also used in the nomenclature of tumors. In the 1830s and 1840s, tumors were thought to arise in extracellular lymph (see Section 2.3.1). Cohnheim [19] referred to tumors as "blastomas" because he thought they all derive from "leftover" embryonic cells, which by then were being referred to as "blasts" (see Section A1.2.5). Borst [35] widened the application of this idea and used "blastomas" for all kinds of pathological lesions which include cellular proliferation.

Currently, different types of tumors are known as "blastomas," but with different meanings. Thus tumors which appear in fetal life or early infancy are derived from embryonic cells and are called "-blastomas" as in "retinoblastoma." In adults, tumors which simply have some morphological similarities to embryonic cells may be called "blastomas." This convention is followed especially in the nomenclature of tumors of the brain in adults, where terms such as "glioblastoma" and "astroblastoma" are

used [36]. The other kinds of tumors in adults in which "blast" is commonly used are those tumor cells which resemble proliferating precursor cells of the adult tissue. This terminological tradition is especially marked in naming of leukemias, where the normal precursor cells have been called "blasts" (as in "myeloblast") and leukemias comprising cells which mainly resemble these precursors are referred to as "blastic" [37].

Thus the term "blast" does not denote any particular theory of tumors, but rather the tumor cells resemble the "blast" cells in normal cell processes.

2.3.6 "Histogenesis" Applied to Tumors

"Histogenesis" in biology refers to the embryonic development of specialized forms of organs and cells from unspecialized tissue [38]. In the late nineteenth century, the term was applied to the idea that individual tumor types arise from individual kinds of normal cells, rather than from a hypothetical generalized tumor precursor cell [39]. Currently, the term is mainly used in discussions of the kind of normal cell from which a particular tumor type may arise. This is particularly significant in theorizing on the origins of tumors with mixed histological appearances, such as carcinosarcoma [40–42].

2.3.7 Discussion of the "One Process Fits All" Theories

Despite their wide popularity, however, it should be remembered that the concept of any single process such as "neoplasia" does not explain the marked distinctions which exist benign and malignant categories of tumors, or any of the other complex differences between the various tumor types (Section 8.2). Most prominently, there is no explanation in the concept of a "neoplastic process" of how tumors are mainly either benign or malignant, as well as other phenomena (see Sections 8.2–8.4).

2.4 THE FIRST GENOMIC THEORIES RELATED TO ABNORMAL CHROMOSOMES

2.4.1 Hansemann's Theory

The first proposal that tumors depend on a disturbance of the intrinsic hereditary material of the cell was made by David Paul Hansemann (1858–1920; see Ref. [8] and Figure 2.3). In his first paper on the topic, Hansemann (1890) suggested that the primary disturbance in tumor cells is in the "balance" of their chromosomal numbers. He suggested ways in which these abnormalities could possibly be induced by carcinogens and might be related to the nuclear changes, altered specialization, and lineage fidelity in tumor cells. Hansemann initially supposed that the chromosomal abnormalities were an integral part of a deviant, öogenesis-like process in a normal cell. Hansemann called the process "anaplasia," which he defined as a process resulting in the phenotypic effects (although that term was not used at the time) of:

i. "dedifferentiation" (loss specialization from öocyte to ovum) together with
ii. an "increased capacity for independent existence" (meaning metastatic potential).

Within 10 years, Hansemann abandoned the öogenic aspect of his theory and in 1904 suggested that tumors form from *two*, possibly particular, re-assortments or damage events in chromosomes in parent cells (see Ref. [8], p. 287–289).

Regardless of the öogenic and chromosomal aspects of this theory, Hansemann's concept of degrees of anaplasia was useful to histopathologists. This was because previous classifications of tumors had used two discrete categories based on Lobstein's ideas [8]. In the main, this meant dividing tumors into "homeoplastic" and "heteroplastic." This classification offered no way of indicating degrees of abnormalities. Hansemann's terms "anaplasia" and "(de)differentiation" became universal because they could

be used for phenomena which are continuously variable (as discussed in Section 8.3).

2.4.2 Other Chromosomal Theories and Opposition to Them

In 1914, Boveri [43] described a theory of tumors based on observations of chromosomes in cells of doubly fertilized (i.e., triploid) sea urchin eggs. No other cell biological abnormalities in tumors were considered [44,45]. The idea was essentially a modification of Hansemann's ideas of imbalances in chromosomal numbers causing tumor formation. He gave little discussion of issues such as specialization in cells, which Hansemann's theory addressed.

Subsequently in the twentieth century, several phenomena were discovered which suggested lesser importance of altered chromosomal numbers in the general biology of cells.

i. The overall numbers and size of chromosomes do not fully correlate with the complexity/"evolutionary level" of the corresponding species. Thus the wheat plant has an enormous genome (in fact, is tri-diploid, with17 trillion bases—approximately five times the size of that of the human)—while the rice plant has only 430 billion nucleotides.
ii. The number of chromosomes can vary enormously between morphologically similar species. For example, the Chinese muntjac deer has its genetic material in 46 small chromosomes, while the Indian muntjac has its genetic material in 6 large chromosomes [46].
iii. Somatic chromosomal abnormalities are inducible in animal and human cells (e.g., by vitamin C exposure in lymphocytes) without necessarily any carcinogenic effect [47].
iv. Chromosomal polymorphisms (including "variant," "marker" and supernumerary chromosomes) occur in phenotypically normal humans [48].

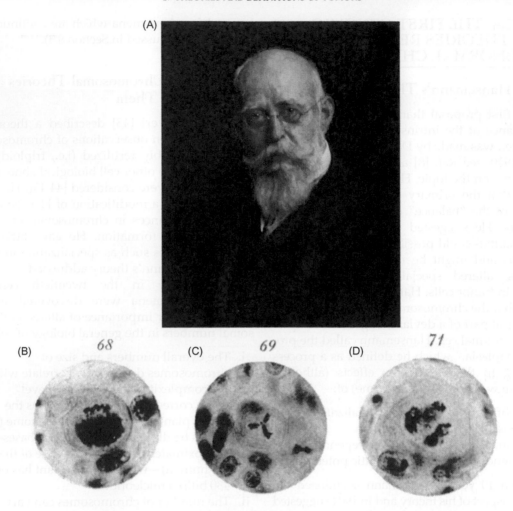

FIGURE 2.3   David Paul Hansemann (1856–1820).

(A) Portrait (courtesy of Herr Wolfgang von Hansemann, grandson).

(B-D) Photomicrographs of abnormal mitotic figures in tumor cells published by Hansemann in 1893.

Hansemann's main contributions to oncology were:

- The idea that tumors may occur through and endogenous changes in the hereditary material of normal cells
- The suggestion that these changes might come about through abnormalities of chromosomes (whole or in parts)
- The words 'de-differentiation' and 'anaplasia', which he used to denote the morphological and behavioral abnormalities in tumor cells brought about by the chromosomal lesions. The words have become almost universally used in medicine for these abnormalities, but without reference to chromosomal abnormalities.
- Descriptions of many other general aspects of tumors.

In relation specifically to cancer cells, various authors, including Koller [49], studied chromosomes in cancer, but no single, constant chromosomal lesion could be found in all types of tumors.

Only in 1961 was any specific chromosomal lesion linked with a particular neoplastic process (chronic myeloid leukemia), as is discussed in Sections 2.9.7 and 9.5.

2.5 CONTRIBUTIONS OF R.A. WILLIS

This author wrote three insightful books on tumors and their relationships to normal biological phenomena [50–52]. Here, two ideas in his *Pathology of Tumors* [50] are discussed.

2.5.1 The Un-Mendelian-Type Phenotypic Changes in Tumors Are Not Consistent with Causation via a Single Somatic Genomic Event

After reviewing the abnormal morphologies and behaviors of tumor cell populations in the various common types of tumors, Willis was forced to discount Mendelian genetics [53]—as the subject was understood at the time—as having any role in tumor formation. His implicit standpoint was that these histopathological features are phenotypic alterations in normal parent cells. This idea has been little discussed subsequently and will be explained here.

When Willis was forming his view (1940s), Mendelism was understood as follows. Mendel had worked before chromosomes were discovered, and his laws were based on the observations of only a limited number of traits in one test organism (the sweet pea, Figure 2.4). His experimental model had the following features:

i. The phenotypic changes were "all-or-nothing," i.e., not showing variability.
ii. The phenotypic changes were always fully expressed in one generation (for dominant traits) or two (for recessive traits—in the F2 generation).
iii. These effects do not change further in subsequent generations without another genomic event.

As his first law (1868), Mendel concluded from his results that:

• "Characters"/traits are determined by parentally derived paired "factors"/alleles; one from each parent.

• Each allele, regardless of from which parent, is either "dominant" or "recessive" (except in sex-linked disorders). Two "dominant" alleles have the same effect on the trait as one dominant allele. Only two recessive alleles change the trait.

Mendel also proposed (as his "second law") that the "factors" are distributed independently during production of gametes. In the 1900s however—after the discovery of chromosomes—this second "law" was found to be incorrect. In fact, all alleles on the same chromosome are distributed together, except when "crossing over" occurs [54]. Later it was found that some genes lie so close together on a chromosome (are "linked") that they form a "complex locus," so that all the traits dependent on the relevant genes can be altered in a single genomic event [55]. Continuous variation was initially thought to be inexplicable by Mendel's Laws, but later, it was suggested that all variation could be explained by polygenism—i.e., multiple genes for the same trait [56]. Polymorphism of genes (i.e., quantitatively different phenotypic effects through different genomic events in the same allele) was discovered only through sequencing of DNA [57].

Somatic genomic events as the mechanism of tumors were mooted by several early authors, including Whitman in 1919 [58]. In the 1928, Muller [59] published observations that radiations—which by that date were known to be carcinogenic—are capable of producing germline mutations. This led to renewed discussion of the question of somatic mutation as the basis of tumors, especially by Bauer in 1928 [60] and Lockhart-Mummery in 1934 [61].

Nevertheless, despite these discoveries, the possible relevance of Mendelian genetics to tumors continued to be rejected because the features of tumors are unlike those used in studies based on Mendel's work (see Figure 2.4). The point was made by Willis [50], when he described the differences between tumors

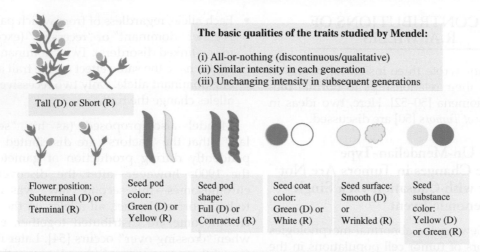

Tall (D) or Short (R)

The basic qualities of the traits studied by Mendel:

(i) All-or-nothing (discontinuous/qualitative)
(ii) Similar intensity in each generation
(iii) Unchanging intensity in subsequent generations

Flower position: Subterminal (D) or Terminal (R)

Seed pod color: Green (D) or Yellow (R)

Seed pod shape: Full (D) or Contracted (R)

Seed coat color: Green (D) or White (R)

Seed surface: Smooth (D) or Wrinkled (R)

Seed substance color: Yellow (D) or Green (R)

In tumors, these basic qualities are not observed.
Tumors show
(i) Continuous/quantitative variabilities in almost all cellular and molecular features consistent with viability of a cell (see Section 8.3). Thus, in a single case of carcinoma of the colon, all grades between the following patterns of cellular abnormalities can be seen.

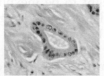
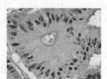

Small size lining small cystic spaces

Large size lining small cystic spaces

Small cells lining large spaces

Very small cells forming no structures

In addition,
(ii) Different cases of the same tumor type often have different intensities of abnormalities.
(iii) Later generations of cells may have different intensities of characteristics compared to their parents ("progression").

FIGURE 2.4   The un-Mendelian features of tumors.

and the phenomena which Mendel studied, as follows:

Benign and malignant tumors are *not* (original emphasis) sharply distinct; in any given class, the individual tumors show every possible gradation of behaviour .... and to explain this, in terms of gene mutation, we would have to suppose as many mutations as there are tumors. Moreover, experimental carcinogenesis .... and structural study of the mode of origin of tumors .... both show plainly that neoplastic change does **not** (original emphasis) take place suddenly, but in a gradual or cumulative

manner, hyperplasia often passing insensibly into neoplasia or benign non-invasive neoplasia into malignant invasive neoplasia without any sudden changes in cell structure or behaviour. These plain facts are incompatible with the mutation hypothesis.

This essentially means that the basic features of tumors are contrary to the Mendelian qualities of phenotypic changes noted in (i), (ii), and (iii) listed at the start of this section. Tumors do not begin with "all-or-nothing" phenotypic changes which are fully expressed in one generation of daughter cells, and in addition, the phenotypic features frequently change thereafter. Moreover, tumors show enormous variability as is discussed in Section 8.3:

- The tumor types vary from one another.
- Different cases of the same type of tumor vary from one another.
- Even different foci of the same case of tumor vary from one another. Finally, many individual cases of tumor change their morphological and behavioral features with time (Section 8.4).

2.5.2 Willis' Definition of Tumors

Willis' other major contribution was a definition of tumors which took into account the advances made in the 1930s and 1940s in the study of experimental carcinogenesis (Section 4.1.3). Willis [50] amended the definition of tumors based only on abnormal growth by including the concept that the tumor cell population is permanently altered from a normal cell by the carcinogen(s). His definition was:

> A tumor is an abnormal mass of tissue, the growth of which exceeds and is uncoordinated with that of the normal tissues, and persists in the same excessive manner after cessation of the stimuli which evoked the change.

This definition indicates that tumors arise from specific causative events, but is perhaps not otherwise greatly different from the earlier definitions. The idea of growth persisting after cessation distinguishes tumors from hyperplasias, as indicated in (iv), (v), and (vi) in Section 2.1.

The definition does not indicate any pathogenetic mechanism of the tumorous change nor does it include the most fundamental characteristic of tumors: that they occur in many different types. The definition is consistent with the idea that the change in the normal cell is a genomic event (Section 1.2.2). However, Willis discounted this mechanism

(see above) and declared in relation to the issue of pathogenesis of tumors "*Ignoramus*"— "We do not know" [50].

Willis' definition has probably been the most popular over the last half century. Foulds (1969, 1976) [18] gave a full account of the difficulties in defining tumors and concluded with a version of Willis' definition. Willis' definition has been used by many subsequent authors, e.g., Refs. [62–64].

2.6 ACCEPTANCE OF THE SOMATIC GENOMIC EVENTS THEORY

2.6.1 Viruses Carry Oncogenes

In the first half of the twentieth century, specific "filterable agents" in extracts of a small number of animal tumors were found to be able to transmit the corresponding disease to other individuals of the same species [65]. Little progress could be made without knowledge of the components of viruses or the identity of the hereditary material in any kind of viruses or any kind of cell.

In the 1950s, three critical fundamental developments were made. These were:

i. The method for sequencing amino acids in proteins [66].
ii. The realization that altering the sequence of nucleotides in DNA might be a mechanism of phenotypic changes [67].
iii. The discovery of the "malignant transformation *in vitro*" model of tumor formation (see Section 14.2.3). In this regard it was found that many cultured cell lines may produce "transformed" cells as a result of infection with viruses (Figure 2.5).

For a time, it was thought possible that a large number of human tumors might be due to direct infection by viruses [68,69], but ultimately it was realized that except for a small number of types of human tumors, the

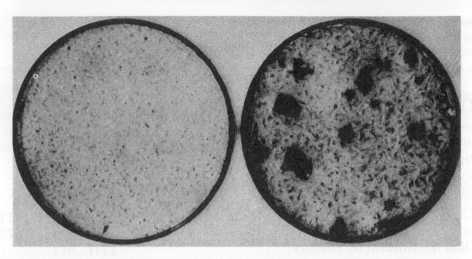

FIGURE 2.5   "Transformation of human fibroblasts *in vitro* by SV40 virus. The left petri dish shows a control culture not infected with SV40. The one on the right shows a culture three weeks after infection with SV40. Both have been fixed with formalin and stained with 1% hematoxylin. Each of the large dark staining areas is a transformed colony produced by SV40". Originally published in 1976 [69]. *Reproduced with permission.*

activated genes associated with the tumorous growth are genes of the human genome, and not transfected viral genes (see Section 4.5).

2.6.2 Uni- or Oligonucleotide Error Genomic Models

Prior to the 1950s, the fact that some chemicals and radiations can cause tumors in humans and experimental animals was established. By the 1960s, it had also been shown that some chemicals and radiations can cause "transformation *in vitro*" of a variety of cultured cells. The relationships between these phenomena and genomic events in these cells, however, remained unclear. Chromosomes were known to contain the hereditary material of the cell, but no studies had shown any constant relationship between any particular chromosomal abnormalities induced by these agents and carcinogenesis [49] (see Section 2.4.2). Also, it was known that loss of function in a gene can lead to loss of a protein, especially an enzyme, and thence to a disease, as in phenylketonuria [70].

In 1959, Ingram [71] showed that the inherited defect in sickle cell anemia results in a single amino acid error in the beta globin chain of the hemoglobin complex. This hemoglobin functions normally in normal circumstances. However, in some situations—mainly hypoxia—the protein complex partially loses its physiological function. In mild hypoxia, the amino acid substitution allows the tertiary structure of the complex to alter, causing changes in its oxygen dissociation curve. In more severe hypoxia, the collapse of the tertiary structure of the hemoglobin complex is so great that the erythrocytes shrink into sickle-shaped forms and become nonfunctional.

After this signal discovery in molecular biology, it was thought possible that amino acid errors in other proteins might be capable of causing "neoplasia" in cells and hence be responsible for many, if not all, tumors.

This "proto-model" could only be investigated when (i) DNA could be efficiently sequenced [72] and (ii) sequencing was made more practicable by the discovery of the polymerase chain reaction [73]. The model has been the basis of much of the work on the genomic bases of tumors ever since (see Chapters 4–7).

2.6.3 Theory of Multiple Uni- or Oligonucleotide Error Genomic Events Producing a Monoclonal Tumor Cell Population

Studies of tumor cell genomes have not so far discovered any single nucleotide or other genomic event (previous section) which is present in the genomes of all tumors. However, the idea that the induction of tumorous change in cells might require more than one genomic event [74] has become more popular. This is a departure from a previous idea that "initiation" may be due to a genomic event, and "promotion" is due to some other phenomenon (see Section 4.1.5).

The multigenomic event theory may be seen to accord with certain general results of experimental carcinogenesis, which have shown:

i. The general need for multiple applications of carcinogens for most tumor formation (Section 4.1.4a).
ii. The particular character of the "two-step" models of experimental carcinogenesis (see Section 4.1.5b).
iii. The necessity for a somatic "hit" to occur in a particular cell for a tumor to form in individuals who carry germ-line events which predispose them to tumors (Section 7.2.2).
iv. The necessity for two transfections and two-phase transformations of cells *in vitro* (Section 14.2.3).

In recent decades, multihit models of oncogenes have been proposed for the adenoma–carcinoma sequence in the human colon (e.g., Ref. [75]).

The genomic event theory of tumors is now almost universally accepted.

2.7 THE SUGGESTED LIMITED POLYCLONALITY OF TUMOR CELL POPULATIONS

In the 1960s–1970s, somatic genomic alterations were thought to be relatively rare, and only a few could be sustained by a cell without loss of viability. The feature of the Mendelian experimental model—that all descendant cells being alike—continued to be applied to tumor cell populations. That is to say, tumor cell populations were thought to be uniclonal descendants of original cells, which had suffered only one or a few genomic events. As a corollary to this, tumor cell populations were thought to genomically homogenous—i.e., have no differences from each other in their nucleotide sequences.

In the late 1970s–1980s however, it was established that metastases of individual cases of malignant tumors can comprise cells with markedly different properties from those of the primary tumor (see Section 8.4). In the same era, expansions of populations of cells (as "clones"), e.g., of "committed" antibody-producing cells, were discovered [76]. This theory of clones rapidly became an accepted phenomenon in biology.

"Clones" of tumor cells in the same population are currently defined in two circumstances:

i. Where two or more different cell lines are cultured *in vitro* from the same case of tumor.
ii. When different cells in the same case of tumor show different protein expressions (by immunoperoxidase stains) or gene copy numbers (by *in situ* hybridization methods—see Section 9.2.6).

Polyclonality is currently often referred to as "heterogeneity" (see Section 8.3.6).

There are several possible origins for polyclonality [77,78]. The theories for the origins of multiple clones include the following (Figure 2.6).

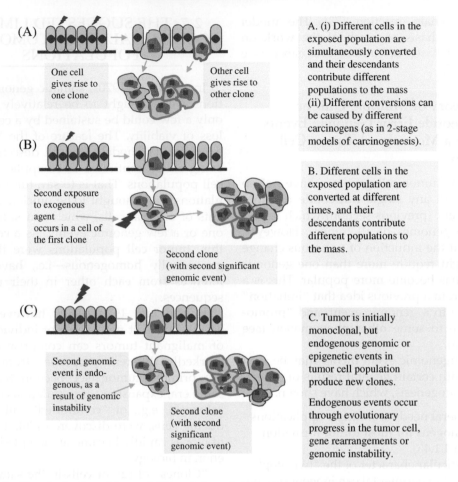

FIGURE 2.6   Theories for the existence of different clones in the one tumorous lesion.

2.7.1 Multiple Clones Arise from Simultaneous Genomic Events in Multiple Adjacent Originally Normal Cells

This theory is based on the Mendelian preconditions that one or more parent cells become mutant, and that the descendants of originally mutant cells are monoclonal, unless acted on by another dose of carcinogen.

According to this theory, the clones arise from an original carcinogenic exposure which has affected multiple different cells in the same location (Figure 2.6A). The tumor is

"polyclonal" because each original cell gives rise to a different clone. The clones are similar to one another because they arise from the same kind of cell, but differ from one another perhaps because of slightly different carcinogen-induced genomic events [78,79].

2.7.2 Multiple Clones Arising from Different Applications of Carcinogen and Having Synergistic Effects on Each Other

This idea is that the initiation of the tumor is due to one carcinogen, but that other carcinogens (i.e., a multifactorial suggestion) then cause different genomic events in different cells in the normal population, creating different, but related, clones [80] (Figure 2.6A).

2.7.3 Multiple Clones Result from Surviving Subpopulations After Additional Genomic Events Occur Endogenously, Possibly in Specified Steps of Its "Evolution"/"Progression"

This idea is essentially a polyclonal version of the Vogelstein–Kinzler model [75]. According to it, successive genomic events produce clones of greater malignancy (Figure 2.6B). However, the earlier clones survive, so that the ultimate population comprises clones in various stages of progression to full malignancy [81].

2.7.4 Theories of Endogenous Further Genomic Alterations

Rearrangements of immunoglobulin genes are a feature of lymphocytes, and it may play a role in lymphoid malignancies [82]. The term "gene rearrangement" is sometimes used for other genomic abnormalities, especially neogene formation via chromosomal aberrations, e.g., Ref. [83]. This process may be considered an endogenous mechanism for producing new clones of cells (Figure 2.6C). It is unclear whether lineage-inappropriate activations of abnormal versions of the normal lymphoid gene rearrangement mechanisms in other cell types might contribute to malignancies in those other kinds of cells.

2.7.5 Additional Points Concerning Clones in Tumor Cell Populations

Two additional points can be made on this topic, where clones are identified by colonies grown *in vitro*.

i. Many cases of tumor produce no colonies when cultured *in vitro* or grafted into experimental animals.
   There is no experimental technique which infallibly provides growing cells outside the human body ("*ex-corpore*") from every human tumor. This fact implies that the current methods may only demonstrate those cells in the tumor cell population which can survive the *ex-corpore* conditions. These cells may not be the biological significant cells *in-corpore* in all cases.

ii. The cell lines which have been grown *ex-corpore* from cases of human tumor are individually unique. Most strikingly, of cases of squamous cell carcinomas of the uterine cervix which have been cultured, only one has resulted in the HeLa cell line. Currently, there are approximately 4,000 human cell lines in the American Type Collection [84] of which approximately 30 are highly characterized "panel" tumor cell lines.

The Collection gives further support to the view that only some of the populations of cells in human tumors include any cells which can survive in culture *ex-corpore*. It also supports the view that the tumor cells which do survive in culture are unique to the extent of not being similar to any other culturable cell lines from the same particular case of tumor, or other cases of the same type of tumor.

2.8 THE "MUTATOR PHENOTYPE" (HETEROGENEOUSLY HETEROGENIZING CELL POPULATION) THEORY

This theory essentially opposed to the view that tumor cell populations comprise small numbers of clones which individually are genomically homogenous (Figure 2.7). It suggests instead that tumor cell populations are

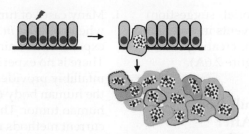

This theory suggests that the initial genomic event causes both excess growth and genomic instability. The cells in the resultant population have heterogeneously heterogenizing genomic events in subsequent generations. The theory is 'aclonal' because it suggests that every cell in the tumor mass is genetically different from every other cell in the mass.

The event which caused the genomic instability ('×' in diagram) remains in at least some daughter cells. A small proportion (red symbols) of the different secondary genomic events cause different alterations in phenotype of the affected cells; i.e. cause the morphological and behavioral heterogeneities in the cell population. See also 5. 1. 6.

FIGURE 2.7   The mutator phenotype theory of tumors.

essentially heterogeneously heterogenizing, in the sense that every cell in the population has a tendency to acquire different genomic abnormalities. These abnormalities are acquired through heterogeneous chromosomal aberrations (some occurring in mitosis), heterogeneous synthesis errors in DNA (in every S phase of cell division), and as a result of mis-repairs of spontaneous damage to DNA (in any phase of the cell cycle except mitosis—see Chapter 5).

The fundamental concept was implied by Hansemann in 1890 [8] in his descriptions of the diverse nature of mitotic abnormalities in tumors and his maxim: "The absence of principle is the principle of tumor cells"—see Section 2.2). The idea was even more clearly expressed by Winge in1930 [85] on the basis of detailed study of chromosomal abnormalities in experimental tumors. However in general, the notion was hardly considered until the last few decades. In the current era, the most significant proponent of the idea has been Loeb, who began experimental work on the idea (discussed in Section 5.1.6) in the 1970s. Loeb's

original suggestion was the genomic instability arises through infidelity of DNA replication during S phase. Later, other mechanisms, such as mis-repairs of damaged DNA and also possibly chromosomal aberrations became included in the idea of "mutator phenotype" (Figure 2.7).

Loeb's ideas have been supported by the discoveries since the 1990s that the genomes of most types of tumors show much larger numbers of genomic events than had been thought possible [86]. These numbers are greatly in excess of the number of events which could be inflicted by units of carcinogens. Hence, they appear to be necessarily the result from one or more kinds of genomic instability (see Chapter 5).

In particular, the theory implies that the original genomic event in the parent cell:

i. Causes the cell to have an excessive growth rate and/or other features (see Chapters 8 and 9).
ii. Renders the descendant tumor cell population genetically unstable. Thus the process is of heterogeneous heterogenization

of the tumor cell population, because with cell divisions, different daughter cells will acquire different "secondary" genomic events resulting in heterogeneity without clonality. More details of this process are given in Section 5.1.6.

2.9 OTHER THEORIES AND CONCEPTS

2.9.1 Stem Cells and Transit-Amplifying Cells in the Origins of Tumors

The significant biological differences between "stem cells" in embryology and "local tissue stem cells" in adult histology are described in Section A1.3.2. Early ideas of the role of embryonic phenomena and/or cells are mentioned in Section 2.3.3a. This section is concerning embryonic and local tissue stem cells as the cells in which tumors originate.

(a) Embryonic Stem Cells (See Figure 2.2B and C)

The ideas of involvements of these cells in tumors are based on several observations:

i. Embryonic proteins may be found in tumors arising in adults [87]. These "embryonic reversions" are found in only a small number of types of tumors.
ii. Genes for embryonic development, e.g., the HOX family of genes, may be altered in certain kinds of tumors [88]. The general issue of whether or not any particular alteration in a gene has any role in either the origin of the tumor or is an epiphenomenon of genomic instabilities is discussed in Sections 9.6.6 and A2.1.5.
iii. Differentiation of embryonic cells may be used as a model for differentiation/specialization of cells in adult tissues [89]. The validity of considering embryonic differentiation as essentially the same as local tissue differentiation might be

treated with caution in view of the greater complexity of germ cell tumors in humans compared with other tumors in adults. Thus in the testis, there are a variety of tumor types [90]. The seminoma comprises cells very like spermatogonia. Another group of types shows cells resemble epithelial cells—and hence have apparently achieved yolk sac, ectodermal or endodermal differentiation. The third group is teratomas, which comprise cells which have completed differentiation to adult cell types.

In the ovary, tumors comprising öocytes are rare, as are primary choriocarcinomas. Teratomas, however, are relatively common [91].

Another suggestion invokes inappropriate activation of the epithelial–mesenchymal transition which occurs in the normal development of the embryo [92–96]. As described in Section A1.1.1, when the embryonic plate forms between the amniotic and yolk sac spaces in the inner cell mass of the blastocyst, mesenchymal cells appear between the primitive ectoderm and endoderm. Because carcinoma cells may, with progression in the body or in *in vitro* culture, become more and more like mesenchymal cells—and even like sarcoma cells of various types—it has been suggested that this embryonic transition phenomenon is being activated in adult cells to produce tumors.

(b) Local Tissue Stem Cells in Adults (See Figure 2.2D)

These theories are based on direct descent of tumor cells from local tissue stem cells. General support for the idea comes from the following points [97]:

i. The local tissue stem cells are the only cells to live long enough to accumulate genomic events, according to the multihit theories.
ii. These cells are the only ones left in original population of initiator-treated cells in two-stage skin models.
iii. These cells have the greatest number of mitoses in their life spans.

iv. These cells could exhibit blocked specialization by a small number of genomic events.

(c) Any Dividing Cell in Adults, Including Transit-Amplifying Cells (See Figure 2.2A and E).

These ideas are essentially an alternative to the stem cell theories. The idea is valid because it is quite possible that an original genomic event of tumor formation could be in a transit-amplifying cell and include changes which confer immortal growth in ways unrelated to the immortality of normal tissue stem cells. Thus essentially, the cell could lose its obsolescence (death-in-terminal differentiation) genes (see Section 10.1) giving rise to immortal populations.

2.9.2 Theories Involving Telomeres and the Immortality of Tumor Cell Populations

Telomeres are made up of repeated sequences of 6–8 nucleobases. They are found at the ends of the arms of chromosomes. Under normal conditions in humans, they are shortened during each division of the nucleus. Large numbers of divisions of cells would cause complete loss of telomeric DNA and after that, loss of nearby genetic DNA. Ultimately, this could cause death of the cell, unless a mechanism for lengthening telomeres was present.

In most multicellular organisms, the germ cells, local stem cells, and some leukocyte series have a particular polymerase called telomerase, which can add repeats to the ends of the telomeres, thus correcting this loss. In many types of human tumors, telomerase has been found to be activated [98]. It has been proposed that this activation step is an essential part of immortalization of tumor cell populations, and hence a necessary contributory biological event in the development of malignant tumors. In cases in which telomerase is not activated, it is possible that telomeres could be replenished by chromosomal crossing over ("alternative telomere lengthening pathway") [99].

As far as is known, telomeres have no direct role in excessive rates of cell division, alterations in specialization, or other features of tumors (Chapter 8).

2.9.3 Theories Involving Plasma Membrane and Cytoskeleton

These theories have been developed to explain situations, such as "solid" carcinogenesis (Section 4.8) and malignant transformation in vitro (Section 14.2.3), in which tumors arise where there does not seem to be a mechanism for any carcinogen to directly access the genome. Generally, the idea is based on supposed cytoskeletal connections between the plasma membrane and the nuclear membrane of cells (Figure 2.8 and Section A1.2).

Plasma membrane comprises complexes of lipids and proteins. The proteins can form associations with the cytoskeleton to determine cell shape and regulate interactions with the environment [100]. The cytoskeleton also interacts with nuclear membrane proteins [101].

However, in pro(sub)phase, the nuclear membrane of the cell dissolves and reassembles as the mitotic spindle. Thus potentially, disturbances in plasma membrane might affect these supposed cytoskeletal relationships, which might then disturb the components of the nuclear membrane. This could mean that during mitosis, the spindle fibers might malfunction and hence cause chromosomal maldistribution and structural errors, which in turn might ultimately result in tumor formation.

The carcinogenic action of in vitro culture conditions (Section 14.2.3) could then simply be excessive demands on production of the proteins of the plasma membrane and cytoskeleton (possibly though their being sequestered in the focal adhesion sites) and hence disturbance in the nuclear membrane/spindle for normal chromosome distributions during anaphase.

Hyperploid cells would then be created (the hypoploid sister cells would be nonviable) and

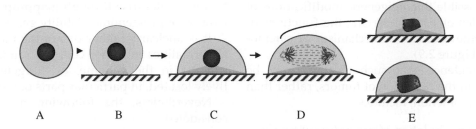

A. Normal cell not in contact with a surface has weak trans-cytoplasmic connections between nuclear membrane and inner plasma membrane.
B. Contact with a surface enforces concentration of the connections on the side of the surface.
C. As the cell flattens, the nuclear membrane loses some integrity, possibly because of enforced proximity to the inner cell membrane.
D. During mitosis, the nuclear membrane does not contribute satisfactorily to the mitotic spindle, and chromosomes become abnormally distributed and/or structurally abnormal.
E. After cell division, the daughter cell with too few chromosomes probably dies. In the cell with excess chromosomes, this or further genomic events may occur which immortalize, and ultimately transform the descendant cells (see also Section 11.9).

FIGURE 2.8 One of various cytoskeletal models for the induction of tumorous cells. Note: in the scheme proposed by Barrett in 1989 (see Section 4.8.1), undegradable particles may be phagocytosed, and come into contact with the outer nuclear membrane. At that site, they may interfere with the nuclear membrane-spindle transitions, resulting in chromosomal abnormalities.

the excess chromosomes would carry excess copies of oncogenes and/or their regulators, resulting in excess growth. The permanent reduction in adhesiveness could be explained by concomitant losses of cell-membrane protein genes.

The most valid test of this scheme might be directed at—if the technologies were available—whether or not nucleotide instability precedes or comes after chromosomal instability in the cells of the culture.

The idea might also be relevant to carcinogenesis by large molecules such as polyaromatic hydrocarbons (PAHs). As noted in Sections 4.2 and 4.4, these particular molecules are multiaromatic ring structures. There is uncertainty about the actual entry of these large molecules into the genomic compartment, to cause genomic events via PAH adducts on DNA [102]. However, these substances are lipid soluble, and could enter the cell membrane, and possibly have effects on the interactions between the proteins of the inner plasma

membrane and the cytoskeletal proteins, with secondary effects on the nuclear membrane proteins. If any of these effects persisted, then they could possibly cause mitotic abnormalities when the nuclear membrane proteins disassemble and reassemble to form the mitotic spindle.

2.9.4 Epigenetic Phenomena and Tumor Formation

Chemical modification of nucleobases and nucleoproteins such as histones are widely suggested to be involved in the regulation of gene expression (Section A2.2.4). The most numerous kind of alteration in DNA is methylation of cytosines, especially in foci ("islands") of concentration of CpGs (cytosine proximate to guanine in the one strand of DNA) [103,104]. In general, tumor cell genomes show foci of hypermethylation in the context of overall hypomethylation of DNA [105]. The commonest kind of alteration to nucleoproteins is acetylation of histones.

It is possible that epigenetic modifications of nucleotides or protein serving the functions of the genome could be a mechanism of carcinogenesis (Figure 2.9).

The evidence that such changes may be involved in the causation of tumors, rather than an epiphenomenon, includes:

i. Hypermethylation of promoters of tumor suppressor genes.
ii. Hypomethylation of promoters of oncogenes.
iii. Acetylation of histones is associated with increased gene activity.
iv. Deacetylation of histones is associated with reduced gene activity.

It is also possible that the acetylation profile of histones performs some kind of localization function (see Section A2.3), as they may do in the phenomenon of genomic imprinting [106].

Genomic instabilities (see Chapter 5) and tumor cell heterogeneity so commonly seen in tumors could arise through inappropriate inactivation of pro-genome-stability genes.

It is unclear whether or not the methylation abnormalities in tumor cells are randomly distributed in the genome, or "targeted"/selectively located, in particular parts of the genome.

Nevertheless, the following may also be considered:

i. There is little evidence of specific patterns of epigenetic changes between different tumor types.
ii. There is also no conclusive evidence of abnormal patterns of epigenetic changes in the genomes of individuals who have an inherited predisposition to tumors.
iii. Chemical damage to DNA nucleoproteins occurs in many nontumorous conditions, especially inflammation. There are apparently no known significant differences between patterns of epigenetic changes in tumors and in nontumorous pathological conditions such as inflammation [107].

A   B   C   D

A. A particular pattern of epigenetic modifications prevent a repairase complex removing a nucleotide - adduct lesion at a particular site, contributing to a failure of repair.

B. A particular pattern of epigenetic modifications + adduct of carcinogen causes failure of attachment of repairase complex, hence failure of repair.

C. Adduct on repairase causes failure of functions of the repairase complex and hence failure of repair of damage produced by another agent (♦)

D. (Possibly) adduct on epigenetic marker causes failure of attachment of repairase complex and then failure of repair of damage produced by another agent (♦)

= adduct on DNA   = repairase complex   = modifications of DNA

FIGURE 2.9   Some "in-principle" mechanisms by which the inherited epigenomic modifications to DNA could contribute to tumor formation.
Note: The chemical modifications in these theories are the same as some of those referred to as 'damage' in DNA-damage theories (see Section 5.1.3 and Figures 1.6 and 5.3).

The issue then becomes, why don't those changes cause tumor formation?

iv. There seems to be no established relationships between histone modifications and tumor type [108,109].

2.9.5 Theories Involving Immunity

There has been no widely considered theory that an abnormality of the immune system can be a cause of tumorous conversion of normal cells. There have, however, been suggestions that immunological deficiency may cause tumorous cells created by other agents to become clinically appreciable lesions. The main proposal of this type was Burnet's immune-surveillance theory [110] (Figure 2.10) which had several components as follows:

i. In normal life, hyper-proliferative cells expressing non-self-antigens are being continuously formed.

ii. Almost all of these cells do not form identifiable masses of cells because they are destroyed in the body by the ordinary processes of immunological rejection. Because no antibodies to patients' tumor cells ever appear in the patients' blood, the rejection process is thought to be T-cell mediated.

iii. Those few cells which do not express non-self-antigens or which for some other reason "escape" immune surveillance grow and form tumors.

The necessary component of this idea would be that it is common for abnormal cells to arise which produce antigens to which the immune system react and kill the cells before they can proliferate. Although discussed at length [111], this idea has been difficult to prove or disprove. The first major problem is that many antigens identified as "tumor-associated" may be embryonic or lineage-unfaithful expressions of normal antigens. Second, the theory does not explain why mice with severe combined immunodeficiency, which have poor T-cell- and B-cell-mediated immune responses, show no increased tendency to spontaneous tumors [112].

Tumors of a small number of types only occur more commonly in humans with immune deficiencies. In acquired immunodeficiency syndrome (see Section 4.5), the predispositions are to lymphomas and Kaposi's sarcoma. Here,

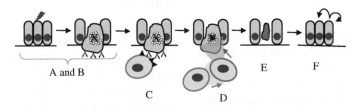

The components of the theory are:
A. Tumorous conversion is common.
B. It is associated with the expression of new antigens on the surfaces of the tumor cells.
C. The new antigens are detected by "afferent arm" immune cells, including dendritic cells.
D. The afferent arm cells communicate with efferent arm cells (T-cells and/or B-cells, orange margin), which produce cytotoxic substances against the tumor cells.
E. The tumor cell dies and does not grow into a mass.
F. An adjacent cell divides, replacing the lost cell. The tissue returns to normal.
   Only rare tumorous cells escape this immuno-surveillance to form tumors.

FIGURE 2.10   Immune-surveillance in cancer.

the human immunodeficiency syndrome virus may be directly carcinogenic or alternatively may serve as an enhancer of tumors because other viruses are not prevented from accessing relevant cells.

In common variable immune deficiency—which so far as is known is not due to a viral infection—there are predispositions to lymphomas and lymphoid hyperplasia of the stomach, but no other types of tumors [113]. Here, the lymphoid lesions could be due to other viruses or agents being able to access the body.

Increased rates of tumors in association with immunosuppressive treatment may be due to this mechanism, and/or to the immunosuppressive agents themselves being carcinogens (see Section 4.2.1).

Antigens in tumors may be used in therapy, as discussed in Section 8.4.1.

2.9.6 Field Theory

From the late nineteenth century, the origin of tumors was controversial [8]. Some authors believed that a single cell converted from a normal cell in one "event," while others thought that a whole field of cells necessarily becomes abnormal, and that the invasive tumor develops in those fields. Two possible ideas were present:

i. The development of the tumor depended on some stochastic second event in a single cell in the predisposed field.
ii. There might be some communication between cells such that some pro-tumorous influence accumulated in certain parts of the field, resulting in a tumor at that site.

Both suggestions would explain multifocal tumors, while the second might better explain why invasive foci tend to occur in the middle of fields.

These ideas were described as "field effect" by Willis [50] and by Slaughter et al. [114] (reviewed in Ref. [115]).

"Field theory" has also been used for the concept of disordered tissue organization as the basis of tumors [116] and as an alternative to the somatic mutation theory of tumors.

2.9.7 Renewed Interest in Chromosomal Theories

The possible role of chromosomal abnormalities in tumor formation was first suggested in the 1890s but the idea was discounted (Section 2.4). In the mid-twentieth century, chromosomal aberrations began to be considered as having possible pathogenetic roles in diseases after the discovery of trisomy 21 as the basis of Down's syndrome by Lejeune in 1959 [117] and the discovery of the Philadelphia chromosome in most cases of chronic myeloid leukemia by Nowell and Hungerford in 1961 [118]. The cytogenetics of tumors has again become the basis of a field of theory, e.g., Refs. [119–122], as well as becoming part of the diagnostic methodology of tumors (see Section 9.5).

2.10 CURRENT DEFINITIONS

2.10.1 Various Twentieth Century Authors

Various authors since Willis (Section 2.5.2) have concerned themselves in different ways with the problem of defining tumors. Some notable examples are as follows.

Wright in 1954 [123] suggested that it was not possible to define tumors in the sense of precisely defining the boundaries of what is, and what is not, a tumor.

Berenblum in 1964 [124] used the concepts of stem cells and maturation which are still discussed today:

> A tumor is an actively-growing tissue, composed of cells derived from one that has undergone an abnormal type of irreversible differentiation; its growth is progressive due to a persistent delay in

maturation of its stem cells. The essential nature of the irreversible differentiation, whether in biological or chemical terms and whether necessarily determined by a virus, is still unknown.

Meissner and Warren in 1971 [125] included the existence of different types of tumors in their definition:

> Neoplasms are disturbances of growth characterised primarily by an unceasing, abnormal and excessive proliferation of cells. Countless varieties arise from essentially all types of human tissues and cells.

This is one of the few definitions until the last decade, to include that fact that tumor types are a phenomenon of tumors overall (see Section 8.2).

Ruddon in 2007 [126]) included the important phenomenon of the abnormal cytokinetics in tumor cell populations (see Section 6.3.2). This author defined a cancer (malignant tumor) as:

> An abnormal growth of cells caused by multiple changes in gene expression leading to dysregulated balance of cell proliferation and cell death and ultimately evolving into a population of cells that can invade tissues and metastasise to distant sites, causing significant morbidity and, if untreated, death of the host.

In this definition, the first part could apply to both benign and malignant tumors, but its second part excludes benign tumors.

The author's own definition, given at the beginning of Chapter 1, incorporates uncontrolled growth together with their occurrence in recognizably different types, which are characterized by variable combinations of variable differences from their respective kinds of cell of origin.

2.10.2 Hanahan and Weinberg's "Hallmarks" of Cancer

In the last decade or so, some authorities have turned to defining malignancy in tumors in cell biological terms. The approach often does not concern itself with benign tumors and underlines the fact that "malignancy" appears to be a complex phenomenon, involving a variety of biochemical and biological traits.

In 2000, Hanahan and Weinberg [127] described the necessary properties of cancer cells in terms of six "hallmarks":

i. Self-sufficiency in growth signals
ii. Insensitivity to growth-inhibitory signals
iii. Evasion of apoptosis
iv. Limitless replicative potential
v. Sustained angiogenesis
vi. Invasion and metastasis.

They also considered the issue of the different tumor types and suggested that these might arise through different disturbances in cellular regulatory circuits (see Chapter 6).

In a revised version of this article [128], these authors reiterated the six hallmarks and added four new hallmarks:

1. Abnormal metabolic pathways
2. Evading the immune system
3. Chromosome abnormalities and unstable DNA
4. Inflammation

   Emphasis was given also to contact inhibition, and mention was made of autophagy, necrosis, as well as replicative senescence. In addition, "emerging hallmarks" were described as:

5. Genome instability and mutation
6. Tumor-promoting inflammation
7. Reprogramming of energy metabolism

Moreover, aspects of the tumor microenvironment, cancer cells, and cancer stem cells, and heterotypic signaling orchestrating the cells of the tumor microenvironment were described.

These "hallmarks" are important descriptors of tumor cells. However, these are not exclusive to tumors. As pointed out by Lazebnik [129], five of the hallmarks, including unceasing growth, are also shown by benign tumors. This omission of benign tumors from concepts of "true" tumors reduces the possible breadth of discussion, because benign tumors can resemble malignant ones in their histology, and also some kinds of malignant tumors commonly arise from benign tumors (see Section 8.4.2). The clear implication of these phenomena is that benign and malignant tumors may well have, in principle, genomic as well as cell biological features in common.

It can also be noted here that these "hallmark" characterizations of tumors do not address the question of certain tumorous phenomena being very similar to nontumorous biological phenomena in certain normal and pathological processes (see Sections A1.4 and A1.5). This is especially true of invasion and metastasis, which various cells in the body perform as part of their physiological functions (Section A1.4 and A1.5).

2.10.3 Definitions Provided by Major Health Agencies

Current (2014) definitions by major health agencies stress the main features of tumors, and the fact that cancers exist in many different types.

Thus the National Cancer Institute (Washington, DC) [130] defines cancer as:

> A term for diseases in which abnormal cells divide without control and can invade nearby tissues. Cancer cells can also spread to other parts of the body through the blood and lymph systems. Cancer is not just one disease but many diseases. There are more than 100 different types of cancer. Most cancers are named for the organ or type of cell in which they start—for example, cancer that begins in the colon is called colon cancer; cancer that begins in melanocytes of the skin is called melanoma (http://www.cancer.gov/dictionary, accessed 13 Nov 2013).

The World Health Organization's definition [131] likewise includes the issue of the types of cancers:

> Cancer is a generic term for a large group of diseases that can affect any part of the body. Other terms used are malignant tumors and neoplasms. One defining feature of cancer is the rapid creation of abnormal cells that grow beyond their usual boundaries, and which can then invade adjoining parts of the body and spread to other organs. (http://www.who.int/cancer/en/ accessed 13 Nov 2013).

2.11 LACK OF THEORIES FOR THE PRECISE GENOMIC EVENTS AND THE THOUSAND OR SO DIFFERENT TYPES OF TUMORS— AND ALL THEIR COMPLEX FEATURES—ALL FROM THE ONE GENOME

Throughout almost all the history of theories of tumors the phenotypic changes in tumors have been attributed to one process—the "neoplastic" process—being instigated or activated in a parent cell. The mechanism of instigation or activation of this hypothetical process has been assumed to be an alteration in the genetic material of the parent cell. This mechanism, however, does not explain all the complexities of the tumor types, and how they can be evinced from the one genome.

The studies of hereditary predispositions to tumors have shown that the germ-line genomic event which predisposes to each type of tumor is located separately from the site of all the events for all the other predispositions. This represents evidence for the idea that each tumor type depends on a genomic event of specific length, in a specific part of the genome, of a particular kind of cell, at a specific phase of its embryological development or histological specialization. Some aspects of this are discussed in Section 8.5.

References

[1] Schwann TH. Microscopical researches into the accordance in the structure and growth of animals and plants, 1839. In the same volume: Schleiden MJ. Contributions to phytogenesis [Smith H, Trans.]. London: Sydenham Society; 1847.

[2] Virchow R. Cellular pathology as based upon physiological and pathological histology. 2nd German edn, 1858 [Chance F, Trans.]. With a new introductory essay by L. J. Rather. New York, NY: Dover; 1971.

[3] Müller J. Ueber die feineren Bau und die Formen der Krankhaften Geschwülste. G. Reimer, Berlin 1838 [West C, Trans.]. On the nature and structural characteristics of cancer and of those morbid growths which may be confounded with it, Sherwood. London: Gilbert and Piper; 1840.

[4] Carpenter WB. The microscope and its revelations. 7th edn, enl. and rev. by WH Dallinger. Philadelphia, PA: Blakiston; 1891.

[5] Albrecht E. Ueber hamartome/On the hamartoma. Verh dtsche pathol Ges 1904;7:153–7.

[6] Ober WB. Selected items from the history of pathology: Eugen Albrecht, MD (1872–1908): hamartoma and choristoma. Am J Pathol 1978;91(3):606.

[7] Willis RA. The borderland of embryology and pathology, 2nd ed. London: Butterworths; 1962. p. 351.

[8] Bignold LP, Coghlan BL, Jersmann HP. David Paul Hansemann: contributions to oncology: context, comments and translations. Birkhäuser Basel 2007:124. 1890 article Ueber asymmetrische Zelltheilung in Epithelkrebsen und deren biologische Bedeutung. Virchows Archiv 119:299–326, translated p. 123–144. Lobstein p. 58; histogenesis p. 99, 243; field theory p. 292, 297; Fibiger, p. 313–14.

[9] Beattie JM, Dickson WEC. A textbook of general pathology for the use of students and practitioners. London: William Heinemann; 1918. p. 239.

[10] Ewing J. Neoplastic diseases: a treatise on tumors, 4th ed. Philadelphia, PA: WB Saunders; 1940.

[11] Anderson W. Boyd's pathology for the surgeon, 8th ed. Philadelphia, PA: Saunders; 1967.

[12] King RJB, Robins MW. Cancer biology (3rd ed.). New York, NY: Pearson/Prentice Hall; 2006.

[13] Buffon GLL. Histoire naturelle, 10. London: Abridged edition printed by JS Barr; 1792. p. 1–26.

[14] Wolff J. Die Lehre von der Krebskrankheit von den Aeltesten Zeiten bis zur Gegenwart/The science of cancerous disease from earliest times to the present (1907). [Ayoub B, Trans.] Introduced by P. Sarco; Science History Publications, Canton, MA, 1989. Le Dran p. 51; Bichat p. 75; Royer-Collard p. 102; Boll p. 273; Broussais p. 80; infection theories, p. 431–590.

[15] Rindfleisch GE. A manual of pathological histology. Translated by E.B. Baxter. London: New Sydenham Soc.; 1872–1873.

[16] Bignold LP, Coghlan BL, Jersmann HP. Virchow's "cellular pathology" 150 years later. Semin Diagn Pathol 2008;25(3):140–6.

[17] Lancereax E. Traité d'anatomie pathologique, Vol. 1. Paris: Anatomie Pathologique Génèrale. Adrien Delahaye; 1875.

[18] Foulds L. Neoplastic development, 2 vols. London: Academic Press; 1969,1976.

[19] Cohnheim J. Lectures on general pathology. 3 vols. [McKee AB, Trans.].London: New Sydenham Society, 1889–1890.

[20] Lotem J, Sachs L. Epigenetics and the plasticity of differentiation in normal and cancer stem cells. Oncogene 2006;25(59):7663–72.

[21] Figg W, Folkman J. Angiogenesis: an integrative approach. From science to medicine. New York, NY: Springer; 2008.

[22] Klagsbrun M, D'Amore P, editors. Angiogenesis: biology and pathology. Cold Spring Harbor, NY: CSHL Press; 2011.

[23] Ribatti D. History of research on tumor angiogenesis. New York, NY: Springer; 2009.

[24] Martin S, Cliff M, editors. Angiogenesis protocols (Methods in molecular biology) (2nd ed.). New York, NY: Humana Press; 2008.

[25] Nicholson GW. Cancer and causation. Br Med J 1935;i(3873):650–1.

[26] Haddow A. On the biological alkylating agents. Perspect Biol Med 1973;16(4):503–24.

[27] Kundu JK, Surh Y-J. Inflammation: gearing the journey to cancer. Rev Mut Res 2008;659:15–30.

[28] Elinav E, Nowarski R, Thaiss CA, et al. Inflammation-induced cancer: crosstalk between tumours, immune cells and microorganisms. Nat Rev Cancer 2013;13(11): 759–71.

[29] Woller N, Kühnel F. Virus infection, inflammation and prevention of cancer. Recent Results Cancer Res 2014;193:33–58.

[30] Maher SG, Reynolds JV. Basic concepts of inflammation and its role in carcinogenesis. Recent Results Cancer Res 2011;185:1–34.

[31] Coussens LM, Werb Z. Inflammation and cancer. Nature 2002;420(6917):860–7.

[32] Virchow R. Über metaplasie/On metaplasia. Virchows Arch. 97:410–430. Abstract in Br Med J 1884;ii(1235):408.

[33] Castellanos A, Vicente-Dueñas C, Campos-Sánchez E, et al. Cancer as a reprogramming-like disease: implications in tumor development and treatment. Semin Cancer Biol 2010;20(2):93–7.

[34] Stolt CM, Klein G, Jansson AT. An analysis of a wrong Nobel Prize—Johannes Fibiger, 1926: a study in the Nobel archives. Adv Cancer Res 2004;92:1–12.

[35] Borst M. Die Lehre von den Geschwulsten mit einem Mikroskopischen Atlas/The Science of Tumors with an Atlas of Microscopy, 2 vols. Wiesbaden: JF Bergmann; 1902.

[36] Burger PC, Scheithauer BW. Tumors of the central nervous system. AFIP Atlas, Ser 4, Fasc 7, 2007.

[37] Swerdlow SH, Campo E, Harris NL, editors. WHO classification of tumours of haematopoietic and lymphoid tissues (4th ed.). Lyon: IARC; 2008.

[38] Dorland's Medical Dictionary for Health Consumers. Philadelphia, PA: Saunders/Elsevier; 2007.

[39] Rather LJ. The genesis of cancer. A study in the history of ideas. Baltimore, MD: The Johns Hopkins University Press; 1978.

[40] Thompson L, Chang B, Barsky SH. Monoclonal origins of malignant mixed tumors (carcinosarcomas). Evidence for a divergent histogenesis. Am J Surg Pathol 1996;20(3):277–85.

[41] McCluggage WG. Malignant biphasic uterine tumours: carcinosarcomas or metaplastic carcinomas? J Clin Pathol 2002;55:321–5.

[42] van Deurzen CH, Lee AH, Gill MS, et al. Metaplastic breast carcinoma: tumour histogenesis or dedifferentiation? J Pathol 2011;224(4):434–7.

[43] Boveri T. Zur Frage der Entstehung der Malignen Tumoren/On the origin of malignant tumors. G Fischer, Jena. Translation with annotations by H. Harris. J Cell Sci 2008;121(Suppl. 1):1–84.

[44] Bignold LP, Coghlan BL, Jersmann HP. Hansemann, Boveri, chromosomes and the gametogenesis-related theories of tumours. Cell Biol Int 2006;30(7):640–4.

[45] Bignold LP, Coghlan BL, Jersmann HP. Cancer morphology, carcinogenesis and genetic instability: a background. EXS 2006;vol. 96:1–24.

[46] Wurster DH, Benirschke K. Indian Momtjac, *Muntiacus muntjak*: A deer with a low diploid chromosome number. Science 1970;168:1364–6.

[47] Ashby J. The genotoxicity of sodium saccharin and sodium chloride in relation to their cancer-promoting properties. Food Chem Toxicol 1985;23(4–5):507–19.

[48] Graf MD, Christ L, Mascarello JT, et al. Redefining the risks of prenatally ascertained supernumerary marker chromosomes: a collaborative study. J Med Genet 2006;43:660–4.

[49] Koller PC. The genetic component of cancer. In: Raven RW, editor. Cancer, vol. 1. London: Butterworth & Co; 1957. p. 335–403.

[50] Willis RA. Pathology of tumours, 4th ed. London: Butterworths; 1967. [Ignoramus] p. 207.

[51] Willis RA. The spread of tumours in the human body, 2nd ed. London: Butterworths; 1952.

[52] Willis RA. The borderland of embryology and pathology, 2nd ed. London: Butterworths; 1962.

[53] Carlson EA. Mendel's legacy: the origin of classical genetics. Cold Spring Harbor, NY: CSHL Press; 2004.

[54] Morgan TH. A critique of the theory of evolution. Princeton, NJ: Princeton University Press; 1916. p 132.

[55] Morgan TH. The theory of the gene, 2nd ed. New Haven, CT: Yale University Press; 1928.

[56] Falconer DS, Mackay TFC. Introduction to quantitative genetics, 4th ed. Harlow, Essex, UK: Addison Wesley Longman; 1996.

[57] Komas AT, editor. Single nucleotide polymorphisms: methods and protocols (Methods in molecular biology) (2nd ed.). New York, NY: Humana Press/ Springer; 2009.

[58] Whitman RC. Somatic mutation as a factor in the production of cancer; a critical review of v. Hansemann's theory of anaplasia in the light of modern knowledge of genetics. J Cancer Res 1919;4:181–202.

[59] Muller HJ. The production of mutations by X-rays. Proc Nat Acad Sci USA 1928;14:714–26.

[60] Bauer KH. Mutationstheorie der Geschwülst-Entstehung/Mutation theory in tumour formation. Berlin: Julius Springer; 1928.

[61] Lockhart-Mummery JP. The origin of cancer. London: J & A Churchill; 1934.

[62] Wyllie AD, MacSween RNM, Whaley K, editors. Muir's textbook of pathology (13th ed.). London: Edward Arnold; 1992. p. 364.

[63] Clayson DB. Toxicological carcinogenesis. Boca Raton, FL: CRC Press/Lewis Publishers; 2001. p. 2.

[64] Kumar V, Abbas AK, Aster JC, editors. Robbins and Cotran's pathologic basis of disease (7th ed.). Philadelphia, PA: Saunders/Elsevier; 2004. p. 270.

[65] Robertson ES, editor. Cancer associated viruses. Boston, MA: Springer US; 2012. p. 1–43.

[66] Bailey K, Sanger F. The chemistry of amino acids and proteins. Annu Rev Biochem 1951;20:103–30.

[67] Watson JD, Crick FH. Genetical implications of the structure of deoxyribonucleic acid. Nature 1953;171:964–7.

[68] Huebner RJ, Todaro GJ. Oncogenes of RNA tumor viruses as determinants of cancer. Proc Nat Acad Sci USA 1969;64(3):1087–94.

[69] Dosic H, Todaro GJ. Viruses, genes and cancer. In: Lynch HT, editor. Cancer genetics. Springfield, IL: Charles C Thomas; 1976. p. 101–10.

[70] Følling A. Über Ausscheidung von Phenyl-benztraubensäure in den Harn als Stoffwechselanomalie in Verbindung mit Imbezillität/ About elimination of phenyl pyruvic acid in the urine as a metabolic abnormality in conjunction with imbecility. Hoppe-Seyler's Zeitschrift für physiologische Chemie 1934; 227(1–4):169–181.

[71] Ingram VM. Abnormal human haemoglobins. III. The chemical difference between normal and sickle cell haemoglobins. Biochim Biophys Acta 1959;36:402–11.

[72] Sanger F, Nicklen S, Coulson AR. DNA sequencing with chain-terminating inhibitors. Proc Nat Acad Sci USA 1977;74(12):5463–7.

[73] Bartlett JMS, Stirling D. A short history of the polymerase chain reaction. PCR Protocols 2003;226:3–6.

[74] Weinstein IB. The origins of human cancer: molecular mechanisms of carcinogenesis and their implications for cancer prevention and treatment—twenty-seventh G.H.A. Clowes memorial award lecture. Cancer Res 1988;48(15):4135–43.

[75] Vogelstein B, Kinzler KW. The multistep nature of cancer. Trends Genet 1993;9(4):138–41.

[76] Forsdyke DR. The origins of the clonal selection theory of immunity as a case study for evaluation in science. FASEB J 1995;9(2):164–6.

[77] Teixeira MR, Heim S. Cytogenetic analysis of tumor clonality. Adv Cancer Res 2011;112:127–49.

[78] Parsons BL. Many different tumor types have polyclonal tumor origin: evidence and implications. Mutat Res 2008;659(3):232–47.

[79] Leedham SJ, Wright NA. Expansion of a mutated clone: from stem cell to tumor. J Clin Pathol 2008;61(2):164–71.

[80] Rubin H. Selective clonal expansion and microenvironmental permissiveness in tobacco carcinogenesis. Oncogene 2002;21(48):7392–411.

[81] Greaves M, Maley CC. Clonal evolution in cancer. Nature 2012;481(7381):306–13.

[82] Dadi S, Le Noir S, Asnafi V, et al. Normal and pathological V(D)J recombination: contribution to the understanding of human lymphoid malignancies. Adv Exp Med Biol 2009;650:180–94.

[83] Shaw AT, Hsu PP, Awad MM, et al. Tyrosine kinase gene rearrangements in epithelial malignancies. Nat Rev Cancer 2013;13(11):772–87.

[84] American Type Culture Collection. Cell lines. Available at: http://www.atcc.org/en/Products/Cells_and_Microorganisms/Cell_Lines.aspx

[85] Winge Ö. Zytologische Untersuchungen über die Natur maligner Tumoren/Cytological investigations of the nature of tumors. Z. fur Zellforschung (now Cell Tissue Res) 1930;10:683–735.

[86] Stoler DL, Chen N, Basik M, et al. The onset and extent of genomic instability in sporadic colorectal tumor progression. Proc Nat Acad Sci USA 1999;96:15121–6.

[87] Bignold LP. Embryonic reversions and lineage infidelities in tumour cells: genome-based models and role of genetic instability. Int J Exp Pathol 2005;86(2):67–79.

[88] Shah N, Sukumar S. The HOX genes and their roles in oncogenesis. Nat Rev Cancer 2010;10(5):361–71.

[89] Sell S. Cancer stem cells and differentiation therapy. Tumour Biol 2006;27(2):59–70.

[90] Ulbright TM, Young RH. Tumors of the testis and adjacent structures. AFIP Atlas, Series 4, Fascicle 18, 2013, p. 65–230.

[91] Scully RE, Young RH, Clement PB. Tumors of the ovary, maldeveloped gonads, fallopian tube and broad ligament. Series 3 Fascicle 23, 1996, p. 239–306.

[92] Kalluri R, Weinberg RA. The basics of epithelial–mesenchymal transition. J Clin Invest 2009;119(6):1420–8.

[93] Lamouille S, Xu J, Derynck R. Molecular mechanisms of epithelial–mesenchymal transition. Nat Revs Molec Cell Biol 2014;15:178–96.

[94] Nisticò P, Bissell MJ, Radisky DC. Epithelial–mesenchymal transition: general principles and pathological relevance with special emphasis on the role of matrix metalloproteinases. Cold Spring Harb Perspect Biol 2012;4:a011908.

[95] Hollier BG, Evans K, Mani SA. The epithelial-to-mesenchymal transition and cancer stem cells: a coalition against cancer therapies. J Mammary Gland Biol Neoplasia 2009;14(1):29–43.

[96] Foroni C, Broggini M, Generali D, et al. Epithelial–mesenchymal transition and breast cancer: role, molecular mechanisms and clinical impact. Cancer Treat Rev 2012;38(6):689–97.

[97] Alison MR, Burkert J, Wright NA, et al. Stem cells and tumourigenesis (2nd ed.). In: Alison MR, editor. The cancer handbook, vol. 1. Chichester, UK: John Wiley & Sons; 2007. p. 22–35.

[98] Shay JW, Wright WE. Role of telomeres and telomerase in cancer. Semin Cancer Biol 2011;21(6):349–53.

[99] Gocha AR, Harris J, Groden J. Alternative mechanisms of telomere lengthening: permissive mutations, DNA repair proteins and tumorigenic progression. Mutat Res 2013;743-744:142–50.

[100] Luna EJ, Hitt AL. Cytoskeleton–plasma membrane interactions. Science 1992;258:955–64.

[101] Starr DA, Fridolfsson HN. Interactions between nuclei and the cytoskeleton are mediated by SUN-KASH nuclear-envelope bridges. Annu Rev Cell Dev Biol 2010;26:421–44.

[102] Baird WM, Hooven LA, Mahadevan B. Carcinogenic polycyclic aromatic hydrocarbon–DNA adducts and mechanism of action. Environ Molec Mutagen 2005;45:106–14.

[103] Hackett JA, Surani MA. DNA methylation dynamics during the mammalian life cycle. Philos Trans R Soc Lond B Biol Sci 2013;368(1609):20110328.

[104] Futscher BW. Epigenetic changes during cell transformation. Adv Exp Med Biol 2013;754:179–94.

[105] Ishida M, Moore GE. The role of imprinted genes in humans. Mol Aspects Med 2013;34(4):826–40.

[106] Ehrlich M, Lacey M. DNA hypomethylation and hemimethylation in cancer. Adv Exp Med Biol 2013;754:31–56.

[107] Karatzas PS, Gazouli M, Safioleas M, et al. DNA methylation changes in inflammatory bowel disease. Ann Gastroenterol 2014;27(2):125–32.

[108] Bojang Jr P, Ramos KS. The promise and failures of epigenetic therapies for cancer treatment. Cancer Treat Rev 2014;40(1):153–69.

[109] Barneda-Zahonero B1, Parra M. Histone deacetylases and cancer. Mol Oncol 2012;6(6):579–89.

[110] Burnet FM. Cancer—biological approach: I. The processes of control. II. The significance of somatic mutation. BMJ 1957;i(5022):779–86.

[111] Dunn GP, Old LJ, Schreiber RD. The three Es of cancer immunoediting. Ann Rev Immunol 2004;22:329–60.

[112] Bosma MJ, Carroll AM. The SCID mouse mutant: definition, characterization, and potential uses. Annu Rev Immunol 1991;9:323–50.

[113] Mellemkjaer L, Hammarstrom L, Andersen V, et al. Cancer risk among patients with IgA deficiency or common variable immunodeficiency and their relatives: a combined Danish and Swedish study. Clin Exp Immunol 2002;130(3):495–500.

[114] Slaughter DP, Southwick HW, Smejkal W. Field cancerization in oral stratified squamous epithelium; clinical implications of multicentric origin. Cancer 1953;6:963–8.

[115] Chai H, Brown RE. Field effect in cancer—an update. Ann Clin Lab Sci 2009;39:331–7.

[116] Soto AM, Sonnenschein C. Emergentism as a default: cancer as a problem of tissue organization. J Biosci 2005;30:103–18.

[117] Neri G, Opitz JM. Down syndrome: comments and reflections on the 50th anniversary of Lejeune's discovery. Am J Med Genet A 2009;149A(12):2647–54.

[118] Nowell PC, Hungerford DA. Chromosome studies in human leukemia. II. Chronic granulocytic leukemia. J Natl Cancer Inst 1961;27:1013–35.

[119] Sandberg AA. Chromosomes in human neoplasia. Curr Probl Cancer 1983;8(2):1–52.

[120] Nowell PC, Nowell PC. Cancer, chromosomes, and genes. Lab Invest 1992;66(4):407–17.

[121] Pathak S, Multani AS. Aneuploidy, stem cells and cancer. EXS 2006;96:49–64.

[122] Rasnick D. The chromosomal imbalance theory of cancer. Boca Raton, FL: CRC Press; 2011.

[123] Wright GP. An introduction to pathology, 2nd ed. London: Longman, Green and Co; 1954. p. 413.

[124] Berenblum I. The nature of tumour growth. In: Florey HW, editor. General pathology (3rd ed.). London: Lloyd-Luke; 1964. p. 528–35.

[125] Meissner WA, Warren S. Neoplasms. In: Anderson WAD, editor. Pathology (6th ed.). St. Louis, MO: Mosby; 1971. p. 529.

[126] Ruddon RW. Cancer biology, 4th ed. Oxford, UK: Oxford University Press; 2007. p. 4.

[127] Hanahan D, Weinberg RA. The hallmarks of cancer. Cell 2000;100(1):57–70.

[128] Hanahan D, Weinberg RA. Hallmarks of cancer: the next generation. Cell 2011;144:646–74.

[129] Lazebnik Y. What are the hallmarks of cancer? Nat Rev Cancer 2010;10(4):232–3.

[130] National Cancer Institute. What is cancer? Available at: http://www.cancer.gov/cancertopics/cancerlibrary/what-is-cancer.

[131] World Health Organization. Fact Sheet: Cancer. Available at: http://www.who.int/mediacentre/factsheets/fs297/en/.

Incidences, Mortality, and Classifications of Tumors

L. P. Bignold: Principles of Tumors.
DOI: http://dx.doi.org/10.1016/B978-0-12-801565-0.00003-2

63

As mentioned in Section 1.1.3, tumors as a category of disease are extremely common, and malignant tumors cause up to 25% of all deaths in developed countries. However, the total of "cancer burden" is made up of the individual "burdens" of each tumor type. The different tumor types occur with different incidences and have different implications with respect to clinical aggressiveness and treatment.

The main reasons for noting differences in incidences, prevalence, and of mortality according to tumor type are:

i. The circumstances of the population in which the incidence of each tumor type is highest may provide insights into causation for the purposes of prevention (Chapter 14).
ii. Projections for health care needs are affected by the incidences of tumor types.

Another aspect of importance is the trends in incidences and mortality of tumor types over long periods, at least many decades, of time. When any significant changes occur over time, it may reflect circumstances of:

i. Causation, with implications for prevention.
ii. Availability and efficacy of treatments.

iii. Diagnostic and other factors related to the collecting of data.

In this chapter, Sections 3.1 and 3.2 discuss matters relating to the reported incidences of tumors. Section 3.3 is concerned with some issues in the mortality data of tumors. Section 3.4 provides detail on the histopathological subtyping of tumors.

Epidemiological methods relating to possible hereditary factors in tumor causation are outlined in Section 7.9 and methods relating to efficacies of treatments of malignancies are discussed in Section 13.2. Epidemiological studies for identifying possible environmental carcinogens are described in Section 14.1.

3.1 GENERAL ASPECTS OF THE REPORTED INCIDENCES AND PREVALENCE OF TUMOR TYPES

3.1.1 Background

The first point to note here is that the data collected by the WHO with the IARC, as well as by the Centers for Diseases Control

(CDC) and other organizations, include only malignant tumors. These are then brought together according to organ of origin (see Section 1.1.3).

Generally in Western countries over the last fifty years, there appears to have been trends indicating (see Figure 3.1) [1–3]:

i. A rise, and then fall, in mortality from lung cancer.
ii. Falls in the mortality from carcinomas of the stomach and uterine cervix.
iii. Rises in the mortality from carcinomas of the liver and pancreas.

However, in the details for the incidences of the types of malignancies in different countries, there have been some quite marked changes in incidence in recent decades (Figures 3.2–3.5).

Factors involved in these changes may be:

i. Changes in the incidence of the tumors, especially in association with the ageing of the population.
ii. Improvements in therapy for the tumors.
iii. Changes in the data collection in relation to tumors.

These data in fact are known to be difficult to interpret, e.g., Ref. [2]. The latest (2012) document by Cancer Research UK affirms this in the statement:

> Caution should be taken when interpreting trends over time for cancers worldwide because changes probably also reflect changes in data recording [3].

Much of this chapter is concerned with the nature of, and factors involved in, these changes in data recording.

3.1.2 Terminology

(a) "Diagnosis"

This refers to the tumor type indicated by, or inferable from, what is written in death certificates. For epidemiologic data recording, the histopathologically identifiable subtypes of tumors of each organ, such as subtypes of non-Hodgkin's lymphoma of lymph nodes, are usually grouped together. The reasons for this are often that the main subtype is also the commonest, and that the numbers of cases of the rarer subtypes are too small for statistically meaningful analysis. Whether or not particular variants may represent "new" tumor types is rarely clearly established.

Histopathological details, however, can affect the overall data for the reported malignant tumors of an organ, as is discussed in Section 3.3.

(b) "Incidence"

This refers to the number of new cases occurring in a defined population in a defined period of time [4].

$$\text{Incidence} = \text{number of new cases in a specified population in a specified time period}$$

Raw data can be used to calculate "incidence rate (population size)" and "incidence rate (time)." Thus:

i. "Incidence rate (population size)" is used to compare incidences in populations of different sizes, e.g., North Americans versus Scandinavians.

$$\frac{\text{Incidence rate}}{\text{(population size)}} = \frac{\text{(number of new cases)}}{\text{(number of people in the population)}}.$$

In studies of causation (see Section 14.1), this can be modified when not all individuals were part of the defined population for the whole period of the study. It is then calculated as the number of new cases in the specified time, divided by the total number of persons

(A)

Trends in Age-adjusted Cancer Death Rates* by Site, Males, United States 1930–2011

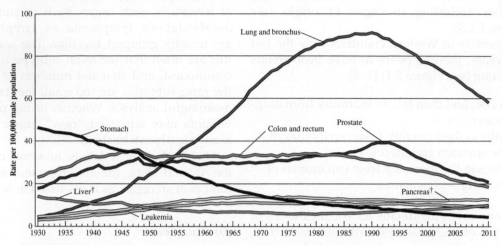

*Per 100,000, age adjusted to the 2,000 US standard population.   †Mortality rates for pancreatic and liver cancers are increasing.

Note: Due to changes in ICD coding, numerator information has changed over time. Rates for cancer of the liver, lung, and bronchus, and colon and rectum are affected by these coding changes.

Source: US Mortality Volumes 1930–1959 and US Mortality Data 1960–2011, National Center for Health Statistics, Centers for Disease Control and Prevention.

©2015, American Cancer Society, Inc., Surveillance Research

(B)

Trends in Age-adjusted Cancer Death Rates* by Site, Females, United States 1930–2011

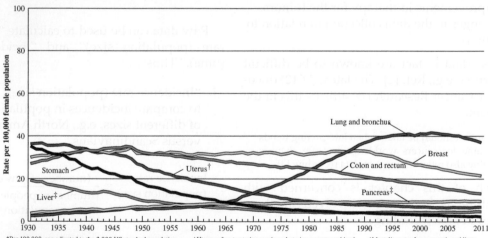

*Per 100,000, age adjusted to the 2,000 US standard population.   †Uterus refers to uterine cervix and uterine corpus combined.   ‡Mortality rates for pancreatic and liver cancers are increasing.

Note: Due to changes in ICD coding, numerator information has changed over time. Rates for cancer of the liver, lung, and bronchus, and colon and rectum are affected by these coding changes.

Source: US Mortality Volumes 1930–1959 and US Mortality Data 1960–2011, National Center for Health Statistics, Centers for Disease Control and Prevention.

©2015, American Cancer Society, Inc., Surveillance Research

FIGURE 3.1   Trends in incidence and mortality of common malignant tumors.
Note: The rise and fall of mortality due to lung cancer is associated with changes in tobacco smoking in the population. The fall in mortality due to stomach cancer is marked, and may be due to changes in the diet of the population. The rise and fall of mortality due to colo-rectal cancer, and that it has a different time pattern in men compared to women, is not explained. *Reproduced with permission from the American Cancer Society.* Cancer Facts and Figures 2015. *Atlanta: American Cancer Society, Inc.*

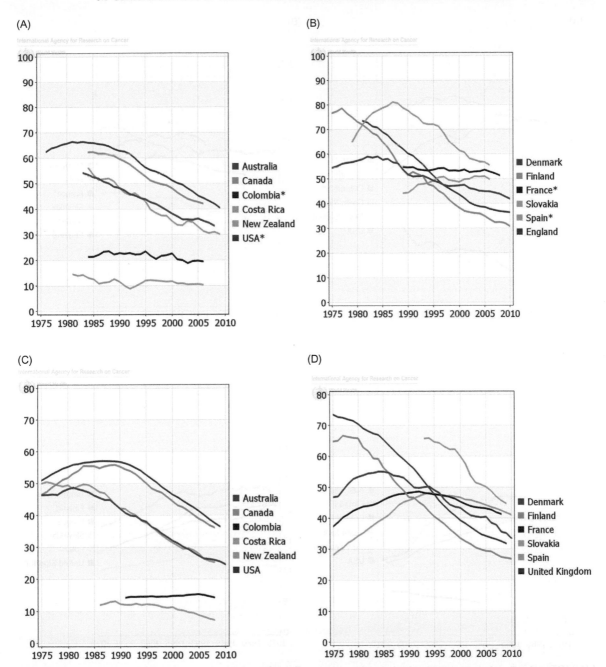

FIGURE 3.2 Lung carcinoma: trends in incidence and mortality, various countries, men. The differences between the different countries appear complex. In some countries, incidences show declines only. In other countries incidences rise and fall with different peaks at different times. *Reproduced with permission from Ferlay J, Soerjomataram I, Ervik M, Dikshit R, Eser S, Mathers C, Rebelo M, Parkin DM, Forman D, Bray, F. GLOBOCAN 2012 v1.0, Cancer Incidence and Mortality Worldwide: IARC CancerBase No. 11 [Internet]. Lyon, France: International Agency for Research on Cancer; 2013. Available from: http://globocan.iarc.fr, accessed on 23 June 2015. All asterisks refer to data which has been collected in the corresponding geographical region.*

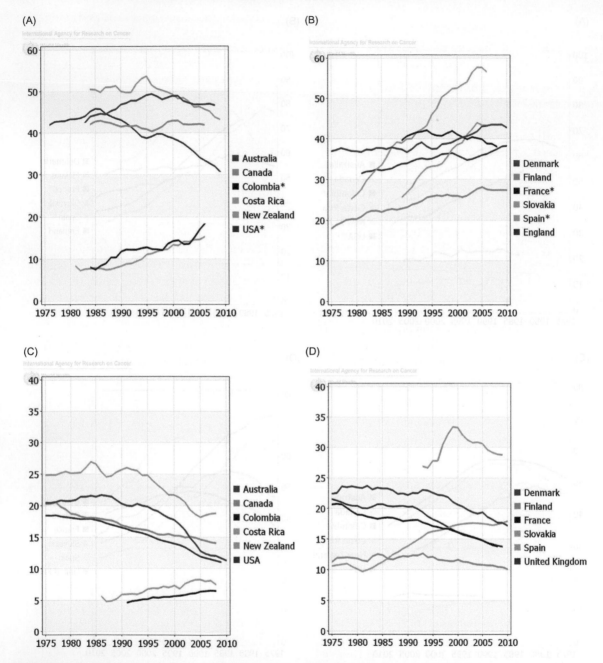

FIGURE 3.3 Trends in incidence of colorectal carcinoma. The differences between the different countries appear complex. Some countries show declines only, others show rises only, and yet others show rises and falls with different peaks at different times. *Reproduced with permission from Ferlay J, Soerjomataram I, Ervik M, Dikshit R, Eser S, Mathers C, Rebelo M, Parkin DM, Forman D, Bray, F. GLOBOCAN 2012 v1.0, Cancer Incidence and Mortality Worldwide: IARC CancerBase No. 11 [Internet]. Lyon, France: International Agency for Research on Cancer; 2013. Available from: http://globocan.iarc.fr, accessed on 23 June 2015. All asterisks refer to data which has been collected in the corresponding geographical region.*

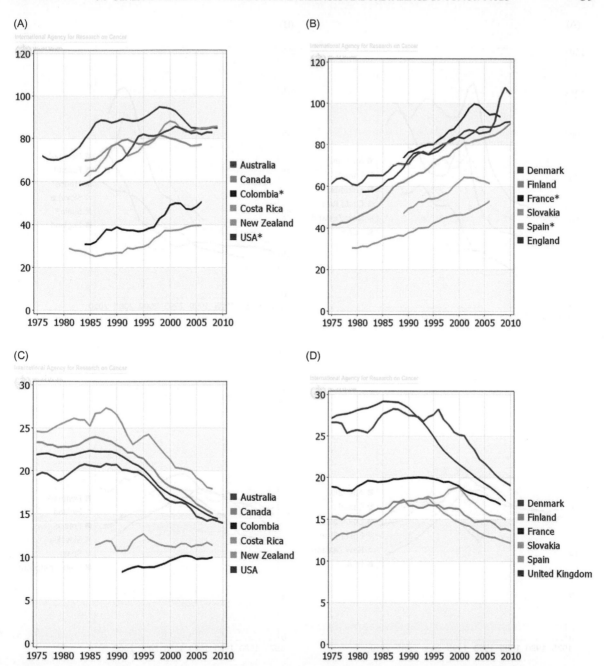

FIGURE 3.4 Trends in incidence of breast carcinoma in females. The differences between the different countries appear complex. Some countries show declines only, others show rises only, and yet others show rises and falls with different peaks at different times. *Reproduced with permission from Ferlay J, Soerjomataram I, Ervik M, Dikshit R, Eser S, Mathers C, Rebelo M, Parkin DM, Forman D, Bray, F. GLOBOCAN 2012 v1.0, Cancer Incidence and Mortality Worldwide: IARC CancerBase No. 11 [Internet]. Lyon, France: International Agency for Research on Cancer; 2013. Available from: http://globocan.iarc.fr, accessed on 23 June 2015. All asterisks refer to data which has been collected in the corresponding geographical region.*

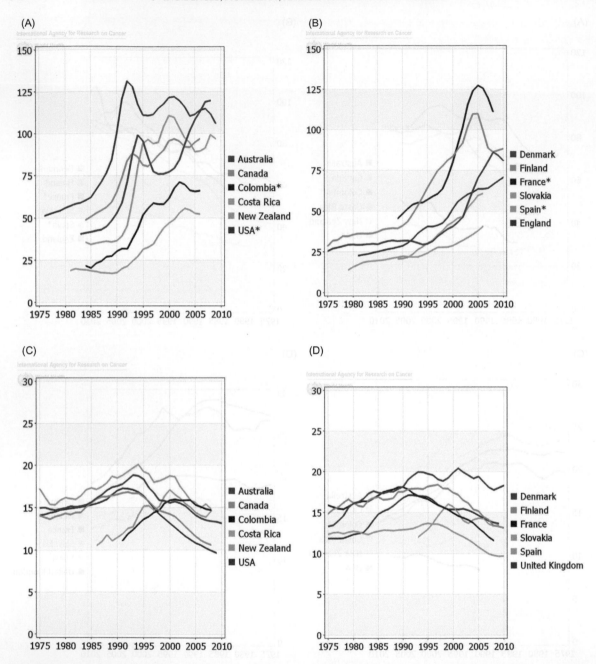

FIGURE 3.5 Trends in incidence of prostatic carcinoma. The differences between the different countries appear complex. Some countries show declines only, others show rises only, and yet others show rises and falls with different peaks at different times. *Reproduced with permission from Ferlay J, Soerjomataram I, Ervik M, Dikshit R, Eser S, Mathers C, Rebelo M, Parkin DM, Forman D, Bray, F. GLOBOCAN 2012 v1.0, Cancer Incidence and Mortality Worldwide: IARC CancerBase No. 11 [Internet]. Lyon, France: International Agency for Research on Cancer; 2013. Available from: http://globocan.iarc.fr, accessed on 23 June 2015. All asterisks refer to data which has been collected in the corresponding geographical region.*

multiplied by the average time the individuals participated in the study.

$$\frac{\text{Incidence rate}}{\text{(suspected cause)}} = \frac{\text{(Number of New Cases)}}{\text{(Person-Time at Risk)}}$$

ii. "Incidence rate (time)" is used for comparisons of incidences in equal time periods, e.g., per annum:

$$\text{Incidence rate (time)} = \frac{\text{(Number of New Cases)}}{\text{(number in whole population)}/\text{(unit of time)}}$$

iii. "Risk" is the same as incidence rate, but the term is used more commonly in epidemiological studies of causation. For "relative risk," see Section 14.1.2.
iv. "Odds ratio" is the number of individuals who developed the disease in the period in comparison with the number which did not develop the disease.

$$\text{Odds ratio} = \frac{\text{(Number of new cases)}}{\text{(number in whole population} - \text{number of new cases)}}$$

For comparisons between communities, the "cumulative incidence" (incidence proportion) is important. It is the number of new cases occurring in defined age-groups in a defined population in a specified period of time as a proportion of the whole population.

3.1.3 Total Incidence

For this discussion (Figure 3.6), it is important to recall that:

i. All cases of cancer have a period of being asymptomatic (especially when the tumor is small).
ii. Especially elderly individuals with a tumor in the asymptomatic phase may die of another disease.

iii. The different types of tumors tend to have markedly different asymptomatic periods, depending on their sites in the body and their inherent degree of aggressiveness.

Thus the total incidence includes all the diagnosed cases and undiagnosed cases of a type of tumor in a community. These relative proportions may vary from country to country.

When the diagnosed incidence and hence the total incidence exceeds the actual incidence of the disease, overdiagnosis is said to have occurred [5].

3.1.4 The Diagnosed Incidence Rate

This is the rate at which cases are diagnosed and hence liable to be reported to cancer registries.

The diagnosed incidence rate can be seen to comprise the following:

a. *The incidence rate of tumors diagnosed through medical investigation of clinical manifestations*
    In societies where medical services are available, this represents the largest part of the total incidence of the tumor type.
b. *The incidence rate of tumors found by medical investigation of unrelated conditions*
    This means essentially the rate at which tumors are accidentally discovered in their asymptomatic early phase during investigations for other illnesses. Tumors may be as small as a few mm across. Sources of such diagnoses include high-resolution imaging methods such as magnetic resonance imaging (MRI) scans.
    In otherwise healthy individuals, these diagnoses are clinically relevant, because without early detection, the tumors would become clinically apparent later in the patients' lives.
    Autopsies are another means of detection of unsuspected tumors (see Section 3.1.3).

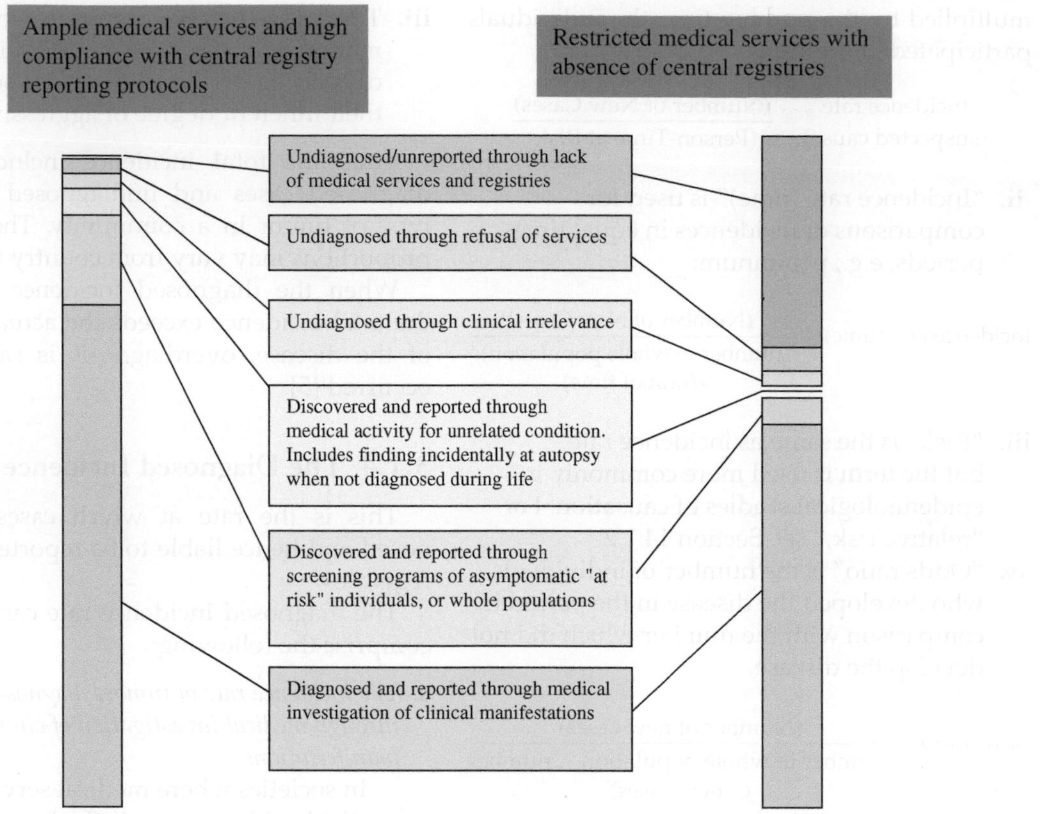

FIGURE 3.6 Potential components of the total incidence of a tumor type according to availability of medical services and registries.

c. *The incidence rate of tumors found by screening methods for prevention*

These are discussed in Sections 3.3.3 and 14.3. Here it can be noted that the sensitivities and specificities of the screening method for the particular tumor type are all important.

3.1.5 The Undiagnosed Incidence Rate

This rate has three components:

a. *Undiagnosed due to clinical irrelevance*

These may occur in all societies and refer to patients who have significant comorbidities (other medical conditions)

which limit their life spans, such that early cancers do not produce symptoms or signs before the patient dies of the other disease. An example of this group would include slow growing tumors in elderly persons with advanced dementia. This neurological disease is liable to limit reporting of symptoms.

b. *Undiagnosed due to refusal of services*

This is mentioned above as the reason for the clinically symptomatic tumors not being diagnosed.

c. *Undiagnosed rate due to lack or inaccuracy of services*

This is the rate at which cases occur but are not detected due to lack of medical services, and hence are not reported to cancer registries. This rate tends to be significant in societies where medical services are limited, and or in subpopulations of communities which do not access medical services. It can be divided into:

i. Undiagnosed clinically apparent cases (above).
ii. Cases which would have been diagnosed incidentally (above) except that the imaging techniques were not available.

3.1.6 Prevalence

(a) General

This term refers to "the number or percent of people alive on a certain date in a population who previously had a diagnosis of the disease. It includes new (incidence) and preexisting cases, and is a function of both past incidence and survival" [6]. It excludes those who:

i. Have never had the disease.
ii. Have had the disease before the time of the survey, but have spontaneously recovered or have been cured by treatment.
iii. Have died with or of the disease before the time of the survey.

The "prevalence proportion" is the number of affected individuals expressed as a proportion of the population. Raw data can be age specific, or age adjusted as for Incidence (see Section 3.2.1).

The overall prevalence of the various tumor types is also difficult to ascertain because different types and cases of tumors can go into remission for different periods of time. The longer the duration of the disease, the more likely the case will be included in prevalence data. On the other hand, if in a particular patient, the tumor is thought to be cured, the patient will not be counted in the prevalence data.

(b) Factors Affecting Prevalence of Cancer

The incidence rate and the average disease duration are the main factors affecting the prevalence of any type of tumor. Early diagnosis and effective treatment of cases resulting in cures may result in reduction of prevalence. Early diagnosis of cases without effective treatment will increase the time period between diagnosis and death, and hence will tend to increase the detected rate of prevalence. Noncurative, but life-extending treatment increases the prevalence of a disease [7].

3.2 MEDICAL ACTIVITY AND THE INCIDENCE AND PREVALENCE RATES OF TUMOR TYPES

Various factors may play roles in apparent differences in incidence data from the same country over time, or comparing incidences in different countries in the same period of time [8].

3.2.1 Availability of Resources

Epidemiological studies for incidences of tumors are more difficult than corresponding studies of many other diseases. The clinical manifestations of tumors can be slow in onset, diagnosis may require complex laboratory studies, and mortality may be difficult to monitor and confirm (Figure 3.6).

In Western societies, data on diagnosed incidence rates are usually accurate, owing to the availability of medical services and the establishment of cancer registries. However, variations in compliance rates of practitioners with respect to reporting cases to registries may possibly occur. In poorer countries, both medical services and reporting to registries may be incomplete. Even within the one country, medical services may differ, e.g., between urban and rural populations [9]. Health differences between urban and rural populations in advanced countries can have a variety of causes [10]. Generally, these may be adjusted accordingly when providing estimates of incidence of tumors [8].

3.2.2 Age Standardization

When comparing populations, either in different geographical locations or in the same location in different periods of time, overall incidence of a disease may be influenced by the age distributions of individuals in the population. Thus if a tumor is common only in elderly people, the unadjusted incidence rate will be lower in populations with low average life expectancies and higher in populations with high average life expectancies. Various methods have been used to adjust data to "international standard populations" [11], but since the 1980s, some studies of cancer epidemiology have considered only tumors arising in middle age [12]. In some countries, data on the age of the population may be difficult to assess. In many countries, birth and death records may be incomplete.

3.2.3 Changes in Histopathological Criteria for Diagnosis of Malignancy

Aspects of changing classifications of tumors into types are discussed in Section 3.4. These matters probably do not significantly affect the incidences of total malignancies of a particular organ. However, there are two main areas by which alterations in histopathological diagnostic criteria may affect total incidence data of tumor types as follows:

(a) Reclassification of indolent lesions as malignant

A number of tumor types have been recognized as having very long courses if left untreated [13]. The main way they were discovered was in autopsy series, as incidental tumors without any metastases. The main examples were low-grade carcinomas of the breast [14], prostate [15],

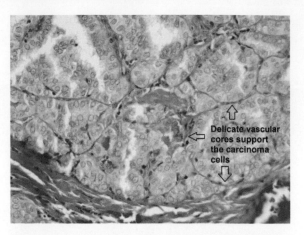

FIGURE 3.7 Example of a common "indolent" lesion resembling malignancy but not having clinically significant effects: "occult" sclerosing papillary carcinoma of thyroid gland.

and occult sclerosing/"papillary micro"— carcinomas of the thyroid [16] (Figure 3.7). These were often omitted from final diagnoses on autopsy reports, and also often not reported to cancer registries.

The practice of treating these lesions in the same way as other cancers has affected cancer registrations when practitioners report lesions as malignancies which previously would have been regarded as being clinically irrelevant.

(b) Including *in situ* malignancies with invasive malignancies

An *in situ* tumor is a lesion which is confined to its original location and does not result in a macroscopically visible swelling, but which is composed of cells that have the cytological features of tumor cells (Figure 3.8). These cells are commonly observed to give rise to invasive tumors of the corresponding cell type. For example, "progression" of these lesions to invasive malignancies occurs in the cervix [17]. *In situ* atypical areas of epidermis [18] give rise to invasive squamous cell carcinomas in 1–10% of cases. In the colon, adenomas show continuous graduations of cytological

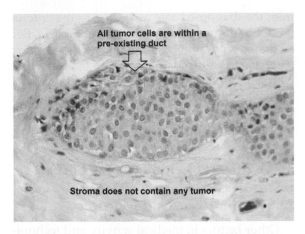

All tumor cells are within a
pre-existing duct

Stroma does not contain any tumor

FIGURE 3.8 Example of a common *in situ* lesion: *in situ* ductal carcinoma of the breast.

atypia (cytostructural abnormality) [19], and carcinomas tend to be found in the most cytologically atypical ones.

In the female, breast *in situ* lesions show all degrees of cytological atypia [20] and are associated with invasive tumor in many cases.

In the urinary tract, the lining epithelium gives rise to cytological papillary overgrowths. The cytologically normal tumors are described as "papillary adenomas" and all of those which show atypia—of any degree in whole spectrum of cytological atypia [21]—are often referred to as carcinomas.

To various degrees in different places and times, all of these lesions may or may not be included in reported cases of cancer.

3.2.4 Changes in Clinical, Imaging, and Biochemical Criteria

(a) Clinical

Some cases of tumors can be diagnosed by clinical examination. For example, diagnoses made by physical examination of advanced, ulcerated carcinomas of the breast and of proctoscopically identified rectal carcinomas are likely to be accurate. In most parts of the world, however, biopsies are taken so that diagnosis

rests with the pathologist. In developing countries, however, biopsies might not be taken in unresectable advanced cases.

(b) Imaging

Some cases of carcinoma of the lung and some other internal organs can be accurately diagnosed by appearances in plain X ray radiographs. Even with more complex imaging techniques, false-positive and false-negative diagnoses of cancer on the basis of imaging are not uncommon [22]. In countries where the relevant diagnostic services are available, biopsy is usually carried out to ensure the diagnosis. Where resources are limited, diagnoses of cancer may be based on radiology alone, with associated possible inaccuracies occurring.

(c) Changes in Biochemical Test Criteria

A few tumors may be diagnosed simply by high levels of particular substances (called in this context "biomarkers") in the blood or other fluid of the body. A long-established example is the presence of high levels of catecholamines in the urine as an indication of pheochromocytoma.

Currently, the most commonly used "biochemical marker" for a tumor is the level of prostate-specific antigen (PSA) in the diagnosis of carcinoma of the prostate. This test has two limitations:

i. Nontumorous diseases of the prostate can raise the serum PSA.
ii. The range of levels of PSA in men with carcinoma of the prostate overlaps the PSA levels in healthy men and in men with nontumor prostate disease.

After this spectrum of PSA levels was identified, changes in the diagnosis rates occurred when the arbitrary "threshold" value for diagnosis of prostate carcinoma was lowered [23]. The effect was a sharp rise in the reported incidence of this tumor [24]. In countries in which serum PSA was not common, there was no change in the reported incidence of the tumor.

These fluctuations in reported incidences are not known to have any biological basis.

3.2.5 Screening Programs

(a) General

The impetus to prevent cancers (Chapter 14) has given rise to many attempts to "screen" whole populations for early cancers. Removal of these early tumors potentially saves patients' lives. In any society, screening is usually associated with the availability of services to treat the cases discovered by screening.

Common, although not universally used, screening programs are mainly:

i. Papanicolaou smears of cervical cells for squamous cell carcinoma of the cervix.
ii. Regular examinations of the breasts of females, including by physical examination or radiological or other techniques.
iii. Testing feces for the presence of blood, as evidence of the presence of bowel carcinoma (older males and females). Sigmoidoscopic examinations are another method of screening for this tumor (see Section 14.3.3).
iv. Regular examinations of the skin for possibly malignant pigmented lesions.
v. Monitoring prostate antigens in the serum for increases associated with prostatic cancer (see above).
vi. Testing blood samples for a protein (CA125) which may indicate ovarian cancer.

(b) Effects of Screening Programs on Reported Incidences

In general, the effects of these programs have been to increase the rates of diagnosed tumor incidences by increasing the number of "incidental" cases, as is especially shown in programs for breast carcinoma [25].

This is most significant if it applies to clinically relevant cases. In some studies, cases from the clinically irrelevant group may be excluded from the diagnosed rate according to the criteria used for administering the screening process. For all tumor types with no later change in screening criteria, early diagnosis is likely to be associated with a peak in the rate of new cases, followed by a fall in the diagnosis rate, as the cases which would have been diagnosed later are diagnosed early to provide for the phase of rising incidence.

3.2.6 Effects of Other Changes in Intensities and Quality of Medical Activity on Reported Incidences

Other factors in medical activity and technologies affect the diagnosis rates of tumor types.

a. *Improved screening technology may further increase the rate of diagnosis of early tumors*
   For example, in screening for breast cancer, mammography usually will detect more lesions than physical examination [26].
b. *Increased use and quality of imaging technologies leads to discovery of more unsuspected lesions*
   Thus computed tomography and MRI scans may detect more "incidental" tumors (i.e., tumors detected when a patient is being investigated for an unrelated medical complaint) than plain X rays, e.g., in endocrine glands [27]. These investigations are not usually included among data arising from screening programs.
c. *Declining autopsy rates*
   Fewer autopsies reduce the number of discovery of incidentally diagnosed cases (Section 3.4.5).

3.3 MORTALITY

Mortality data are important to several issues in epidemiology. These include assessing the incidence of malignant tumors, the assessment of the effects of therapy (Section 12.6), as well as in epidemiological investigations of causation and prevention (Chapter 14).

3.3.1 Terminology

"Mortality of a disease" refers to deaths due to the disease being studied, in a defined population in a defined period of time [28].

Mortality = number of deaths due to the disease in a specified population in a specified time period

Cancer mortality may also be expressed as a rate (population) and a rate (time). Thus

$$\frac{\text{Mortality rate}}{(\text{population size})} = \frac{(\text{number of deaths due to the disease})}{(\text{number in whole population})}.$$

$$\frac{\text{Mortality rate}}{(\text{time})} = \frac{\text{Number of deaths due to disease}}{\text{Number in population}} \times (\text{unit of time})$$

Mortality rates may be adjusted for age distributions for comparisons of populations [29].

For tumor types which, if untreated, will cause the death of the patient, the mortality rate is the overall incidence rate minus the cure rate.

3.3.2 Cancer Registry Data and Death Certificate Data

Most of the available data on incidence are the reports of cases to cancer registries. Most of the data on mortality are supplied in death certificates. Death certificates rarely indicate a disease from which the individual did not die—i.e., a tumor which was diagnosed, treated, and cured in life.

For tumor types which cannot be effectively treated, such as lung cancer, the diagnosed incidence (cancer registry data) and the mortality (death certificate) incidence rates are likely to be equal. For tumors which are often cured by treatment, such as carcinoma of the thyroid, and to lesser extents, breast, colorectal carcinomas, and melanomas, diagnosed/cancer registry incidence rates are greater than mortality rates.

3.3.3 Changes in Mortality Rates Over Time

An important aspect of mortality data is whether or not in a given community, the rate changes over time.

(a) Causes of falls in Reported Death Rates when Total Incidences Remain Constant

These are mainly:

i. Greater numbers of cures, due to therapy including earlier detection and treatment, so more cases are at a lower stage of spread in the body (see Section 1.3.2).
ii. Therapy which prolongs life but does not cure cases. This fall in incidence should be temporary, and—if all other factors are equal—be followed by a rise in incidence in later years.
iii. Changes in classifications of tumor types.
iv. Changes in practices involved in reporting deaths.

(b) Causes of Fall in Death Rates per Number of Diagnoses

These are as for the causes in (a), with the additional possibility of:

i. Changes in diagnostic practices resulting in overdiagnosis of tumors. This possibility arises only in relation to a few types of tumors (see Section 3.2.4).

3.3.4 Death Certificates

It is general in Western countries for death certificates, signed by a medical practitioner, to be issued before a patient's body can be disposed of. Any inaccuracies in death certificates affect registry data for both incidence (for tumors not diagnosed) and mortality (if the tumor has caused death, but this is not recorded) [30–32].

The topic is under continuing study. The most important tool has become linking of cancer registry data with death certificates [33]. The linkage gives the most accurate survivals over a fixed interval (say, 10 years). However, tumors which can have longer postdiagnosis survivals (e.g., carcinoma of the breast) may not be accurately assessed.

The longer the patient survives, the more likely he or she will move outside the Registry/data collecting area.

3.3.5 Role of Autopsies

Autopsies play roles in cancer management and statistics in three main ways [34]:

a. *May increase accuracies of diagnoses of tumor types*

This is especially if premortem diagnostic studies were inconclusive.

b. *May occasionally reveal undiagnosed cancers, particularly in the colon and breast*

These cases may have been clinically overlooked because many cases of these tumors are slow growing and do not cause symptoms or signs early in their course. The fall in autopsy rates over recent decade means that fewer "incidental" cancers are no longer found by this means.

Autopsies, with notification of tumors found to the appropriate cancer registry, may result in increased numbers of clinically undiagnosed cases in incidence data.

c. *May assist in assessing the efficacy of treatment*
See Section 12.6.

3.4 CLASSIFICATIONS OF TUMORS

It has already been noted that there are over a thousand distinguishable tumor types. The general classification of these tumor types according to organ, cell, and behavior is also described (Section 1.1.2).

For statistical purposes, they are usually reduced to 30 or so categories (Table 3.1).

However, the existence of the different subtypes, and the fact that their names have changed over time, must be discussed because:

(i) Different subtypes may require different treatments.
(ii) The existence of so much diversity in tumors—all from the one genome—is a phenomenon of cellular phenotypes which genomic theories of tumors should take into account (Section 4.1.3).

Because of these reasons, the issues in classifying and subtyping tumors are discussed in some detail.

3.4.1 Pre-1940s

Since the late nineteenth century (see Section 2.1), all classifications of tumors have been based primarily on their clinicopathological features. From that time until the middle of the twentieth century, almost all classifications of tumors were produced by individuals or small groups of clinicians and pathologists. There were considerable variations between these classifications. This was partly because the histogeneses (the parent kinds of cells) and actual clinical behaviors of many types of tumors were not fully clarified. The variations were also because some terminologies were based on traditions developed before cellular pathology was developed (see Section 2.1) and hence did not depend on histopathological concepts. Moreover, most classifications were not fully satisfactory because many cases could not be diagnosed definitively into one type or another. As a result, these early classifications tended to include categories of "tumor not otherwise specified ('NOS')," "sarcoma NOS," etc. for these cases of tumor.

3.4.2 Armed Forces Institute of Pathology: *Atlas of Tumor Pathology*

In 1949, the Armed Forces Institute of Pathology (AFIP) (Washington, DC) began to

TABLE 3.1 Abbreviated Classifications of the Main Subtypes and Patterns of Malignant Tumors

Epidermal carcinoma

Squamous cell carcinoma

Basal cell carcinoma. Subtypes: superficial spreading, nodular, ulcerative and morphoeic. Mixed examples are common.

Malignant melanoma

Subtypes: nodular, superficial spreading, arising in Hutchinson's melanotic freckle, acral lentiginous. Mixed examples are common.

Lung: Nonsmall cell carcinoma

Subtypes: squamous, adeno- and anaplastic large cell patterns

Lung: Small cell carcinoma

Subtypes: small cell; mixed small and large cell; occasional cases are mixed with nonsmall cell patterns (squamous or adenocarcinoma)

Female breast

Invasive ductal; invasive lobular carcinomas, both with variable proportions of papillary, cribriform, desmoplastic ("scirrhous"), and other patterns; medullary. Mixed examples are common.

Stomach

Adenocarcinoma with variable proportions of papillary, tubular, mucinous, and poorly cohesive patterns. Sometimes subtyped as "intestinal" and "diffuse." Mixed examples are common.

Large bowel (colon and rectum)

Adenocarcinoma with variable tubular, solid, mucinous, undifferentiated, and uncommonly other patterns. Mixed examples are common.

Pancreas

Generally as for large bowel. Mucinous variants constitute a higher proportion. Mixed examples are common.

Leukemia

Acute lymphoblastic (mainly B-cell)

Acute myelocytic

Chronic lymphocytic

Chronic myeloid leukemias

Tumorous proliferative lesions of other kinds of cells in the bone marrow.

Note: Many broad groups are subdivided according to chromosomal abnormalities. (Mixed examples among the hematopoetic kinds of cells are common.)

Lymph nodes: Hodgkin's disease

Subtypes diffuse lymphocyte predominant, nodular lymphocyte predominant, nodular sclerosing, mixed cellularity, lymphocyte depleted. Mixed examples are common.

(Continued)

TABLE 3.1   (Continued)

Lymph nodes: Non-Hodgkin's lymphoma

   Current classifications rely on combinations of abnormalities in morphology, antigen expression, and genomic lesions resulting in definition of individual types without little hierarchical taxonomic relationships being clear (see main text). Frequencies of combined examples are difficult to assess.

Uterine cervix

   Squamous cell carcinoma of variable degrees of loss of differentiated.

Endometrium

   Adenocarcinoma with variable proportions of papillary, solid, and undifferentiated patterns.

Ovary

   Adenocarcinoma originating from surface epithelium.

   Subtypes include cystic/solid; serous, mucinous, "clear cell," endometrioid, and other patterns.

   (Mixed examples are not very common.)

Testes

   Seminoma (like spermatogonia); other germ cell tumors (i.e., showing differentiation to early embryonic, or to adult tissue kinds of cells). Mixed examples are common.

Prostate

   Adenocarcinoma of variable, tubular, solid, and undifferentiated patterns.

   Mixed examples are common.

Brain

   Glial cells, ranging from well-differentiated ("astrocytic") to poorly differentiated (glioblastomatous) cases; usually in "pure" forms.

Kidney

   Adenocarcinoma of tubular epithelium with variable mixtures of papillary, tubular, solid, and undifferentiated patterns. Mixed examples are common.

Bladder

   Urothelial tumors showing variable degrees of invasiveness. Noninvasive tumors are usually purely papillary; invasive tumors consist of variable mixtures of papillary, tubular, solid, and undifferentiated patterns.

Bones

   Osteogenic sarcoma. Many cases showing variable additional components of cartilaginous, fibrous, and other tumors.

   Chondrosarcomas, consists of chrondrocytes with variable degrees of loss of differentiation. Usually "pure" forms.

Source: From Ref. [1].

publish an *Atlas of Tumor Pathology* issued as Series of Fascicles [35].

Importantly, each fascicle dealt with a particular organ, not a whole organ system (see Section A1.1.2). For example, separate fascicles were written for the tumors of salivary glands, oral cavity, esophagus and stomach, and intestines. The fascicles were intended to be, and indeed have largely become (the fascicles are now being produced in their fourth series), major texts for the classifications and descriptions of tumors.

Of note is that in Series 1 of the Fascicles, the description of each type of tumor was preceded by a list of "synonyms and related terms." In Series 2, the list was called "Synonyms." In Series 3 and 4, this part of the text has been omitted. This is presumably because many synonyms are now obsolete. In other cases, it may be because equating one term in one system of nomenclature with another term in another system may be controversial.

3.4.3 World Health Organization: *International Classification of Diseases* and *International Histological Classification of Tumours*

As a separate project, since its inception in 1946, the World Health Organization (WHO) has sponsored international classifications of diseases. The first was the *Statistical Classification of Diseases, Injuries, and Causes of Death* [36]. Subsequently, the WHO has produced a "family" of classifications. Currently, the main one relevant to tumors is the *International Classification of Diseases*, which is now in its 11th edition [37].

Beginning in 1967, and in support of the part of this classification which deals with tumors, the WHO has published a series of diagnostic handbooks entitled *International Histological Classification of Tumours* [38]. Part of the

purpose has been to achieve higher degrees of intercountry consistency in diagnosis, to better serve international epidemiological studies.

In recent series, each of these volumes has tended to describe all the tumors of each organ system of the body rather than being limited to an individual organ (see above). These volumes are shorter, and in some aspects less detailed, than the Fascicles of the *Atlas of Tumor Pathology*. There are also some differences between the classifications in the *Histological Classification* compared with the *Atlas*.

In the previous sections of this chapter, the possibilities of difficulties in pathologic diagnoses have been mentioned. This section outlines these matters in some detail.

3.4.4 General Aspects of Identification of New Types and Subtypes of Tumors

A major reason for efforts to identify new types of tumors or subtypes of recognized categories of tumors—which are then included in classifications—is because the newly discovered types or variants might have different prognoses or respond differently to new therapeutic regimens (Chapter 10).

Thus the general tendency in tumor classifications has been to detail more and more types. This process of "splitting" into smaller and smaller groups, as opposed to "lumping" into larger and perhaps more manageable groups, leads to difficulties in comparing data over time concerning what may well be the same tumor types. For examples of increases with time in diagnosable types/subtypes, see Table 3.2.

3.4.5 Classifying Cases Showing Mixed Variants/Patterns of a Tumor Type

There are difficulties associated with using some of the classifications in practice. They mainly arise from the fact that many cases of

TABLE 3.2 Increasing Numbers of Diagnosed Tumor Subypes

1. AFIP: *Atlas of Tumor Pathology*

Lung	Bronchogenic carcinoma	1980: 5 types; 5 subtypes	1994: 6 types; 7 subtypes
Prostate	Epithelial malignancies	1973: 5 subtypes.	2011: 7 subtypes.
Breast	Invasive carcinoma	1968: 1 type; 10 variants	2009: 3 "major variants"; 25 variants of which 15 are uncommon
Colo-rectum		1967: 1 types; 1 variant	2003: 1 type; 10 variants

2. WHO: *International Histological Classification of Tumours*

Lung	Bronchogenic carcinoma	1967: 4 types; 10 subtypes;	2004: 26 types and variants
Prostate	Invasive carcinoma	1980 4 types; 8 variants	2004: 3 types; 10 subtypes
Breast	Invasive carcinomas	1968: 2 types; 5 rare types; 6 variants	2012 10 types; 20 variants; 10 rare types; 3 variants
Colo-rectum	Epithelial malignancies	1989: 7 types; no subtypes or variants	2010: 5 types; 6 subtypes

Source: Data from these two sources enable different subdivisions of cases into types, subtypes, and variants.

These numbers are assessed from the contents pages of the various volumes. The numbers are not completely definite, because the criteria used by authors for making the divisions into types, subtypes, and variants are not always clear in the texts.

Note: The increases in numbers of types of tumors reflect increased knowledge of the natural history of the lesions, as well as of the genomic abnormalities which are identified in lesions.

tumor may show more than one histological pattern. When the histological pattern becomes a "type," then it may become difficult in practice to decide which diagnosis should be applied to the whole mass.

Usually, a case is diagnosed according to the pattern which is most extensive in the microscopy slides. "Extent" is usually expressed as a percentage. However, in relation to assessments of percentages, it must be remembered that in histopathological practice, the proportion of a whole tumor which is made into histological slides is usually very small (typically less than 1% of the tumor bulk may be in the slides). Hence sampling error (i.e., the sample is too small to reliably reflect the whole) is likely to affect many of these assessments of percentages of variant histological subtype.

Examples of tumors exhibiting multiple patterns/types are as follows.

(a) Carcinoma of the Breast

Multiple histological patterns are commonly seen in individual cases of carcinoma of the breast (Figure 3.9). This is seen especially in *in situ* tumors [39]. It is also seen in invasive tumors. For example, the patterns of "invasive ductal" and "invasive lobular" may occur together in individual tumors. In a study of 12,206 cases of breast carcinoma, 6.2% were classified as pure lobular, 70.5% as pure ductal, and 23.2% were identified as invasive ductal mixed with invasive lobular or other variant [40]. As another example, a study of 100 cases of carcinomas of the breast showed that mucinous carcinoma was mixed with other kinds of breast carcinoma in 55 of the cases. The other 45 cases—which were classed as "pure"—showed 0–10% of other patterns in the sections examined [41].

An exemplary case is shown in Figure 3.9. Much of the lesion comprised cystic

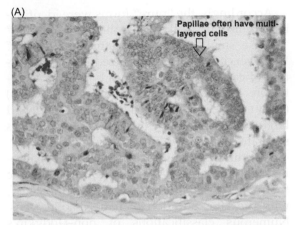

(A) Papillae often have multi-layered cells

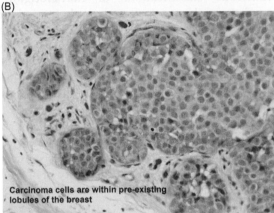

(B) Carcinoma cells are within pre-existing lobules of the breast

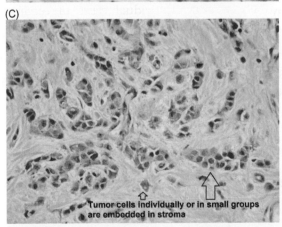

(C) Tumor cells individually or in small groups are embedded in stroma

FIGURE 3.9 Mixtures of subtypes/variants of carcinoma of the breast. These three different appearances occurred in the same case: (A) Papillary intra-cystic appearance; (B) Intralobular; (C) Scirrhous.

spaces lined by papillary tumor, and others appeared as cylinders filled with tumor cells (Figure 3.9A). Other areas, however, consisted of intralobular carcinoma (Figure 3.9B) and infiltrating ductal breast carcinoma of the "scirrhous" pattern characterized by small regular cells in lines one cell thick ("Indian files") between dense fibrous tissues (Figure 3.9C). The usual way to allocate such a case to a particular type is by assessing the percentages of all the types and to diagnose the case as the type which is represented by the highest percentage.

(b) Adenocarcinomas of the Large Bowel (Colon and Rectum)

Eighty percent of carcinomas of the colon are well or moderately differentiated [42]. The remainder includes well-differentiated papillary carcinomas and undifferentiated "anaplastic" types. The phenomenon of mixed variants is seen in most of these tumors because of the focally marked accumulations of mucus. The amounts of mucinous tumor in a case can vary from 0% to 100%. In a typical series [26], 6% of 462 cases of adenocarcinoma of colon were mucinous, and a further 28% had some mucinous foci. To make the diagnosis of the mucinous variant, the recommended thresholds for mucinous tumors versus nonmucinous tumors is greater than 50% of the mass being pools of mucus [27].

(c) Malignant Lymphoma

i. Hodgkin's lymphoma

Hodgkin's disease has traditionally been diagnosed by the presence of a particular kind of cell originally described independently by Sternberg [43] and Reed [44]. A later classification of the disease [45] was into "paragranuloma" (no inflammatory cells and only mildly abnormal lymph node cells), "granuloma" (inflammatory cells, and even fibrosis, with mildly–moderately abnormal lymph node cells), and "sarcoma" (wholly comprising markedly abnormal lymph node cells). In 1966, Rappaport renamed the first as "lymphocyte predominant," the last as "lymphocyte depleted," and divided the second into "nodular sclerosing" and "mixed cellularity" [46]. This classifications has been preserved in the WHO classification [47] with the addition of a type referred to as "nodular lymphocyte predominant Hodgkin's lymphoma." However, in individual cases, the features of nodular sclerosing and mixed cellularity may be mixed, and combinations of Hodgkin's and non-Hodgkin's disease may occur [48].

ii. Non-Hodgkin's lymphoma

These have long been the most difficult tumors to classify [49,50]. In the era before immunoperoxidase stains for specific lymphoid antigens, these tumors were classified on morphology alone [46]. The features were:

i. By the architecture of the tissue: whether the cells formed follicles ("follicular lymphoma") or not ("diffuse lymphomas").

ii. By the size of the tumorous lymphoid cells: whether they were small or large. Small cells were considered lymphoid in origin and large cells histiocytic in origin.

Accordingly, the "pure" forms were identifiable: "follicular small cell," "follicular large cell," "diffuse small cell," and "diffuse large cell." However, it was also recognized that a proportion of cases were "mixed" in both their architecture ("mixed follicular and diffuse" cases) (Figure 3.10) and kind of cell ("mixed small and large cell" cases), and these were included as separate categories in the classification [46].

As immunostaining and molecular pathological techniques have become available, the first new division was into "T-" and "B-cells", with the later addition of natural killer cells. Numerous classifications of non-Hodgkin's lymphomas have appeared and tended to become successively more complex. The current WHO classification [47] uses mainly morphological and antigen expression (immunoperoxidase) profiles as criteria. Genomic abnormalities (such as chromosomal translocations) are taken into account for a few types.

Only some tumors comprise purely one or other kind of lymphocyte. The entity "T-cell/histiocyte-rich large B-cell lymphoma" is recognized [47]. "Mixed" cases both in uniformity of antigen expressions in different foci of the same tumor,

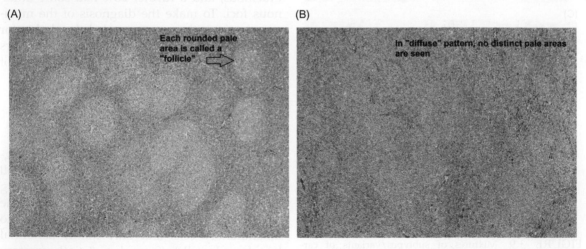

(A)   (B)

FIGURE 3.10 Mixtures of subtypes/variants of non-Hodgkin's malignant lymphoma. Malignant lymphomas are often classified as either 'follicular' or 'diffuse'. Not uncommonly, as in this case, both patterns were present in the one affected lymph node. (A) Follicular pattern; (B) Diffuse pattern.

and also mixed positivities in staining for antigens between cases are common in practice. Thus ease and consistency in diagnosis of the various subtypes have been difficult to achieve [51,52].

(d) Testis

In the testis, many cases of nonseminomatous germ cell tumors show more than one of its major pathological patterns (embryonal, yolk sac, and teratomatous), so that a term "mixed germ cell tumor" may be introduced alongside diagnostic categories of "pure" cases showing only one pattern [53].

(e) Thyroid

In the thyroid gland, low-grade cancers are often classified as papillary and follicular types [31]. However, many cases show both papillary and follicular foci. Arbitrarily, certain nuclear abnormalities have been deemed to be more important than architecture, so that a category of "follicular variant of papillary carcinoma" has been provided [54].

3.4.6 Classifying Tumors Showing Continuous Spectra of Morphological Features

Another kind of difficulty arises when the individual diagnostic categories are really parts of continuous spectra of cytological abnormality (as opposed to overall composition of patterns discussed in Section 3.4.3), and not discrete clusters of low and high degrees of abnormalities (see Section 8.2.6).

One example is *in situ* epithelial lesions of the female breast [39]. Another example is transitional cell carcinomas of the urinary tract, in which the cases of tumor completely fill the spectrum from "benign" to "malignant" [55]. A further example is the epithelial tumors of the endometrium [56]. In this type, there is an uninterrupted range of patterns from "adenomatous hyperplasia," through "atypical adenomatous hyperplasia" and "papillary/grade 1"

adenocarcinomas, to invasive solid adenocarcinoma patterns.

3.4.7 Relevance to Grading

The fact that in individual cases, the degrees of abnormality may vary from focus to focus affects the process of allocating a grade to cases of malignancy (Section 1.3.3). The phenomenon is often described as focally x and focally y degrees of "differentiation" (see Section 2.4.1). The question extends to what should be called "malignant" and what should be called "atypia"? Arbitrary percentages may be recommended, e.g., more than 10% high-grade atypia requires the whole lesion to be called "high grade."

This problem is evident in relation to various types of tumors, including ductal carcinomas *in situ* [57], carcinoma *in situ* of the cervix [58], carcinomas of the prostate [59], premalignant lesions of the gastrointestinal tract [60], and renal cell carcinomas [61].

3.4.8 Resulting Interhistopathologist Differences in Diagnoses

In relation to diagnoses made of types and subtypes, the difficulties presented above are probably the main factors in the commonly described wide ranges of incidences of subtypes of tumors.

One example is the "medullary" variant of carcinoma of the breast. Originally applied to soft carcinomas (as opposed to the more common firm, "scirrhous" variant), it became a diagnosis based on histopathological criteria. The criteria, however, were difficult to apply, and reported incidences of this variant have ranged from less than 1% to 7% [62].

Another example is the diagnosis of "dysplastic naevus," which can be made entirely on clinical observations or entirely on histopathological features. Different authors have offered different clinical and pathological criteria, so that the diagnosis "dysplastic naevus" has varied in different series between 5% and 20% of the total number of naevi [63].

3.4.9 Nonhistological Bases for Classifications

Cases of tumors are usually divided according to stage and grade, but this is not a classification in the usual sense.

In clinical practice, however, cases are classified according to biomarker or other characteristics. An example of these relates to expressions of estrogen receptors, progesterone receptors, as well as receptors for human epidermal growth factor 2 in breast carcinoma. If a tumor is negative for expression of all these receptors, it is classified as "triple negative" for the planning of therapy [64].

References

[1] American Cancer Society. Cancer facts and figures 2014. Available at: http://www.cancer.org/research/cancerfactsstatistics/cancerfactsfigures2014/.

[2] Ellis L, Woods LM, Estève J, et al. Cancer incidence, survival and mortality: explaining the concepts. Int J Cancer 2014;135(8):1774–82.

[3] Cancer Research UK. Trends in world mortality. Available at: http://www.cancerresearchuk.org/cancer-info/cancerstats/world/mortality/#Trends.

[4] dos Santos Silva E. Cancer epidemiology: principles and methods. Lyon: IARC/WHO; 1999. p. 60.

[5] Klotz L. Cancer overdiagnosis and overtreatment. Curr Opin Urol 2012;22:203–9.

[6] National Cancer Institute. Cancer statistics. Available at: http://www.cancer.gov/statistics.

[7] National Cancer Institute - Surveillance Research. Care and cure prevalance. Available at: http://surveillance.cancer.gov/prevalence/carecure.html.

[8] Curado MP, Edwards B, Shin HR, editors. Cancer incidence in five continents, vol. IX. Lyon: WHO and IARC Scientific Publication No 160; 2012.

[9] Shi L. Health care in China: a rural–urban comparison after the socioeconomic reforms. Bull World Health Organ 1993;71(6):723–36.

[10] Nasca PC, Mahoney MC, Wolfgang PE. Population density and cancer incidence differentials in New York State, 1978–82. Cancer Causes Control 1992;3(1):7–15.

[11] Ahmad OB, Boschi-Pinto C, Lopez AD, et al. Age standardization of rates: a new WHO standard.

World Health Organization; 2001. GPE Discussion Paper Series: No.31 EIP/GPE/EBD. Available at: http://www.who.int/healthinfo/paper31.pdf

[12] In: Ref. [4]. p. 81.

[13] Esserman LJ, Thompson Jr. IM, Reid B. Overdiagnosis and overtreatment in cancer: an opportunity for improvement. JAMA 2013;310:797–8.

[14] Santen RJ, Yue W, Heitjan DF. Modeling of the growth kinetics of occult breast tumors: role in interpretation of studies of prevention and menopausal hormone therapy. Cancer Epidemiol Biomarkers Prev 2012;21:1038–48.

[15] Zlotta AR, Egawa S, Pushkar D, et al. Prevalence of prostate cancer on autopsy: cross-sectional study on unscreened Caucasian and Asian men. J Natl Cancer Inst 2013;105:1050–8.

[16] Sakorafas GH, Giotakis J, Stafyla V. Papillary thyroid microcarcinoma: a surgical perspective. Cancer Treat Rev 2005;31:423–38.

[17] Kurman RJ, Ronnett BM, Sherman ME, et al. Tumors of the cervix, vagina, and vulva. AFIP Atlas, Ser 3, Fasc 4; 2010. p. 75–110.

[18] Patterson JW, Wick MR. Nonmelanocytic tumors of the skin. AFIP Atlas, Ser 4, Fasc 4; 2006. p. 28.

[19] Riddle RH, Petras RE, Williams GT, Sobin LH. Tumors of the intestines. AFIP Atlas, Ser 3, Fasc 32; 2003. p. 110.

[20] Tavassoli FA, Eusebi V. Tumors of the mammary Gla nd. AFIP Atlas Ser 4, Fasc 10; 2009. p. 75–90.

[21] Murphy WM, Grignon DJ, Perlman EJ. Tumors of the kidney, bladder and related urinary structures. AFIP Atlas Ser 4, Fasc 1; 2004, p. 253–82.

[22] De Boo DW, van Hoorn F, van Schuppen J, et al. Observer training for computer-aided detection of pulmonary nodules in chest radiography. Eur Radiol 2012;22(8):1659–64.

[23] Canto EI, Shariat SF, Slawin KM. Biochemical staging of prostate cancer. Urol Clin North Am 2003;30(2):263–77.

[24] McDavid K, Lee J, Fulton JP, et al. Prostate cancer incidence and mortality rates and trends in the United States and Canada. Public Health Rep 2004;119(2):174–86.

[25] Bach AG, Abbas J, Jasaabuu C, et al. Comparison between incidental malignant and benign breast lesions detected by computed tomography: a systematic review. J Med Imaging Radiat Oncol 2013;57(5):529–33.

[26] Nelson AL. Controversies regarding mammography, breast self-examination, and clinical breast examination. Obstet Gynecol Clin North Am 2013;40(3):413–27.

[27] Vardanian AJ, Hines OJ, Farrell JJ, et al. Incidentally discovered tumors of the endocrine glands. Future Oncol 2007;3(4):463–74.

[28] Hjartåker A, Weiderpass E, Bray FI. Cancer mortality International encyclopedia of public health. Philadelphia, PA: Elsevier; 2008. p. 452–64.

[29] Kirch W, editor. Encyclopedia of public health. New York, NY: Springer; 2008. p. 996–7.

[30] Hu CY, Xing Y, Cormier JN, Chang GJ. Assessing the utility of cancer-registry-processed cause of death in calculating cancer-specific survival.. Cancer 2013; 119(10):1900–7.

[31] Johansson LA, Björkenstam C, Westerling R. Unexplained differences between hospital and mortality data indicated mistakes in death certification: an investigation of 1,094 deaths in Sweden during 1995. J Clin Epidemiol 2009;62(11):1202–9.

[32] Johnson CJ, Hahn CG, Fink AK, et al. Variability in cancer death certificate accuracy by characteristics of death certifiers. Am J Forensic Med Pathol 2012;33(2):137–42.

[33] Johnson CJ, Weir HK, Fink AK, et al. (Accuracy of cancer mortality study group). The impact of national death index linkages on population-based cancer survival rates in the United States. Cancer Epidemiol 2013;37(1):20–8.

[34] Chute CG, Ballard DJ, Nemetz PN. Contributions of autopsy to population-based cancer epidemiology: targeted intervention to improve ascertainment. : IARC Sci Publ No 112; 1991. p. 207–16.

[35] Henry RS. The Armed Forces Institute of Pathology - Its First Century. US Printing Office. Available at: http://archive.org/stream/TheArmedForcesIns tituteOfPathology-ItsFirstCentury/ArmedForces InstituteOfPathologyItsFirstCentury-Henry_djvu.txt

[36] Moriyama IM, Loy RM, Robb-Smith AHT. History of the statistical classification of diseases and causes of death. Rosenberg HM, Hoyert DL, editors. National Center for Health Statistics. Hyattsville, MD; 2011.

[37] World Health Organization. International Classification of Diseases (ICD). Available at: http://www. who.int/classifications/icd/en/.

[38] Sobin LH. The international histological classification of tumours. Bull World Health Organ 1981;59(6):813–9.

[39] Rosen PP. Rosen's breast pathology. Philadelphia, PA: Lippincott-Raven; 1997. p. 235.

[40] Pestalozzi BC, Zahrieh D, Mallon E, et al. Distinct clinical and prognostic features of infiltrating lobular carcinoma of the breast: combined results of 15 international breast cancer study group clinical trials. J Clin Oncol 2008;26(18):3006–14.

[41] Ranade A, Batra R, Sandhu G, et al. Clinicopathological evaluation of 100 cases of mucinous carcinoma of breast with emphasis on axillary staging and special reference to a micropapillary pattern. J Clin Pathol 2010;63(12):1043–7.

[42] Riddle RH, Petras RE, Williams GT et al. Tumors of the intestines. AFIP Atlas, Ser 3 Fasc 32; 2003. p. 145.

[43] Sternberg C. Über eine eigenartige unter dem Bilde der Pseudoleukamie verlaufende Tuberculose des lymphatischen apparates. Ztschr Heilk 1898;19:21–90.

[44] Reed D. On the pathological changes in Hodgkin's disease, with special reference to its relation to tuberculosis. Johns Hopkins Hosp Rep 1902;10:133–96.

[45] Jackson H, Parker F. Hodgkin's disease: II. Pathology. N Engl J Med 1944;231:35–44.

[46] Rappaport H. Tumors the hematopeotic system. AFIP Atlas, Scr 1, Fasc 8; 1966. p. 156–60. Non-Hodgkin's lymphoma. p. 91–54.

[47] Swerdlow SH, Campo E, Harris NL, et al. T cell rich B cell lymphoma. In: WHO classification of tumours of the haematopoeitic and lymphoid system. 4th ed.; 2007. p. 321–34.

[48] Warnke RA, Weiss LM, Chan JKC, et al. Tumors of the lymph nodes and spleen. AFIP Atlas, Ser 3, Fascicle 14; 1997. p. 329–40.

[49] Campo E, Swerdlow SH, Harris NL, et al. The 2008 WHO classification of lymphoid neoplasms and beyond: evolving concepts and practical applications. Blood 2011;117(19):5019–32.

[50] Taylor CR, Hartsock RJ. Classifications of lymphoma; reflections of time and technology. Virchows Arch 2011;458(6):637–48.

[51] Cartwright R, Brincker H, Carli PM, et al. The rise in incidence of lymphomas in Europe 1985–1992. Eur J Cancer 1999;35:627–33.

[52] Chan JK, Kwong YL. Common misdiagnoses in lymphomas and avoidance strategies. Lancet Oncol 2010;11(6):579–88.

[53] Ulbright TM, Young RH. Tumors of the testis and adjacent structures. AFIP Altas Ser 4, Fasc 18; 2013. p. 211–38.

[54] Rosai J, Carcangiu ML, DeLellis RA. Tumors of the thyroid gland. AFIP Atlas Ser 3, Fasc 5; 1990. p. 21–122. Follicular variant of papillary. p. 100–8.

[55] Murphy WM, Grignon DJ, Perlman EJ. Tumors of the kidney, bladder and related urinary structures. AFIP Atlas, Ser 4, Fasc 1; 2004. p. 246–87.

[56] Silverberg SG, Kurman RJ. Tumors of the uterine corpus and gestational trophoblastic disease. AFIP Atlas, Ser 3, Fasc 3; 1991. p. 15–90.

[57] Tavassoli FA. Ductal carcinoma in situ: introduction of the concept of ductal intraepithelial neoplasia. Mod Pathol 1998;11(2):140–54.

[58] Baak JP, Kruse AJ. Use of biomarkers in the evaluation of CIN grade and progression of early CIN. Methods Mol Med 2005;119:85–99.

[59] Brimo F, Montironi R, Egevad L, et al. Contemporary grading for prostate cancer: implications for patient care. Eur Urol 2013;63(5):892–901.

[60] Guindi M, Riddell RH. The pathology of epithelial pre-malignancy of the gastrointestinal tract. Best Pract Res Clin Gastroenterol 2001;15(2):191–210.

[61] Delahunt B. Advances and controversies in grading and staging of renal cell carcinoma. Mod Pathol 2009;22(Suppl. 2):S24–36.

[62] Tavassoli FA, Eusebi V. Tumors of the mammary gland. AFIP Atlas, Ser 4, Fasc 10; 2009. p. 200–1.

[63] Elder DE, Clark Jr WH, Elenitsas R, et al. The early and intermediate precursor lesions of tumor progression in the melanocytic system: common acquired nevi and atypical (dysplastic) nevi. Semin Diagn Pathol 1993;10(1):18–35.

[64] Davis SL, Eckhardt SG, Tentler JJ, et al. Triple-negative breast cancer: bridging the gap from cancer genomics to predictive biomarkers.. Ther Adv Med Oncol 2014;6(3):88–100.

Etio-pathogenesis I: Causative Agents of Tumors

L.P. Bignold: Principles of Tumors.
DOI: http://dx.doi.org/10.1016/B978-0-12-801565-0.00004-4

As outlined in 1.2, the overall causation of diseases—their etio-pathogenesis—includes the etiological agents and the mechanisms by which they have their observable effects. For tumors, both these aspects of formation are complex. This is particularly because

i. There are many different kinds of etiological agents,

ii. There are several potential targets in the cell for these agents,

iii. There are many possible mechanisms by which the causative agents may act on these targets, and

iv. The phenotypic outcomes are highly diverse, a thousand or so, different tumor types—all from the one genome.

This and the next three chapters are concerned with these complex issues. The general basis for these accounts is that tumorous conversion occurs through genomic events in cells that are partly or wholly specialized into one or other kind of cell in the adult organism (see 1.1.2).

In this chapter, Section 4.1 introduces general issues in the formation of tumors. Section 4.2 compares general aspects of the known etiological agents. Sections 4.3 through to 4.8 deal with each category of carcinogens and how they might react with potential targets in cells.

Chapter 5 discusses the known genomic events in tumor cells and how the known etiological agents might cause them. Chapter 6 describes what is known of particular genes in relation to the major features (trait-like phenotypic changes) which are seen in tumor cell populations. Chapter 7 describes hereditary factors in tumor formation. Possible mechanisms of combinations of features, as define the tumor types, are discussed in Chapter 8.5.

In support of these chapters, Appendix 2 outlines relevant aspects of the normal genome, and Appendix 3 describes factors in the resistance of cells and tissues to noxious agents ("noxins") generally—including carcinogens.

4.1 GENERAL ISSUES IN TUMOR FORMATION

4.1.1 Terminology

(a) "Mutation" and "Genomic Event"

Originally in germ-line genetics, the term "mutation" was applied to a change in phenotype, especially if having Mendelian characteristics (see 2.5.1). In the 1930s, it was recognized that some phenotypic changes can be due to particular abnormalities in chromosomes [1]. Since the 1960s, it has been understood that certain phenotypic changes (such as for sickle cell anemia—see 2.6.2) can be due to single abnormalities in sequences of nucleobases in DNA. Gradually after that, "mutation" came to be used also for the assumed lesion in the hereditary material which causes a change in phenotype. Subsequently, however, it was established that not all chromosomal abnormalities and not all DNA sequence changes necessarily produce a detectable change in phenotype (see in Chapters 5, 7 and 9). Nevertheless, the term "mutation" has been maintained to refer to all changes in the genome, whether or not they produce any phenotypic change. As an example, the term "silent mutation" may be used in germ-line genetics for genomic events, which do not alter the phenotype of the organism, but in molecular biology for alterations in nucleotide sequences in exons, which do not alter amino acid sequence [2]. It is not necessary to avoid "mutation" in many areas of tumor genomics, especially in hereditary predispositions, because the meaning is usually clear. However, in some parts of this book, the phrases "phenotypic change" and "genomic event" are used to indicate the relevant precise meaning.

(b) Non-Genopathic and "Genopathic"

The injuries and unphysiological deaths of cells (see Chapter 10) comprise:

i. Those resulting from interference with the structural or metabolic activities of the cell membrane or cytoplasm, without necessarily affecting the nucleus or the genome. These are non-genopathic injuries or cell deaths.

ii. Those resulting from damage to the structure, or enzymatic processes involving the genome (A2.3). These are termed "genopathic" phenomena. The term has no implication for whether or not the agent reacts directly chemically with DNA (see 4.4.1 and (c) below). In all genopathic events, the cells are not killed immediately—and not necessarily killed at all—but are assumed to have altered functional capacity in some feature of their normal biology because of alterations in protein production in comparison with the original cell. Also, genopathic events occurring in a cell may cause defects which only manifest in daughter cells.

(c) Ambiguity of "Genotoxic"

The term "genotoxic" has been used with different meanings in different texts. Some authors use "genotoxic" to indicate either:

i. Direct covalent combination of agents with DNA, as for chemicals, or
ii. Direct induction of covalent chemical changes in DNA, as for radiations [3,4] (discussed in 5.1.2).

Using this definition, "non-genotoxic carcinogen" refers to a carcinogen which does not react directly with DNA [5]. In association with this usage, "indirect genotoxin" has been used for substances which do not react with DNA, but which might cause the production of other molecules—such as reactive oxygen species—which can damage DNA [6]. (In this book the term "indirect genopathic effect" is used for this phenomenon.)

Other authors have used "genotoxic" to refer to an ability to inflict any manifestation of genomic damage in living organism. The manifestations include nuclear morphological changes, chromosomal aberrations, DNA strand breaks, changes in DNA nucleotide sequences or in fact, any inheritable phenotypic change, whether or not the agent reacts covalently with DNA [7].

(d) "Locus" and "Complex Locus"

The term "locus" was introduced for "the site of a gene" and "complex locus" for "site of a cluster of genes" [8]. Recently, authors have used terms such as "contiguous" [9] for genes in complex loci. In this book, lesions of various sizes are discussed, some of which are large enough to contain multiple genomic elements including those which do not encode for protein. Most frequently in this book, general words such as "site" for locus, and "regions" for large sites/complex loci, are used.

(e) Clastogens

The topic of chromosomal aberrations is discussed in 5.3. The induction of chromosomal aberrations is one kind of genopathic event. Agents that are capable of inducing chromosomal aberrations are referred to as "clastogens."

4.1.2 Spontaneity versus Causation by Exogenous Factors

As mentioned in 1.2.1, many types of tumors occur without known exogenous influences. In addition, among cases of types of tumor for which an etiological agent is known, some cases arise without known exposure to the agent. An example is seen in the relationship between tobacco smoking and carcinoma of the lung. Most cases occur in smokers, but a few cases arise in individuals who have never been smokers themselves and who have rarely inhaled smoke exhaled by others ("passive smoking").

This gives rise to the idea that tumors might be "spontaneous," in the sense of having no specific external causative factor (see Figure 4.1). Alternative scenarios for "spontaneity" are:

i. That the individual is predisposed from birth by his/her specific genomic makeup to the tumor. Attempts to establish this kind of hereditary component to tumor formation has been made in studies of identical twins [10]. However, twins are usually raised together, so that some common environmental factor affecting both twins to develop a tumor of a particular type may be the basis of any association found.

Also, it can be noted that the spontaneous tumor incidences vary markedly among different strains of laboratory animals living under the same environmental conditions [11,12].

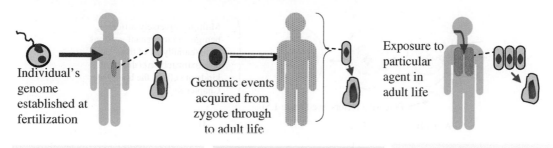

A

Spontaneous due to individual's genome

Some aspect of the individual's genome which is formed at fertilization of a human egg, predisposes to tumor formation in the particular kind of cell— with only a few exceptions— later in life.

B

Spontaneous due to accumulated multi-factorial low-impact genomic events

From zygote through to adult life, the genome unavoidably acquires genomic events which, infrequently and randomly, cause tumors, but not in any particular kind of cell.

C

Exogenous

"Exogenous" suggests that with few exceptions, causative agents ( ) for tumors act on particular kinds of cells in adult life.

FIGURE 4.1 Conceptual differences between spontaneous occurrence and exogenous causation in relation to tumors.

This suggests that there may be differences between humans too in their liability to develop tumors without exogenous cause.

ii. That during the lifetime of the individual, the general environment of individuals inflicts large numbers of genomic events on the somatic cells. These, in the right combinations in the right genes, lead to tumors in occasional cells of the body.

This idea is supported by the observation that tumors arise as a concomitant of increasing age of the individual (see 3.2.2).

Nevertheless, in all these situations, "spontaneity" cannot be proved without excluding all the possibilities for unrecognized exposures to known or so far undiscovered causative environmental agents. This is a very important issue: Do chronic, low-level "environmental" exposures to putative carcinogens (as opposed to occupational exposures) cause the cases of tumors which appear to have no cause? The issue has been argued in many books, including two works for the general reader [13,14], without definitive conclusions drawn.

4.1.3 Five Steps in the Etio-Pathogenesis of Tumors by Exogenous Agents

For many diseases, the general steps in their etio-pathogenesis may be relatively straight forward.

i. The exogenous causative agent accesses a cell.

ii. The agent acts with one or more targets in one or more kinds of cell.

iii. The effects of the action of the agent on the target result in the manifestations of the disease.

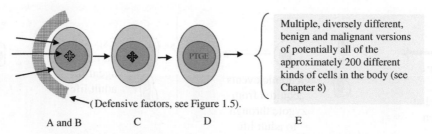

(Defensive factors, see Figure 1.5).

A and B   C   D   E

A. The exogenous causative factor(s) enters, or in some cases only impinges on (see Sections 4.8 and 4.11) a somatic cell.

B. The exogenous agent(s) acts on a critical structural or biochemical target(s) in the cell (✜).

C. A particular function of the target is disturbed (✜).

D. The disturbance in the function in the target has a pro-tumor genomic effect (PTGE).

E. The relevant effect(s)/genomic event(s) produces the features of thousand or so different types of tumors—from different parent cells, but nevertheless, all from one genome.

FIGURE 4.2   The five steps of tumor etio-pathogenesis involving exogenous agents.

With respect to tumors, however, these steps in exogenous causation—that is to say in the full sequence between exposure to carcinogen and the thousand or so different types of tumors—are more complicated than in this scheme.

Thus, for tumors, the steps can be envisaged (Figure 4.2) as follows:

A. The exogenous causative factor(s) enters, or in some cases only impinges on (see 4.8 and 4.11), a somatic cell. This contact or entry can only occur if the cell lacks effective defensive mechanisms against the carcinogen (A3.1). Multiple modifying factors, with possibly different targets, may be involved.

B. The exogenous agent(s) act on a critical structural or biochemical target(s) in the cell. The critical target of exogenous carcinogens is almost universally thought to be the DNA of the cell. Possibly, alternative or additional targets may be involved (4.2.6).

C. A particular function of the target is disturbed. In the sickle cell anemia model (A4.2.5), this effect is analogous to the change in the oxygen-binding curve of the

hemoglobin complex. In tumor formation, the disturbance in function in DNA is that it becomes less reliable as a template for synthesis and transcription.

D. The disturbance in the function in the target has a pro-tumor effect. This effect for tumors is almost universally accepted to be a change in the nucleotide sequence/genomic event particular genes in the DNA of the cell. (The reasons for calling these changes "genomic events" are explained in 4.1.1).

E. The relevant effect/genomic event(s) produce the features of tumors. In considering this step, it must be remembered that

   i. Different genes or regulators of genes are probably responsible for each feature of tumor cells.

   ii. The thousand or so different tumor types manifest as complexly different combinations of these features.

   iii. All these tumor types arise from events in the one genome, because the genome is the same in all proliferating cells in the body of the individual.

4.1.4 Time Factors

The periods of time between exposures to carcinogens and appearance of tumors are important aspects of tumor causation because they have implications for many of the proposed mechanisms of etio-pathogenesis.

(a) Repeated or Continuous Exposures

For induction of almost all human and experimental tumors, normal cells must be exposed to carcinogens repeatedly or continuously, either directly or indirectly, over long periods of time.

All human tumors with a known cause, including the well-known examples of chimney sweeps' cancer, shale oil-workers' tumors, aniline dye workers' bladder tumors, and tobacco-smokers' lung cancers (see 1.2), arise only after these repeated exposures. Even when the agent is permanently located in the tissues, for example with inhaled asbestos fibers, the associated tumors usually appear only after many years, even five decades or more [15].

Similarly, in experimental skin carcinogenesis, at least months of application are usually necessary to produce tumors, as was demonstrated by Ichikawa and Yamagiwa [16] in experiments producing tumors in rabbits' ear epidermis with coal tar (see 4.1.3b).

For radiations, it is well known that long periods of time are required for tumors to appear, for example, for skin cancers to result from ultraviolet light (UVB).

With viruses, long periods of exposure are well established, as discussed in 4.5.3.

Only a few chemical carcinogens, such as certain nitrosamine analogues, are able to produce tumors by a single administration. This model (the Faber model [12,17]) is restricted mainly to rat livers. The model requires that, at the time of administration of the carcinogen, the liver is regenerating after partial hepatectomy [18]. This implies that cells which are undergoing division, including mitosis, may be particularly susceptible to the carcinogen. Few examples of carcinogenesis involving only a single exposure to carcinogen are known to have occurred in humans.

(b) Delays and "Latencies"

These terms relate to timing and are applied in many situations. Generally, "delays" refer to the fact that tumors may appear long after cessation of the application of the carcinogen. A clear example was shown by Doll and Hill [19] who showed that the increased liability to lung carcinoma persists for up to 20 years after cessation of smoking. This observation has been confirmed in larger studies (Figure 4.3) [20]. Emphasizing this observation is the fact that, at the present time in some countries, carcinomas are common in ex-smokers than in "current-smokers" [21]. Also, for radiation-induced tumors, cases of sarcomas have been reported as occurring 30 years after the completion of courses of radiotherapy (always given in multiple doses) for bone lesions [22]. There are also delays of different durations in the appearance of tumors associated with inherited predispositions (see in Chapter 7).

The term "latency" has been used for the same circumstances as above, as well as for certain slightly different additional situations as follows:

i. The interval between the first exposure to carcinogen—in a situation of continuous applications—and the appearance of tumors [23],
ii. Periods of time between the surgical removal of primary tumors and the appearances of clinically appreciable metastases (see in Chapter 1), and
iii. Malignant progression (see 8.4).

4.1.5 Multi-Factorial Causation of Tumors

The fact that more than one cell-damaging agent might be involved in the formation of many kinds of tumors has been understood

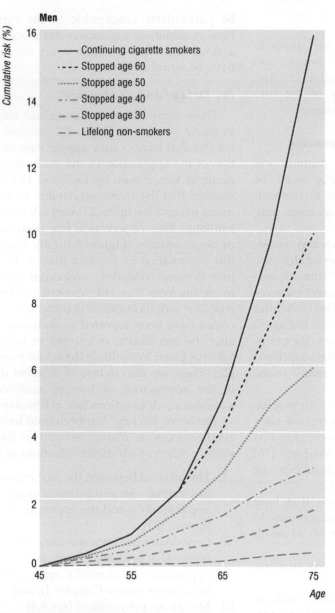

Age of appearance of lung cancer in 75-year-old persons according to age of cessation of smoking.

The figure shows that 15 years after cessation of smoking, the risk of lung cancer is still two-thirds of that of continuing smokers, and more than ten times that of never-smokers.

FIGURE 4.3   Delays between starting smoking and development of carcinoma of lung. Effects of stopping smoking at various ages on the cumulative risk (%) of death from lung cancer up to age 75, at death rates for men in United Kingdom in 1990. Non-smoker risks are taken from a US prospective study of mortality. Reproduced with permission from BMJ 2000;321:323.

since the early twentieth century [24,25]. This is often considered in terms of multiple different chemicals being required for tumor formation. However, it is also possible that different classes of causative agents may be involved in some tumors. Examples include the possibilities of syncarcinogenesis between chemicals and radiations, as well as between chemicals and viral infections.

The main component issues in multi-factoriality have been studied with different chemicals, as noted here:

(a) Essential and Modifying Factors

Investigation of multi-factorial causation of tumors only became possible with methods for causing tumors in experimental animals. Coal tar was already known to be a mixture of chemicals when Ichikawa and Yamagiuchi [15] (see above) developed the experimental method for demonstrating its tumorigenic effect. A major area of study then became: which chemicals in coal tar are the carcinogens? When pure compounds were isolated from coal tar, several were shown to be carcinogens [26].

After that, studies were undertaken on the effects of multiple individual chemicals applied to the same tissue (usually, area of skin of an appropriate experimental animal). The possibility that multiple carcinogens might be necessary to produce tumors appears to be broadly consistent with the facts that:

i. Environmental carcinogens often exist in mixtures with non-carcinogens (as in coal tar), and
ii. Humans are exposed to great numbers of different chemicals on a daily basis.

The broad results of the studies were that:

i. Some chemicals ("complete carcinogens") cause tumors without the application of any other substance (e.g. aromatic amines) [27].
ii. Other chemicals were found which only had an effect in the presence of other chemicals, which were called "co-carcinogens." By definition, chemicals in the second category were not carcinogenic in their own right.

Ultimately, causative agents were divided into those which are essential for tumor formation and those factors which have modifying roles (either increasing or reducing incidences of tumors). The essential agents are often referred to as "primary carcinogens" or "initiators" (see next subsection). The modifying factors that have enhancing effects are called "enhancers" or "promoters." Modifying factors that decrease the incidence of tumors in this context are usually called "inhibitors" [28].

Modifying factors might also include those that affect the activation(s) necessary to convert pro-carcinogens to ultimate carcinogens (see A3.1.3).

(b) Exposures to Carcinogens in Specific Sequences

In the studies of multiple carcinogens mentioned in the previous subsection, the chemicals were often administered in mixtures. However, when the chemicals were administered separately, it was found that in some protocols (Figure 4.4) tumors only appeared in the exposed skin when one chemical (later called an "initiator") was applied before the second chemical (called the "promoter") [28–30]. Skin treated with the initiator alone did not develop tumors. Moreover, the promoter could not cause tumors on its own. It was only able to produce a tumor in skin which had been pretreated with the initiator.

These results led to the popular "two-stage" concept of carcinogenesis with "initiation" and "promotion" being necessary phases of tumor formation [31,32]. Subsequently, many additional complex phenomena were discovered. For example, "promotion" was found to be reversible, but "initiation" was found to be irreversible. Moreover, some promoters would only be effective if another promoter had preceded it (Figure 4.4) [33]. No particular chemical molecules or chemical groups in analogues could be identified to be specific to any particular carcinogenic role. Moreover, many of the protocols demonstrated the different effects of agents only in certain strains of rodents [34]. At present, the biochemical basis of the model, especially of "promotion", is not fully explained [35,36].

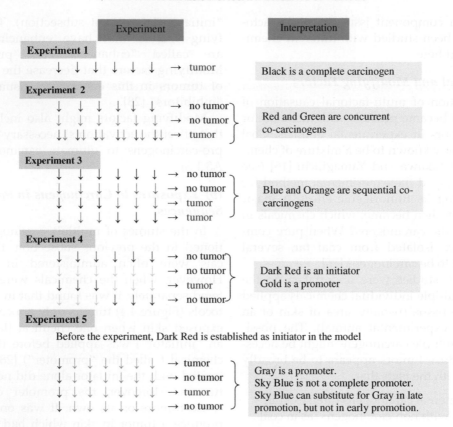

FIGURE 4.4 Interpretation of effects of different agents in different combinations in the skin model for initiation and promotion in carcinogenesis.

4.2 GENERAL ASPECTS OF ETIOLOGICAL AGENTS AND THEIR EFFECTS

4.2.1 Many Carcinogens Have Multiple Other Noxious Effects

In the general consideration of the biology of tumor causation, it is important to recognize that carcinogenesis is only one of many possible noxious effects of external agents on tissues and cells. Radiations in particular have major additional effects, such as inflammation of tissues, deaths of cells, immune-suppression, and non-carcinogenic genomic damage, such as chromosomal aberrations and induction

of micronuclei [37] (see also 10.2.4). Among the various categories of carcinogenic chemicals, many substances cause other pathological effects. For example, polyaromatic hydrocarbons (found in all tars and in tobacco smoke) provoke inflammation, cell death, and fibrosis [38] as well as tumors. Most kinds of asbestos have similar effects, of which fibrosis is the most marked [39].

In general, however, in considering possible relationships between these non-tumorous effects and possible carcinogenic potencies, it must be noted that almost all the agents which cause most of the non-tumor pathological effects are not carcinogens. For example, chronic inflammation due to tuberculosis, and

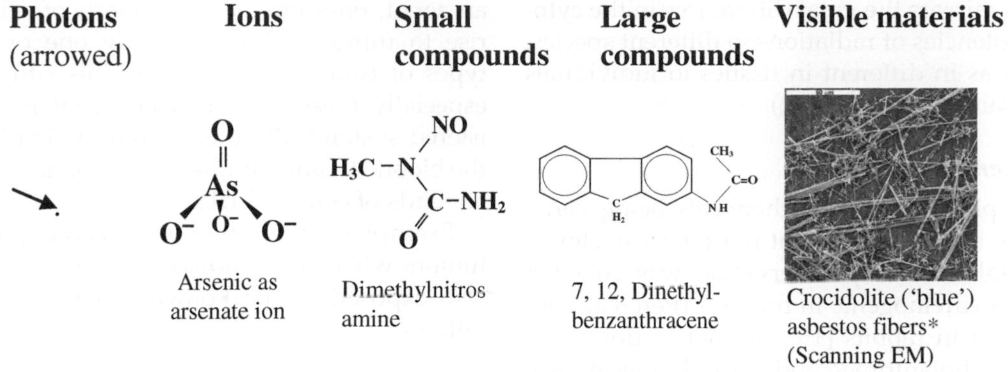

Photons (arrowed)	Ions	Small compounds	Large compounds	Visible materials

Arsenic as arsenate ion / Dimethylnitros amine / 7, 12, Dinethyl-benzanthracene / Crocidolite ('blue') asbestos fibers* (Scanning EM)

FIGURE 4.5 Differences in physicochemical structure of carcinogens (not to scale). *Left to right:* Arsenate; Dimethylnitrosamine; 2-Acetylaminofluorine; Crocidolite ('blue') asbestos fibers. Image from the Gallery of the Denver Microbeam Laboratory of the US Geological Survey, reproduced with permission from the Geological Survey Department of the Interior/USGS (Dr H Lower).

lung fibrosis due to silica, are not associated with carcinogenesis.

A general classification of noxious effects in cells is given in Chapter 10. These additional biological effects may also be important in the experimental assessment of potential carcinogens for humans as discussed in 14.2.

4.2.2 Diverse Physicochemical Natures of Etiological Factors

One of the most striking features of carcinogens is the diversity in their physicochemical natures (see Figure 4.5 and later in this chapter). The major carcinogens include:

i. Photons of wavelengths which are capable of either exciting electrons in atoms (UVB) or ejecting electrons so that ionized radicals are created (ionizing radiations).
ii. Chemicals including ions (such as As3+, Cr3+), small molecules such as single-ring aromatic amines, and larger molecules up to 5-ring polyaromatic hydrocarbons (PAHs, e.g., 1,2;5,6 dibenzanthracene).
iii. Certain viruses and some other biological agents.
iv. Some apparently insoluble solid materials, such as asbestos fibers. Even glass or plastic

surfaces may be considered carcinogens, because cells cultured on them often undergo malignant transformation (15.4).

The fact that such diverse agents are able to cause, or contribute to, the same biological result—i.e., conversion of normal cells to tumor cells—is something of a challenge for any theory of tumor etio-pathogenesis. In fact, it may support the possibility that different classes of agent may act through different mechanisms (see 2.6–2.8).

4.2.3 Species Differences in Susceptibilities to Carcinogens

This is a major phenomenon in carcinogenesis, and has significant implications for testing of the carcinogenic properties of environmental agents (15.3). Species differences in susceptibilities occur in relation to both radiations and chemicals.

(a) Radiations

The potencies of ionizing radiations as carcinogens are quite different in different animals. Species that are particularly susceptible are mice [40] and beagle dogs [41]. In mice, there are considerable differences between different strains [42].

This mirrors the great differences in the cytotoxic potencies of radiations in different species, as well as in different in tissues in individuals of the same species (12.1.2).

(b) Chemicals

The phenomenon of chemicals being carcinogenic in one animal but not others is almost universal. An early observation was coal tar, which is carcinogenic in humans (Percival Pott, 1775) and in rabbits [15], but not in dogs [43]. 3-methylcholanthrene and several other hydrocarbons, which are among the most potent of skin carcinogens in rodents, have little potency in monkeys [44]. Among rodents there are also many differences in susceptibility to other carcinogens. For example, dimethylarsinic acid (DMA(V)) is carcinogenic to the rat urinary bladder, but not in mice [45]. The issue is discussed both in terms of general principles and individual carcinogens in Reference [46].

Finally, some widely used chemicals are carcinogenic in animals but not in humans. A particular example is saccharin, which produces tumors of the urinary bladder in male rats but is not carcinogenic in humans [47,48]. Other examples are D-limonene—a major oil in the rind of citrus fruit—which causes kidney tumors in male rats but not in humans [48,49], and phenobarbital, which produces liver tumors in mice but not in humans [50].

4.2.4 Differences between the Kinds of Parent Cells in Susceptibilities to Carcinogens

Some examples of formation of tumors by known agents involve direct application of carcinogen to a particular kind of cell in a human or animal. However, in many situations, the carcinogen is applied to tissues (see A1.1.2) rather than to just a single kind of cell. Frequently in these situations, when different kinds of cells appear to be equally exposed to

an agent, only one or a few kinds of cells give rise to tumors. Moreover, only one or a few types of tumors are induced. This situation is especially observed when an agent is administered systemically (i.e., so that it circulates in the blood stream) and hence potentially reaches all kinds of cells (A3.1.1).

Examples of only one cell giving rise to tumors when many adjacent kinds of cells are also exposed to the known carcinogen are as follows:

a. *Human chronic arsenic toxicity.*

Most events involving inorganic arsenic—as arsenite or arsenate—occur when the substance is inadvertently consumed via contaminated drinking water [51]. The tumors that are induced, however, do not occur in the mouth, gastrointestinal tract, liver, or kidneys, in which the arsenic is deposited. In fact, arsenic-related tumors arise mainly in the epidermis of the skin [52]. In this situation, exposure to the solar UVB rays might be a co-carcinogen (see 4.1.3) [53,54]. However, melanocytes and dermal cells are not affected.

If arsenic-containing dusts are inhaled, pulmonary carcinomas may arise [55]. In these cases too, a wide range of other cells, especially those of the reticulo-endothelial system, are in close contact with the arsenic. Tumors are not induced in these latter cells.

b. *Human and experimental skin tumors induced by coal tar and shale oil.*

In humans exposed to these agents, tumors develop from the keratin-producing cells but not melanocytes or dendritic cells. In certain strains of mice, tumors of melanocytic cells may be produced [56] but few other experimental animals are known to be susceptible.

c. *Human mesothelioma caused by asbestos*

In this disease, the asbestos fibers are deposited in the lungs and cause tumors of the lung-covering cells (the mesothelium), as well as possibly carcinomas of the bronchus [see 57,58]. However, asbestos fibers also come into direct contact with alveolar lining cells, fibroblasts, and lung capillary endothelial cells. No tumors develop from these kinds of cells.

d. *Human vinyl chloride exposure*

This substance is used in the plastics industry, mainly in the production of polyvinyl chloride. The monomer is either inhaled or swallowed. The main tumor type caused is angiosarcoma of the liver. Associations are described with hepatomas, glial tumors of the brain, and lymphomas. Tumors are not provoked in lung capillary endothelial cells, macrophages, hematopoietic, or any other kinds of cells [59].

e. *Human thorium exposure*

Thorium is a naturally occurring radioactive element, emitting alpha particles (helium nuclei). Thorium dioxide is radio-opaque, and so as fine particulate suspension ("Thorotrast"), and was widely and commonly used as a "contrast" agent in radiology. The particles are taken up by the reticulo-endothelial system (see A1.1.1) to "image" these organs. Its radioactivity was not involved in this use. Patients who had been given Thorotrast, however, after many years have developed angiosarcomas and carcinomas of the liver, as well as leukemias [60]. Various types of tumors at other sites have been reported in patients receiving Thorotrast, but whether or not these were causative or spurious associations is difficult to establish. Thorotrast was also associated with hepatic cirrhosis which, whatever its cause, is a predisposing factor to liver carcinoma. The mechanism of toxicity of thorium dioxide is thought likely to be its radioactivity, rather than any chemical toxicity.

f. *Experimental nitrosamine carcinogenesis*

Cell type specificity in susceptibility to carcinogens has been confirmed in experimental studies of the carcinogenic potencies of nitrosamines in rats. The agent is administered in drinking water, but causes tumors of liver cells, but not other cells in the liver, in kidney, or any other cells in the body—the Faber model [16].

g. *Perspective on these differences*

Although there is no proof, overall it would seem that defensive factors in tissues and cells against exogenous noxious agents may play at least partial roles in these phenomena, as is discussed in A3.1 and A3.2. Such mechanisms do not readily explain the differences in types of tumor where no such factors can be operative. This is particularly the case where radiations are involved.

4.2.5 Dose and Dose Rates of Carcinogens

A major issue in all experimental and epidemiological studies of the effects of carcinogens is that particular amounts of the carcinogen may be necessary for tumor formation to occur. "Dose" usually refers to the total amount of carcinogen delivered to the subject, whether in parts or as a single administration. In relation to tobacco smoking and lung cancer, the term "pack years" is used to indicate the individual's total dose of tobacco smoke exposure.

Total dose also has considerable implications for assessing the possible effects of unavoidable exposures to carcinogens, e.g., "background" or "environmental" exposures to radiations.

"Dose rate" refers to the dose (usually average) exposure per unit time. This is more difficult to assess and accommodate in human studies.

4.3 RADIATIONS

Radiation carcinogenesis is part of the much larger field of radiobiology, which encompasses all the noxious (non-genopathic and genopathic—Chapter 10) effects of radiations. UVB and ionizing radiations have long been known to produce tumors. The details of doses, periods of administration, etc. can be found in References [61–63].

This section is concerned with the physics and damage effects to DNA and proteins of UVB and ionizing radiations. The absorptions of energy from radiations by tissues are essential aspect of radiotherapy, and are discussed in 12.1.

4.3.1 Background

Radiations with wavelengths longer than approximately 10^{-7} m (mainly radio waves, microwaves, infrared, and visible light) agitate atoms, which manifests as heat. Molecular agitation disturbs noncovalent inter-atomic bonds, including those of electric charge, hydrogen bonds, van der Waals force, and hydrophobic interactions [64]. Disturbance of these bonds can lead to loss of water molecules from structures, as well as disturbances in noncovalent bonds, leading to coagulation of macromolecules [65] (Figure 4.6). Both mechanisms abolish function in most of the highly structure-critical macromolecules. The components of each macromolecule—especially the amino acids and nucleotides—are additional important factors in the susceptibility of the macromolecule to radiation damage.

Visible light has no noxious effects on living things, except in the presence of a photosynthesizer pigment such as methylene blue or porphyrins[66]. In this situation, "singlet oxygen"—a reactive but unionized form of O_2—can be generated with tissue-damaging effects. (The phenomenon is used in a form of cancer therapy known as "phototherapy" [67].)

Radiations of wavelengths shorter than 10^{-7} m (UVB, X-rays, and gamma rays) have these same heat-producing effects, but in addition, can disturb the electrons of atoms with potential effects on covalent bonds, as discussed in the following subsections.

4.3.2 Ultraviolet Light

(a) Physics

In addition to generating heat in tissues, ultraviolet radiations (wavelengths 10^{-8} m to 10^{-7} m) have the effect of exciting electrons so that they move to higher orbital sub-shells. The affected atoms become more reactive with those nearby [68] (Figure 4.6). In addition, singlet oxygen can be generated [69]. Generally, existing covalent bonds are not broken, but formation of new covalent bonds can be facilitated in macromolecules as follows.

(b) Damage to DNA from Ultraviolet Light

In DNA in aqueous solutions, UVB can cause a variety of transient modifications as well as permanent oxidations to purines and pyrimidines. However, the most characteristic modification is dimerization of adjacent pyrimidine bases to cyclobutane [68]. In living cells, these errors are usually corrected by nucleotide excision repairs. Errors in these repairs may cause transversions, mainly C→T and CC→TT. Another kind of transversion is T→G, which is rarely induced by a wide range of other mutagens including UVB, but is commonly found with UVA [70,71].

(c) Damage to Proteins from Ultraviolet Light

Ultraviolet light damages proteins both directly (causing alterations mainly in the side chains of the amino acids) or indirectly through energy transfers from activated chromophores attached to the proteins [68]. The consequences of the actions of ultra violet light on proteins are mainly through alterations in the three-dimensional structure of proteins.

The functional results include inactivation of enzymes, liability of the abnormal proteinaceous material to be attacked by lysosomal and other intracellular proteolytic enzymes, and tendency to aggregation. In relation to aggregation, ultraviolet-damaged proteins can interact

A. Atomic (oxygen atom with vacant vertical *2p* orbital in resting state)

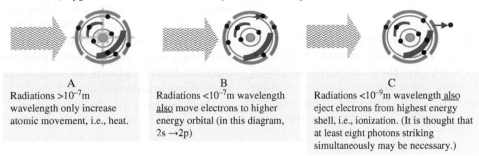

A	B	C
Radiations >10^{-7}m wavelength only increase atomic movement, i.e., heat.	Radiations <10^{-7}m wavelength <u>also</u> move electrons to higher energy orbital (in this diagram, 2s →2p)	Radiations <10^{-9}m wavelength <u>also</u> eject electrons from highest energy shell, i.e., ionization. (It is thought that at least eight photons striking simultaneously may be necessary.)

B. Molecular and inter-molecular results leading to reductions in function

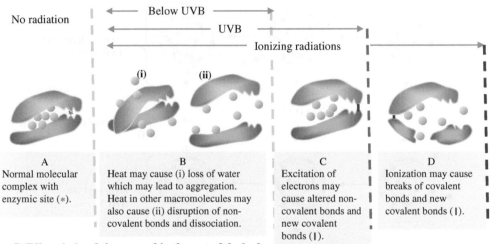

A	B	C	D
Normal molecular complex with enzymic site (∗).	Heat may cause (i) loss of water which may lead to aggregation. Heat in other macromolecules may also cause (ii) disruption of non-covalent bonds and dissociation.	Excitation of electrons may cause altered non-covalent bonds and new covalent bonds (I).	Ionization may cause breaks of covalent bonds and new covalent bonds (I).

C. Effects in local tissues, and in the rest of the body

> Heat, UV, and ionizing radiations are themselves inflammogenic. Also denatured fragments of macromolecules in tissues are inflammogenic. When denatured fragments produced by almost any mechanism enter the blood circulation they cause systemic effects such as fever and malaise.

FIGURE 4.6   Atomic, molecular, and inter-molecular effects of photonic radiations.

illicitly with other macromolecules. Breaking of the amino acid chain is a relatively uncommon effect of ultraviolet light, and intra-protein or inter-protein cross-linking may occur [68,71].

All of these studies are consistent with the possibility that all radiations may have direct or indirect effects on the functional sites on proteins.

(d) Protein–DNA Cross-Linking

By altering the state of DNA nucleobases and the side chains of proteins, UV can result in ring-opening and cross-links causing aggregation of proteins, but little breakdown of the bonds between amino acids. Cross-links with DNA and RNA also can cause aggregation, and hence loss of function in these molecules [68,72].

4.3.3 Ionizing Radiations

(a) Physics

Radiations of wavelengths shorter than approximately 10^{-9} m can induce heat and excite electrons in their orbitals—affecting hydration and noncovalent bonds—similar to the effects of UVB. However, these shorter wavelengths can also eject electrons from the outer shells of atoms, creating ions from un-ionized atoms (Figure 4.6). Most importantly, these radiations are capable of inducing free radicals ($^\bullet$H and $^\bullet$OH) from water, which in turn react with other water molecules to produce hydroxyl radicals as well as superoxide and peroxide anions (Figure 4.7) [73].

Dislodgement of electrons shared between atoms can break many types of existing

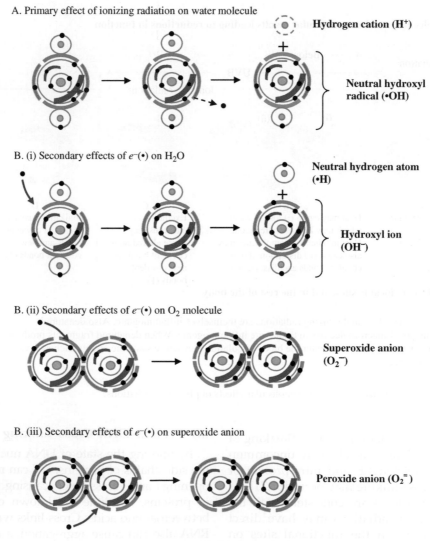

FIGURE 4.7 Some of the radicals which are produced in H_2O and in O_2 molecules by ionizing radiations.

covalent bonds, and lead to formation of new bonds between different, but adjacent atoms. In relation to macromolecules, common examples are as follows.

(b) Damage to DNA from Ionizing Radiations

Damage to DNA in living organisms is thought to occur mainly through radicals generated in water (above) [73–75]. These alterations in turn may cause the various kinds of damage as discussed in Chapter 1 (including Figure 1.6), in Chapter 5 (Figures 5.2, 5.3, and 5.4), and in Appendix 2 (Figure A2.2).

In particular, they may cause:

i. Various nucleobase lesions, including fissions in the rings in the pyrimidine and purine bases [74,75]. However, radiations do not increase the spontaneous rate of depurination, especially loss of cytosines. Only heat and acidic conditions have this effect [76].

ii. Breaks of the phosphodiester bonds, as well as in the ribose rings of the DNA "backbone" [74,75]. The doses required for these breaks in DNA in aqueous solution are much higher than for breaks to occur in living cells, as is discussed in Reference [77].

iii. DNA-DNA cross-links may occur [78].

(c) Damage to Proteins from Ionizing Radiations

Radiations cause denaturation in proteins, indicated by loss of solubility in water, and/or loss of biochemical activity (mainly of enzymes) [79]. As for DNA, these changes occur through the combined actions of heat and free-radical damage. High doses of ionizing radiations can cause breaks in covalent bonds randomly through the amino acid chain, and conformational changes with alterations of bioactivities can follow [79,80].

It has also been suggested that radiations may cause cross-links between proteins when ionized sites are created in adjacent molecules at the same time [81]. This allows the sites to interact with each other, forming new cross-links without any "molecular bridge" ("zero-length type" cross-linkage, see [82]).

(d) Protein–DNA Cross-Links

Radiation-induced cross-linking between proteins and DNA is described [83]. The chemical sites of these changes are altered by the presence of water, but not the types of primary damage.

Overall, little has been published on whether or not formation of cross-links between any proteins and other macromolecules due to radiations correlates with carcinogenicity.

4.3.4 The Basis of Cumulative Ionizing and Non-ionizing Radiation Damage over Long Periods of Time

Biologically significant radiation damage consists of trillions of individual events over long periods of time. The DNA in long-lived kinds of cells, such as local tissue stem cells, would be expected to receive very large numbers of potentially damaging "hits." It is known that the epidermis becomes atrophic with UV exposure. This suggests several possible mechanisms:

i. The local tissue stem cells are reduced in number,

ii. These produce fewer transit amplifying cells, and/or

iii. The transit amplifying cells undergo fewer divisions before full specialization.

All these changes may be due, at least in part, to vascular damage to the dermis of the skin.

Overall, these changes are anti-proliferative in all the cells, but paradoxically, only a few cells in the affected population of cells give rise to tumors. These facts suggest that there is presumably some storing/accumulation of damage in susceptible cells, which in turn implies that cellular repair mechanisms are overwhelmed (A3.3). A likely mechanism of the anti-proliferative effect of radiations is cumulative unrepaired genomic damage, so that loss-of-function events reduce replicative and transcriptional efficiencies in affected cells [84–86].

In this situation, the tumorigenesis may be caused by much rarer combinations of gain-of-function events in cells which have suffered relatively little genomic loss of function, as mentioned in Section 4.1.2.

4.4 CHEMICALS

4.4.1 Many Carcinogens Must Be Activated in the Body to Have Their Effect

The first demonstration of activation of a chemical to an active form was by Miller and Miller in 1948 [87], who showed that 4-diame-thyl-aminoazobenzene ("butter yellow") is activated in the liver in association with it causing tumors in that organ.

It is now recognized that many carcinogens, including polycyclic aromatic hydrocarbons, nitrosamines, aflatoxin, aromatic and heterocyclic amines, and estrogens, are activated in the tissues of the body [88]. Complete activation may involve more than one enzymatic step. Thus, carcinogens may sometimes result from modification of "pro-carcinogens" to "ultimate carcinogens", via intermediary "proximate carcinogens" [89].

Activating enzymes include oxidases, carbonyl reductases, and hydrolases [88]. The most important are the cytochrome P450 group of enzymes (CPY category, [88,90], see also A3.1). However, the details of the involvement of CPY enzymes in tumor formation may be complicated. For example, for polycyclic hydrocarbons, assessments of metabolic outcomes can include the possible involvement of several

enzymes of the CYP families; formation of multiple metabolites, and secondary metabolites; and de-activation of the ultimate carcinogen by further modifications [88].

Further complicated aspects of carcinogen activation include alteration of expression and activity of CYPs induced by the pro-carcinogenic hydrocarbon; tissue-specific differences in expression patterns of CPYs and factors relating to intracellular localization of both the enzyme (membrane-bound or free) and the target molecules [90,91]. Variations in these factors may contribute to the different susceptibilities between different individuals to environmental carcinogens [92].

4.4.2 Chemical Carcinogens Have Different Chemical Reactivities

As detailed in publications, especially the IARC *Monographs on the Evaluation of Carcinogenic Risks to Humans* [93], numerous chemicals, in many different classes, are known to be carcinogenic. This section, however, is concerned with the diversity of chemical reactivities of chemical carcinogens (see also 1.2.1 and 5.1). Carcinogens in their native or activated forms are classified according to the kinds of their interaction(s) with DNA and with proteins. Actions on lipids are not discussed, but are mentioned in relation to solid state carcinogenesis (4.6). The possible relevance of the cell membrane to some aspects of tumors induced by polyaromatic hydrocarbons are mentioned in 2.9.3.

a. *Carcinogens which react covalently only with or are incorporated into DNA*

Few chemicals react covalently only with DNA. The most studied is bromodeoxyuridine (BDU), which replaces thymidine during DNA synthesis. BDU causes a variety of experimental genopathic effects, but is not considered a carcinogen in either IARC or NIH lists of carcinogens.

b. *Carcinogens which react covalently with both DNA and proteins*

Most carcinogens, especially after endogenous activation, react with both DNA and proteins. For example, polyaromatic hydrocarbons, which have been activated by the actions of CPY enzymes (A3.1.2), then interact with DNA primarily at N7 position on purine bases as well as with proteins. As additional examples, aromatic amine-, amide-, and nitroso-carcinogens, which are known to react with DNA, also interact with the amino acids of proteins, including with methionine, tryptophan, cysteine, and tyrosine [94].

Cross-linking of proteins to proteins and proteins to DNA is common among carcinogens and non-carcinogens [95,96].

c. *Carcinogens that react covalently only with proteins*

Only a few carcinogens react with proteins but not DNA. Ethionine (an analogue of methionine) presumably has its effect by competing with methionine for incorporation into proteins [97].

Certain heavy metals can interact with proteins by substituting for zinc ions in peptide chains [98], presumably altering the functions of these polypeptides.

Arsenic, nickel, ferric, and cobalt ions do not react directly with DNA but probably do react with some proteins [99]. According to Finnegan and Chen [100], "AsIII(arsenite) is a dithiol reactive compound that binds to and potentially inactivates enzymes containing closely spaced cysteine residues or dithiol co-factors."

At the present time, this field of "toxico-proteomics" in relation to carcinogens and other chemicals is being investigated with many new techniques [101] but is beyond the scope of this book.

d. *Carcinogens which react with neither DNA nor proteins*

There are only a few chemicals in this category. They include various herbicides, for example, the 2,3,7,8-tetrachlorodibenzo-p-dioxin (TCDD), which is thought to be a "promoter" of carcinogenesis [102]. Others are ethylbenzene [103] and naphthalene [104].

The mechanism of action of these substances is not clear, but may be by "intercalation" (see Figure 4.8), which is defined as having an action through noncovalent competitive binding to sites on macromolecules which influence the activities of those molecules [105]. Indirect mechanisms via redox cycling, oxidative damage by the 8-oxo-2'-deoxyguanosine base modification, as well as possibly by other metabolic effects and gene activations, are discussed in 5.1.2(b).

4.4.3 Chemical Structure Does Not Perfectly Indicate Carcinogenic Potency

The essential observation here is that, as well as no consistency of structure *across all the classes* of carcinogens, there are no consistencies among analogues *within each class* of chemical carcinogen among analogues.

This major issue in chemical carcinogenesis has been recognized since the 1930s, when pure preparations of hydrocarbon analogues became available [106]. For example, 3,4-benzypyrene is a potent carcinogen while 1,2-benzypyrene has only a weak carcinogenic effect (Figure 4.9). The difference between the molecules is in the position of one benzene ring [32].

Among aromatic amines, N,N-dimethyl-4-aminoazobenzene (DAB, "butter yellow") is a potent carcinogen, while 3'-methyl-dimethyl-aminoazobenzene is not carcinogenic. Here, the addition of one methyl group abolishes carcinogenicity of DAB [32].

And among metal ions, beryllium, chromium, cadmium, nickel, and arsenic ions have been found to be carcinogenic, while copper, manganese, and stannous ions are probably not carcinogenic. The carcinogenicity of Fe^{3+} is unclear, while Fe^{2+} is not carcinogenic [107].

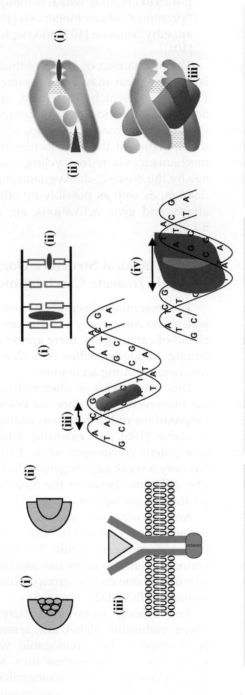

A

In-principle chemical concept. The space in the recipient molecule may be occupied by many small structures (i) or by one large structure (ii). An example of (ii) is the insertion of ligands into receptors (iii). See also Fig 1.8.

B

Potentially, DNA has many possible sites for intercalation.
(i) Between nucleotide pairs
(ii) Between nucleotides of a pair
(iii) In the minor groove of the double helix
(iv) In the major groove of the double helix.

C

Proteins, especially protein complexes, and even DNA–protein complexes, may have many sites of intercalation.
(i) Into an enzymic site
(ii) Into a site of covalent bonds
(iii) Into a space normally occupied by water molecules

FIGURE 4.8 Intercalation.

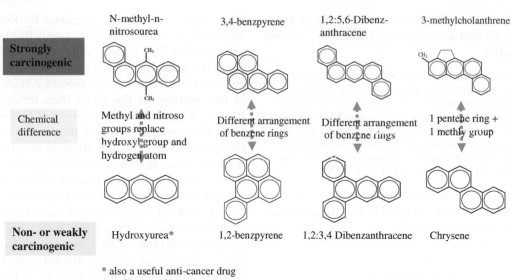

FIGURE 4.9   Small chemical differences affect carcinogenic potency.

With the various polycyclic hydrocarbons (PAHs), several attempts have been made to link carcinogenicity with structure [89,108–110], but these are not applicable to other classes of carcinogens.

All of these differences in carcinogenic effect may be due to different efficacies in defensive factors which determine whether or not the particular agent reaches the genome and the compartment in which the genome processes occur in the cell (see A3.1).

4.4.4 Adducts on DNA Are Not Always Associated with Tumors in the Relevant Cells

Because induction of chemical changes in DNA—either directly or indirectly—is the most widely studied possible pathogenetic mechanism of carcinogens (see 5.1), the issue of a possible role of adducts on DNA is of considerable importance.

In the 1950s, studies of alkylating agents revealed no correlation between adduct formation and carcinogenicity. In particular, in relation to ethylenimines,

> … neither polyfunctionality nor cross-linkage, nor the formation of a polyreactive polymer or micelle is necessary for carcinogenic action in ethyl-enimine derivatives, whatever may be their significance for other radiomimetic effects. [111]

Subsequent studies of agents including polycyclic hydrocarbons, aromatic amines, and mycotoxins [112,113] have had various findings. In particular, adducts can be found in organs which do not develop tumor, as well as organs in which tumors arise [114]. This suggests that the detection of adducts in a tissue does not necessarily indicate a specific tumorigenic risk for that tissue. Another finding has been that individual carcinogens, especially "bulky" ones are often associated with more than one adduct type [115].

The strengths of the association between particular types of adducts and mutations vary unpredictably by several orders of magnitude, so that a "consensus statement" in 1996 concluded:

> As a general conclusion, the Panel suggested that the current technological capabilities for detection of DNA adducts exceed our ability to define the biological significance of adducts as it relates to toxicity or health outcome. DNA adducts are likely to play an important role in human risk for cancer induction and progression, but the quantitative aspects of this relationship remain to be determined. [114]

A later review has confirmed the concept that adducts may be related to carcinogenic risk only in a general way [4].

4.5 VIRUSES

4.5.1 Background

The fact that viruses might be causes of tumors was established when particular tumor types were found to be transferable from one animal to another by subcellular tumor extracts ("filterable agents"). In this way, Peyton Rous in 1911 identified the sarcoma virus which bears his name of chickens [116]. Koch's postulates for an infectious disease were satisfied by showing that virus taken from a tumor in one animal could cause tumors in other animals [117]. Nevertheless, very few other animal tumors could be transmitted from one animal to another, and no human tumor yielded any viral agent which could produce a tumor in animals. Experiments involving attempts to transmit tumors from one human to another are unethical [118]. This means that it is not possible to test the Koch's Postulate, which requires that the pure agent be isolated from a lesion in one individual, and then shown to cause a tumor when administering to a healthy individual.

However, further research into viral oncogenesis developed from studies using embryonic animal tissues, and *in vitro* models of tumors. Growth of Rous sarcoma virus (RSV) in chick eggs had been described by Keogh in 1938 [119]. In the 1958, Temin and Rubin [120] showed that RSV could cause cultured chicken fibroblast cells to become new stable cell lines, i.e., transformation *in vitro* (see 14.2.3). Analysis of the genome of the virus then revealed the particular gene which causes the transformation. The gene was called the *src* gene. Later work, however, [121,122] showed that the tumorigenic gene in the virus had, in fact, originated from the chicken genome. The gene had become inserted into the viral genome during infection of one chicken, and had been altered to a constitutionally active form, such that when descendants of that virus were transmitted to other chickens, the altered gene caused tumors of corresponding cells in those birds.

These studies led to the search for "oncogenes," being any gene of either viral or tumor origin which, when transfected into cells *in vitro* caused malignant transformation of those cells (see 14.2.3). Numerous "oncogenes" have since been discovered (see 6.2). Many of the pieces of DNA in viruses which can cause tumors are in fact unchanged, or mutant versions of, normal cellular genes [122]. The viral forms of the genes were termed *v-(oncogenes)* and the cellular genes from which they are derived were called *c-(oncogenes)*. The term "proto-oncogene" is sometimes used for the first-discovered viral version, and sometimes for the wild-type cellular version of the gene.

So far as humans are concerned, however, only a few oncogenic viruses have been found, and these are associated with only a small number of tumor types (4.5.3). An avenue of investigation for these viruses involves the lifestyle factors, which result in transmission of these organisms from one human to another.

4.5.2 General Aspects of Viral Infections

Before discussing the viruses which are associated with tumors in humans, it is necessary to mention some general aspects of viral infections. Genomic issues in viral carcinogenesis are considered in 5.7.

(a) Classifications of Viruses

Two main classifications of viruses are used: that of the International Committee on the Taxonomy of Viruses [123], and the (simpler) Baltimore Classification [124]. Both classifications divide viruses first according to their hereditary material—DNA or RNA—and second, according to whether the nucleic acid is double or single stranded. At the taxonomic level of family, the names given to viruses are usually related to a particular disease caused by at least one type member (e.g. papilloma). Morphological and other criteria may then be applied for further sub-divisions.

RNA viruses possess reverse transcriptases, so that the viral RNA can be retro-transcribed to DNA [125]. This DNA derived from the viral RNA then can be replicated in the infected mammalian cells. There is no need for this step with the DNA viruses.

There is no specificity in pathological or clinical features of either RNA versus DNA viral infections. For example, hepatitis B virus has DNA while hepatitis C virus has RNA as its hereditary material. Both viruses cause chronic inflammation in the liver, and are implicated in the causation of hepatocellular carcinomas.

(b) Tropisms

Most viruses are highly specific as to the species, and also the kinds of cells, which they infect [126]. This "tropism" is thought to depend on host factors and viral factors. The host factors include the "pre-target" defenses as discussed in A3.1. In particular, for viruses which enter the body and come into contact with cells, the specificity of the cells invaded could be based in different expressions of surface molecules in different kinds of cells.

The viral factors involved in the tropism may include possession of the appropriate ligands for the host's cell surface molecules, as well as the virus's capacity to commandeer host processes for synthesis of nucleic acids and proteins.

Some viruses cause tumors in a multiple kinds of cells, and in more than one kind of experimental animal. These are called "polyoma" viruses.

(c) Morphologically Detectable Pathogenetic Effects

Generally, for a virus to cause an infection, a virion must enter into a cell and proliferate there [127]. In cells, different viruses have different pathogenetic mechanisms. Some incorporate their DNA into the host cell genome. Others replicate their DNA in the cytoplasm of the host cells. However, regardless of their primary proliferative mechanism, each virus produces only one or a few kinds of lesions.

These invasions and proliferations of virions (see Figure 4.10) may cause:

i. No pathological lesion.
ii. Inflammation of tissues and deaths of infected cells. Common viruses such as rhinoviruses, adenoviruses, and influenza viruses, all of which infect cells of the respiratory tract in humans, cause inflammation, and sometimes cell death. However, none of these viruses have been linked to any proliferative lesion or tumor of those or any other kinds of cells.
iii. Release of virions. This is usually transient, but with some viruses such as for hepatitis B and C, it occurs on a chronic basis (even for years).
iv. For some viruses only, particular cytopathic effects may be seen microscopically, especially rounded, stainable "inclusion" bodies in the cytoplasm or nucleus.
v. Host cell proliferation and, for some viruses, tumor formation.

In this, a few viral infections cause excess, but transient, proliferative activity in host cells, without being inflammogenic or carcinogenic.

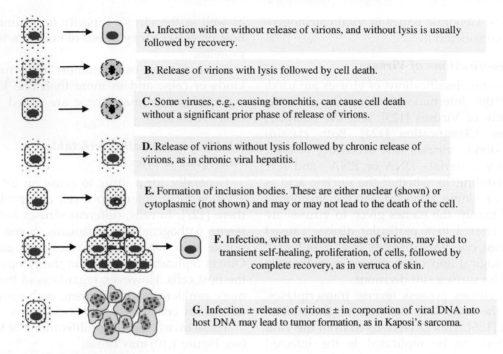

A. Infection with or without release of virions, and without lysis is usually followed by recovery.

B. Release of virions with lysis followed by cell death.

C. Some viruses, e.g., causing bronchitis, can cause cell death without a significant prior phase of release of virions.

D. Release of virions without lysis followed by chronic release of virions, as in chronic viral hepatitis.

E. Formation of inclusion bodies. These are either nuclear (shown) or cytoplasmic (not shown) and may or may not lead to the death of the cell.

F. Infection, with or without release of virions, may lead to transient self-healing, proliferation, of cells, followed by complete recovery, as in verruca of skin.

G. Infection ± release of virions ± in corporation of viral DNA into host DNA may lead to tumor formation, as in Kaposi's sarcoma.

FIGURE 4.10   Principal outcomes of viral infections.

For example, particular strains of the human papilloma virus infect the epidermal cells of the skin causing warts (verrucae vulgaris, plantaris, etc.). These wart lesions are not inflamed, are self-healing, and never undergo "malignant transformation." Similarly, the molluscum contagiosum virus causes a self-healing tumor-like lesion [128].

(d) Different Effects In Different Species

In general, if a virus, or class of viruses, infects more than one species, then similar pathological lesions are produced in the different species. Examples include certain herpes viruses (see HHV8 and HTLV below). For tumor formation, however, some viruses which are non-tumorigenic in one species and may cause tumors in others. An example is the adenovirus which causes respiratory infections in humans, but also tumors in Syrian hamsters [129].

4.5.3 Human Tumor Types Associated with Viral Infections

Except where specifically indicated (see Table 4.1), the following is taken from Reference [130].

(a) Epstein-Barr Virus (EBV)

In all populations, there is a high incidence of antibodies to this virus (human herpes virus 4) in persons who have had no remembered illness. This suggests a high rate of subclinical infection with the virus. In a minority of individuals, EBV causes a non-tumorous illness known as infectious mononucleosis/"glandular fever." This disease is characterized by a transient proliferative effect on one particular kind of leukocyte, with a variety of other clinical manifestations. In Western parts of the world, individuals who have suffered clinically evident infectious mononucleosis are no

TABLE 4.1 Virus-Associated Malignant Tumors in Humans

Virus	Type	Distribution of infection	Rate of tumors in infected individuals	Tumor type
Epstein-Barr virus (EBV)	DNA	Universal	Very low	Burkitt's lymphoma (Africa)
				Nasopharyneal carcinoma (East Asia)
Hepatitis B viruses (HBV)	DNA	Worldwide in subpopulations[a]	Low	Hepatocellular carcinomas
Hepatitis C viruses (HCV)	RNA	Mainly in countries with poor hygiene facilities	Low	Hepatocellular carcinomas
Human Herpes virus-8 (HHV8)	DNA	Worldwide in subpopulations[a]	High	Kaposi's sarcoma
Human Immunodeficiency Virus 1 (HIV1)	RNA	Commonest in Africa and Caribbean, other countries in sub-populations[a]	High	Lymphomas, carcinomas
Human Papilloma Viruses (mainly types 16 and18)	DNA	Worldwide	Low	Squamous carcinomas mainly of cervix
Human T-cell Lymphotrophic virus type-1 (HTLV-1)	RNA	Most common in Japan and certain indigenous populations in many countries	Low	T-cell leukemia
Merkel cell polyoma virus (MCV or MCCPyV)	DNA	Universal, all ages	Very low	Epidermal tumors in elderly

[a]*Those individuals who are exposed to infection through human–human transfers of body fluids.*

more likely than never-infected individuals to develop any tumor of these leukocytes, or any other type of tumor in later life.

In African populations, the Epstein-Barr virus is especially associated with Burkitt's lymphoma. In East Asian populations, there is an association with nasopharyngeal carcinoma. In Western societies, there are associations with classical Hodgkin's disease, lymphomas in immunosuppressed individuals, including cases of HIV infection and T-cell and NK cell lymphomas.

This discrepancy between the populations has been linked to more successful immune responses to the virus in the former, compared to the latter, populations.

(b) Hepatitis B Viruses (HBV)

This family of DNA viruses has been divided into eight genotypes, each comprising a number of sub-genotypes. Infections are worldwide, and transmission is by passage of blood or other body fluids from an infected individual to an uninfected person. Incidences in different parts of the world reflect lifestyle factors in the particular population.

Viral infection causes primarily liver cell necrosis and reactive fibrosis of liver connective tissue cells (A1.1.3). The incidence of hepatocellular carcinomas in asymptomatic infected individuals varies between populations from approximately 0.1% to 1.0% per year. Factors in these differences may be the viral load, the relative proportions of different genotypes, and co-carcinogens such as alcohol intake and hepatitis C infection [131].

(c) Hepatitis C Viruses (HCV)

This family of RNA viruses comprises six major genotypes. They have the same global range, due to the same factors, as hepatitis B viruses. Most cases are asymptomatic; approximately 25% go on to cirrhosis. The risk of hepatocellular carcinoma is probably smaller than in hepatitis B infection, but the issues of viral load, differences in genotype, and cofactors probably play similar roles. The mechanism of tumorigenesis by hepatitis C is unclear [132].

(d) Human Herpes Virus-8 (HHV8) and Kaposi's Sarcoma

This virus has a worldwide distribution, with incidences similar to the hepatitis viruses. This family of herpes viruses with similarities to HHV8 is extensive in Africa, including populations of non-human primates. The range of human infection is worldwide, with local incidences being influenced by the modes of transmission, which are the same as for the hepatitis viruses. For patients on anti-retroviral therapy, the rate of development of Kaposi's sarcoma in HHV8-infected individuals is approximately 1.4/1000 patient years [133]. For untreated patients, the incidence is several-fold higher. This suggests that viral load is a major factor in the causation of this tumor.

Other tumors associated with HHV8 infection include primary effusion lymphoma and multi-centric Castleman's disease.

(e) Human Immunodeficiency Virus 1 (HIV1)

The genome of this virus consists of two copies of single stranded RNA. Its distribution and incidences are similar to those of the hepatitis viruses and of HHV8. The virus has non-tumorigenic effects of death of T lymphocytes, as well as probably other cell types.

It is thought that the immunosuppression caused by HIV1 allows infection with HHV8 to occur, and hence the association of Kaposi's sarcoma with HIV1 infection.

HIV1 infection is associated with a variety of human tumors, including lymphomas and carcinomas. The HIV virus may not be directly causative for all these tumors, but perhaps permissive by way of causing immune depression.

(f) Human Papilloma Viruses

The IARC volume [130] cites 118 types of papilloma viruses. All types infect only squamous epithelium. All sexually active individuals are liable to contract at least one infection in their life time. Most infections are self-limiting and have no pathogenetic effects. Ten to ninety percent of cases show in situ epithelial changes known as "dysplasia." In the uterine cervix this is referred to as "cervical in situ neoplasia" ("CIN"). In the glans penis and prepuce it is much less common, and has a variety of names, including "Bowen's disease" and "erythroplasia of Queyrat" [134].

HPV of types 16 (50%) and 18 (20%) account for 70% of invasive cervical cancers which have HPV DNA in their cells. The clinical aggressiveness of the tumors seems to be the same for both viral types. The carcinogenic effect appears to depend on the presence of viral protein in the cells.

(g) Human T-Cell Lymphotrophic Virus Type-1 (HTLV-1)

This is one of a family of RNA viruses which cause leukemias in a wide variety of mammals. HTLV-1 infects various lymphoid cells, but causes transformation in only T lymphocytes. Transmission can be via milk from mother to baby, sexual contact, and transfers of body fluids as in blood transfusions. Among carriers of the virus, the rate of clinical disease (adult T-cell leukemia/lymphoma) is 2–4%, with latency periods of 40–60 years.

(h) The Merkel Cell Polyoma Virus (MCV or MCCPyV)

Merkel cell tumor is rare, occurs mainly in the elderly, arises in the epidermis, and is highly malignant. MCV is a small, double-stranded DNA virus, and hence included among the polyoma family. Its presence in Merkel cell tumors was discovered in 2008. To date, it has not been found in any other tumors [135].

4.5.4 Potential Bases for Associations between Infections and Tumor Formation

At the beginning of this chapter, when referring to Koch's Postulates, it was pointed out that relationships between microbiological agents and pathological lesions can be complex.

The facts that virus-associated tumors may have low incidence rates compared to the infection rates (shown especially by EBV [136]), and that long periods of time may elapse between infection and tumor (e.g., hepatitis viruses), suggest various possible bases for the association [137].

Bases for associations of viral infections with tumors can be summarized as follows:

a. *The viral infection is the essential sole causative agent of the tumor formation, but*
  i. Its effect is weak, and a critical "viral load" is necessary for tumorous effect,
  ii. The immune system prevents most infected cells from becoming tumorous,
  iii. More than one effect of the viral infection must occur in a cell, e.g., a critical site of insertion of viral DNA into the host cell genome, as well as production of a pro-tumor viral protein.
b. *The viral infection is the essential causative agent, but tumor formation requires a second agent*
  Examples of possible second agents may be a chemical from local bacterial flora, or another viral infection.

c. *Another factor is the essential causative agent, and the virus is a second agent.*
  The infection plays a contributory but not initiating role in the formation of the tumors, i.e., the tumor formation has another initiating factor, including possibly another virus (see above).
d. *The infection is not an etiological factor for the tumor, but a co-incidental infection which is more likely in the individuals also exposed to the actual essential causative agent.*
  This could be the situation in the high incidence of Epstein-Barr virus in lymphomas in patients with AIDS.
e. *The virus has been a passenger in the genome of the individual.*
  In this situation, the virus and has no etiological role, but is merely expressed as part of dysregulation of genes (see 5.6) in the cell population of the particular type of tumor.
f. *The virus is an inter-current infection of the tumor cell population.*
  In this, the virus has no etiological role, but occurs merely because the virus has selective trophism (4.5.1 above) for the cells of the particular type of tumor.
g. *The virus induced the tumorigenic change in the original cell, but its presence is not required for the descendant cells to form tumors.*
  This "hit and run" hypothesis has very wide implications, but at the present time, it is unproven [138]. Possible mechanisms of genomic damage by viruses are discussed in 5.7.

4.5.5 Animal and *In Vitro* Studies

Until a decade or so ago, these were discussed in most general accounts of viral carcinogenesis. With current methods of studying human tumors, these have given less prominence in some texts [130]. However, experimental models are essential for investigating the various different possible associations of viral infections with tumors.

4.6 OTHER MICROORGANISMS AS CARCINOGENS

4.6.1 Bacteria

In general, the vast majority of chronic bacterial infections, such as tuberculosis and chronic osteomyelitis, are not associated with tumor formation. Currently, the main bacterium being widely studied in relation to carcinogenesis is *Helicobacter pylori* in relation to gastric carcinoma and gastric lymphoma. It is also being investigated for other gastro-intestinal tract (GIT) malignancies, as well as non-GIT tumors[130]. The role of this organism in causing inflammation of the stomach was shown by Marshall and Warren in 1984 [139], but roles of the organism in causation of tumors are still unclear [130,140,141].

4.6.2 Parasites

Only a few parasites are associated with tumors in humans [130,142]. The bile duct flukes *Clonorchis sinensis*, and *Opisthorchis viverrini* are associated with tumors of bile duct cells as well as liver cells. However, other liver flukes such as *Fasciola* species are not.

Schistosoma haematobium selectively colonizes the mucosa of the urinary bladder and predisposes the urothelium first to squamous metaplasia, and then to squamous cell carcinoma. However, other schistosomes, for example, *Schistosoma mansoni* and *Schistosoma japonicum*, which locate in the wall of intestine, do not predispose to tumors there.

4.7 HORMONES AS CARCINOGENS

Hormones are only relevant to tumors of the secondary sex organs. In the male, this is the prostate gland, while in the female, they are the breasts and endometrium. It is well known that removal of the primary sex organs (testes;

ovaries) prevents the development of tumors in the secondary organs. Excess testosterone is not known to be associated with carcinoma of the prostate, although there is concern that testosterone therapy for older men might cause prostatic carcinoma, or increase the growth of pre-existing tumors [143].

Greatly increased serum estrogens, especially in the absence of progestogens, can cause hyperplasia of the endometrium and possibly carcinoma [144]. Whether estrogens cause carcinoma of the breast is unclear, but the use of anti-estrogens to prevent breast carcinoma in older women is widely suggested [145]. In the case of oral contraceptive use (which constitutes excess estrogens and progestogens), no definite effect on the incidence of carcinoma of the breast has been confirmed [146].

In considering the roles of these hormones, it would seem that physiological hormones in physiological concentrations may be considered contributory agents to the formation of some tumors, because their normal function is to stimulate growth in a particular kind of cell in the body (see above), and this enhancing effect applies also to tumorous derivatives of those kinds of cells. By definition, a hormone does not produce any irreversible effect in a cell, and its physiological action (as opposed to any different action it might have in supraphysiological doses) should not be considered as an initiating carcinogenic action.

However, analogues of normal hormones may be carcinogenic. The use of the synthetic oestrogen diethylstilboestrol is important [147]. This agent, when given to pregnant mothers—with the aim of preventing miscarriages—led to the occurrence of adenocarcinomas of the lower genital tracts in the daughters born of the pregnancies. The stilboestrol given to the mother had the effect of causing glandular epithelial cells to line the vaginas of the daughters, not squamous epithelium. The adenocarcinomas arose from this precondition of "vaginal adenosis." The mechanism of the tumors could then be that the agent is the primary causative agent, and the adenosis is only a phase which the tumor cells went through before becoming malignant. Alternatively, the mechanism may have been that the adenosis was the entire direct action of the stilboestrol, and that the subsequent carcinomas occurred through the action of some unknown carcinogen acting on the adenosis cells.

4.8 "SOLID" CARCINOGENS *IN VIVO*

So far in this book, tumor formation has been addressed mainly according to the DNA nucleobase damage model (4.2.4): carcinogen → damage to nucleobases of DNA → uni- or oligonucleotide genomic event → instigation/ activation of the characteristics of one type of tumor. This scenario involves carcinogens gaining access to the intranuclear compartment of the cell.

In some kinds of tumorigenesis, however, it would seem that the carcinogens do not enter the cells at all. Instead, they appear to induce tumors by virtue of being a non-physiological surface to which the cells attach. They have been described in terms of "solid state" carcinogenesis [148].

This section describes the situations *in vivo* which appear to be based on this phenomenon. Suggestions for their mechanisms are discussed in 2.9.3.

4.8.1 Asbestos Fibers

"Asbestos" is the term used for a variety of hydrated silicates [106,149] which have in common a fine fibrous structure (Figure 4.5). The fibers may be as thin as a chromosome ($3\,\mu$). The main kinds are:

i. Chrysotile (white asbestos), a magnesium silicate. It is the only asbestos comprising

flexible threads and has been the most widely used type. It is the only member of the Serpentine class.

The remaining kinds are known collectively as Amphiboles.

ii. Amosite (brown asbestos), a hydrated iron silicate

iii. Crocidolite (blue asbestos), a hydrated sodium-ferric and ferrous silicate (illustrated in Figure 4.5).

Three others (tremolite, actinolite, and anthophyllite) consist of mixed calcium-magnesium, calcium, magnesium iron, and magnesium iron hydrated silicates. Several additional minerals as classed as "asbestiform."

Crocidolite has greater carcinogenic potency than other asbestos fibers, especially chrysotile. For carcinogenic effect, fiber size, durability, and surface properties appear to be important factors [150]. In regard to possible mechanisms of carcinogenesis, it has been noted that asbestos fibers differ from many carcinogens in that they do not induce gene mutations in *Salmonella* assays or other phenotype-changing assays [150] and 14.2. However, in other genopathic assays (see 4.1.1 and 14.2), asbestos induces chromosomal abnormalities (numerical and structural, see 5.3) in a wide variety of mammalian cells including mesothelial cells. Asbestos also induces transformation of cells in culture including mesothelial cells and fibroblasts [151] and 14.2.

On the basis of these observations, Barratt and coworkers [152,153] proposed a mechanism for cell transformation, which is dependent on fiber dimension. The suggestion is that fibers are phagocytized by the cells and accumulate in the perinuclear region of the cells. When the affected cell undergoes mitosis, the physical presence of the fibers interferes with chromosome segregation and results in anaphase abnormalities. The transformed cells show aneuploidy and other chromosome abnormalities. This is similar to the mechanism described in 2.9.3 is a variation on this theory.

Another possible mechanism is that contact with asbestos generates electrophilic or metabolically activate secondary products that can alter DNA [154]. This is essentially the same proposal as the "indirect genopathic" mechanism in 4.1.1.

4.8.2 Other Fibers and Dusts

Since the prohibition of the use of asbestos in many industries, various other mineral fibers have been introduced as alternatives [155,156]. These include glass fibers (as in fiber glass), and various "mineral wools" including "ceramic fiber wool." These have little or no carcinogenic potency.

High exposures to wood dusts in industrial situations cause tumors of the nose and nasal sinuses and other sites [107]. No particular species of trees seem to be responsible.

Biological fibers, such as cotton and sugar cane, are potentially inhaled by workers in the relevant industries and cause pulmonary fibrosis, but no carcinogenic effects of these are proven.

4.8.3 Plastic Film and Miscellaneous "Solid" Carcinogen-Induced Tumors

(a) In Vivo

In the 1940s, it was found that implanting Bakelite or cellophane film into the subcutaneous tissues of rats led to sarcoma formation at the sites within months to years [157,158]. During the next 25 years, there were numerous papers describing the potencies of different chemical types of film and the possible roles of their physicochemical properties [148,159]. It was not certain that the films did not contain some carcinogenic filler, plasticizer, or other additive which might have acted as a slow release carcinogen.

In 1956, metallic gold and silver films were reported to have this effect [159] but this may not be relevant in human situations. Especially, the silver in dentists' amalgam has never been found to be carcinogenic in man. In the normal situation, the silver is only in contact with acellular dentine and enamel. However, even when amalgam is accidently injected into gum tissue ("amalgam tattoo"), it has no known carcinogenic effect.

The topic receives little attention at the present time [160]. The issue is conceivably relevant to the biomaterials industry, but to date, carcinogenicity of the inert materials used for joint prostheses and Stent devices has not been recorded [161].

(b) In Vitro

Solid carcinogenesis in vitro is essentially the phenomenon of cells of a limited number of species undergoing malignant transformation when cultured in vitro (see 14.2.3).

References

[1] Painter TS. A new method for the study of chromosome rearrangements and the plotting of chromosome maps. Science 1933;78:585–6.

[2] Alberts B, Johnson A, Lewis J, editors. Molecular biology of the cell (5th ed.). New York, NY: Garland; 2008.

[3] United States EPA. Search page. Term = genotoxic. Available at: http://nlquery.epa.gov/epasearch/.

[4] Poirier MC. Chemical-induced DNA damage and human cancer risk. Discov Med 2012;14(77):283–8.

[5] Hernández LG, van Steeg H, Luijten M, et al. Mechanisms of non-genotoxic carcinogens and importance of a weight of evidence approach. Mutat Res 2009;682(2-3):94–109.

[6] Hsu C-H, Stedeford T. Cancer risk assessment: chemical carcinogenesis, hazard evaluation, and risk quantification. Hoboken, NJ: John Wiley & Sons; 2010. p. 668.

[7] US Department of Health and Human Services, Food and Drug Administration Guidance for Industry: S2(R1) Genotoxicity Testing and Data Interpretation for Pharmaceuticals Intended for Human Use; 2012. Available at: http://www.fda.gov/downloads/Drugs/Guidances/ucm074931.pdf.

[8] Carlson EA. Comparative genetics of complex loci. Q Rev Biol 1959;34(1):33–67.

[9] Strachan T, Read A. Human molecular genetics, 4th ed. New York, NY: Garland Science; 2010. pp. 427–8.

[10] Lichtenstein P, Holm NV, Verkasalo PK, et al. Environmental and heritable factors in the causation of cancer--analyses of cohorts of twins from Sweden, Denmark, and Finland. N Engl J Med 2000;343(2):78–85.

[11] Lee G-H, Nomura K, Kanda H, et al. Strain specific sensitivity to diethylnitrosamine-induced carcinogenesis is maintained in hepatocytes of C3H/HeN↔C57BL/6N chimeric mice. Cancer Res 1991;51:3257–60.

[12] Bannasch P. Strain and species differences in susceptibility to liver tumour induction Modulators of experimental carcinogenesis. : IARC Sci Publ. No 51; 1983. pp. 9–38.

[13] Proctor RN. Cancer wars. New Your: Basic Books; 1995.

[14] Davis D. The secret history of the war on Cancer. New York: Basic Books; 2007.

[15] Bianchi C, Giarelli L, Grandi G, et al. Latency periods in asbestos-related mesothelioma of the pleura. Eur J Cancer Prev 199; 6(2):162–6.

[16] Yamagiwa K, Ichikawa K. Experimental study of the pathogenesis of carcinoma. J Cancer Res 1918;3:1–29.

[17] Solt DB, Medline A, Farber E. Rapid emergence of carcinogen-induced hyperplastic lesions in a new model for the sequential analysis of liver carcinogenesis. Am J Pathol 1977;88(3):595–618.

[18] Craddock VM, Frei JV. Induction of liver cell adenomata in the rat by a single treatment with N-methyl-N-nitrosourea given at various times after partial hepatectomy. Br J Cancer 1974;30(6):503–11.

[19] Doll R, Hill AB. Mortality in relation to smoking: ten years' observations of British doctors. Brit med J 1964:1460–7. i, 1399–1410.

[20] Peto R, Darby S, Deo H, et al. Smoking, smoking cessation, and lung cancer in the UK since 1950: combination of national statistics with two case-control studies. BMJ 2000;321(7257):323.

[21] Mong C, Baron EB, Fuller C, et al. High prevalence of lung cancer in a surgical cohort of lung cancer patients a decade after smoking cessation. J Cardiothoracic Surg 2011;6(1):19.

[22] Weiss SW, Goldblum JR. Enzinger and weiss' soft tissue tumors, 5th ed. Philadelphia, PA: Mosby/Elsevier; 2008.

[23] National Cancer Institute. Dictionary of Cancer Terms. Available at: http://www.cancer.gov/dictionary?CdrID=693544.

[24] Bauer KH. Mutationstheorie der Geschwülst-Entstehung/Mutation theory in tumour formation. Berlin: Julius Springer; 1928.

[25] (No authors listed). Multistage and multifactorial nature of carcinogenesis. IARC Sci Publ. No 116; 1992, pp. 9–54.

[26] Kennaway EL. Further experiments on cancer-producing substances. Biochem J 1930;24(2):497–504.

[27] Neumann HG, Hammer R, Hillesheim W, et al. Role of genotoxic and nongenotoxic effects in multistage carcinogenicity of aromatic amines. Environ Health Perspect 1990;88:207–11.

[28] Ruediger HW. Antagonistic combinations of occupational carcinogens. Int Arch Occup Environ Health 2006;79(5):343–8.

[29] Berenblum I. The mechanism of carcinogenesis. A study of the significance of cocarcinogenic action and related phenomena. Cancer Res 1941;1:807–14.

[30] Mottram JC. A developing factor in experimental blastogenesis. J Pathol Bacteriol 1944;56:181–7.

[31] Berenblum I. The study of tumours in animals. In: Florey HW, editor. General pathology (3rd ed.). London: Lloyd-Luke; 1964. pp. 622–58.

[32] Becker FF. Recent concepts of initiation and promotion in carcinogenesis. Am J Pathol 1981;105(1):3–9.

[33] Süss R, Kinzel V, Scribner JD. Cancer experiments and concepts. New York, NY: Springer Verlag; 1973. pp. 53–62.

[34] Reiners Jr JJ, Nesnow S, Slaga TJ. Murine susceptibility to two-stage skin carcinogenesis is influenced by the agent used for promotion. Carcinogenesis 1984;5:301–7.

[35] Slaga TJ, DiGiovanni J, Winberg LD, et al. Skin carcinogenesis: characteristics, mechanisms, and prevention. Prog Clin Biol Res 1995;391:1–20.

[36] Abel EL, Angel JM, Kiguchi K, et al. Multi-stage chemical carcinogenesis in mouse skin: fundamentals and applications. Nat Protocols 2009;4:1350–62.

[37] Nais AWH. An introduction to radiation biology, 2nd ed. Chichester, UK: John Wiley and Sons; 1998. pp. 217–40.

[38] (No authors stated). Chemical, environmental and experimental data. Polynuclear aromatic compounds. Part 1. IARC Monographs 32; 1983. pp. 1–477.

[39] Mossman BT, Shukla A, Heintz NH, et al. New insights into understanding the mechanisms, pathogenesis, and management of malignant mesotheliomas. Am J Pathol 2013;182(4):1065–77.

[40] Rivina L, Schiest R. Mouse models for efficacy testing of agents against radiation carcinogenesis - a literature review. Int J Environ Res Public Health 2012;10(1):107–43.

[41] Fisher DR, Weller RE. Carcinogenesis from inhaled (239)PuO$_2$ in beagles: evidence for radiation homeostasis at low doses? Health Phys 2010;99(3):357–62.

[42] Imaoka T, Nishimura M, Iizuka D, et al. Radiation-induced mammary carcinogenesis in rodent models: what's different from chemical carcinogenesis? Radiat Res 2009;50(4):281–93.

[43] Hanau A.N. (1889), Silzungsber. d. Gesellsch. d. Aerzte in Zurich vom 9 März, 1889 / Report on the Proceedings of the Society of Medical Practitioners on 9th March, 1889. Fortschritte der Med., 1889, vii, 338. Cited by Woglom H. The Study of experimental Cancer. NY: Columbia University Press; 1913. p. 21. See also Shimkin MB, Cancer, 1960, 13:221.

[44] Gold LS, Manley NB, Slone H, et al. Supplement to the carcinogenic potency database. Environ Health Perspect 1999;Suppl. 4:529.

[45] Cohen SM, Ohnishi T, Arnold LL, et al. Arsenic-induced bladder cancer in an animal model. Toxicol Appl Pharmacol 2007;222(3):258–63.

[46] D'Amato R, Slaga TJ, Farland WH. Relevance of animal studies to the evaluation of human Cancer risk Progr Clin Biol Res, vol. 374. New York, NY: Wiley-Liss; 1992.

[47] Cohen SM. Saccharin: past, present, and future. J Am Diet Assoc 1986;86(7):929–31.

[48] Dybing E. Development and implementation of the IPCS conceptual framework for evaluating mode of action of chemical carcinogens. Toxicology 2002; 181-182:121–5.

[49] (No authors listed). d-limonene. IARC Monogr Eval Carcinog Risks Hum. No 7; 1999. pp. 307–27.

[50] McClain RM, Keller D, Casciano D, et al. Neonatal mouse model: review of methods and results. Toxicol Pathol 2001;29(Suppl):128–37.

[51] United States EPA. Basic information about arsenic in drinking water. Available at: http://water.epa.gov/drink/contaminants/basicinformation/arsenic.cfm.

[52] Martinez VD, Vucic EA, Becker-Santos DD, et al. Arsenic exposure and the induction of human cancers. J Toxicol 2011;2011:431287.

[53] Rossman TG, Uddin AN, Burns FJ. Evidence that arsenite acts as a cocarcinogen in skin cancer. Toxicol Appl Pharmacol 2004;198(3):394–404.

[54] Burns FJ, Uddin AN, Wu F, et al. Arsenic-induced enhancement of ultraviolet radiation carcinogenesis in mouse skin: a dose-response study. Environ Health Perspect 2004;112:599–603.

[55] Järup L. Hazards of heavy metal contamination. Br Med Bull 2003;68:167–82.

[56] Takizawa H, Sato S, Kitajima H, et al. Mouse skin melanoma induced in two stage chemical carcinogenesis with 7,12-dimethylbenz[a]anthracene and croton oil. Carcinogenesis 1985;6(6):921–3.

[57] Mossman BT, Lippmann M, Hesterberg TW, et al. Pulmonary endpoints (lung carcinomas and asbestosis) following inhalation exposure to asbestos. J Toxicol Environ Health B Crit Rev 2011;14(1-4):76–121.

[58] Gibbs A, Attanoos RL, Churg A, et al. The "Helsinki criteria" for attribution of lung cancer to asbestos exposure: how robust are the criteria? Arch Pathol Lab Med 2007;131(2):181–3.

[59] United States EPA. Toxicological review of vinyl chloride. Available at: http://www.epa.gov/iris/toxreviews/1001tr.pdf.

[60] Agency for Toxic Substances and Disease Registry, US Public Health Service and US EPA. Toxicological profile for Thorium. Available at: http://www.atsdr.cdc.gov/toxprofiles/tp147.pdf.

[61] (No authors stated). Radiation. IARC Monographs on the evaluation of Carcinogenic Risks to Humans. Lyon: IARC – WHO. Vol 100D; 2012.

[62] Boice Jr. JD. Ionising radiation. In: Schottenfeld D, Fraumeni Jr JF, editors. Cancer epidemiology and prevention. New York, NY: Oxford University Press; 2009. pp. 259–93.

[63] United States EPA. Health effects. Available at: http://www.epa.gov/radiation/understand/health_effects.html.

[64] Lodish H, Berk A, Kaiser CA, editors. Molecular Cell Biology (7th ed.). New York, NY: W. H. Freeman; 2013.

[65] Ooi T. Thermodynamics of protein folding: effects of hydration and electrostatic interactions. Adv Biophys 1994;30:105–54.

[66] Kiesslich T, Gollmer A, Maisch T, et al. A comprehensive tutorial on in vitro characterization of new photosensitizers for photodynamic antitumor therapy and photodynamic inactivation of microorganisms. Biomed Res Int 2013;2013:840417.

[67] Ross K, Cherpelis B, Lien M, et al. Spotlighting the role of photodynamic therapy in cutaneous malignancy: an update and expansion. Dermatol Surg 2013;39(12):1733–44.

[68] Pattinson DI, Davies MJ. Actions of ultraviolet light on cellular structures. EXS 2006;96:131–58.

[69] Ravanat JL, Douki T, Cadet J. Direct and indirect effects of UV radiation on DNA and its components. J Photochem Photobiol B 2001;63(1-3):88–102.

[70] Pfeifer GP, You YH, Besaratinia A. Mutations induced by ultraviolet light. Mutat Res 2005;571(1-2):19–31.

[71] Cadet J, Wagner JR. DNA base damage by reactive oxygen species, oxidizing agents, and UV radiation. Cold Spring Harbor Perspect Biol 2013;5:a012559.

[72] Budowsky EI, Abdurashidova GG. Polynucleotide-protein cross-links induced by ultraviolet light and their use for structural investigation of nucleoproteins. Prog Nucl Acid Res Mol Biol 1989;37:1–65.

[73] O'Neill P, Wardman P. Radiation chemistry comes before radiation biology. Int J Radiat Biol 2009;85(1): 9–25.

[74] Von Sonntag C. The chemical basis of radiation biology. London: Taylor & Francis; 1987.

[75] Hütterman J, Teoule R, Kohnlein W. Effects of ionizing radiation on DNA: Physical, chemical and biological aspects. New York, NY: Springer; 1978.

[76] Sinden RR. DNA structure and function. San Diego, CA: Academic Press / Elsevier; 1994. p. 34.

[77] Bignold LP. Mechanisms of clastogen-induced chromosomal aberrations: a critical review and description of a model based on failures of tethering of DNA strand ends to strand-breaking enzymes. Mutat Res Rev Mutat Res 2009;681:271–98.

[78] Dextraze ME, Gantchev T, Girouard S, et al. DNA interstrand cross-links induced by ionizing radiation: an unsung lesion. Mutat Res 2010;704(1-3):101–7.

[79] Schulman SG. Fundamentals of interaction of ionizing radiations with chemical, biochemical, and pharmaceutical systems. J Pharm Sci 1973;62(11):1745–57.

[80] Kempner ES. Damage to proteins due to the direct action of ionizing radiation. Q Rev Biophys 1993;26(1):27–48.

[81] Yamamoto O. Ionizing radiation-induced crosslinking in proteins. Adv Exp Med Biol 1977;86A:509–47.

[82] Kunkel GR, Mehrabian M, Martinson HG. Contact-site cross-linking agents. Mol Cell Biochem 1981;34(1):3–13.

[83] Cress AE, Bowden GT. Covalent DNA-protein crosslinking occurs after hyperthermia and radiation. Radiat Res 1983;95(3):610–9.

[84] Agency for Toxic Substances and Disease Registry. Toxicological profiles. Available at: http://www.atsdr.cdc.gov/toxprofiles/tp149-c3.pdf.

[85] Schöllnberger H, Stewart RD, Mitchel RE, et al. An examination of radiation hormesis mechanisms using a multistage carcinogenesis model. Nonlinearity Biol Toxicol Med 2004;2(4):317–52.

[86] Little JB. Radiation carcinogenesis. Carcinogenesis 2000;21(3):397–404.

[87] Miller JA, Miller EC. The carcinogenicity of certain derivatives of p-dimethylaminozobenz in the rat. J Exp Med 1948;87(2):139–56.

[88] Penning TM. Metabolic activation of carcinogens. In: Penning TM, editor. Chemical Carcinogenesis. New York, NY: Springer; 2011. pp. 135–58.

[89] Shimada T. Xenobiotic-metabolizing enzymes involved in activation and detoxification of carcinogenic polycyclic aromatic hydrocarbons. Drug Metab Pharmacokinet 2006;21(4):257–76.

[90] Bolt HM, Roos PH. Generation of reactive intermediates by cytochromes P450. In: Ioannides C, editor. Cytochromes P450. Cambridge UK: RSC Publishing; 2008. pp. 47–97.

[91] Henderson RF. Species differences in the metabolism of olefins: implications for risk assessment. Chem Biol Interact 2001;135-136:53–64.

[92] Bartsch H, Hietanen E. The role of individual susceptibility in cancer burden related to environmental exposure. Environ Health Perspect 1996;104(Suppl. 3):569–77.

[93] International Agency for Research on Cancer. Monographs on the evaluation of carcinogenic risks to humans, IARC & WHO. Lyon 1972, vols. 1–106, plus supplements.

[94] Luch A. The mode of action of organic carcinogens on cellular structures. EXS 2006;96:65–95.

[95] Fornace Jr AJ, Little JB. DNA-Protein cross-linking by chemical carcinogens in mammalian cells. Cancer Res 1979;39:704–10.

[96] Wong SS, Lameson DM. Chemistry of protein and nucleic acid cross-linking and conjugation, 2nd ed. Boca Raton FL: CRC Press; 2011.

[97] Hoffman RM. Altered methionine metabolism, DNA methylation and oncogene expression in carcinogenesis. A review and synthesis. Biochim Biophys Acta 1984;738(1-2):49–87.

[98] Sarkar B. Metal replacement in DNA-binding zinc finger proteins and its relevance to mutagenicity and carcinogenicity through free radical generation. Nutrition 1995;11(5 Suppl):646–9.

[99] Song WJ, Sontz PA, Ambroggio XI, et al. Metals in protein-protein interfaces. Annu Rev Biophys 2014;43:409–31.

[100] Finnegan PM, Chen W. Arsenic toxicity: the effects on plant metabolism. Front Physiol 2012;3:182.

[101] Liebler DC. Protein damage by reactive electrophiles: targets and consequences. Chem Res Toxicol 2008;21(1):117–28.

[102] McGregor DB, Partensky C, Wilbourn J, et al. An IARC evaluation of polychlorinated dibenzo-p-dioxins and polychlorinated dibenzofurans as risk factors in human carcinogenesis. Environ Health Perspect 1998;106(Suppl. 2):755–60.

[103] Henderson L, Brusick D, Ratpan F, et al. A review of the genotoxicity of ethylbenzene. Mutat Res 2007;635(2-3):81–9.

[104] Schreiner CA. Genetic toxicity of naphthalene: a review. J Toxicol Environ Health B Crit Rev 2003;6(2):161–83.

[105] Strekowski L, Wilson B. Noncovalent interactions with DNA: an overview. Mutat Res. (Mutation Research/Fundamental and Molecular Mechanisms of Mutagenesis) 2007;623(1-2):3–13.

[106] Hartwell JL, Shubik P. Survey of compounds which have been tested for carcinogenic activity. Washington, DC: Public Health Service publication No 149, US Govt Printing Office; 1957.

[107] (No authors cited). Arsenic, metals, fibres and dusts. Lyon: IARC Monograph 100c; 2012.

[108] Benigni R, Passerini L. Carcinogenicity of the aromatic amines: from structure-activity relationships to mechanisms of action and risk assessment. Mutat Res 2002;511(3):191–206.

[109] Luch A. On the impact of the molecule structure in chemical carcinogenesis. EXS 2009;99:151–79.

[110] Felton JS, Knize MG, Wu RW, et al. Mutagenic potency of food-derived heterocyclic amines. Mutat Res 2007;616(1-2):90–4.

[111] Walpole AL. Carcinogenic action of alkylating agents. Ann N Y Acad Sci 1958;68:750–61.

[112] (No authors cited). DNA adducts: identification and biological significance. Lyon: IARC Sci Publ No 125; 1994.

[113] Seo KY, Jelinsky SA, Loechler EL. Factors that influence the mutagenic patterns of DNA adducts from chemical carcinogens. Mutat Res 2000;463(3):215–46.

[114] Nestmann ER, Bryant DW, Carr CJ. Toxicological significance of DNA adducts: summary of discussions with an expert panel. Regul Toxicol Pharmacol 1996;24(1 Pt 1):9–18.

[115] Loechler EL. The role of adduct site-specific mutagenesis in understanding how carcinogen-DNA adducts cause mutations: perspective, prospects and problems. Carcinogenesis 1996;17(5):895–902.

[116] Rous P. A sarcoma of the fowl transmissible by an agent separable from the tumor cells. J Exp Med 1911;13:397–411.

[117] Evans AS. Causation and disease: the henle-koch postulates revisited. Yale J Biol Med 1976;49(2):175–95.

[118] Gross L. Transmission of cancer in man. Tentative guidelines referring to the possible effects of inoculation of homologous cancer extracts in man. Cancer 1971;28(3):785–8.

[119] Keogh EV. Ectodermal lesions produced by the virus of Rous sarcoma. Brit J Exptl Pathol 1938;19:1–8.

[120] Temin HM, Rubin H. Characteristics of an assay for Rous sarcoma virus and Rous sarcoma cells in tissue culture. Virology 1958;6(3):669–88.

[121] Rubin H. The early history of tumor virology: Rous, RIF, and RAV. Proc Natl Acad Sci USA 2011;108(35):14389–96.

[122] Javier RT, Butel JS. The history of tumor virology. Cancer Res 2008;68(19):7693–706.

[123] King AMQ, Adams MJ, Carstens EB, editors. Virus taxonomy: classification and nomenclature of viruses: ninth report of the international committee on taxonomy of viruses. San Diego, CA: Elsevier; 2012.

[124] Murphy FA, Fauquet CM, Bishop DHL, editors. Virus taxonomy: sixth report of the international committee on taxonomy of viruses. New York, NY: Springer-Verlag; 1995, [Chapter 41].

[125] Fan H. Retroviruses Encyclopedia of microbiology, 3rd ed. Philadelphia, PA: Elsevier; 2009. pp.519–34.

[126] Nathanson N, Holmes KV. Cellular receptors and trophism Nathanson N, editor. Viral pathogenesis and immunity (2nd ed.). San Diego, CA: Academic Press; 2007. pp. 27–40.

[127] Taralo KP. Foundations of microbiology, 6th ed. New York, NY: McGraw Hill; 2007. pp. 174–6.

[128] White GM, Cox NH. Diseases of the skin, 2nd ed. Philadelphia, PA: Elsevier/ Mosby; 2006. pp. 533–9.

[129] Wold WS, Toth K. Syrian hamster as an animal model to study oncolytic adenoviruses and to evaluate the efficacy of antiviral compounds. Adv Cancer Res 2012;115:69–92.

[130] (No authors stated). Biological agents, vol. 100b. Lyon: IARC Monographs on the Evaluation of Carcinogenic Risk; 2012.

[131] Shlomai A, de Jong YP, Rice CM. Virus associated malignancies: the role of viral hepatitis in hepatocellular carcinoma. Semin Cancer Biol 2014;26: 78–88.

[132] Koike K. The oncogenic role of hepatitis C virus. Recent Results Cancer Res 2014;193:97–111.

[133] Franceschi S, Maso LD, Rickenbach M, et al. Kaposi sarcoma incidence in the Swiss HIV Cohort Study before and after highly active antiretroviral therapy. Br J Cancer 2008;99:800–4.

[134] Viral-associated nonmelanoma skin cancers: a review Am J Dermatopathol 2009;31(6):561–73.

[135] Hughes MP, Hardee ME, Cornelius LA, et al. Merkel cell carcinoma: epidemiology, target, and therapy. Curr Dermatol Rep 2014;3:46–53.

[136] Klein G, Klein E, Kashuba E. Interaction of Epstein-Barr virus (EBV) with human B-lymphocytes. Biochem Biophys Res Commun 2010;396(1):67–73.

[137] Butel JS. Viral carcinogenesis: revelation of molecular mechanisms and etiology of human disease. Carcinogenesis 2000;21(3):405–26.

[138] Niller HH, Wolf H, Minarovits J. Viral hit and run-oncogenesis: genetic and epigenetic scenarios. Cancer Lett 2011;305(2):200–17.

[139] Marshall BJ, Warren JR. Unidentified curved bacilli in the stomach of patients with gastritis and peptic ulceration. Lancet 1984;i(8390):1311–5.

[140] Chung HW, Lim JB. Role of the tumor microenvironment in the pathogenesis of gastric carcinoma. World J Gastroenterol 2014;20(7):1667–80.

[141] Hussein NR. Helicobacter pylori and gastric cancer in the middle east: a new enigma? World J Gastroenterol 2010;16(26):3226–34.

[142] Fried B, Reddy A, Mayer D. Helminths in human carcinogenesis. Cancer Lett 2011;305(2):239–49.

[143] Morgentaler A. Testosterone therapy for men at risk for or with history of prostate cancer. Curr Treat Options Oncol 2006;7(5):363–9.

[144] Kim JJ, Chapman-Davis E. Role of progesterone in endometrial cancer. Semin Reprod Med 2010;28(1):81–90.

[145] Advani P, Moreno-Aspitia A. Current strategies for the prevention of breast cancer. Breast Cancer (Dove Med Press) 2014;6:59–71.

[146] Moorman PG, Havrilesky LJ, Gierisch JM, et al. Oral contraceptives and risk of ovarian cancer and breast cancer among high-risk women: a systematic review and meta-analysis. J Clin Oncol 2013;31(33):4188–98.

[147] Schrager S, Potter BE. Diethylstilbestrol exposure. Am Fam Physician 2004;69(10):2395–400.

[148] Bischoff F, Bryson G. Carcinogenesis through solid state surfaces. Prog Exp Tumor Res 1964;5:85–133.

[149] (No authors stated). Workshop on mechanisms of fibre carcinogenesis and assessment of chrysotile asbestos substitutes. 8–12 November 2005. Lyon, France. Lyon: IARC/WHO. Available at: www.who.int/ipcs/publications/new_issues/summary_report.pdf.

[150] Kane AB. Mechanisms of mineral fibre carcinogenesis. IARC Sci Publ. No 140; 1996. pp. 11–34.

[151] Huang SX, Jaurand MC, Kamp DW, et al. Role of mutagenicity in asbestos fiber-induced carcinogenicity and other diseases. J Toxicol Environ Health B Crit Rev 2011;14(1-4):179–245.

[152] Barrett JC, Lamb PW, Wiseman RW. Multiple mechanisms for the carcinogenic effects of asbestos and other mineral fibers. Environ Health Perspect 1989;81:81–9.

[153] Barrett JC, Shelby MD. Mechanisms of human carcinogens. Prog Clin Biol Res 1992;374:415–34.

[154] Barrett JC. Cellular and molecular mechanisms of asbestos carcinogenicity: implications for biopersistence. Environ Health Perspect 1994;102(Suppl. 5):19–23.

[155] National Toxicology Program. 13th Report on carcinogens. Available at: http://ntp.niehs.nih.gov/ntp/roc/twelfth/profiles/GlassWoolFibers.pdf.

[156] (No authors listed). Man-made vitreous fibres. Lyon: IARC Monograph No 81; 2002.

[157] Turner FC. Sarcomas at sites of subcutaneous implanted Bakelite disks in rats. J Natl Cancer Inst 1941;2:81–3.

[158] Oppenheimer BS, Oppenheimer ET, Stout AP. Sarcomas induced in rats by implanting cellophane. Proc Soc Exp Biol Med 1948;67(1):33–4.

[159] Nothdurf H. Experimental development of sarcoma by means of enclosed foreign bodies. Strahlentherapie 1956;100(2):192–210.

[160] Baker SG, Kramer BS. Paradoxes in carcinogenesis: new opportunities for research directions. BMC Cancer 2007;7:151.

[161] Delaunay C, Petit I, Learmonth ID, et al. Metalonmetal bearings total hip arthroplasty: the cobalt and chromium ions release concern. Orthop Traumatol Surg Res 2010;96(8):894–904.

Etio-Pathogenesis II: Genomic Events and Processes Potentially Caused by Etiological Agents

L.P. Bignold: *Principles of Tumors.*
DOI: http://dx.doi.org/10.1016/B978-0-12-801565-0.00005-6

The previous chapter outlined five steps in the etiology and pathogenesis of tumors and described the various known carcinogens and how they might interact with targets in cells (A and B of those steps). The next step (C) to discuss is how interaction(s) of agents with an intracellular target(s) might cause genomic events.

This area in carcinogenesis is not fully clarified mainly because:

i. There are many kinds of genomic abnormalities (Figure 5.1).
ii. It is not known whether one or all of them might be critical to tumor formation.
iii. It is also not known whether there might be, on the one hand, one or several specific genomic events for all tumors, or on the other hand, different kinds of genomic events for different types of tumor.
iv. It is not known precisely how the actual genomic lesions are created.

In this chapter, Sections 5.1–5.3 give accounts of the different kinds of genomic lesions—including genomic instabilities involving these events—and how they might occur. Section 5.4 describes additional kinds of genomic instabilities in tumors.

Subsequent to these sections, Section 5.5 outlines potential effects of genomic instabilities. Section 5.6 mentions possible abnormalities of gene regulation in tumor cell populations. Section 5.7 deals with the genomic lesions potentially caused by viruses and other agents.

Events in specific protein-coding genes in tumor cells (step D) are discussed in Chapters 6 and 9, and the possible bases of combinations of features (step E) in Section 8.5.

5.1 UNINUCLEOTIDE EVENTS AND PROCESSES

From the late 1950s, the main theory of tumor formation has been that one or more uni- or oligonucleotide alterations in particular sequences in DNA. In the main, these genomic events were assumed to be "fixed"—i.e., unchanging in subsequent generations of the cell—according to "mutations" in germline genetics. A particular example was the sickle cell anemia model of genetic disease (see Section 2.6). However, it is now thought to be probable that tumors comprise cell populations in which the individual cells are liable to develop different additional genomic events

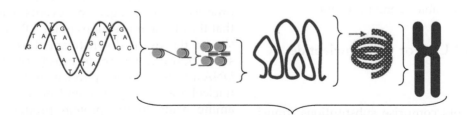

Uni-nucleotide events. These are mainly only detected by sequencing of nucleotides.

Larger lesions comprise:
(i) Abnormalities of copy numbers (amplifications and deletions): These can be detected in relation to any length of protein-coding as well as much of the nonprotein-coding DNA (see Section 9.2.6).

(ii) Re-positional changes (inversions and transpositions): These are more difficult to detect (see Sections 9.4 and 9.5).

FIGURE 5.1   The different kinds of genomic events.

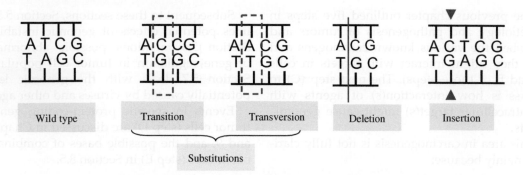

FIGURE 5.2   The kinds of uni- and dinucleotide errors.

in subsequent generations—i.e., the nucleotide errors and chromosomal complements are unstable after the original genomic event.

This section describes this kind of genomic event, emphasizing the distinction between "fixed/"stable" genomic events and changing/"unstable" genomic events.

5.1.1 Fixed Uni- or Oligonucleotide Errors

(a) Kinds

These errors comprise substitutions (transitions and transversions) as well as insertions and deletions (Figure 5.2). In many experimental studies, most errors are substitutions of nucleobases. Insertions and deletions of nucleotides are less common [1]. Nucleotide substitutions may result in substitutions in amino acids in the protein product of the gene. This is the case in sickle cell anemia (see Section 2.6.2).

However, uni- or oligonucleotide insertions and deletions may have other effects.

i. Frameshift mutations.

In exons, all numerical gains or losses of nucleotides which are not a multiple of three produce frameshift genomic events. This means that the all the nucleotides in the DNA from the site of the event downstream during synthesis or transcription are incorrect. The length of DNA affected can be up to tens of thousands of nucleobases in sequence and hence also many amino acids in the protein product (Section A2.1.5). Similar small insertions and deletions may have substantial effects on noncoding (RNA) genes [2].

ii. Creation of an inappropriate ("hidden") stop codon.

Substitution of a nucleotide may cause a previous amino acid–coding triplet to be converted to a "stop codon"/triplet. The overt synthesis stop codons are TAG, TAA, TGA. The overt transcription stop codons are UAG, UAA, UGA.

As well as by substitution, inappropriate stop codons can be created in frameshift mutations. These are called "hidden stop codons," because the nucleotide sequence is not altered, but two of a triplet for one amino acid, with another from the triplet of another amino acid, is "read" in the frameshift as a triplets which is a stop codons.

Inappropriate stop codons cause "truncated" and often functionally deficient, protein products.

(b) No Particular Error of This Type Has Been Found so far Found in All Tumors

It was noted in Section 2.6.2 that all cases of sickle cell anemia are caused by a single nucleotide substitution in the globin gene, leading to a single amino acid substitution in the globin molecule. In tumors however, no universal error of uni- or oligonucleotide genomic event has been found in the exons of any protein-coding gene so far studied.

Nevertheless, it remains possible that a genomic event of any of these kinds—frameshift or not—is involved in tumor formation by being:

i. In an exon, but has not been detected.
ii. Is in a noncoding structure such as an RNA gene (see Section A2.2.5).
iii. Is in an intronic regulator of a gene(s).

5.1.2 Mechanisms of Nucleotide Errors: The Nucleobase Damage Model

Nucleobase damage is the most widely accepted, long-standing suggestion for the mechanism of nucleotide genomic events, especially those which might cause tumors [3,4]. It relates to steps 2, 3, and 4 in the pathogenesis of tumors (see Section 4.1.3), i.e., those which occur between the interaction of an exogenous agent with a cellular target and the induction of alterations in its genome (see also Section 1.2 and Figure 4.2). It is particularly applied to studies of radiation and chemical carcinogenesis (Sections 4.3 and 4.4). The principles of this model are as follows.

(a) Nucleobases as the Targets of Carcinogen

In this nucleobases are considered the exclusive target of the carcinogens.

(b) The Nucleobases Are Chemically Altered/"Damaged"

Chemical alteration may be direct or indirect (see Section 1.2.2 and Figure 1.3) and may involve structural changes within the nucleotides themselves. The damage is of three kinds:

i. Nonadduct damage

This is particularly seen with ultraviolet B and ionizing radiations, as described in Section 4.3.

ii. Direct adduct formation by the chemical agent—or an activated metabolic derivative of it.

This means that the exogenous agent or the activated derivative becomes covalently bound to nucleobases of the DNA (Figures 1.3 and 5.3).

iii. Indirect adduct damage arising when the agents cause production in tissues of reactive endogenous secondary molecules which covalently bind to DNA.

The most prominent example is the production of reactive oxygen species (ROS) by ionizing radiations (Section 4.3.3). Chemical carcinogens too can cause endogenous substances to release chemicals which react with DNA. These include ROS [5] as well as aldehydes such as 4-hydroxy-2-nonenal, malondialdehyde, acrolein, crotonaldehyde, and methylglyoxal [6].

This general mechanism has been studied in relation to those metal ions which do not react directly with DNA [7]. These metal ions are thought to form complexes with other atoms or molecules, which in turn act on other molecules

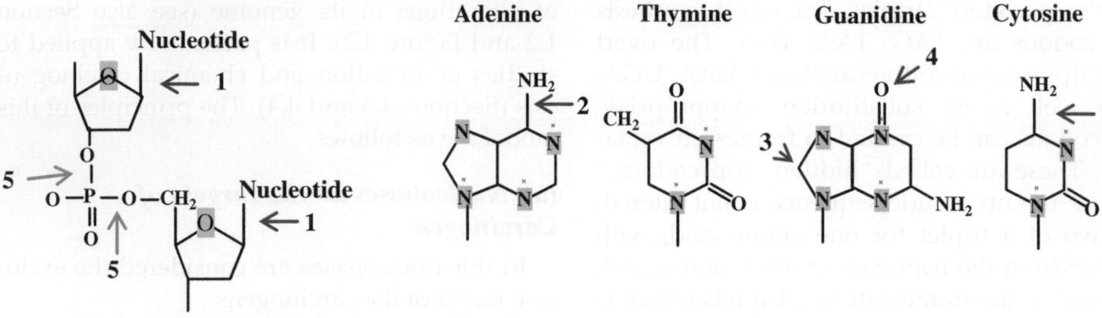

1. Breaking of covalent bond by hydrolysis causes loss of nucleotide.
2. Breaking of covalent bond by hydrolysis causes loss of amino group.
3. Addition of oxygen to carbon atom.
4. Addition of an alkyl group to an oxygen atom.
5. Breaks in the phospho-diester bond between the riboses (required for DNA and chromosome breaks) rarely occur spontaneously.

FIGURE 5.3   Examples of common sites of damage on DNA (refer also to Figure 1.3).

to produce endogenous substances which react with DNA. For example, arsenic, nickel, ferric, and cobalt ions do not react directly with DNA and are believed to affect DNA indirectly via redox cycling, oxidative damage by the 8-oxo-2′-deoxyguanosine base modification, other metabolic effects, and gene activations [8].

(c) The Chemical Alterations in Nucleobases Cause Failures of Accurate DNA Repairs and/or Synthesis

The next step in the mechanism involving nucleobase damage is that the chemical changes cause sequence errors in the DNA of the cell. Most frequently, the errors observed in experimental studies with carcinogens are substitutions. This effect must be an indirect one, because no known etiological agent or indirect reactive molecule can substitute one nucleobase for another in DNA in physiological, cell-free conditions.

Substitutions and other nucleotide errors are thus believed to occur as an indirect effect during repairs to, or synthesis of, DNA via two steps:

i. *Mis-repairs in G0 and G1 (see Section A1.3.4)*

Repairs of the nucleotides in the DNA require various complexes of proteins [9,10]. Repairs to DNA can occur in all phases of the cell cycle except mitosis (Section A1.3.3) [11]. The base-excision repair enzymes and nucleotide-excision repair enzymes seem to be most relevant to alterations in nucleotide sequences. In general, the potential for mis-repairs by these enzymes to cause sequence alterations are thought to be considerable. This is particularly because spontaneous damage to DNA—especially losses of nucleobases causing abasic sites—is thought to occur at rates of tens of thousands of bases per nucleus per day (see Section A2.1.4).

A mis-repair in DNA in G0 or G1 may be corrected prior to transcription (transcription-coupled repair, [12]). If that step too were to fail, or inappropriately correct the error, an error may persist for the life of the cell.

ii. *Mis-repairs and matching errors in S phase*

Almost all nucleotide damage inhibits DNA synthesis. The damage is often corrected immediately before the separation of the strands by helicases prior to synthesis [13]. Failed repairs here may lead to errors in template strands of DNA and hence potentially a permanent nucleotide error.

The next step at which errors may arise is in the pair matching by the DNA polymerase complex. Errors of nucleotide pairing at this point are mainly corrected by the "checking" functions referred to as "proofreading" and "mismatch repair" (Section A1.3.2, Figure 5.4) [14].

In the early years of study of DNA synthesis, it was assumed that the mis-pairing errors might arise because enzymatic sites are sufficiently "elastic" to accommodate adducts and thus to permit the matching errors [15]. However, now it is recognized that most of the DNA polymerases are extremely "stringent"/precise in their matching of nucleotides on new DNA to those on template DNA [14]. Most mismatches therefore are thought to result from deployment of "low stringency" of DNA polymerases (mainly Y POL family) which can "synthesize over" the adduct-nucleobase complex (i.e., "trans-lesional synthesis" [16]).

However, for the error to persist, the subsequent "checking" (proofreading and mismatch repairs) must also fail (Figure 5.5).

The fact that transitions and transversions comprise alterations in the nucleotide sequences in *both* strands of DNA indicates that the original "correct" nucleotide is replaced with one appropriate to the new incorrect nucleotide at this second step (Figure 5.5).

iii. *These errors may be commoner in metabolically active or dividing cells*

It is worth noting at this point that for both mechanisms, the numbers of DNA mis-repairs and mis-syntheses per nucleus may be greater in cells which are synthesizing DNA (i.e., are proliferating) or are synthesizing RNA (i.e., are transcriptionally active). In both these situations, relatively more DNA is in single-strand configuration than is present in nondividing or metabolically less active cells (see Section A3.3).

5.1.3 Discussion of the Nucleobase Damage Model

While the direct and indirect nucleobase damage theory has received much attention, there are still some phenomena which are difficult to explain by it as follows.

(a) Species Differences in Radiosensitivities to Genomic Damage

Species differences are a major phenomenon in carcinogenesis. In relation to chemical carcinogens, (see Section 4.2.3), many species differences in susceptibilities may be due to defensive factors in cells (see Sections A3.1 and A3.2). However, such factors are not involved in susceptibilities of different species to the carcinogenic effects of radiations. Radiations deposit their energy at random wherever there is water. No significant differences are known of between the water contents of parts of cells in different species. The DNA is chemically the same in all species—the same ratios of the nucleobases are observed. This indicates that the radiations are acting on something else. The main alternatives would seem to be the proteins in the cell. The issue would then arise: what proteins might, if their functions were altered, cause tumors? It would seem likely that dysfunctions in the proteins which act on DNA—for synthesis, transcription, unraveling, and repairs could have genomic effects. Reduced stringency of nucleobase selection by the synthetic sites of DNA polymerases has long been proposed as a mechanism of nucleotide sequence errors (see Section 5.1.6), but for lack of sufficiently sensitive technology, has been difficult to prove.

1. Mechanism of transition during repair of base damage

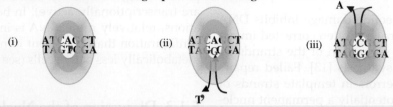

(i) Damage to a nucleotide is detected.
(ii) First repair: Mis-pairing occurs when the damaged nucleotide is replaced by a pair-incorrect nucleotide.
(iii) Second repair: The pair for the damaged nucleotide is changed to a nucleotide which is correct for the wrongly inserted replacement for the damaged nucleotide, creating a permanent change in the nucleotide sequence.

2. Mechanism of transversion during repair of base damage

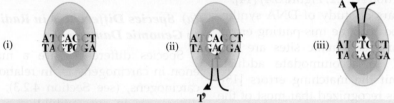

(i) Damage to a nucleotide is detected.
(ii) First repair: The damaged nucleotide is replaced by a pair-incorrect nucleotide.
(iii) Second repair: The original undamaged nucleotide is changed to a nucleotide which is pair-correct to the wrongly inserted replacement for the damaged nucleotide, creating a permanent change in the nucleotide sequence.

3. Mechanism of transition or transversion during synthesis

For transition
(i) At synthesis site, mis-pairing occurs when a nucleotide is mispaired to the template nucleotide.
(ii) At proof reading or mis match repair, mis-pairing is corrected, but by replacing the template nucleotide with a correct pair for the incorrect new nucleotide, creating a permanent change in the nucleotide sequence.

For transversion, the mispairing at synthesis is of an A with an A, followed at proof reading or mismatch repair, by replacement of the template nucleotide with a correct pair for the incorrect new nucleotide.

FIGURE 5.4   Nucleotide substitution errors may arise when primary errors in repairs or synthesis are followed by errors in later repairs (examples only).

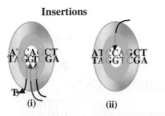

Insertions   Deletions

(i) First repair: Mis-pairing occurs when an additional nucleotide is erroneously inserted proximate to the correct replacement nucleotide.

(ii) Second repair: A pair nucleotide is inserted into the other strand for the extra nucleotide, creating a permanent addition of a nucleotide pair to the sequence.

(i) First repair: Mis-pairing occurs when no nucleotide is inserted to replace a damaged nucleotide.

(ii) Second repair: The pair of the damaged nucleotide is excised and not replaced, creating a permanent loss of nucleotide pair in the sequence.

FIGURE 5.5   Mechanisms of insertions and deletions of single nucleotides.

(b) Genomic Damage Caused by Noncarcinogens

In relation to direct DNA adducts and tumor formation, it is important to recognize that there are substances which react with DNA but are not carcinogenic. Notable examples of the latter are the psoralens [17] which cross-link DNA strands but do not cause tumors (provided ultraviolet light is absent) and analogues of benzo(a)pyrene [18] which are not carcinogenic. The supravital stain Hoechst 33342 [19] binds to the minor groove of DNA especially A-T-rich sites, but is not known to be carcinogenic.

(c) Similar Genomic Damage in Nontumorous Pathological Processes

In relation to indirect DNA adducts (oxidations, etc.), it may also be noted that activation of ROS have been detected in many nontumorous conditions, including respiratory and cardiovascular diseases [20], ageing, and inflammation generally [21] (see also Chapter 11). Mechanisms by which these secondary mechanisms might cause tumors in one situation and one or more nontumorous

pathological conditions in other situations are unclear.

(d) Lack of Correlation Between Potencies in Carcinogenicity and Other Genopathic Effects of Many Agents

This is important to the consideration of testing of potentially carcinogenic chemicals, as is discussed in Section 14.2.7.

5.1.4 The "One-Off" Impaired Replicative Fidelity of DNA Polymerase Theory of Nucleotide Genomic Events

This possible mechanism for nucleotide errors has been recognized since the 1960s [22]. It was suggested as early as 1965 [23] and was investigated from the mid-1970s [24–26]. The essential concept is that "one-off" episodes of DNA replicative hypofidelity in S phase occur because a DNA polymerase complex is caused to unfaithfully pair nucleotides during synthesis of DNA for at least part of one-half replication "bubble" (see Section A2.3 and Figure 5.6), or because low-fidelity polymerases have been deployed by the cell [27].

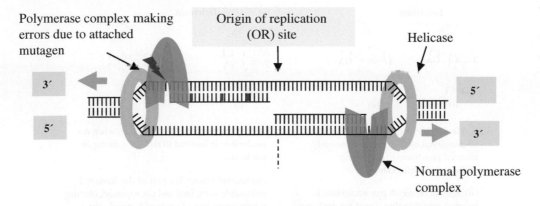

According to this model, attachment of mutagen (⚡) to the polymerase complex results in an increased rate of mis-pairings (▮) in the daughter strand of DNA from the origin of replication sites (30–300 kb). The other strand is not affected. If the complex were to be used in additional "bubble" of synthesis it may cause additional foci of replicative infidelity in the one S phase.

Note: For simplicity, the diagram illustrates the principle using the double helicase model of synthesis. The helicase is the MCM type [97].

FIGURE 5.6   Principle of the concept of mutagenesis by "one-off" impaired replicative fidelity of DNA in S phase in eukaryotes (see Section A2.3).

The phenomenon is to be distinguished from mutator phenotype. In this phenotype, there is an inherited tendency to nucleotide mis-pairings during DNA synthesis (see Section 5.1.6), so that the whole genome tends to be affected.

Many proteins, especially those for the synthetic, proofreading, and mismatch repair steps, might be affected by carcinogens and hence could be responsible for replicative infidelity of DNA synthesis.

5.1.5 "Mutational Spectra" (High Incidences of Nucleotide Errors at Specific Locations)

A common, but little understood phenomenon is that nucleotide sequence errors (substitution and less frequently insertions and deletions) in some parts of the genes than others can be clustered in particular locations in the genome [28].

Different carcinogens may induce different spectra [29]. It means that some parts of the genes are genomically altered more frequently than others either:

i. by one carcinogen alone (carcinogen-specific spectra), e.g., Ref. [30] or
ii. by many different carcinogens (carcinogen nonspecific spectra; "site-specific" mutations) [31].

Different mutational spectra also occur in different tumor types (see Section 9.3.3) and in different viral infections (Section 5.7).

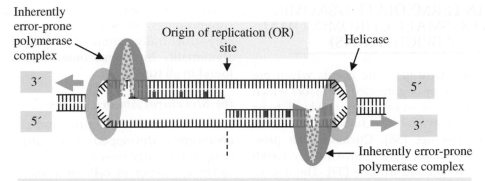

There is an inherent defect in the replicative fidelity of all DNA polymerase complexes because of a gene-dictated structural abnormality at the synthesis site.
– Every DNA polymerase complex makes a higher than normal number of random errors.
– All replication bubbles in the whole genome are affected.
– The errors are not necessarily site- or nucleotide-specific.
– With every division of the cell, random errors accumulate.
The phenotype might also arise by any inherent defect in DNA repair proteins, rather than the polymerases.

FIGURE 5.7 Principle of accumulating nucleotide errors due to inherently defective DNA polymerases ("mutator phenotype"). Note: Double helicase model of DNA synthesis shown as in Figure 5.6.

5.1.6 Nucleotide Sequence Instability ("Mutator Phenotype"; "Point Mutation Instability")

This phenomenon is mentioned in Section 2.9 and refers to the possibility that in tumor cells, the polymerase complexes inherently tend to mis-pair nucleotides during synthesis, proof-reading, or mismatch repairs [32–37]. The cause of the tendency is assumed to be a somatic genomic event in the gene(s) for the polymerase protein(s), or alternatively, a metabolic alteration in the cell which affects the pairing accuracy of the polymerase complex (Figure 5.7).

The situation at present seems to be that the nucleotide sequence instability in tumor cell populations may arise from one or more of the following:

i. The DNA polymerases, proofreading, and/or mismatch repair enzymes in tumor cells are intrinsically prone to making pairing errors.
ii. Endogenous factors are produced in tumor cells which cause these enzymes to malfunction with the same effects.
iii. The G0, G1, G2 repair enzymes are inherently liable to create mis-repairs.
iv. The same enzymes are affected by endogenous factors to produce the same effect.
v. The numbers of endogenous DNA damaging events is increased, because the cells of the tumor cell population are under "stress," and this overwhelms the capacities of the normal repair enzymes.
vi. The numbers of exogenous DNA damaging events increase because the normal toxico-kinetic barriers to exogenous mutagens (see Appendix 3) are defective in the cells of the population. As a result, these new exogenous damage events overwhelm the normal repair enzymes.

The important implication of this idea is that the cells in single cases of tumors would likely to be different from each other and from their cell of origin, through successive generations, without additional insults from known exogenous agents.

5.2 INTERMEDIATE GENOMIC EVENTS (SMALL CHROMOSOMAL ABERRATIONS)

There are many genomic events which are larger than uni- or oligonucleotide errors which are too small, or of an inappropriate kind, to be appreciated by available technology for visualizing chromosomes [38]. Currently, the presence of pieces larger than approximately 100 kb may be demonstrated in cells [39]. These techniques show changes in copy number (deletions and amplifications). However, inversions and transpositions—within a chromosome or to another chromosome—cannot be detected.

The various issues of these techniques are described further in Chapter 9.

5.3 FIXED AND UNSTABLE CHROMOSOMAL ABNORMALITIES IN TUMOR CELL POPULATIONS

5.3.1 Fixed Chromosomal Abnormalities

(a) Number

In normal cells in G0 or G1 (Section A1.3.4), the chromosome number is constant, excepting in normal multinucleate cells, such as megakaryocytes. At the end of S phase and in mitosis, the normal number of chromosomes for the cell type is doubled. No normal human cell varies from this.

A few germ-line numerical abnormalities in chromosomes are known. These are especially 45X0 (Turner's syndrome) and 47XXY (Klinefelter's syndrome) [40]. Loss of gain of any whole autosomal chromosome is not recorded in any viable postpartum human.

In all forms of experimental damage to the chromosomes, as well as in transformation *in vitro* (Section 14.2.3) and in tumors, numerical changes are always associated with structural changes, and *vice versa*.

Tumor cells tend to have excess numbers of chromosomes, of which many are undoubtedly structurally abnormal as well. No constant abnormality in chromosome number has been found in all tumors.

(b) Structural Abnormalities

The terminology of structural chromosomal aberrations developed from studies beginning in the early twentieth century of damage to chromosomes in cells of *Tradescantia* which have small numbers of readily identifiable chromosomes [41] (Figures 5.8–5.10).

Later authors [42,43] described the kinds of aberrations in terms of:

i. Breaks with separation of a fragment (deletions). The deleted fragment could sometimes be seen in the metaphase cell, as one or more fragment ("acentrics," "minutes"), or not visible, when it was assumed to be lost from the cell.

ii. Discontinuities in staining. This is indicated where there is an apparent break in the chromosome, but the broken part retains its position with the part of the arm from which it apparently broke off. These lesions are also called "gaps" or "achromatic lesions" [42].

iii. Breaks with rejoinings ("changes," "interchanges"— now understood as large transpositions). These were subdivided into breaks and rejoinings involving

iiia. Chromatid arm interchanges.

These could be in the same chromatid (intraarm changes): in different arms of the same chromatid (intrachromosome changes) and between different chromosomes (interchanges. Each kind of exchange could be "balanced"/unbalanced."

iiib. Ring forms, with zero, one, two, or rarely three centromeres (acentric, dicentrics, and tricentric, etc. ring forms).

iiic. Multiarm forms with multiple centromeres (triradials, quadradials, etc.).

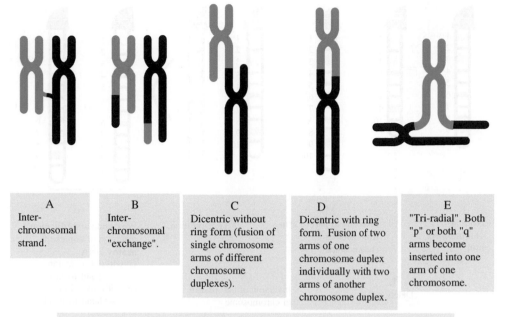

| A
Inter-chromosomal strand. | B
Inter-chromosomal "exchange". | C
Dicentric without ring form (fusion of single chromosome arms of different chromosome duplexes). | D
Dicentric with ring form. Fusion of two arms of one chromosome duplex individually with two arms of another chromosome duplex. | E
"Tri-radial". Both "p" or both "q" arms become inserted into one arm of one chromosome. |

Note: When any two breaks in whatever morphological form occur in two genes and join, the result may be a "fusion" neo-gene (see Section 9.3.3).

FIGURE 5.8 Interchromosomal aberrations seen by ordinary staining of squash preparations of cells in meta(sub)phase (see also Section A1.3.5).

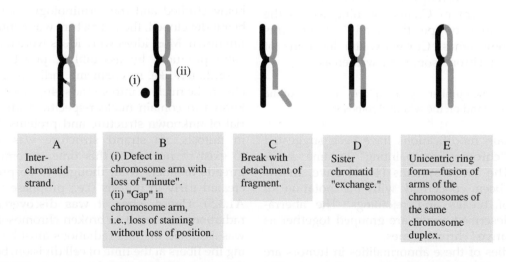

| A
Inter-chromatid strand. | B
(i) Defect in chromosome arm with loss of "minute".
(ii) "Gap" in chromosome arm, i.e., loss of staining without loss of position. | C
Break with detachment of fragment. | D
Sister chromatid "exchange." | E
Unicentric ring form—fusion of arms of the chromosomes of the same chromosome duplex. |

FIGURE 5.9 Intrachromosomal lesions seen with ordinary staining in squash preparations of meta(sub)phase.

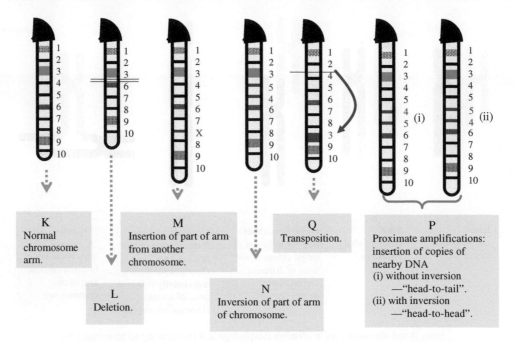

FIGURE 5.10   Intrachromosomal aberrations mainly identifiable by Giemsa staining (see Section 9.3).

In the 1960s, these terms were amended slightly according to a theory proposed by Revell [43] by using the word "exchanges" broadly instead of "changes" or "interchanges." This usage is currently the most popular.

The advent of Giemsa staining led to the identification of specific "bands" in the arms of chromosomes. Currently, the locations of lesions on chromosomes are written as:

Chromosome number/"p"or "q" arm/number of the Giemsa band on the whole chromosome

Certain modifications have been suggested since "chromosome painting" became available. The ISCN systems [44,45] have, in the main, been concerned with the notation of sites of breaks and rejoinings. The aberrations described above are grouped together as "derivative" chromosomes.

Studies of these abnormalities in tumors are discussed in Chapter 9.

5.3.2 Theories of Mechanisms of Chromosomal Abnormalities Involving Primary DNA Strand Breaks

At the time that chromosomes were first being studied and the terminology above was being developed, their structure was completely unknown. Most ideas were ideas were that they were produced by assembling possibly by a coagulation-like mechanism, individual particles in the nucleus into strands [46]. Nuclei were known to contain nuclear-specific acidic material of unknown structure and proteins. Except in mitosis, no strand structure was known to exist in nuclei. At this time, chromosomal threads or fibers were thought to be probably created early in mitosis (i.e., prophase Section A1.3.5). Hence when it was discovered that radiations can result in broken chromosomes, it was assumed that the radiations must be breaking the fibers at the time of cell division, because they were not recognized to exist at other times.

The major change in fundamental concept occurred only in the 1970s, when it was established that each chromosome has at its core only one continuous double-strand of DNA (the "unineme" hypothesis [47]). Thus the core of the chromosomes—the DNA—exists as 46 separate molecules in every human nucleus. The fiber is there to be broken at any time in the cell cycle (Section A1.3.4).

From the 1970s therefore, attention shifted to how agents which cause chromosomal aberrations can break DNA strands. No satisfactory explanation of chromosomal aberrations based on direct breakage of strands by radiations or chemicals emerged [48–51]. This was mainly because:

i. Chemicals which produce chromosomal aberrations in living cells cannot themselves break the "backbone" phosphodiester and ribose ring bonds of DNA (which are necessary for these aberrations) when the DNA is in cell-free solutions.
ii. Ionizing radiations may break these bonds under these conditions, but only in doses which are lethal to any living organism, see Section 9.2.7.

5.3.3 The "Tether Drop" Theory of Chromosomal Aberrations

This theory is based on data that has accumulated in the last two decades or so concerning the function of the enzymes and associated proteins which serve DNA (Section A2.3). The essential points are that DNA strands are broken during normal enzymatic processes relating to DNA such as unraveling, transcription, nucleotide-excision repair, and in synthesis. For all of these processes, it is necessary that the broken ends be held in place ("tethered," or "clamped" [10,52]), ready for the re-ligation of the ends of the DNA to complete the relevant process. Protein conformation is much more sensitive to noxious agents than chemical bonds of the backbone of DNA. On this basis, the tetherings of DNA ends to enzyme complexes could fail causing a "tether drop" of the broken ends. Alone such a "drop" would simply result in a break in the chromosome. However, if two broken ends were close to one another in the same nucleus, rejoining could occur if unaffected tether proteins were present.

On the basis of this concept (Figure 5.11), there are three possible mechanisms of action of clastogens (substances which cause chromosomal aberrations, Section 4.1.1).

Mechanism 1 would be the failure of DNA–protein tethering due partially or completely to the quality of DNA to which it was tethering (i.e., the DNA being too "slippery" by virtue of damage).

In mechanism 2, there would be failure of binding sites on proteins: the clastogen may have bound to and altered the stereology of the DNA-binding site, or the accessory protein-enzyme-binding sites so that normal DNA is inappropriately released. (The last of these is not relevant to topoisomerases.)

In mechanism 3, substances may simply position themselves in the binding sites, and thus reduce the function of the tether without affecting the structure of either of the DNA or the enzyme/accessory tethering site. Whether or not these agents must bind to the major or minor grooves of the DNA duplex or intercalate between the bases is unclear, and it is seems likely that the various noncovalently binding clastogens act at different sites (see Sections 3.1 and 4.2).

Assumptions of the suggestion include:

i. Single- and double-strand breaks can be created variously by unraveling, synthesis and repair "anywhere, anytime" in the cell cycle [11].
ii. Different accessory tether proteins (or none in the case of topoisomerases) may be involved in different types of DNA synthesis and different types of DNA repairs.

The essential points of this theory [48] are:
(i) That the DNA strand breaks, which underlie the separation of DNA ends, are those created by physiological genomic processes.
(ii) That the separation of ends is due to induced failure of function of proteins which hold the DNA strands in place during the brief period of existence of the physiological breaks.

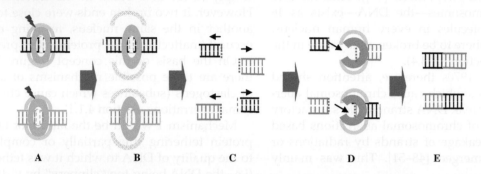

A. Clastogens (   ), see Section 4.1.1, impinge on the tether proteins for genome-related processes (Section A2.3.3) which physiologically cause *double* strand breaks. The possible processes include those of topoisomerase II unraveling and synthesis

B. The functions of the tether proteins are reduced so that their attachments to the DNA and other proteins of the complex fail.

C. With the additional influence of normal mechanical distractive forces (◄ ►) on the DNA, the DNA ends separate and are left bare.

D. These bare ends encounter ligase (   ) and fresh tether protein molecules.

E. The ligases ligate the ends of DNA, but often illicitly (inappropriate nonhomologous end-joining), resulting in chromosomal arms in continuity with arms of other chromosomes.

FIGURE 5.11   The tether drop theory of chromosomal aberrations for double-strand physiological breaks.
Note: Tether drops of physiological single strand breaks in combination with double strand breaks can explain chromosomal formations such as tri-radials.

iii. DNA synthesized on damaged template DNA by translesional synthesis is often poorly complementary to its template strand, and this is more likely to dissociate from the same template strand if "tether drop" occurs.

iv. The individual DNA ends resulting from repair-created single-strand breaks can rejoin with ends resulting from repair-created double-strand breaks.

v. (Related to the appearances of minutes in cytogenetic preparations): separated fragments of single strands of DNA can "hairpin" and bind to itself and subsequently condense to form "minutes" at metaphase.

On the basis of the morphology of chromosomal aberrations and these assumptions, it is possible to describe the basic and complex chromosomal aberrations, including duplications, inversions, and chromosomal stickiness, as is indicated in Figures 5.8 and 5.9.

By considering the variety in proteins at all the levels of their structure, it is possible to explain additional features of chromosomal aberrations, including the existence of clasto-gens which do not react with DNA, variations in effects according to species, the necessary proximity of chromosomal fibers/DNA strands allowing for "exchanges," and "G2 aberrations" (see Ref. [48]).

5.3.4 Unstable Chromosomal/Karyotypic Abnormalities

(a) "Karyo-Instability" Not "Aneuploidy": The Distinction Between Stable and Unstable Genomic Abnormalities in Cell Populations

"Ploidy" was used in 1894 in a German lan-guage work by Strasburger and copied without change in an English translation in 1908 [53]. The condition of abnormally low numbers in a cell was called "hypoploidy" and high numbers are referred to as "hyperploidy."

"Aneuploidy" came to be used for any fixed abnormality in numbers—other than multiples of the number in gametes—of chromosomes in the cells of an individual. A medical example is the 45 chromosomes in all cells of patients with Turner's syndrome (45XO).

"Aneuploidy" was subsequently applied to tumor cell populations, but often without the distinction being made between the "stable" (fixed) and "unstable" (cell-to-cell variable) con-ditions. "Stable" aneuploidy refers to cell popu-lations in which all cells have exactly the same abnormality in their chromosome number, and which does not change with cell division as in germ-line disorders. "Unstable aneuploidy," on the other hand, refers to populations in which the constituent cells have different abnor-malities, and new abnormalities are liable to appear with each new cell division, as clearly described by Hansemann (see Section 2.4.1). Another point is that germ-line "aneuploidy" is not associated with structural abnormalities in chromosomes (such as partial deletions, inver-sions, amplifications, etc.). In contrast, these kinds of chromosomal abnormalities are almost invariable in tumor cell populations, espe-cially malignant ones. Because of this, the term "karyo-instability"—which can be qualified as "numerical" and/or "structural" if required—is appropriate (Figure 5.12).

(b) Near-Universality of Hyperploidy and Karyo-Instability in Tumor Cell Populations

These abnormalities of chromosomes are seen not only in cases of human tumors, but also in viable tumor cell lines. Thus in the American Type Tissue Collection [54], most cell lines range between triploid and tetraploid (i.e., between 69 and 92). The chromosomes in HeLa cells notably range from 70 to 164.

In all these tumor cell populations, hyper-ploid cells are much more common than hypoploid cells. It is generally assumed that the hypoploid forms are either dying or vegeta-tive cells (see Chapters 11 and 12). The prolif-erative components of tumor cell populations are thought to be the cells with ploidies in the range $2n$–$6n$. Extremely hyperploid cells ($6n$ and above) are thought to be viable, but like-wise incapable of division. Almost all the cell lines which are near diploid are in fact hemat-opoietic malignancies (mainly leukemias and lymphomas) [54].

As already mentioned, the chromosomes in tumor cells with numerical instability are usu-ally also structurally abnormal. Many chromo-somes in tumor cells are larger or smaller than those in normal cells. The chromosomes may break up. Those parts without centromeres are liable to be lost from the relevant daughter cell. Both chromosomes of a pair may go to one daughter cell (i.e., asymmetric distribution at anaphase).

These observations imply that the tumor cell population, on average, is hyperploid but is not replicating its cells exactly with each cell division. In fact, most malignant tumor cell populations are constantly generating new, vari-able, hyperploid cells. This constantly changing hyperploidy occurs because of repetitions of

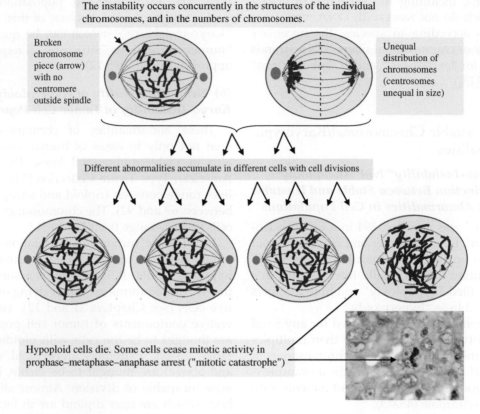

FIGURE 5.12   The main known principles of karyo-instability (unstable aneuploidy).

asymmetric distributions of chromosomes. After each of these asymmetric divisions, the cells with the surplus of chromosomes tend to survive and the chromosome-deficient cells usually either die or remain vegetative (Section 5.3.4).

5.3.5 Mechanisms of Karyo-Instability

Because of its nature, karyo-instability is assumed to be due to event(s) in a genomic element(s) which is/are necessary for preservation of mitotic and chromosomal stability, or due to event(s) which generate(s) intracellular conditions which indirectly disturb mechanisms for preservation of the same [55,56].

At present it seems that genomic events in a large number of candidate genomic elements could be responsible, although perhaps, only one need be affected in each case. The directly relevant particular gene(s) could be for a variety of proteins including those:

  i. involved in condensation of chromatin—perhaps especially of centromeric chromatin;
 ii. involved in binding to spindle fibers (kinetochore proteins);
iii. for the tubulin of the spindle fibers themselves;
 iv. for a possible pathogenetic neoprotein which interferes with the same processes.

It can be noted that karyo-instabilities appear also to be the presumably indirect effects of the following:

i. All proteins which are involved in regulation of checkpoints in the cell cycle [57] and cellular responses to DNA damage [58]
ii. Chromatin remodeling factors [59]
iii. Telomerase [60].

5.3.6 Imperfect Correlations Between Chromosomal Abnormalities and Carcinogenesis

In some experimental models, both "permanent-small number" chromosomal lesions and karyo-instability can result from carcinogens, including irradiation and alkylating agents, but also from noncarcinogens, such as caffeine [61].

5.4 OTHER KINDS OF GENOMIC INSTABILITY IN TUMORS

5.4.1 "Microsatellite" Instabilities

(a) Nature of the Phenomenon

"Microsatellite instabilities" refers to the fact that in certain conditions, there is defective correction of errors created during synthesis in the numbers of oligonucleotide repeats making up the microsatellites (for oligonucleotide repeat sequences, see Section 5.4.1. Most frequently, the actual "mis-repair" is the failure to excise or replace copies of cytosine–adenine repeats in DNA which had been amplified or deleted at the synthetic site. The effect is that the cells in the tumor cell population have different microsatellite lengths compared to the nontumor cells in the body—hence the term "microsatellite instability" [62].

(b) Microsatellite Instability in Colorectal Carcinoma

The origins of names of the genes involved are complex [63]. In colorectal tumors, the relevant repair genes are known as *MSH2*, *MLH1*, *MSH6*, and *PMS2*. Both alleles must be inactivated before expression of the protein fails. The disorder can be diagnosed:

i. by demonstrating different patterns of sizes of microsatellites in the tumor versus the patient's normal cells [64];
ii. by using immunohistochemistry, showing that the tumor cells do not have a mismatch repair protein [65,66].

The latter test is much cheaper than the former, but is susceptible to issues in immunohistochemical studies [66], as also discussed in Section 9.6.

Using these methods, it has been found [67,68] that:

i. Approximately 85% of colorectal tumors show no abnormality of mismatch repair proteins.
ii. Approximately 12% show abnormal mismatch repair protein expression without detectable loss of alleles of relevant genes. These are considered due to hypermethylation of the promoter of the *MLH1* gene.
iii. The remaining cases have a germ-line deficiency in an allele of one of the relevant genes (Lynch syndrome).

(c) Unexplained Aspects of Tandem Nucleotide Repeat Variabilities in Cancer

i. The repeats are not parts of genes, so that the mechanism of the genomic lesion and its biological effects in the tumor cell population are unclear.
ii. Existence of other nucleotide repeat disorders which are not associated with cancer.

Two or three nucleotide repeat mutations are characteristics of "expansion of repeats," which are associated with certain "degenerative" conditions (e.g., FXTAS [69]). These conditions are not associated with any marked predisposition to any cancer type or to cancer generally.

iii. In patients with Lynch syndrome, only a few kinds of cells have increased liability to tumors, while at the same time, all cells in the body carry the same germ-line genomic event.

5.4.2 Irregular and Possibly Partial Endo-Reduplication

The possibility that episodic DNA synthesis (other than during DNA repairs) may occur without the cell proceeding to mitosis and cell division has been frequently overlooked in tumor cell biology [70,71]. Thus, if in S phase of the cell cycle, the DNA were to be partly replicated but not proceed to chromosomal assembly (prophase), then a partially hyperploid/hyperchromatic cell (between $2n$ and $4n$) persisting in the interdivision period of the cell cycle (Section A1.3.4) would result.

5.4.3 Other Mechanisms and Considerations

(a) Inappropriate Meiotic Crossing-Over as a Kind of Genomic Instability

Before metaphase 1 of meiosis (Section A1.1.1), the homologous parental DNA duplex strands are aligned and more or less at exactly the same point in their DNA sequence. By the action of particular enzymes (recombinases), each duplex is broken and the ends swapped. That is to say, the broken end of the proximal arm of each chromatid is joined to the broken end of the distal piece of the other. Chromosome breaks and rejoinings are common in malignant tumor cell populations. Whether or not the meiotic mechanisms might be activated in cancers is unclear [72].

(b) Inappropriate Gene Rearrangements in Lymphocytes as Physiological, Limited Kind of Somatic Cell Genomic Instability

When "naïve" lymphocytes are exposed to antigens (often in association with presentation by macrophages), there is first a proliferative reaction, in which each new lymphocyte produces randomly a different antibody-type protein to express on its surface. The different proteins are produced by randomly reshuffling parts of the few hundred or so VDJ genes. This "reshuffling" involves breaking and rejoining of DNA duplexes, but only in these genes. All other genes in the lymphocytes are not affected.

The second phase is selection of the lymphocytes exhibiting a protein which binds to the antigen and stimulus of this cell to form a clone.

Whether or not the gene rearrangements—which may be associated with fusion genes (see Section 9.5.1)—of malignant tumor cell populations might be due to an inappropriate activation of these mechanisms in nonlymphoid cells is unclear [73].

(c) Abnormalities in the Nuclear "Matrix"

An abnormality of the nuclear matrix might cause a generalized failure of genetic integrity—and hence a form of genetic instability [74].

(d) Tetraploidy and Subtraction of Chromosomes

This form of genomic instability [75] depends on a supposed general tendency of cultured cells to become tetraploid. A population of cells which are tetraploid in the nondividing period could arise by:

i. Binucleation with subsequent fusion of nuclei, without cell division.
ii. A complete genome replication to metaphase without telophase, i.e., dissolution of chromosomes at metaphase without nuclear division, also called "endomitosis" [76].

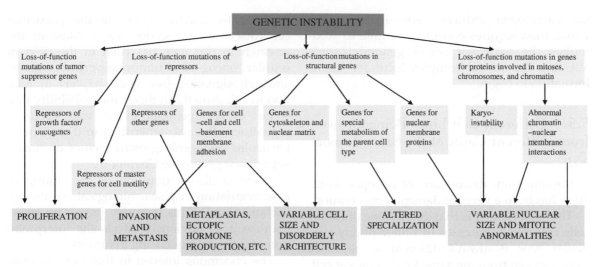

FIGURE 5.13 Potential for genomic instability to contribute to the diversity of abnormalities in tumor cell populations.

iii. "Complete endo-reduplication," i.e., complete S phase without progression to M phase.

Occasional losses of chromosomes could then occur in later cell divisions, without replenishment by further reduplication or other.

5.5 POTENTIAL EFFECTS OF THE GENOMIC INSTABILITIES IN TUMOR CELL POPULATIONS

5.5.1 Unstable Genomic Abnormalities as a Source of Variabilities and Heterogeneities

Variabilities and heterogeneities in normal cell populations are mentioned in Section A1.2.5 and in tumor cell populations in Section 2.4 (see Ref. [77]). At the present time, there seems little explanation of these aspects of the complexities of tumors other than genomic instabilities (Figure 5.13). As discussed in Chapter 2, the features of the tumor types (the "phenotypes" of the parent kinds of cells) are much too complex

to be explained by "one mechanism fits all" tumors. Again, stable genomic lesions ("mutations" in the original sense) do not explain the complexities of tumors, because stable mutations should produce stable phenotypes in descendant populations, which, as shown in Chapter 8, may suit many "benign" tumor types, but is not the case in the common malignant tumors.

5.5.2 As the Mechanisms of Tumor Progression

This issue is discussed in Section 8.4. Its mechanism(s) were little discussed until the new era of chromosome research in the 1960s, based on metaphase squash preparations. Numerous authors have made this suggestion [78].

5.5.3 As a Mechanism of Delays in Carcinogenesis

It has been noted (Section 4.1.2) that in carcinogenesis, almost always, multiple applications of carcinogen overly long periods of time is necessary. A mechanism of this could be that the carcinogen induces genomic instability, which then requires considerable time to accumulate the particular set of genomic events affecting the critical complex locus for tumor formation to begin.

5.5.4 Different Cell Lines Being Grown from Different Cases of the Same Tumor Type

Missing from discussions of genomic instability has been a very fundamental observation.

This observation is that every clone of tumor cells cultured from cases of tumor of any particular type is always different to all other clones grown from the same kind of parent cell. While each case of tumor begins with essentially the same morphological features of its cells, it (each case of tumor) is capable of producing different clones of cells in itself, and these clones are different from the clones developing from all other cases of tumor.

The most prominent example is the HeLa cell, which was grown from a case of carcinoma of the uterine cervix [79]. This line of cells has never been grown from any other case of carcinoma of the cervix or any other organ. All the established cell lines derive from single cases of tumor, and not one of them is like any other. This phenomenon suggests a heterogenizing process in all cases of tumor so that the cell lines from them are all different from one another.

5.6 ABNORMAL REGULATION OF GENES IN TUMOR CELL GENOMES

5.6.1 General

Interest in the regulation of genes, rather than just the numbers of copies of genes, has increased along with the realization that many features of tumors—excess growth and other features—might result from derepression of normally inactive genes in the particular cell (discussed in Section 8.1.1). Most of the research in this area has been in the field of cellular oncogenes, tumor suppressor genes, and cell signaling (see next chapter) in relation to cell growth. In this is the possibility that some "driver" genomic events may occur in regulators of genes rather than in the exons of particular oncogenic growth factors or tumor suppressor genes (see Chapter 6).

There is also a question of the timing of gene regulation in phasic biological events (see Section A2.2.1). This is a very important issue, because it so obviously affects the interpretation of genome sequencing in tumors.

The enormous interest in this topic is indicated by the current databases of gene expression patterns [80–83].

An account of studies of the various factors influencing protein-coding gene expression is given in Chapter 9. There is too little information on mechanisms of control of noncoding genes, especially "long RNA genes" for comment here.

5.6.2 Cancer Epigenetics

The general suggestion of possible epigenetic regulation of normal protein-coding gene expression is discussed in Section A2.2.4. Without a clear conception of the normal significance of chemical modifications of DNA for gene regulation, it is difficult to assess the literature of epigenetic changes in tumor cell populations [84,85]. In cancer, it has been suggested that the alterations on the nucleotides and histones alter their interrelationships in the primary chromatin formation (the nucleosome), leading to the inactivation of genomic material, especially if the affected DNA contains a promoter of a tumor suppressor gene [86]. In tumors, however, the overall patterns which have been detected are of demethylation of repeat sequences (Section A2.2.6), pericentromeric DNA, and in genes studied individually.

In general, several issues can be identified as follows [87]:

i. The changes in cancer could represent "hijacking" of DNA methylation, which is necessary for normal mammalian embryogenesis.
ii. There is often more hypomethylation than hypermethylation of DNA during carcinogenesis, leading to a net decrease in the genomic 5-methylcytosine content.
iii. The exact methylation changes between different cases of the same type of cancer are not the same.
iv. While the transcription-silencing role of DNA hypermethylation at promoters of many tumor-suppressor genes is clear, the biological effects of cancer-linked hypomethylation of genomic DNA are less well understood.
v. Widespread DNA hypomethylation might contribute to destabilizing chromosomal integrity.

5.7 GENOMIC LESIONS POTENTIALLY INDUCIBLE BY VIRUSES AND OTHER AGENTS

The general principles of viral infections were described in Section 4.5. For viruses, that pathogenetic step between interaction with a cellular target and the induction of genomic events is complicated because:

i. The whole living cell and its metabolism is the target of the virus,
ii. The virus itself contains genomic material.

Nevertheless, there are several possible mechanisms by which viruses alter the host genome as described in the next sections (Figure 5.14).

5.7.1 Insertion of Viral Genes into the Genome

This may have pro-tumor effects in several ways.

(a) Expression of a Viral Protein with Growth-Promoting Properties

The main suggested mechanism of tumor formation by viral DNA in cells is that the protein products of the viral DNA have pro-tumorous effects on the cell cycle. Viral oncogene products may directly enhance proliferation of cells (see Ref. [138] in Chapter 4), as has been shown for the Kaposi's sarcoma–associated herpes virus [88].

(b) Expression of Viral Protein Which Inhibits a Tumor Suppressor Gene

Viral gene products may inactivate host tumor suppressor genes, such as *RB* and *P53* (see Section 5.3) [89].

(c) Loss of Function Genomic Event in a Host Gene: "Common Insertion Sites"

The position of the viral DNA might inactivate host tumor suppressor genes, e.g., by separating the exons from the promoter.

Of importance here is that insertion of viral genome may occur preferentially at specific "hot-spots," also known as "common insertion sites." This concept means in essence that certain viruses show preferential sites of insertion of viral nucleic acid into the host genome [90]. The existence of these sites, however, has proved difficult to identify and they remain controversial [91,92].

(d) Secondary "Hot-Spots" of Genomic Events

There are reports that specific viral infections can cause genomic-event "hot-spots" in infected cell lines which are not directly associated with the inserted viral genome [93]. These preferences for sites might be mediated by viral proteins. If the preferred sites relate to any pro-growth genomic element, they could serve as a tumorigenic mechanism of the virus.

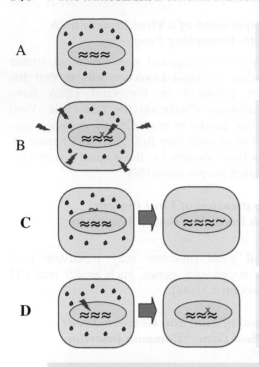

A. A continuing population of virions in the cytoplasm produces non-DNA pro-growth molecules/growth factors.

B. A continuing population of virions in the cytoplasm produces factor(s)which allows other carcinogens (⤺) to enter cell/genome compartment and cause oncogenic genomic events.

C. A transient population of virions causes viral DNA to become inserted into host genome, with oncogenic effects

D. A transient population of virions produces proteins which cause a pro-growth genomic event in host genome, which continues after the viral infection has finished ("hit and run" model).

x = somatic genomic event caused by exogenous agent, including possibly viral protein(s). It is not viral DNA, but a lesion induced in the host DNA.

If the genomic event is in a genomic element supporting genomic stability, then the cell's descendants will show genomic *instability*.

FIGURE 5.14   Some possible genomic mechanisms in viral carcinogenesis.

5.7.2 Viral Infections Producing Genomic Transfection Events Between Genomes of Adjacent Host Cells

There are several possible mechanisms to account for tumors in which viral DNA is present in tumor cells, but no transcription or other positional or other genomic mechanism of action is apparent (i.e., the DNA is "passenger") (see Ref. [138]) [94].

Such events might be at the level of nucleotide sequences or at the level of chromosomal aberrations (Sections 5.2 and 5.3).

One mechanism suggested has been that the viral particle in a nontumorous but infected cell

takes up a "cellular oncogene" and selectively transfects that gained function human gene to another human cell which is thus made tumorous. It would then be unnecessary for the viral DNA in tumor cell genome to be transcribed.

5.7.3 Genomic Abnormalities Associated with Other Known Etiological Agents

(a) Helicobacter pylori

Gastric infection with this bacterium is associated with gastric carcinoma and other tumors (see Section 4.6.1). Some evidence for direct effects of its products on the host genome has been reported [95].

(b) Parasites

Genomic effects of parasite infestations are only beginning to be studied [96].

(c) Solid-Phase Carcinogens

How these agents induce lesions in the genome is discussed in Section 4.8.

References

[1] Friedberg EC, Walker GC, Siede W, et al. DNA repair and mutagenesis (2nd ed.). Washington, DC: ASM Press; 2006. p 614.

[2] In Ref. [1]. pp. 73–5.

[3] Horsfall Jr. FL. Current concepts of cancer. Can Med Assoc J 1963;89:1224–9.

[4] D'Andrea AD. DNA repair pathways and human cancer. In: Mendelsohn J, Howley PM, Israel MA, editors. The molecular basis of cancer (3rd ed.). Saunders/Elsevier; 2008. pp. 39–55.

[5] Klaunig JE, Wang Z, Pu X, et al. Oxidative stress and oxidative damage in chemical carcinogenesis. Toxicol Appl Pharmacol 2011;254(2):86–99.

[6] Voulgaridou GP, Anestopoulos I, Franco R, et al. DNA damage induced by endogenous aldehydes: current state of knowledge. Mutat Res 2011;711(1-2):13–27.

[7] Durham TR, Snow ET. Metal ions and carcinogenesis. EXS 2006;96:97–130.

[8] IARC Working Group on Evaluation of Carcinogenic Risks to Humans. Arsenic, metals, fibres and dusts, IARC Monographs, No 100c, Lyons, 2012.

[9] In Ref. [1]. pp. 107–461.

[10] Iyama T, Wilson III DM. DNA repair mechanisms in dividing and non-dividing cells. DNA Repair (Amst) 2013;12(8):620–36.

[11] Essers J, Vermeulen W, Houtsmuller AB. DNA damage repair: anytime, anywhere? Curr Opin Cell Biol 2006;18(3):240–6.

[12] Vermeulen W, Fousteri M. Mammalian transcription-coupled excision repair. Cold Spring Harb Perspect Biol 2013;5(8):a012625.

[13] Brosh Jr. RM. DNA helicases involved in DNA repair and their roles in cancer. Nat Rev Cancer 2013;13(8):542–58.

[14] DePamphilis ML, editor. DNA replication and human disease. Cold Spring Harbor, NY: CSH Press; 2006.

[15] Lehninger AL, Nelson DL, Cox MM, editors. Biochemistry. New York, NY: Worth Publishers; 1970. p. 670.

[16] Lehmann AR, Niimi A, Ogi T, et al. Translesion synthesis: Y-family polymerases and the polymerase switch. DNA Repair (Amst) 2007;6(7):891–9.

[17] Hönigsmann H, Tanew A, Brücke J, Ortel B. Photosensitizing compounds in the treatment of psoriasis. Ciba Found Symp 1989;146:159–67.

[18] Rubin H. Synergistic mechanisms in carcinogenesis by polycyclic aromatic hydrocarbons and by tobacco smoke: a bio-historical perspective with updates. Carcinogenesis 2001;22(12):1903–30.

[19] Immunochemistry Technologies, LLC. Material data safety sheet for Hoechst Stain at ~200 ug/mL. Available at: http://www.biogenesis.com.tw/DataBank/1011120Apoptosis%20Research/Cellular%20imaging%20tools/DNA%20dyes/Hoechst_MSDS.pdf.

[20] Laskin DL, editor. Oxidative/nitrosative stress and disease. Hoboken, NJ: Wiley-Blackwell; 2010.

[21] Preedy V. Aging: oxidative stress and dietary antioxidants. San Diego, CA: Academic Press/Elsevier; 2014.

[22] Bignold LP. Carcinogen-induced impairment of enzymes for replicative fidelity of DNA and the initiation of tumors. Carcinogenesis 2004;25(3):299–307.

[23] Speyer JF. Mutagenic DNA polymerase. Biochem Biophys Res Commun 1965;21(1):6–8.

[24] Loeb LA, Kunkel TA. Fidelity of DNA synthesis. Annu Rev Biochem 1982;51:429–57.

[25] Salnikow K, Zhitkovich A. Genetic and epigenetic mechanisms in metal carcinogenesis and cocarcinogenesis: nickel, arsenic, and chromium. Chem Res Toxicol 2008;21(1):28–44.

[26] Kitchin KT, Wallace K. The role of protein binding of trivalent arsenicals in arsenic carcinogenesis and toxicity. J Inorg Biochem 2008;102(3):532–9.

[27] Pagès V, Fuchs RP. How DNA lesions are turned into mutations within cells? Oncogene 2002;21(58):8957–66.

[28] In Ref. [1]. pp. 523–5.

[29] Lea IA, Jackson MA, Li X, et al. Genetic pathways and mutation profiles of human cancers: site- and exposure-specific patterns. Carcinogenesis 2007;28(9):1851–8.

[30] Hollstein M, Moriya M, Grollman AP, et al. Analysis of TP53 mutation spectra reveals the fingerprint of the potent environmental carcinogen, aristolochic acid. Mutat Res 2013;753(1):41–9.

[31] Georgakilas AG, O'Neill P, Stewart RD. Induction and repair of clustered DNA lesions: what do we know so far? Radiat Res 2013;180(1):100–9.

[32] Loeb LA, Springgate CF, Battula N. Errors in DNA replication as a basis of malignant changes. Cancer Res 1974;34(9):2311–21.

[33] Cheng KC, Loeb LA. Genomic instability and tumor progression: mechanistic considerations. Adv Cancer Res 1993;60:121–56.

[34] Loeb LA. A mutator phenotype in cancer. Cancer Res 2001;61(8):3230–9.

[35] Beckman RA, Loeb LA. Genetic instability in cancer: theory and experiment. Semin Cancer Biol 2005;15(6):423–35.

[36] Preston BD, Albertson TM, Herr AJ. DNA replication fidelity and cancer. Semin Cancer Biol 2010;20(5):281–93.

[37] Lange SS, Takata K, Wood RD. DNA polymerases and cancer. Nat Rev Cancer 2011;11(2):96–110.

[38] Campbell LL, editor. Cancer cytogenetics: methods and protocols (2nd ed.). New York, NY: Humana Press/Springer; 2011.

[39] McDonnell SK, Riska SM, Klee EW, et al. Experimental designs for array comparative genomic hybridization technology. Cytogenet Genome Res 2013;139(4):250–7.

[40] Strachan T, Reed A. Human molecular genetics (4th ed.). New York, NY: Garland Science; 2011. p. 52.

[41] Lea DE. Actions of radiations on living cells (2nd ed.). Cambridge, UK: Cambridge University Press; 1962. pp. 189–281.

[42] Savage JR. Classification and relationships of induced chromosomal structural changes. J Med Genet 1976;13(2):103–22.

[43] Revell SH. A new hypothesis for the interpretation of chromatid aberrations, and its relevance to theories for the mode of action of chemical agents. Ann NY Acad Sci 1958;68(3):802–7.

[44] Shaffer LG, Tommerup N. ISCN 2005: An international system for human cytogenetic nomenclature: recommendations of the international standing committee on human cytogenetic nomenclature. Basel: Karger; 2005. pp. 88–95.

[45] Brothman AR, Persons DL, Shaffer LG. Nomenclature evolution: changes in the ISCN from the 2005 to the 2009 edition. Cytogenet Genome Res. 2009;127(1):1–4.

[46] Lima de Faria A. One hundred years of chromosome research. Dordrecht: Kluwer; 2003. pp. 13–15; 169–76.

[47] Gall JG. Chromosome structure and the C-value paradox. J cell Biol 1981;91:3s–14s.

[48] Bignold LP. Mechanisms of clastogen-induced chromosomal aberrations: a critical review and description of a model based on failures of tethering of DNA strand ends to strand-breaking enzymes. Mutat Res 2009;681(2–3):271–98.

[49] Roukos V, Misteli T. The biogenesis of chromosome translocations. Nat Cell Biol 2014;16(4):293–300.

[50] Williams SE, Fuchs E. Oriented divisions, fate decisions. Curr Opin Cell Biol 2013;25(6):749–58.

[51] Guo Y, Kim C, Mao Y. New insights into the mechanism for chromosome alignment in metaphase. Int Rev Cell Mol Biol 2013;303:237–62.

[52] Mailand N, Gibbs-Seymour I, Bekker-Jensen S. Regulation of PCNA–protein interactions for genome stability. Nat Rev Mol Cell Biol 2013;14(5):269–82.

[53] Strasburger E, Noll F, Schenck H, Karsten G. Lehrbuch der Botanik für Hochschulen. A textbook of botany. [Lang WH, Trans.]. London: Macmillan; 1908.

[54] American Type Culture Collection. Product list. Available at: http://www.atcc.org/products/.

[55] Jefford CE, Irminger-Finger I. Mechanisms of chromosome instability in cancers. Crit Rev Oncol Hematol 2006;59(1):1–14.

[56] Roschke AV, Rozenblum E. Multi-layered cancer chromosomal instability phenotype. Front Oncol 2013;3:302.

[57] Enders GH. Mammalian interphase cdks: dispensable master regulators of the cell cycle. Genes Cancer 2012;3(11-12):614–8.

[58] Voineagu I, Freudenreich CH, Mirkin SM. Checkpoint responses to unusual structures formed by DNA repeats. Mol Carcinog 2009;48(4):309–18.

[59] Papamichos-Chronakis M, Peterson CL. Chromatin and the genome integrity network. Nat Rev Genet 2013;14(1):62–75.

[60] Genescà A, Pampalona J, Frías C, et al. Role of telomere dysfunction in genetic intratumor diversity. Adv Cancer Res 2011;112:11–41.

[61] D'Ambrosio SM. Evaluation of the genotoxicity data on caffeine. Regul Toxicol Pharmacol 1994;19(3):243–81.

[62] Geiersbach KB, Samowitz WS. Microsatellite instability and colorectal cancer. Arch Pathol Lab Med 2011;135:1269–77.

[63] Boland CR, Goel A. Microsatellite instability in colorectal cancer. Gastroenterology 2010;138(6):2073–87.

[64] Vilar E, Gruber SB. Microsatellite instability in colorectal cancer—the stable evidence. Nat Rev Clin Oncol 2010;7(3):153–62.

[65] Shia J, Ellis NA, Klimstra DS. The utility of immunohistochemical detection of DNA mismatch repair gene proteins. Virchows Arch 2004;445(5):431–41.

[66] Hall G, Clarkson A, Shi A, et al. Immunohistochemistry for PMS2 and MSH6 alone can replace a four antibody panel for mismatch repair deficiency screening in colorectal adenocarcinoma. Pathology 2010;42(5):409–13.

[67] Jass JR. hMLH1 and hMSH2 immunostaining in colorectal cancer. Gut 2000;47(2):315–6.

[68] De la Chapelle A, Hampel H. Clinical relevance of microsatellite instability in colorectal cancer. J Clin Oncol 2010;28:3380–7.

[69] Hagerman P. Fragile X-associated tremor/ataxia syndrome (FXTAS): pathology and mechanisms. Acta Neuropathol 2013;126(1):1–19.

[70] Fox DT, Duronio RJ. Endoreplication and polyploidy: insights into development and disease. Development 2013;140(1):3–12.

[71] Walen KH. The origin of transformed cells. Studies of spontaneous and induced cell transformation in cell cultures from marsupials, a snail, and human amniocytes. Cancer Genet Cytogenet 2002;133(1):45–54.

[72] Erenpreisa J, Cragg MS. MOS, aneuploidy and the ploidy cycle of cancer cells. Oncogene 2010;29(40):5447–51.

[73] Radman M, Jeggo P, Wagner R. Chromosomal rearrangement and carcinogenesis. Mutat Res 1982;98(3):249–64.

[74] Lever E, Sheer D. The role of nuclear organization in cancer. J Pathol 2010;220(2):114–25.

[75] Ganem NJ, Storchova Z, Pellman D. Tetraploidy, aneuploidy and cancer. Curr Opin Genet Dev 2007;17(2):157–62.

[76] Zielke N, Edgar BA, DePamphilis ML. Endoreduplication. Cold Spring Harb Perspect Biol 2013;5(1):a012948.

[77] Bignold LP. Variation, "evolution", immortality and genetic instabilities in tumor cells. Cancer Lett 2007;253(2):155–69.

[78] Nowell PC. Tumor progression: a brief historical perspective. Semin Cancer Biol 2002;12(4):261–6.

[79] Masters JR. HeLa cells 50 years on: the good, the bad and the ugly. Nature Reviews Cancer 2002;2:315–9.

[80] Expression Atlas. Differential and baseline expression. Available at: http://www.ebi.ac.uk/gxa/home;jsessio nid=6B3C2DFAD084ABD8679916CDBDE93087.

[81] Human Genome Center. Homepage. Available at: http://hgc.jp/english.

[82] Korea Research Institute of Bioscience and Biotechnology. Gene expression database service. Available at: http://medical-genome.kribb.re.kr/GENT/.

[83] National Center for Biotechnology Information. Gene expression omnibus. Available at: http://www.ncbi. nlm.nih.gov/geo/.

[84] Baylin SB. Epigenetics and cancer Mendelsohn J, Howley PM, Israel MA, editors. The molecular basis of cancer (3rd ed.). Philadelphia, PA: Saunders/Elsevier; 2008. pp. 57–65.

[85] Tannock IF, Hill RP, Bristow RG, et al. editors. The basic science of oncology (5th ed.). New York, NY: McGraw Hill; 2013. pp. 20–5.

[86] Tollefsbol T, editor. Cancer epigenetics. Boca Raton, FL: CRC Press; 2009.

[87] Ehrlich M. Cancer-linked DNA hypomethylation and its relationship to hypermethylation. Curr Top Microbiol Immunol 2006;310:251–74.

[88] Bais C, Santomasso B, Coso O, et al. G-protein-coupled receptor of Kaposi's sarcoma-associated herpesvirus is a viral oncogene and angiogenesis activator. Nature 1998;391(6662):86–9.

[89] Levine AJ. The common mechanisms of transformation by the small DNA tumor viruses: the inactivation of tumor suppressor gene products: p53. Virology 2009;384(2):285–93.

[90] Uren AG, Kool J, Berns A, et al. Retroviral insertional mutagenesis: past, present and future. Oncogene 2005;24(52):7656–72.

[91] Knight S, Collins M, Takeuchi Y. Insertional mutagenesis by retroviral vectors: current concepts and methods of analysis. Curr Gene Ther 2013;13(3):211–27.

[92] Wu X, Luke BT, Burgess SM. Redefining the common insertion site. Virology 2006;344(2):292–5.

[93] McDougall JK. "Hit and run" transformation leading to carcinogenesis. Dev Biol (Basel) 2001;106:77–82. discussion 82–3, 143-60].

[94] McGivern DR, Lemon SM. Virus-specific mechanisms of carcinogenesis in hepatitis C virus associated liver cancer. Oncogene 2011;30(17):1969–83.

[95] Touati E. When bacteria become mutagenic and carcinogenic: lessons from *H. pylori*. Mutat Res 2010;703(1):66–70.

[96] Biron DG, Loxdale HD. Host–parasite molecular crosstalk during the manipulative process of a host by its parasite. J Exp Biol 2013;216(Pt 1):148–60.

6

Etio-Pathogenesis III: Growth, Invasion, and Metastasis

L.P. Bignold: Principles of Tumors.
DOI: http://dx.doi.org/10.1016/B978-0-12-801565-0.00006-8

In Section 4.1.3 of Chapter 4, the pathogenesis of tumors caused by exogenous agents was divided into five steps:

(A) entry into, or possibly contact with, a carcinogen and a cell; (B) action on a biochemical structure of the cell; (C) disturbance in function in that structure; (D) changes in the cell caused by those disturbances; and (E) in different cells, variable phenotypic changes in many different combinations such that the tumors are classifiable into types.

So far as step D is concerned, the clinically significant changes of tumor cells are excessive accumulation of cells, invasion of adjacent tissues, and metastasis to distant sites in the body. These features have therefore been the main aspects of tumors investigated at the biochemical and molecular biological levels.

The other features of tumor cells—especially abnormal specialization, abnormal spatial arrangements, loss of cyto-structural regularity, mitotic abnormalities (see Section 8.1)—are phenotypic features of tumor cells. However, they are almost exclusively used in histopathological diagnosis and are of little direct clinical significance. Because of this, and also because they are difficult topics for experimental investigation, these latter features of tumors have not received as much attention as the three main features. Possible models for some of these features are given in Appendix 1.

This chapter therefore deals with what is known of growth, invasion, and metastasis according to step (D) in the overall scheme of etiopathogenesis. In general, the changes are thought to result from one or more genomic events in specific genes. The main aspects considered here are:

i. What genomic events in which genes may be important for growth, invasion, and metastasis?
ii. What are the cell biological mechanisms of these main features?

For this, Sections 6.1–6.5 discuss the biology and genomics of excess growth. Sections 6.6 and 6.7 deal with what is known of the pathogenesis of invasion and metastasis respectively.

Step (E)—how different combinations of features, i.e., the tumor types may occur together in specific kinds of cells—is discussed after detailing the feature types, as given in Section 8.5.

6.1 GENERAL PERSPECTIVE OF CELL ACCUMULATION

6.1.1 Terminology

"Growth" is used in two senses:

i. "Growth" is applied to the rate of cell division. This is the meaning which is used in the remainder of this chapter.
ii. "Growth" is also used clinically for the rate of accumulation of cells. This is important, because abnormal accumulation is the only cellular feature which is present in all tumors (see Sections 2.2, 8.1.7, and A1.3).

However, accumulation of cells depends on excess production over losses [1–4]. Thus "accumulation" can be considered as "rate of cell division minus rate of losses."

Both processes are thought to be relevant to many types of tumors. This aspect of tumor cell populations is discussed in Section 8.3.2.

6.1.2 Control of Cell Replication

Since the nineteenth century, it has been recognized that in multicellular organisms, the different types of cells function cooperatively for the existence of the whole organism. This means that the growth of individual kinds of cells must be controlled.

i. In adult life, the body is clearly a collection of mutually interdependent organ systems, in which to degrees, parenchymal cellular activity (see Section A1.1.2) in each organ is controlled by one or more others.

ii. In embryonic development, the developing parts appear not to be so much interdependent for their final function, but coordinated in their immediate developmental processes, so that in the ultimate adult the right numbers of cells in the right places are achieved.

iii. In nontumor proliferative lesions too, regrowths of cells after loss appear to be coordinated and controlled. For example in wound healing, when the new epidermis has completed growing on underlying connective tissue, epidermal and dermal cells cease proliferating and return to normal functional states at the same time.

How these phenomena of control and coordination occur is a major topic of developmental biology and pathology. Up until the 1970s, most studies concerned pro-growth factors in serum and tissues [5] as well as the phenomena of contact inhibition (see Section A1.3.1c) and of transformation *in vitro* (Section 14.2.3). With the beginnings of molecular biology, rapid progress was made by interrelated findings in cell biology, virology, and molecular biology ([6–8]; see Section 5.7). Many proteins and the genes which code them were discovered and investigated for their effects on intracellular biochemical events. This is particularly required for the study of membrane receptors, because proteins are not lipid soluble and hence cannot enter cells passively.

The results of these studies revealed that many of these growth factors are participants in cascades of activations of proteins in the cytoplasm of cells (see Section 6.3). The genes coding for the growth factors were called "oncogenes."

The next several sections provide an introduction to growth factors and oncogenes (Table 6.1). How oncogenes may be variants of the normal growth control genes or enhancers of normal pro-growth mechanisms in cells are considered in Section 6.3.

TABLE 6.1 Major General Categories of Growth Factors

1. Exogenous

Viruses

Products of a small number of bacterial species

Mitogens used for certain experimental purposes, e.g., lectins such as Concavalin-A

2. Endogenous

Endocrine growth factors:
For example, somatostatin, trophic hormones from the anterior pituitary gland or primary sex organs

Paracrine growth factors:
Inflammatory and immune-cell cytokines, e.g., interleukins and cytokines

Autocrine and intracrine growth factors:
Receptors and cascade proteins of cell signaling

3. Specifically for cells in culture: supporting viability

Vitamins, nutrients, basement membrane proteins, adhesion molecules, and other plasma proteins.

Unspecified substances provided by "feeder" layers of other cells used in some *in vitro* techniques

6.2 GENERAL ASPECTS OF GROWTH FACTORS AND ONCOGENES

6.2.1 Discovery and Definitions of "Growth Factors"

Various definitions of growth factors are available. In the early days of culturing cells *in vitro*, the term was applied to substances in serum which promote the growth of cells [9]. This definition included substances such as hormones, which have what is called an "endocrine" effect [10] (see Figure 6.1). Subsequently, it was found that cultures of cells may release growth factors into their medium, and that extracts of tissues, including tumors, contain pro-growth factors. These factors which do not circulate in the blood, but are produced locally, were considered to have a "paracrine effect" [10]. A common test for all these factors was the ability of the substance to induce growth in cells which had previously cultured *in vitro*, but held inactive, e.g., by starving them of serum [11].

Subsequently, additional mechanisms by which substances promote the growth of cells have been discovered [10]. These are:

i. The cell produces the factor, which acts on that cell alone without being expressed on the surface (intracrine action).

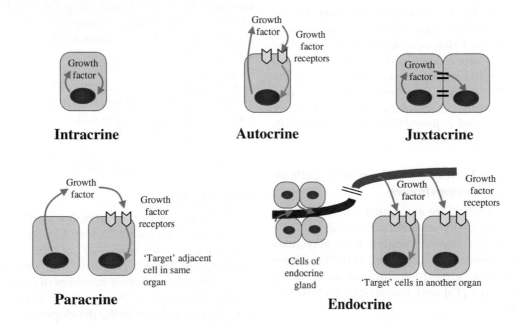

Intracrine   **Autocrine**   **Juxtacrine**

Paracrine   **Endocrine**

Note: "Exocrine" refers to secretion of materials outside the "milieu interieu" e.g., sweat from sweat glands and pancreatic enzymes from the pancreatic acinar cells. It is not usually considered a mechanism of control of cells.

FIGURE 6.1 The "crines" of cell control.
"Eccrine" is used when the secretion by these cells is without structural change in the cell.
"Apocrine" is used when the secretion occurs by separation of specific parts of the cell.
"Holocrine" refers to secretion by release of disintegrating whole cells into the duct of the gland.
None of these exocrine secretions are considered a mechanism of control of cells in the body.

ii. The cell secretes the factor to its surface, but the factor then binds to the cell's own receptors (autocrine action).

iii. Cells produce factors which they then pass only to cells with which they are in plasma membrane-to-plasma membrane contact (juxtacrine action).

"Growth factors" can thus be defined as any substances which induce growth in other cells [12]. Growth factors generally include three physiological categories.

(a) Growth Factors Relating to Immune Responses

These are usually called cytokines, and are proteins made by certain immune and nonimmune cells which have effects on the immune system. Some cytokines stimulate the immune system and others slow it down. They can also be made in the laboratory and used to help the body fight cancer, infections, and other diseases. Examples of cytokines are interleukins, interferons, and colony-stimulating factors (filgrastim, sargramostim) [13].

(b) Growth Factors for Nonimmune Cells in Nontumor Circumstances

These proteins are associated with nontumor growth, especially in embryonic and fetal life [14], although altered versions may play roles in certain aspects of tumor cell behavior. These nontumorous growth factors include nerve growth factor [15] and insulin-like growth factor [16].

(c) Oncogenic Growth Factors

These are, by definition, the pro-growth protein products of oncogenes (see next section). They may have a pathologic effect though being expressed in increased amounts, or by being physiologically abnormal through a genomic event in the exons of the genes causing gain of function.

6.2.2 Discovery and Definitions of Oncogenes

This section describes some of the main oncogenes which have been associated with tumors. Some code for signaling proteins is as described in the next section. The functions of many other cancer-associated genes have not yet been clarified. The abnormalities in all these genes which have been documented in cases of human tumors are discussed in Section 9.3.

(a) Discovery

Since the first studies of oncogenic viruses (Section 4.5.1), many oncogenes have been found by extracting fragments of DNA from human tumors and transfecting them individually into cells *in vitro*. Oncogenes give relative independence of the cultured cells from external growth factors in the culture medium. The occurrence of immortalization or malignant transformation of cells (Section 15.4) is taken as strong additional evidence that the DNA contains an oncogene (Figure 6.2).

Oncogenes are usually classified as Mendelian "dominant," because only a genomic event in one allele may be required for the full effect.

(b) Definition

At the present time, there are multiple definitions of "oncogene." The Genetics Home Reference Guide web site of the National Institutes of Health [17] lists four definitions, which describe different numbers of genes.

One definition is "An oncogene is a mutated gene that contributes to the development of a cancer. In their normal, un-mutated state, oncogenes are called proto-oncogenes, and they play roles in the regulation of cell division" (Figure 6.3).

This definition only refers to about 40 known oncogenes for which protein-coding proto-oncogenes have been established [18,19].

Another definition in Ref. [17], however, is: "A gene, one or more forms of which is associated with cancer." This definition may include

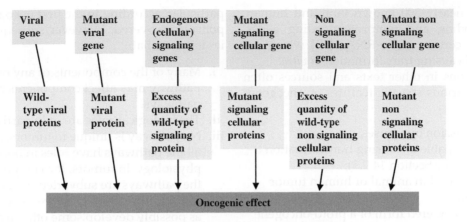

FIGURE 6.2 Conceptual overview of the possible mechanisms involving the genome, the protein products, and the oncogenic effect.

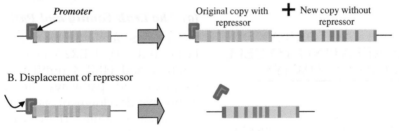

A. Repressed gene amplified with own promoter

B. Displacement of repressor

C. Gain-of-function—or gain of increased function—event in promoter or exon

D. Amplification or transposition of an inactive gene without promoter to promoter region of another gene (Second gene may or not be inactivated)

E. Loss of insulator between normally inactive GFO gene and upstream active gene

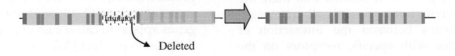

Dark shading of gene indicates inactivity

FIGURE 6.3 Some genomic events which may activate a growth factor–oncogene (GFO).

all of the nearly 300 known viral and human, protein-coding and nonprotein-coding genes, including certain micro-RNAs [20] which are capable of causing transformation *in vitro*.

Definitions in other texts and sources often mention various specific features of oncogenes such as:

i. transmission by viruses;
ii. being capable of causing transformation *in vitro* (see Section 14.2.3);
iii. being found in animal or human tumor genomes;
iv. being an altered form of a proto-oncogene involved in normal cell control;
v. providing relative independence from growth factors in the culture medium;
vi. being Mendelian dominant.

6.3 GROWTH FACTORS AND ONCOGENES RELATING TO CELL SIGNALING PATHWAYS

6.3.1 General

This topic includes not only the identities of the messengers which pass between cells to coordinate growth, but how the messengers, once contacting a cell, induce that cell to divide [21–24]. The general concept of a receptor has been described in Section 1.2.3. The principle is similar to those for steroid hormone-induced cellular responses [25] and the activation of immunocytes [26]. In both of these, a series of enzymes are sequentially activated to produce a final functional product, in principle similar to the blood-clotting system [27].

By the early 1990s, it seemed that there are several different, but apparently discrete, signaling pathways between the interaction of growth factor with specific receptors on the external surface of the cell, and the instigation of cell division for specific biological phenomena (Figure 6.4). The pathways are often named according to one of its components, and in

particular, according to the first oncogenic component discovered. However subsequent studies indicate that:

i. Many of the components of any one pathway may affect components of other pathways.
ii. The biological roles are only partly specific.
iii. No pathway is unique to tumors. Almost all the pathways have roles in normal physiology. In tumors, the components of the pathways are subverted by mutations, causing the cells to grow excessively, as well as possibly develop some other features.

6.3.2 Pathways Involving Tyrosine Kinase Activations

(a) The ErbB Family and Pathways

This large family of tyrosine kinases includes EGFR (ErbB-1), HER2/c-neu (ErbB-2), HER 3 (ErbB-3), and HER 4 (ErbB-4) [21,28,29]. These receptors and pathways are associated with embryonic development, as well as regeneration and repair in adults. The cascade of activations involves first proximate intermediaries (associated with, or activated by, the cytoplasmic domain of the receptor). Different receptors do not necessarily activate different proximate intermediaries.

Next in the pathways are "deep" or "free cytosolic" intermediaries.

Finally, there are effector proteins, which are mainly transcription factors for proteins which initiate the cell cycle (i.e., cause G0→G1).

i. Proximate intermediaries include Src, Shc, Grb2, SOS, P13K, and membrane RAS proteins (Kras, Mras, and Nras). Nucleotide errors at codons 12, 13, or 61 of these latter genes appear to cause gain of function of the protein product [30].

Some additional proximate intermediaries are:

• Jak, which associates mainly with the cytoplasmic domains of surface

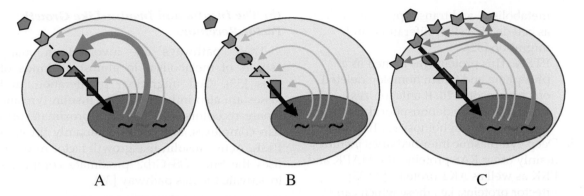

A. Excess oncogene product causes over activity of any part of the signaling pathways/networks (intracrine mechanism).

B. Mutant oncogene produces a hyper-functional variant of the gene product (intracrine mechanism).

C. Production of excess receptors for ambient stimulatory ligand (autocrine mechanism).

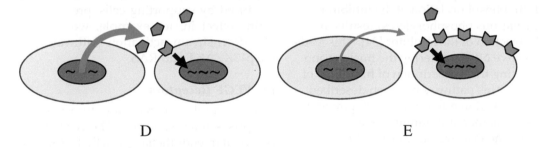

D. Excess production of ligand to stimulate adjacent cells (paracrine mechanism).

E. Induction in adjacent cells of excess receptors for the normal ligand (paracrine mechanism).

FIGURE 6.4   Principles of some mechanisms of action of growth factor–oncogenes.

receptors for the cytokines interferon, erythropoietin, and thrombopoietin [31]. Activated Jak phosphorylates tyrosines on STAT peptides, which then dimerize to become active transcription factors. This pathway therefore allows direct signaling from plasma membrane to nucleus [31].

- P13K. When activated by membrane receptor kinases, this protein activates Akt1 (protein kinase B), which was originally identified as the product of the oncogene in a retrovirus AKT8. Akt8 binds phospholipids and interacts with other pathways to participate in a variety of processes including glucose metabolism, and transcription, as well as cell death, cell proliferation, and migration [32].

- PTEN. This protein was named as a phosphate and tensin homolog deleted on chromosome 10. It antagonizes p13K activity by dephosphorylating the membrane phosphoinositol PIP3 [32].

ii. Deep cytoplasmic intermediaries include mainly other RAS proteins, the MAPK and ERK as well as AKT proteins [32,33].

iii. Effector proteins, i.e., those which can enter the nucleus and act as transcription factors.

They include:

- The family of deep intermediaries known as NFκB. These proteins are normally held inactive by specific inhibitors (IκB). Phosphorylation of the inhibitor (by various phosphorylases) results in destruction of the inhibitor [34]. This leads to active NFκBs entering the nucleus and affecting the transcription of hundreds of genes. This pathway has been described as involved in numerous developmental inflammatory and immunologic phenomena as well as cancer [35].
- MYC family of proteins. These transcription factors enhance transcription of many target genes, with correspondingly numerous biological effects associated with both proliferation and apoptotic death in different cells. Increased levels of myc protein is found in many kinds of tumors [36].
- Proteins of the activating protein-1 family, including *jun* and *fos*. These proteins are involved in several cellular processes, including growth, specialization, and apoptosis [37].

The effector proteins which stimulate growth ultimately act by increasing the synthesis of cyclin-dependent kinases [38]. Cyclins are discussed in Section A1.3.6.

(b) The Insulin and Insulin-Like Growth Factor Receptors

These pathways are involved in many aspects of metabolism in almost all kinds of cells [33]. Cell migration, proliferation, and losses are also influenced. The insulin tyrosine kinase receptors use IRS–Grb2 proximate protein complexes, and then acts mainly through P13K. Active insulin-like growth factor receptor uses the Shc–SOS–Grb2 proximate complexes to activate the Ras pathway [39].

(c) c-Met

This receptor is widely expressed in embryonic cells, and in adults, is mainly expressed on epithelial cells [40]. It thus has roles in wound healing and organ regeneration. The activating ligand is hepatocyte growth factor which is produced by supporting cells, providing a paracrine effect in, for example, wound healing. The downstream activation pathways include the p13K/AKT and RAS.

(d) VEGF Receptors

This family of receptor tyrosine kinases comprises three subtypes. The relevant ligand is vascular endothelial growth factor (VEGF). Its main effect is the production of new capillary blood vessels ("angiogenesis" see Section 2.3.3c), along with proliferative effects on other cells especially inflammatory cells [41]. After activation of the receptor kinase, downstream events include activations of p13K and Jak/STAT (see above).

(e) WNT-Beta-Catenin Pathway

Wnt proteins are bound to extracellular matrix proteins. Defects in Wnt protein are associated with a variety of developmental defects [42,43]. When it binds to the Frizzled membrane receptor and/or the LDL receptor proteins, complexes form including Axin, which is a tyrosine kinase. This kinase degrades beta-catenin. Catenins are ubiquitous proteins which are particularly associated with actin–cell membrane complexes [44] (see also Section 2.9.3). Their concentration in the cytoplasm is normally limited by a proteolytic complex dependent on the Frizzled membrane protein (see above). Wnt, when it binds to Frizzled, prevents the formation of the proteolytic complex, so that beta-catenin concentration increases in the cytoplasm. The protein enters the nucleus where it interferes with the transcription of multiple genes and has pro-growth effects.

The pathway is also important in nervous system pathologies [45] and immunologic phenomena [46].

6.3.3 Pathways Involving Serine–Threonine Kinase Activity

These enzymes transfer phosphate groups between the named amino acids.

(a) TGF-Beta

This pathway has a wide range of activities in relation to development of vertebrates, as well as wound healing and immune responses in adults. Binding of TGF-beta to appropriate receptors in the membrane causes phosphorylation of intracytoplasmic proteins known as Smads [47]. Complexes of Smads 2, 3, and 4 enter the nucleus and have both enhancing and suppressive effects on transcription of various enzymes.

(b) Other Serine–Threonine Kinases

These include many of the cytoplasmic intermediaries, e.g., mos/Raf kinases and MAPKK kinases. These proteins have similar structures to many tyrosine kinases and have no function which is apparently particular to their chemical activity [48].

6.3.4 Guanosine-Phosphate-Dependent Activity

These are called G-protein-coupled receptors and are used in a large number of nontumor physiological regulatory mechanisms, including for hormones and neurotransmitters [49]. They are membrane proteins with seven transmembrane α-helical units. There are approximately 800 of these proteins encoded in the genome. They are switches which are "on" when they bind guanosine triphosphate, and "off" when they bind guanosine diphosphate.

6.3.5 Other Pathways

(a) Pathways Involving Other Enzymatic Activities in Membrane Receptors

They include lipid kinases (e.g., p13 kinase), lipid phosphatases (SHIP, PTEN), phospholipases (CPLA2), and nucleotide cyclases [50].

(b) Hedgehog-Patched Pathway

This pathway does not involve an enzymatic activity. Hedgehog is the ligand which activates Patched, upon which the latter releases a normally bound protein known as Gli into the cytoplasm. Gli is then free to enter the nucleus and act as a transcription-affecting factor [51].

(c) Notch Protein

Following binding by specific proteins known as Delta and Jagged, this transmembrane protein suffers proteolysis outside and inside the cell membrane. An intracellular fragment acts directly as a transcription factor in the nucleus [52].

6.4 OTHER ASPECTS OF CELL SIGNALING, GROWTH FACTORS, AND ONCOGENES

6.4.1 Acquisition of Oncogenic Effect by Proto-Oncogenes

By convention, all acquisitions of the oncogenic capability involving human proto-oncogenes are referred to as "activations." This applies to all kinds of proto-oncogenes,

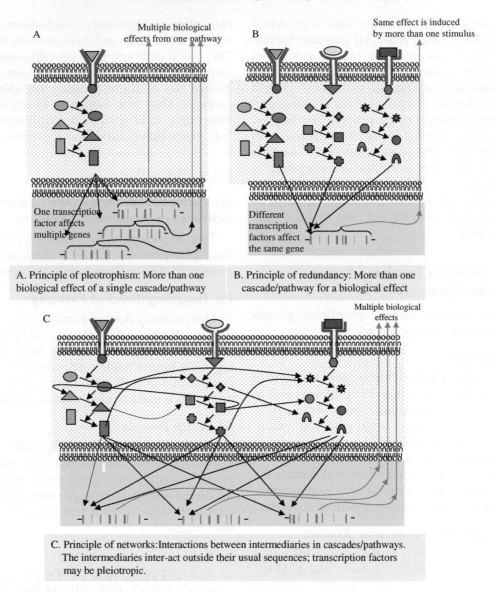

A. Principle of pleotrophism: More than one biological effect of a single cascade/pathway

B. Principle of redundancy: More than one cascade/pathway for a biological effect

C. Principle of networks:Interactions between intermediaries in cascades/pathways. The intermediaries inter-act outside their usual sequences; transcription factors may be pleiotropic.

FIGURE 6.5   Pleotrophisms, redundancies, and networks of oncogenes.

whether they are surface receptors, intermediaries in signaling pathways, or normally secreted products of cells (see next section).

Because these are vitally important to the assessments of genomic abnormalities in human tumors, they are discussed in Chapter 9.

6.4.2 Pleotrophisms and Redundancies in Intracellular Signaling Events

At the present time, there appear to be certain difficulties in determining the relative significance of each cell signaling system (Figure 6.5). The issue is especially important to the development of anticancer drugs (see Sections 13.1 and 13.2). The difficulties derive from the following issues.

(a) Pleiotropy(ism)

In this phenomenon, one agent or pathway leads to more than one biological effect [53,54]. Antagonistic pleiotrophy refers to situations in which the effects are contrary to one another, e.g., lead to proliferation, but also apoptosis [54].

(b) Redundancy

This is the situation in which more than one pathway leads to activation of the same final effector event (usually induction of transcription of genes) [55,56]. It is usually identified as occurring when a whole pathway is ablated—as in an appropriate "knockout" mouse—without any phenotypic change. The difficulty consists of whether or not blocking one signaling pathway might be frustrated by the unaffected efficacy of another.

6.4.3 Networks in Intracellular Signaling Events

Originally, this term was used by developmental biologists to indicate supposed intercellular signals which might control the coordinated behavior of embryonic populations of cells.

In relation to the intracellular cascades of enzyme activations (previous section), large numbers of reports have indicated that there can be interactions across the supposed dividing lines between the major pathways [57]. This implies possible nonspecificity of each activated enzyme for a downstream substrate.

Such cross-pathway actions are often referred to as "cross-talk." Some authors have suggested that the actual system of cell undergoing G0→G1 is through networks of oncogene and enzyme activities, not discrete pathways. The networks together constitute a "circuitry."

Some additional proposed signaling networks include the target of rapamycin network [58].

As the number of identified "cross-talk" reactions increases, and the "circuits" are found to be more detailed, the difficulty consists of whether or not inactivating any particular enzyme might frustrated by bypass circuits in the cell. This area is investigated through mathematical analysis and modeling from gene expression patterns revealed by microarrays (see, for example, Ref. [59]).

6.4.4 The Diffusion-Feasibility Aspect of the Pathway Concept

There is potentially a difficulty in pathway and network concepts of cell signaling, relating to whether or not specific cascades of interactions of proteins are possible in "the viscous soup that is present in the cytoplasm and nucleus" was is alluded to by Weinberg [60].

One concept of how the pathways may be kept separate from one another by "in-complexing" of the enzymes of the particular cascade. In this model, all the enzymes of a given pathway are held in multiprotein complexes, which are held together by "scaffold" and "adaptor" proteins. Thus the enzymes of different pathways may interact ("cross-talk"), but only when permitted by the quaternary structures of the complexes [61].

6.4.5 Oncogene "Cooperation" May Be Necessary for Tumor Formation

As mentioned in Sections 2.6.2 and 4.5, after the discovery of increased growth by simple amplification of gene copy numbers, it was found that transformation of normal cells sometimes requires transfection with more than one gene [62]. The idea was put forward especially to explain the transitions from adenoma to carcinoma of the colon (see Section A2.3.3). In fact, with microarray studies of gene expressions in cancer, it is possible that multiple oncogenes must "cooperate" to induce tumor formation [62,63].

6.4.6 Oncogene "Addiction"

This term was introduced to describe the fact that in some tumor cell populations, activating genomic events in relation to one individual oncogene may play the major role in tumor cell growth. This provides the rationale for offering "single target" ("selective") drug therapies for patients with tumors which show the appropriate overactivity of the oncogene [64,65]. The validity of the concept is still unclear [66,67].

6.5 TUMOR SUPPRESSOR GENES

6.5.1 Discovery and Definitions of Tumor Suppressor Genes

(a) Discovery

The discovery of genes in this class began in the 1970s through investigations of hereditary predispositions to tumors (see Section 7.1). At this time, it was already known that all cells in an individual have the same chromosomal complements, and hence all cells carry a germ-line alteration through adult life. Two major questions arose:

i. Why does the germ-line mutation affect only one or a few kinds of cells in the body?
ii. Why does only a tiny proportion of susceptible kind(s) of cells become tumorous?

Historically, Bauer in 1928 [68] had thought that in familial polyposis coli (FAP), the inherited predisposition permits the cells to undergo abnormal hyperplastic responses to environmental stimuli. Subsequently, numerous authors made similar "two-hit" suggestion without specifying whether the germ-line and somatic hits were on the same genes or on different genes (see Ref. [69]).

In the 1960s, a few authors [70–72] suggested that the somatic hit could be on the second allele of the affected germ-line gene. However,

it was Knudson [73,74] who provided the data. In his work, he compared the incidences of spontaneous retinoblastomas with the incidences in predisposed individuals, and showed that the somatic genomic events are liable to be on the other allele of the affected gene. By conventional thinking at the time, the gene was said to be recessive, because loss of both alleles are required for the somatic phenotypic effect.

Knudson's model was confirmed by later cytogenetic analyses of retinoblastoma tumors [75]. They proved the existence of genes which, through apparent loss of function through loss of both alleles, can allow tumors to form in otherwise normal cells.

(b) Definitions

With this background, the original meaning of "tumor suppressor gene" (see also Section 1.2.2) was an autosomal recessive gene which was assumed to be responsible in biology and pathology for holding the growth of normal cells in check (Section A1.3) [76]. Loss of the suppressor allows uncontrolled cell division to occur.

In recent decades, however, the term has been extended to apply to all substances which have an antitumor effect by any mechanism. Thus the NIH Genetics Home Reference Guide defines these as "Genes in the body that can suppress or block the development of cancer" [77]. For these genes, the recessive classification of the gene is not included in the concept, because one abnormal copy alone may be enough to predispose the individual to tumor [76,78,79].

6.5.2 Loci for the Germ-Line and Somatic Events for Each Predisposition

(a) Sites of Germ-Line Genomic Events

At the time that the two-hit model of tumor formation became accepted, methods for locating genomic events on the chromosomes were being developed. Tumor suppressor genes were initially located by cytogenetic analysis of rare human tumors (see above), but many additional methods have since been developed [80].

The genes for most of the other high penetrance predispositions have been mapped and are in different sites on the chromosomes.

(b) Sites of Somatic Genomic Events

Besides Knudson's bi-allelic model [73,74] (Figure 6.6) for tumor suppressor genes, it was recognized that there are at least two other possible locations of additional (somatic) "hits" as follows:

i. On the same allele as the germ-line genomic event.

In current terms, this means a genomic event which converts a "hypomorphic" allele into a "null" allele). This somatic hit could be on different exons. This sequence, were it to occur, would imply that neither genomic alteration was sufficient alone to produce the loss of function of the tumor inhibitor. According to Mendelian concepts, the gene would be "dominant," because damage to one allele could cause the phenotypic change. However, the genomic events would be "hypomorphic"—being individually unable to cause the change.

Such a sequence is open to other interpretations. For example, if the gene in question is "dominant" in this way, then it could be regarded simply as an oncogene. In this way, the predisposition would by through the germ-line carrying a hypomorphic genomic ("incomplete mutation") which then can be converted to a normo-morphic event (complete mutation) by somatic alteration to another part of the protein product (see sickle cell anemia model, Section 2.6.1, Hb S + C).

Further, in some individuals, the gene may be represented by multiple alleles (poly-alleleism) but only one of these alleles is active (see Section A2.3.1). This would mean that for many individuals (rather than the few in the well-known inherited predisposition

syndromes—see Section 7.1), loss of the one good allele will constitute in a pro-growth event. This concept of "dominant" (i.e., haplo-insufficient) tumor suppressor genes has been widely considered [76,78,79].

ii. The second hit might be on another gene in the genome. In current terms, this means a tumor resulting from additive effects of a new hypomorphic mutation on the effects of a germ-line hypomorphic mutation. This could be a markedly increased functional significance if the second protein forms a complex with the product of the first gene.

6.5.3 Possibilities of Roles of Tumor Suppressor Genes in Spontaneous Tumors

Because of the small likelihood of random inactivating genomic events occurring on the two wild-type alleles in a nonpredisposed individual, these genes are thought less likely than oncogenes to be involved in spontaneous tumors in humans. Nevertheless, several tumor types show high rates of genomic events in their genomes, as is discussed in Chapters 5 and 9.

6.5.4 Mechanisms of Action of Tumor Suppressor Genes

An early idea was that the protein products of tumor suppressor genes would act against growth factor oncogenes, and hence were called "anti-oncogenes" [81]. Most of these genes have therefore been studied in relation to proteins of cell signaling pathways (see Section 6.5). Other mechanisms might include general slowing in the progress of the cell cycle, or promotion of apoptosis (in various senses—see Chapter 12), or even have the effect of reducing contact inhibition (see Section A1.3.1c).

Using the more recent, wider definition of tumor suppressor genes, mechanisms can include genes for proteins which effectively

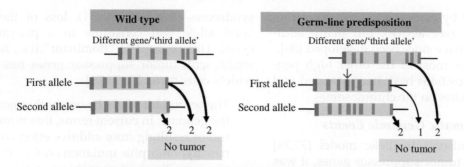

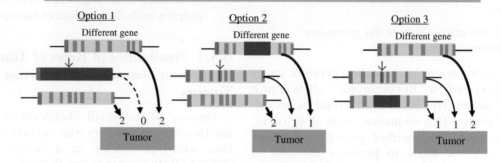

These options necessarily invoke concepts of morphism and trialleleism.

The diagram represents each allele having three possible notional values: wild type (dark green, "2"), hypomorphic (red-green, "1"), and amorphic (red, "0"). To prevent a tumor, the cell must have a total of five or more values.

In the predisposition, the mutant TSG allele could be hypomorphic, but not amorphic.

In option 1, the second hit is suggested to render the mutant allele amorphic, so that tumor forms.

In option 2, there is suggested to be another tumor suppressor gene involved ('trialleleism'). If the second hit renders one allele of this gene hypo- or amorphic, the cell becomes tumorous. The 'other gene' could be an entirely different gene, or conceivably another copy of the same gene.

In option 3, the second hit renders hypomorphic, the second allele of the germ-line-affected TSG, so that the cell becomes tumorous.

Option 3 is now generally referred to as Knudson's model. Knudson's model does not account for the fact that the predisposition has effect only in particular parent kinds of cells, although all cells in the body carry the germ-line mutation (see section 7.1.2)

FIGURE 6.6   Options for second hits in relation to tumor suppression [73,74].

repair DNA. This is because faulty DNA repair may play a role in tumor formation (see Sections 2.7 and 2.8; [82]). It also includes mechanisms such as repression of cell division when the DNA of the dividing cell is damaged. This is assumed to be "tumor suppressive" on the basis that DNA damage is a pro-mutational, and hence tumorigenic, event. In this scenario,

TABLE 6.2   Some of the Mainly Known Tumor Suppressor Genes

Tumor suppressor gene	Location	Function of protein product	Inherited disorders (see also in Chapter 7)
APC	5q21	Signaling through adhesion molecules	Familial adenomatosis polyposis
BRCA1	17q21	Repair double strand breaks	Familial breast cancer
BRCA2	13q12.3	Repair double strand breaks	Familial breast cancer
CDK4	12q14.1	Cyclin-dependent kinase 4	Familial melanoma
DCC	18q21.3	Transmembrane receptor (netrin 1)	Esophageal squamous cell carcinoma
			Colorectal adenocarcinoma
MSH2	2p22-p21	DNA mismatch repair	Mismatch repair syndromes esp Lynch syndrome
MLH1	3p21.3	DNA mismatch repair	Mismatch repair syndromes esp Lynch syndrome
NF1	17q11.2	Catalysis of RAS inactivation	Neurofibromatosis type 1
NF2	22q12.2	Linkage of cell membrane to cytoskeleton	Neurofibromatosis type 2
PTEN	10q23	Regulated cell survival	Cowden's syndrome
P53	17p13	Cell cycle regulation, apoptosis	Li–Fraumeni syndrome
RB1	13q13	Cell cycle regulation	Retinoblastoma (less commonly, sarcomas)
VHL	3p26-p25	Regulation of transcription elongation	Von Hippel Lindau syndrome
WT1	11p13	Transcriptional regulation	Wilms' tumor

Note: This list does not include the many rare and complex subtypes and variants of the main syndromes. These related syndromes may be due to co-mutation of genes proximate to the tumor suppressor gene.

the tumor suppressor is identified as a genomic event suppressor.

Studies of the protein products of tumor suppressor genes have revealed unexpected complexities. As the following examples show, it is not clear how many of them have their normal effect of inhibiting cell division.

It is also worth remembering that the situation may be more complicated than simply loss of function of a single protein according to the sickle cell anemia model (Section 2.6.1). Especially, the clinical features of the syndromes of inherited predisposition (Section 7.1) appear to be too complex to be explained by dysfunction or loss of one protein product of one gene.

6.5.5 Particular Tumor Suppressor Genes

This section deals with the genes and gene products in known germ-line predispositions to tumors (Table 6.2). The clinical and pathological features of the syndromes are discussed in Chapter 7.

(a) The RB Gene and Normal Protein Product

This gene is located on the short arm of chromosome 13. It comprises 27 exons and spans 180 kb of DNA. The exons encode a nuclear protein of m.w. 105 kd known as p105-Rb, or Rb-protein. Normal Rb protein binds and thereby inhibits certain transcription factors. During G1, cyclin-dependent kinases phosphorylate Rb protein, inactivating it, so that transcription of the other proteins relevant to cell division can proceed [83]. Inactivating mutations in the Rb protein removes this control of cell division. Rb protein also binds other cellular proteins and is reported to have many other effects on cellular metabolism [84].

(b) Adenomatous Polyposis Coli (APC) Gene

This gene for FAP is located on the long arm of chromosome 5. It spans 110 kb of DNA, of which 9 kb is divided among 15 exons [85]. The main function of the normal protein is thought to bind components of the Wnt-beta catenin signaling pathway (see Section 6.3.2e) [86,87] but also has roles in maintaining chromosomal stability [88].

(c) Mismatch Repair Genes All Causing Lynch Syndrome (See Section 7.5.1)

There are seven genes of which four account for greater than 90% of cases.

MSH2 is located on 2p, comprising 80 kb, with 16 exons, with a spliced transcript of 3 kb [89].

MLH1 is located on 3p, comprising 58 kb in 19 exons with a spliced transcript of 2.5 kb [90].

MSH6 is located on 2p, comprising 24 kb in 10 exons, with a spliced transcript of 4 kb [91].

PMS6 (PMS2L4) is located on 7q, comprising 350 kb, with 10 kb in 7 exons [92].

All proteins are intranuclear and are thought to be participants in protein complexes which repair errors in numbers of repeats in DNA comprising tandem repeat short sequences.

(d) BRCA1 and -2

BRCA1 is located at 17q and consists of 80 kb, of which approximately 6 kb is distributed in 24 exons. Several isoforms through alternative splicing have been described. The protein product locates in the nucleus and is ubiquitously expressed with the highest levels in ovaries, testis, and thymus. Its main function is thought to be as part of the BASC protein complex, a super complex of BRCA1-associated proteins involved in DNA repairs and may also be a transcription factor [93].

BRCA2 is located on 13q, comprising approximately 84 kb, including 10 kb in 27 exons. The protein locates in the nucleus and has roles in DNA repair, and probably in meiotic recombination [94].

(e) The WT1 (Wilms' Tumor) Genes and Normal Protein Products

WT1 is located on the short arm of chromosome 11, spanning approximately 47 kb. There are nominally 10 exons, but start codons, exons, splices, and RNA editing create the possibility of up to 36 isoforms. These seem to have overlapping but also distinct functions during embryonic development and the maintenance of organ function [95]. The protein(s) are only expressed in embryonic kidneys. It has been reported to suppress transcriptional activity of certain growth factors, such as insulin-like growth factor-2 [33]. The possible roles of WT protein or other genes/proteins in the renal tumors (nephroblastoma) as well as in complex abnormalities of Beckwith–Weidemann syndrome (BWS, see also Section 7.2.3) are unclear.

The WT2 gene is located near WT1 on 11p. Germ-line loss of WT2 causes BWS alone, and co-loss with WT1 gene causes nephroblastoma in BWS.

WT3 is located on 16q and losses of it have been associated with Wilms' tumor [96].

The functions of WT2 and WT3 proteins are unclear [97].

(f) Neurofibromins 1 and 2 (NF, Neurofibromatoses) Genes and Normal Protein Products

The NF1 gene, is located on chromosome 17q, spans approximately 280 kb of DNA of which approximately 84 kb is distributed among 60 exons (57 constitutive, 3 alternative), with approximately 4 alternative splicing variants [98]. The protein products (neurofibromins) contain about 2,800 amino acids and are expressed mainly in neural support cells such as Schwann cells. The proteins appear to function as inactivators of *ras* GTPases (see Section 6.3.2). The gene is often altered in tumors of other kinds of cells. The protein is expressed in all normal cell kinds in adults, not just Schwann cells.

The protein product has chemical similarities to cystoskeletal proteins, all of which are expressed in many other kinds of cells in normal adults [99].

NF2 gene is located on 22q and comprises approximately 120 kb, of which 1.8 kb is distributed among 17 exons and has 1 alternative splice site [100,101]. The protein (also called "merlin") is highly expressed in fetal brain but also expressed in nonneural tissues. Its functions are particularly associated with cell motility and adhesion, as well as with regulation of transcription of other genes [102].

(g) VHL

This gene is located on chromosome 3p, spanning approximately 11 kb, of which nearly 5 kb is distributed among 3 exons [103]. The full protein product is widely expressed in fetal and adult cells. It acts as an E3 ubiquitin ligase and hence is involved in intracellular protein metabolism, especially in the degradation of heat-inducible factor. The protein product also complexes with other proteins in the regulation of VEGF, as well as in the control of a variety of other cellular processes [104,105].

(h) TP53

This gene is located on 17p, spanning 19 kb of DNA, including 1.2 kb in 10 coding exons. [106]. Its protein product (p53, m.w. 53 kDa) is expressed in all cells and appears to function as an inhibitor of transcription for a variety of genes. The role of the gene is complex (Figure 6.7) especially because some of these genes appear to be growth-enhancers, while others of the inhibited genes have no clear function in tumors [107–113]. This is discussed in Section 7.2.6 (see also Figure 6.7).

The subject is enormously complex. In cases of tumor arising without a germ-line TP53 mutation (see Li–Fraumeni syndrome, Section 7.4.9), the gene is mutant in perhaps more types of tumors than any other. In the International Agency for Research on Cancer (IARC) TP53 mutations database, there are 26,325 events which have been detected over a wide range of tumor types. On the IARC web site there is a slide show which outlines the complexities of mutations in this gene [114].

(i) XP genes

These genes are Mendelian recessive, because the disease (xeroderma pigmentosum) only occurs when the individual inherits mutations in both germ-line alleles. The mutant gene products are mainly truncated versions of proteins associated with DNA repair. The gene products are not growth factors, nor are they suppressors of normal growth mechanisms.

There are seven known genes which cause variants in this disease group (see Section 7.2.9) of which four account for the majority [115–118].

The locations of these genes are XPA 9q; XPB 3q; XPC, 3p; and XPD 19q.

6.6 PATHOLOGICAL OBSERVATIONS, CELL BIOLOGY, AND POSSIBLE GENOMIC PATHOGENESIS OF INVASION

In tumor cell populations, invasion—usually in association with metastasis—is the most significant of the cell biological abnormalities [119]. Several general issues may be noted.

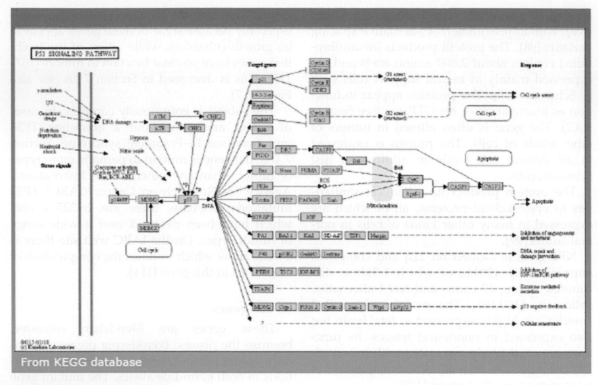

From KEGG database

FIGURE 6.7 The TP53 pathway.

6.6.1 Diversity in Invasions by Tumor Cell Populations

Tumor cells can invade various structures (Figure 6.8), and some of these invasions are commoner among some types of tumors than others. These have clinical significance but also indicate that different mechanisms may be relevant in different circumstances.

(a) Epithelium Invading Supportive Tissues

This is the commonest circumstance of invasion and has received most attention in the experimental literature. Histological studies show that malignant cells expand as a mass and that cells may be found at variable distances from the edges of the main mass. These distances can be up to millimeters, or even

centimeters. This tendency to become distributed over long distances is different in different tumor types. Most characteristically, breast cancers can show separate foci centimeters from the main mass [120], raising always the possibility of separate initial foci (see discussion of "field theory" in Section 2.9.6).

(b) Epithelial and Melanocytic Tumors Invading Other Epithelia

Invasion of cells from the ectodermal cell layer into other cell types from the same layer are less common than invasions of connective tissue. However, they are seen in Paget's disease of the nipple (Figure 6.8A) [121] and in squamous cell carcinoma of the cervix [122]. It is also seen in melanomas, when the epidermis is invaded by single melanoma cells [123] (see Figure 6.8B).

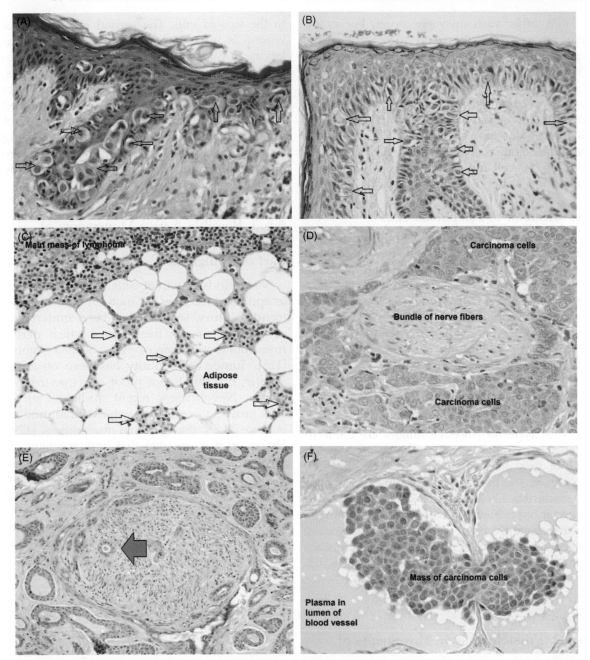

FIGURE 6.8 Tumor invasions. (A) Invasion of epidermis of nipple (Paget's disease of nipple) by breast carcinoma cells ×20. Arrows indicate some of the carcinoma cells. (B) Invasion of epidermis by malignant melanoma ×20. Arrows indicate some of the melanoma cells. (C) Lymphoma (arrows) invading adipose tissue around a lymph node. (D) Perineural (i) and intraneural (ii) invasion (arrowed) by adenoid cystic carcinoma of parotid gland ×10. (E) Squamous cell carcinoma of skin invading through a blood vessel ×20. The carcinoma cells may have gained access through the wall in continuity to a site deep to this particular histological section. Alternatively, this cluster of cells may have broken off from the main mass and after moving in the blood stream, may have impacted in an angulation of the vessel.

Few other such invasions occur. For example, carcinomas of the colon do not invade the epithelium of the small bowel or the anal canal (they may spread under the epithelial layer, but not in it). Carcinomas of the stomach do not invade into the mucosa of the esophagus or duodenum. These examples of the "trait" of invasion without tissue lysis are important, because they raise questions such as exactly what abnormalities of cell–cell adhesion between the normal cells allow tumor cells to move between them?

(c) Behavior of Lymphoma Cells

Under normal conditions, and also in certain inflammations, lymphocytes enter the epithelium of various tissues, including the intestine [124]. However, when B-cells form "germinal centers," they remain strictly localized in supportive tissues. A common aspect of tumorous lymphocytes is diffuse spread of the lymphoma cells through adipose and other tissues (Figure 6.8C) [125]. In the gastrointestinal tract, this includes invading the epithelium of the mucosa [126].

(d) Invasion Without Metastasis

Most types of invasive tumors are liable to metastasize (although some cases are excised before they have done so—see Section 10.2). However, the basal cell carcinoma of the skin characteristically does not metastasize, even when it has achieved a large size [127]. Perhaps more relevantly, basal cell carcinomas are never seen in blood vessel lumina. This emphasizes that invasion of connective tissue may be a different "trait" to invasion of vessels and hence metastasis (see Section 6.8).

6.6.2 Possible Passive Movements of Cells

In normal biology, passive sliding, movement of cells is easily appreciated in two situations. The first is of epithelial cells "sliding" from the bases of crypts in the intestinal mucosa to the tips of villi. The second is seen after a wound is inflicted on an epithelial cell population. In this, the adjacent epithelial cells "despecialize" and form a flattened sheet of cells. The margins of this sheet slides over the defect as a plate, propelled by the cells produced behind them. Individual epithelial cells do not separate from the main mass, and the sliding movement ceases when its edge meets the edge of epithelium advancing from the opposite side (contact inhibition, Section A1.3.1c).

6.6.3 Possible Active Movements of Tumor Cells: Acquisition of Motility as Part of Malignant Tumorous Change

Most kinds of parent cells (Figure 6.9), such as epithelial cells, are not motile. Since the nineteenth century, studies have been undertaken of whether or not the tumors which derive from nonmotile parent cells become motile [128]. Tumor cell motility has been observed *in vitro*, but this is not proof of its occurrence *in vivo*. Nevertheless, recent studies of tumor cells growing *in vivo* have shown that motility is more likely than passive mechanisms [129]. However, it is possible that both mechanisms might occur at the same time in the same case of tumor. The remainder of this section examines some of the possible mechanisms of cell movement (see also Section A1.5.3).

It would appear that in tumors arising from nonmotile cells, motility might be acquired in several ways as follows:

a. *The normal parent cell has an active, but constrained motile apparatus present*

In this option, motility might be considered constrained by adhesion of the cell to local structures, especially cell–cell junctions (desmosomes) and attachments via integrins, e.g., Refs. [130,131] or other adhesion proteins (see above).

Motility in tumor cells arising from normally immotile parent cells might

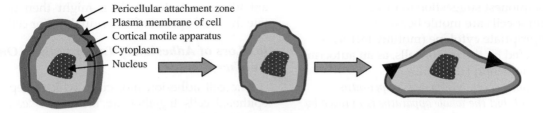

Pericellular attachment zone
Plasma membrane of cell
Cortical motile apparatus
Cytoplasm
Nucleus

A. A genomic event reduces pericellular restraints/attachments so that pre-existing active motile apparatus (green zone) is able to act.

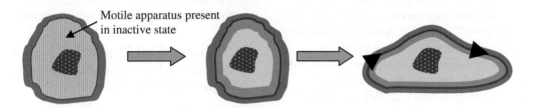

Motile apparatus present in inactive state

B. A genomic event activates motile apparatus which is previously synthesized but inactive. Changes in pericellular restraints are not defined in this theory.

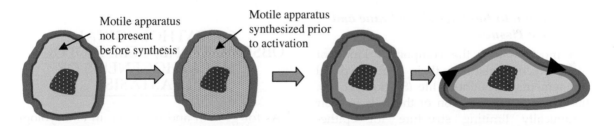

Motile apparatus not present before synthesis

Motile apparatus synthesized prior to activation

C. A genomic event induces synthesis and activation of motile apparatus. Changes in pericellular restraints are not defined in this theory

FIGURE 6.9 Principles of possible mechanisms for motility in cancer cells.

begin with expression (implying gene derepression) by tumor cells of enzymes which lyse basement membrane, thus allowing the inherent motility of cells to express itself [132,133].

b. *The normal parent cell has inactive motile apparatus present*

Essentially, in this concept, the apparatus is inactive state because of lack of an appropriate stimulus. According to the commonest suggestion in this category, tumor cells are motile because the appropriate cytokine (motility factor) is secreted by the tumor cells, as an autocrine phenomenon [134,135].

c. *The normal cell has no motile apparatus present, but the whole apparatus is induced by derepression of a "motile apparatus master gene"*

The last of these mechanisms has been less popular than others. The protein products of many signaling genes (see Section 6.4) may affect cell motility indirectly. However, there is evidence that some genes for embryonic differentiation may serve as motility-specific "master gene" [136,137]. Another possibility is that leukocyte motility mechanisms might be in some way derepressed and then be responsible for malignant motility in nonhematologic cells, e.g., Refs. [138,139].

6.6.4 Adhesion in Relation to Passive and Active Cell Movement

(a) Adhesion to Basement Membrane and Supportive Tissues

In most tissues, the component kinds of cells perfectly maintain their proper spatial arrangements throughout life (Section A1.1.3). This depends on adhesion of the cells to their anatomically limiting structure—for epithelium, the basement membrane. If the basement membrane of the epithelium were destroyed, or made less adhesive for the local cells, or the tumor cells themselves lost their adhesiveness for basement membrane structures, then invasion could occur. The adhesion molecules on cells which bind to collagen and other basement membrane components are known as "integrins" [140,141]. Their possible roles of tumorous invasions have been extensively investigated, e.g., Ref. [142]. Nevertheless, in some circumstances, nonintegrin molecules such as CD44 might also be involved [143]. Also possible is that tumor cell populations may release

antiadhesion factors, which might then facilitate the local spread of individual tumor cells.

(b) Loss of Adhesion of Tumor Cells to One Another

The cell adhesion molecules which keep the epithelial cells together are known as "cadherins" [144]. Loss of their effectiveness could contribute to the spatial irregularities in tumor cell masses (see Section 8.1.3) as well as invasion of supporting tissues, e.g., Ref. [145]. It is possible that the loss of adhesion is by tumor cells secreting relevant antiadhesion molecules, which facilitate the separation of tumor cells from each other.

6.6.5 Tumor Cell Population Movements by Differential Growth According to Gradients of Growth Factors

This mechanism of population movements is mentioned in Section A1.5.2. Whether or not it may play roles in invasion by tumor cells is unclear.

6.7 PATHOLOGICAL OBSERVATIONS, CELL BIOLOGY, AND PATHOGENESIS OF METASTASIS

As for growth and invasion, in embryological and adult tissues, there are normal "metastatic" phenomena which might correspond to metastasis of tumor cells (see Section A1.5). "Metastasis" of tumors refers to secondary growths in tissues at a distance from the primary tumor. The most important avenue of spread before the growth is "transvascular" (i.e., transit through lymphatic or blood vessels) while less common avenues are other spaces, such as the pleural or peritoneal cavities. The general principles which affect the likelihood of a metastasis having formed in a case of malignant tumor are described in Section 1.3.2. A summary is given in Figure 6.11.

This section deals with the physiological cell biological factors involved in metastasis. Discussion is restricted to transvascular spread, since this is the most important (Figures 1.1 and 1.12).

For the topic of "circulating solid tumor cells," see Section 9.7.4.

6.7.1 Tumor-Type Characteristic Patterns of Metastases

A long recognized feature in metastatic growth by tumor is often organ-specific according to tumor type as described by Paget in 1889 [146] and documented in detail by Willis in 1952 [147]. For example, carcinomas of the lung tend to metastasize to bone, brain, liver, and kidneys while melanomas metastasize to skin and mucosa of the intestinal tract, as well as the brain, liver, and kidney.

6.7.2 Excess Transplantability of Tumor Cells Compared to the Cells from Which They Arise; Tissue-Specific Antigens

In considering growth in distant sites in general, it may be noted that tumor cells grow at the site of metastasis better than do corresponding

normal cell of origin of the tumor cell. The most obvious example is growth in lymph nodes. Thus normal epithelia, by whatever preparative method, do not grow when grafted into lymph nodes of the same individual. However, many carcinomas, which derive from these cells, are capable of such growth. This point was noted in the 1890s by Hansemann [148] who described it as "an increased capacity for independent existence." In the first half of the twentieth century, the phenomenon of easier transplantability of tumors compared to normal tissues was studied, for example, by Leo Loeb [149]. The results, however, were not conclusive, probably mainly because the methods did not exclude immunological (graft-rejection) phenomena. Since then, transplantability of tumors has been studied in athymic mice, but the mechanisms are still not fully understood [150].

6.7.3 General Cell Biological Factors Affecting the Likelihood of Metastasis of a Case of Tumor

As general principles, the likelihood of metastasis having occurred before diagnosis (Figure 6.10) depends on the following.

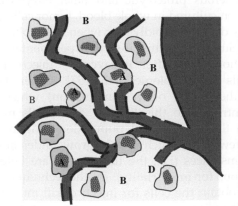

A. Inherent motility and invasiveness of tumor cells

B. Penetrability of intercellular matrix

C. Penetrability of the walls of the vessels:
   i. Density of vascularization
   ii. Penetrability of vessel walls (indicated by dashes)

D. Proportion of new vessels—these are usually more penetrable than established vessels.

FIGURE 6.10 Factors affecting invasiveness of tumors into vessels.

(a) The Inherent Invasiveness of the Tumor

In practice, the inherent invasiveness is assessed, in the surgically removed specimen, by the extent of any spread which has occurred into vessels, around nerves or to local lymph nodes or other structures. Generally, the more de-specialized is a tumor, the greater the apparent invasiveness of its cells. There is some evidence that reduced expression of adhesion molecules is associated with worse prognosis in patients [151]. Invasiveness of vessels may be a separate phenomenon from invasiveness of supportive tissues for basal cell carcinomas of the skin [127]. It is possible that the reverse may occur: some cases of carcinoma of the thyroid appear to be more likely to invade vessels than supportive tissue [152].

(b) The Quantity of Adjacent Structures Which Can Serve as Avenues for the Tumor to Spread

The availability of structures to invade is assumed to be the same for all tumors arising in any one type of tissue. However, the different organs in which tumors originate differ greatly in their vascularity. This may be relevant to lung carcinomas, which arise in the most vascular tissue in the body and metastasize readily.

Entry into local vessels is again more likely if the tumor itself produces factors which induce additional blood vessels in adjacent tissues by primary tumors. This matter of tumors inducing angiogenesis is a whole field of tumor research (see Section 2.3.3c) [153]. It may also be important to note the possibility that elaboration of angiogenic factors may occur more in some tumor types than in others [153], as might be expression of VEGF receptor.

(c) The Degree of Degradation of the Walls of the Vessels into Which the Tumor Cells May Grow

The walls of vessels are a barrier to cells growing into their lumina ("intravasation"). Degradation of intratumoral vessel walls may occur through mechanical forces associated with the growth of the tumor [154]. The damage to the vessels may include loss of endothelial cells and erosion of basement membrane [155]. Hemorrhage from these vessels may occur. The damaged vessels may allow the ingress of tumor cells more than normal vessels. These features are likely to be focal in the individual case of tumor. This means that they may not be reliably used as a prognostic factor when assessed by ordinary histopathological methods.

(d) The Period of Time During Which the Tumor Cells Have Been Near the Relevant Structures

The period of time that a tumor has been present is usually inferred from the size of the primary tumor. However, this may not be valid for all tumor types, because more rapidly growing tumors will reach a given size more quickly than a slowly growing tumor.

6.7.4 Survival in Lymph and/or Bloodstream

Once in vessels, the tumor cells must survive if they are to cause metastases. Survival in lymph, when spreading through lymphatic vessels, is not surprising since that fluid is very similar to interstitial fluid and contains few degradative enzymes (Figure 6.11). However, blood contains numerous proteolytic and other enzymes, and when cultured in whole blood, tumor cells generally die more rapidly than when in lymph.

Supporting this are the experimental observations that after intravenous injection, few cells form colonies [156,157]. Other factors (either alone or in combination) which may contribute to this loss of viable cells may be (a) the original cell population contains only a few cells which would grow under any circumstances (i.e., the tumor cells are heterogeneous for immortality); (b) the techniques used to obtain the cells for injection kill most of the cells; or (c) the sites of impaction of tumor cells is inhospitable to metastatic growths.

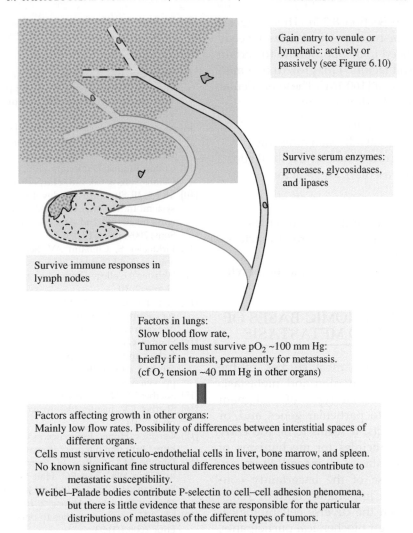

Gain entry to venule or lymphatic: actively or passively (see Figure 6.10)

Survive serum enzymes: proteases, glycosidases, and lipases

Survive immune responses in lymph nodes

Factors in lungs:
Slow blood flow rate,
Tumor cells must survive pO_2 ~100 mm Hg: briefly if in transit, permanently for metastasis.
(cf O_2 tension ~40 mm Hg in other organs)

Factors affecting growth in other organs:
Mainly low flow rates. Possibility of differences between interstitial spaces of different organs.
Cells must survive reticulo-endothelial cells in liver, bone marrow, and spleen.
No known significant fine structural differences between tissues contribute to metastatic susceptibility.
Weibel–Palade bodies contribute P-selectin to cell–cell adhesion phenomena, but there is little evidence that these are responsible for the particular distributions of metastases of the different types of tumors.

FIGURE 6.11 Factors affecting tumor cells once present in vessels and at sites of potential metastatic growth.

6.7.5 Thrombosis and the Impaction Site

Considerable effort has gone in to the investigation of factors in tumor cells which promote thrombosis. This is relevant because, when tumor cells are in the bloodstream, they are more likely to impact in a small vessel of the distant organ if they are part of a thrombus, rather than a single deformable structure. Both tissue factor [158] and platelets [159] may play roles.

6.7.6 Factors in Growth at the Metastatic Site

In principle, a cell which has already entered the small vessels at the site of primary tumor and survived in the circulation has demonstrated that it can survive in conditions other than those of the environment of the parent cell. However, different tumor types show different patterns of "successful" colonization of other tissues (see Section 8.3.3). This suggests that certain tumor-type-specific factors ("seed" qualities) may need to be appropriate to certain local factors ("soil factors") for growth of a metastatic tumor deposit [160,161]. These particular combinations may be drawn from:

a. *The impacted tumor cell*
   i. Compatible cell adhesion molecules
   ii. Adequate growth rate for the local site
   iii. Adequate vascularization-factor production
b. *The local tissue*
   i. Compatible cell adhesion molecules
   ii. Adequate available local growth factors
   iii. Adequate permissiveness of any neovascularization action of the tumor cells.

6.8 POSSIBLE GENOMIC BASES OF INVASION AND METASTASIS

At the present time, most studies of the molecular biology of invasion and metastasis usually involve particular types of malignant tumors in relation to particular genes and/or their protein products. It seems to be too early to infer general principles for mechanisms of invasion and metastasis—if indeed there are any. This is because of the uncertainty concerning the relative importance and/or range of tumors to which the diverse cell biological phenomena, and their biochemical participants, may be relevant. Valuable insights are provided in many current works, e.g., Refs. [162–165].

References

[1] Steel GG. Cell loss as a factor in the growth rate of human tumors. Eur J Cancer 1967;3(4):381–7.
[2] Shackney SE, McCormack GW, Cuchural Jr. GJ. Growth rate patterns of solid tumors and their relation to responsiveness to therapy: an analytical review. Ann Intern Med 1978;89(1):107–21.
[3] Spratt JS, Meyer JS, Spratt JA. Rates of growth of human neoplasms: parts I and II. J Surg Oncol 1995;60:137–46. 1996, 61(1):68–83.
[4] Tubiana M. Tumor cell proliferation kinetics and tumor growth rate. Acta Oncol 1989;28(1):113–21.
[5] Burgess AW. Growth factors: the beginnings. Growth Factors 1988;1(1):1–6.
[6] Weinberg RA. The biology of cancer, 2nd ed. New York, NY: Garland Science; 2014. pp. 71–130.
[7] Vogt PK. Retroviral oncogenes: a historical primer. Nat Rev Cancer 2012;12(9):639–48.
[8] Morange M. A history of molecular biology. [Cobb M, Trans]. Cambridge, MA: Harvard University Press; 2000.
[9] Collins English Dictionary (10th ed.). London: William Collins Sons & Co. Ltd/HarperCollins; 2009.
[10] Morris JF. Structure and development of the endocrine system. In: Hay ID, Wass JAH, editors. Clinical endocrine oncology (2nd ed.). Hoboken, NJ: John Wiley & Sons; 2009. pp. 3–17.
[11] Pirkmajer S, Chibalin AV. Serum starvation: caveat emptor. Am J Physiol Cell Physiol 2011;301(2):C272–9.
[12] Webster's College Dictionary. New York, NY: Random House; 2010.
[13] National Cancer Institute. NCI Dictionary of cancer terms. Available at: http://www.cancer.gov/dictionary?CdrID=46130.
[14] Sherbet GV. Introduction Sherbet GV, editor. Growth factors and their receptors in cell differentiation, cancer and cancer therapy. Amsterdam: Elsevier; 2011. pp. xvii–vix.
[15] Sherbet GV. Nerve growth factors. In: Ref. [14]. pp. 81–5.
[16] Sherbet GV. Insulin-like growth factors. In: Ref. [14]. pp. 87–104.
[17] Genetics Home Reference. Oncogene. Available at: http://ghr.nlm.nih.gov/glossary=oncogene.
[18] Pierotto MA, Frattini M, Sazzi G, et al. Oncogenes. In: Hong WK, Bast RC, Hait WN, editors. Holland–Frei cancer medicine (8th ed.). Shelton, CT: People's Medical Publishing House; 2010. pp. 68–85.
[19] Chial H. Proto-oncogenes to oncogenes to cancer. Nat Educ 2008;1(1):33.
[20] Lawrie CH. MicroRNAs as oncogenes and tumor suppressors MicroRNAs in medicine. Hoboken, NJ: Wiley; 2013. pp. 223–43.
[21] In: Ref. [6]. p. 131–230.
[22] Bradshaw R.A.Dennis EA, editors. Handbook of cell signaling, (2nd ed.) 3 vols. San Diego, CA: Academic Press, Elsevier; 2011.
[23] Hynes NE, Ingham PW, Lim WA, et al. Signalling change: signal transduction through the decades. Nat Rev Mol Cell Biol 2013;14(6):393–8.
[24] Harvey A, editor. Cancer cell signalling. Hoboken, NJ: Wiley-Blackwell; 2013.
[25] Beato M, Truss M, Chávez S. Control of transcription by steroid hormones. Ann NY Acad Sci 1996;784: 93–123.

[26] König R. Signal transduction in T lymphocytes. In: Ref. [22]. pp. 2679–98.

[27] Antovic JP, Blombäck M. Essential guide to blood coagulation. Chichester, UK: John Wiley & Sons; 2013.

[28] Köstler WJ, Yarden Y. The epidermal growth factor receptor family. In: Ref. [22]. pp. 435–41.

[29] Harvey A. Epidermal growth factor receptor family. In: Ref. [24]. pp. 1–24.

[30] Prior IA, Lewis PD, Mattos C. A comprehensive survey of Ras mutations in cancer. Cancer Res 2012;72(10):2457–67.

[31] In: Ref. [6]. pp. 202–4.

[32] In: Ref. [24]. pp. 7–9.

[33] Stephen AG, Esposito D, Bagni RK, et al. Dragging ras back in the ring. Cancer Cell 2014;25(3):272–81.

[34] Westwick JK, Schwamborn K, Mercurio F. NFκB: a key integrator of cell signaling. In: Ref. [22]. pp. 2069–76.

[35] Erstad DJ, Cusack Jr. JC. Targeting the NF-κB pathway in cancer therapy. Surg Oncol Clin N Am 2013;22(4):705–46.

[36] In: Ref. [6]. pp. 306–13.

[37] Angel P, Hess J. The multi-gene family of transcription factor AP-1. In: Ref. [22]. pp. 2059–68.

[38] Bruyère C, Meijer L. Targeting cyclin-dependent kinases in anti-neoplastic therapy. Curr Opin Cell Biol 2013;25(6):772–9.

[39] Thorpe M, Shehu E, Harvey A. Insulin and insulin-like growth factor (IGF) family. In: Ref. [24]. pp. 25–44.

[40] Hiscox S. c-Met receptor signaling. In: Ref. [24]. pp. 115–38.

[41] Leszczynska K, Hillyar C, Hammond EM. Vascular endothelial growth factor and its receptor family. In: Ref. [24]. pp. 139–70.

[42] In: Ref. [6]. pp. 157–8.

[43] Tree D. Wnt signaling. In: Ref. [24]. pp. 67–91.

[44] Saito-Diaz K, Chen TW, Wang X, et al. The way Wnt works: components and mechanism. Growth Factors 2013;31(1):1–31.

[45] Meffre D, Grenier J, Bernard S, et al. Wnt and lithium: a common destiny in the therapy of nervous system pathologies? Cell Mol Life Sci 2014;71(7):1123–48.

[46] Gattinoni L, Ji Y, Restifo NP. Wnt/beta-catenin signaling in T-cell immunity and cancer immunotherapy. Clin Cancer Res 2010;16(19):4695–701.

[47] Stenbeck G. Transforming growth factor-β receptor signaling. In: Ref. [24]. pp. 45–66.

[48] Hubbard SR. Structures of serine/threonine and tyrosine kinases. In: Ref. [22]. pp. 413–17.

[49] In: Ref. [6]. pp. 209–12.

[50] Nishizuka Y, Kikkawa U. Historical overview: protein kinase C, phorbol ester and lipid mediators. In: Ref. [22]. pp. 1033–5.

[51] In: Ref. [5]. pp. 215–16.

[52] In: Ref. [5]. pp. 214–15.

[53] Dinarello CA. Historical insights into cytokines. Eur J Immunol 2007;37(Suppl. 1):S34–45.

[54] Stepanenko AA, Vassetzky YS, Kavsan VM. Antagonistic functional duality of cancer genes. Gene 2013;529(2):199–207.

[55] Fearon ER. Molecular genetics of colorectal cancer. Annu Rev Pathol 2011;6:479–507.

[56] Citri A, Yarden Y. EGF-ERBB signaling: towards the systems level. Nat Rev Mol Cell Biol 2006;7(7):505–16.

[57] Logue JS, Morrison DK. Complexity in the signaling network: insights from the use of targeted inhibitors in cancer therapy. Genes Dev 2012;26(7):641–50.

[58] Guertin DA, Sabatini DM. Cell growth. In: Mendelsohn J, Howley P, Israel M, editors. The molecular basis of cancer (3rd ed.). Philadelphia, PA: Saunders/Elsevier; 2008. pp. 169–75.

[59] Fendler B, Atwal G. Systematic deciphering of cancer genome networks. Yale J Biol Med 2012;85(3): 339–45.

[60] In: Ref. [6]. p. 176.

[61] Zeke A, Lukács M, Lim WA, et al. Scaffolds: interaction platforms for cellular signaling circuits. Trends Cell Biol. 2009;19(8):364–74.

[62] Pedraza-Fariña LG. Mechanisms of oncogenic cooperation in cancer initiation and metastasis. Yale J Biol Med 2006;79(3–4):95–103.

[63] Aguirre E, Renner O, Narlik-Grassow M, et al. Genetic modeling of PIM proteins in cancer: proviral tagging and cooperation with oncogenes, tumor suppressor genes, and carcinogens. Front Oncol 2014;4:109.

[64] Weinstein B, Joe AK. Mechanisms of disease: oncogene addiction—a rationale for molecular targeting in cancer therapy. Nat Clin Pract Oncol 2006;3:448–57.

[65] Weinstein B, Joe A. Oncogene addiction. Cancer Res 2008;68(9):3077–80.

[66] Torti D, Trusolino L. Oncogene addiction as a foundational rationale for targeted anti-cancer therapy: promises and perils. EMBO Mol Med 2011;3(11):623–36.

[67] Sawyers CL. Shifting paradigms: the seeds of oncogene addiction. Nat Med 2009;15(10):1158–61.

[68] Bauer KH. Mutationstheorie der Geschwülst-Entstehung/Mutation theory in tumour formation. Berlin: Julius Springer; 1928. pp. 14–6.

[69] Bignold LP. The cell-type-specificity of inherited predispositions to tumors: review and hypothesis. Cancer Lett 2004;216(2):127–46.

[70] Burch PRJ. Carcinogenesis and cancer prevention. Nature 1963;197:1145–51.

[71] Nicholls EM. Somatic variation and multiple neurofibromatosis. Hum Hered 1969;19:473–9.

[72] Comings DE. A general theory of carcinogenesis. Proc Natl Acad Sci USA 1973;70:3324–8.

[73] Knudson AG. Genetics and the etiology of human cancer. Adv Hum Genet 1977;8:1–66.

[74] Knudson AG. Genetics of human cancer. Annu Rev Genet 1986;20:231–51.

[75] Squire J, Gallie BL, Phillips RA. A detailed analysis of chromosomal changes in heritable and non-heritable retinoblastoma. Hum Genet 1985;70(4):291–301.

[76] In: Ref. [6]. pp. 231–390.

[77] Genetics Home Reference. Tumor suppressor gene. Available at: http://ghr.nlm.nih.gov/glossary=tumor suppressorgene.

[78] Ruddon RW. Cancer biology (4th ed.). New York, NY: Oxford University Press; 2007. pp. 352–66.

[79] Hohenstein P. Tumour suppressor genes—one hit can be enough. PLoS Biol 2004;2(2):E40.

[80] El-Deiry WS. Tumor suppressor genes. vol 1, Part II (Methods in molecular biology, vol. 222). Totowa, NJ: Humana Press; 2003. pp. 295–500.

[81] Knudson AG. Anti-oncogenes and human cancer. Proc Natl Acad Sci USA 1993;90:10914–21.

[82] Sherr CJ. Principles of tumor suppression. Cell 2004;116(2):235–46.

[83] Dick FA, Rubin SM. Molecular mechanisms underlying RB protein function. Nat Rev Mol Cell Biol 2013;14(5):297–306.

[84] Nicolay BN, Dyson NJ. The multiple connections between pRB and cell metabolism. Curr Opin Cell Biol 2013;25(6):735–40.

[85] Tirnauer J. APC (adenomatous polyposis coli). Atlas Genet Cytogenet Oncol Haematol 2005;9(2):132–3.

[86] Neufeld KL. Nuclear APC. Adv Exp Med Biol 2009;656:13–29.

[87] Minde DP, Anvarian Z, Rüdiger SG, et al. Messing up disorder: how do missense mutations in the tumor suppressor protein APC lead to cancer? Mol Cancer 2011;10:101.

[88] Caldwell CM, Kaplan KB. The role of APC in mitosis and in chromosome instability. Adv Exp Med Biol 2009;656:51–64.

[89] Domingo E, Schwartz Jr. S. MSH2 (human mutS homolog 2). Atlas Genet Cytogenet Oncol Haematol 2005;9(4):291–2.

[90] Domingo E, Schwartz Jr. S. MLH1 (human mutL homolog 1). Atlas Genet Cytogenet Oncol Haematol 2005;9(2):120–2.

[91] Banerjee S. MSH6 (mutS homolog 6 (E. coli). Atlas Genet Cytogenet Oncol Haematol 2007;11(3):169–72.

[92] National Center for Biotechnology Information. PMS2P4. Available at: http://www.ncbi.nlm.nih.gov/gene/?term=PMS2L4.

[93] Banerjee S. BRCA1 (breast cancer 1, early onset). Atlas Genet Cytogenet Oncol Haematol 2008;12(3):197–200.

[94] Atlas of Genetics and Cytogenetics in Oncology and Haematology. BRCA2. Available at: http://atlasgeneticsoncology.org//Genes/BRCA2ID164ch13q13.html.

[95] Hohenstein P, Hastie ND. The many facets of the Wilms' tumour gene, WT1. Hum Mol Genetics 2006;15:R196–201.

[96] Guert B, Ratschek M, Harms D, et al. Clonality and loss of heterozygosity of WT genes are early events in the pathogenesis of nephroblastomas. Hum Pathol 2003;34(3):278–81.

[97] Segers H, van den Heuvel-Eibrink MM, de Krijger RR, et al. Defects in the DNA mismatch repair system do not contribute to the development of childhood Wilms tumors. Pediatr Dev Pathol 2013;16(1):14–19.

[98] Wimmer K. NF1 (neurofibromin 1). Atlas Genet Cytogenet Oncol Haematol 2006;10(3):171–2.

[99] Carroll SL. Molecular mechanisms promoting the pathogenesis of Schwann cell neoplasms. Acta Neuropathol 2012;123(3):321–48.

[100] Genetics Home Reference. NF2. Available at: http://ghr.nlm.nih.gov/gene/NF2.

[101] Baser ME, Kuramoto L, Woods R, et al. The location of constitutional neurofibromatosis 2 (NF2) splice site mutations is associated with the severity of NF2. J Med Genet 2005;42(7):540–6.

[102] Cooper J, Giancotti FG. Molecular insights into NF2/Merlin tumor suppressor function. FEBS Lett 2014 pii: S0014-5793(14)00276-2.

[103] Richard S. VHL (von Hippel–Lindau tumor suppressor). Atlas Genet Cytogenet Oncol Haematol 2002;6(2):106–10.

[104] Robinson CM, Ohh M. The multifaceted von Hippel–Lindau tumour suppressor protein. FEBS Lett 2014;558(16):2704–11.

[105] Bader HL, Hsu T. Systemic VHL gene functions and the VHL disease. FEBS Lett 2012;586(11):1562–9.

[106] Soussi T. P53 (protein 53 kDa). Atlas Genet Cytogenet Oncol Haematol 2003;7(1):6–9.

[107] Levine AJ, Lane DP, editors. The p53 family. Cold Spring Harbor, NY: CSH Press; 2010.

[108] Ayed A, Hupp T, editors. p53 (Molecular biology intelligence unit. Austin, TX: Landes/Springer; 2010.

[109] Hainaut P, Olivier M, Wiman KG, editors. p53 in the clinics. New York, NY: Springer; 2013.

[110] Liao JM, Cao B, Zhou X, et al. New insights into p53 functions through its target microRNAs. J Mol Cell Biol 2014;6(3):206–13.

[111] Muller PA, Vousden KH. p53 mutations in cancer. Nat Cell Biol 2013;15(1):2–8.

[112] Golubovskaya VM, Cance WG. Targeting the p53 pathway. Surg Oncol Clin N Am 2013;22(4):747–64.

[113] Rivlin N, Brosh R, Oren M, et al. Mutations in the p53 tumor suppressor gene. Important milestones at the various steps of tumorigenesis. Genes Cancer 2011;2:466–74.

[114] International Agency for Research on Cancer. TP53 genetic variations in human cancers. Available at: http://p53.iarc.fr/Download/SlideShow2013_online.pdf.

[115] Stary A, Sarasin A. XPA (xeroderma pigmentosum, complementation group A). Atlas Genet Cytogenet Oncol Haematol 2001;5(2):100–2.

[116] Stary A, Sarasin A. ERCC-3 (excision repair cross-complementing rodent repair deficiency, complementation group 3). Atlas Genet Cytogenet Oncol Haematol 2001;5(2):88–90.

[117] Stary A, Sarasin A. XPC (xeroderma pigmentosum, complementation group C). Atlas Genet Cytogenet Oncol Haematol 2001;5(2):103–5.

[118] Stary A, Sarasin A. ERCC2 (excision repair cross-complementing rodent repair deficiency, complementation group 2). Atlas Genet Cytogenet Oncol Haematol 2001;5(2):83–7.

[119] Mareel M, Leroy A. Clinical, cellular, and molecular aspects of cancer invasion. Physiol Revs 83:337–76.

[120] Tavassoli FA, Eusebi V. Tumors of the mammary gland. AFIP Atlas, Ser 4, Fasc 10, 2009:149–50.

[121] In: Ref. [120]. pp. 350–60.

[122] Kurman RJ, Ronnett BM, Sherman ME, et al. Tumors of the cervix, vagina, and vulva. AFIP Atlas Ser 4, Fasc 13; 2010. p. 92.

[123] Elder DE, Murphy GF. Melanocytic tumors of the skin. AFIP Atlas Ser 4, Fasc 13; 2010. pp. 278–96.

[124] Islam SA, Luster AD. T cell homing to epithelial barriers in allergic disease. Nat Med 2012;18(5):705–15.

[125] Ben-Ezra J, Burke JS, Swartz WG, et al. Small lymphocytic lymphoma: a clinicopathologic analysis of 268 cases. Blood 1989;73(2):579–87.

[126] Fenoglio-Preiser CM, Noffsinger GN, Lantz PR, et al. Gastrointestinal pathology (3rd ed.). Philadelphia, PA: Lippincott Williams and Wilkins; 2008. pp. 1162–9.

[127] Patterson JW, Wick MR. Nonmelanocytic tumors of the skin. AFIP Atlas, Ser 4, Fasc 4; 2006. pp. 46–70.

[128] Hansemann DP. Bignold LP, Coghlan BL, Jersmann HP, editors. David Paul Hansemann: contributions to oncology: context, comments and translations. Basel: Birkhäuser; 2007. p. 248.

[129] Sahai E. Illuminating the metastatic process. Nat Rev Cancer 2007;7(10):737–49.

[130] Chan KT, Cortesio CL, Huttenlocher A. Integrins in cell migration. Methods Enzymol 2007;426:47–67.

[131] Onodera Y, Nam JM, Sabe H. Intracellular trafficking of integrins in cancer cells. Pharmacol Ther 2013;140(1):1–9.

[132] Liotta LA, Wewer U, Rao NC, et al. Biochemical mechanisms of tumor invasion and metastases. Prog Clin Biol Res 1988;256:3–16.

[133] Egeblad M, Rasch MG, Weaver VM. Dynamic interplay between the collagen scaffold and tumor evolution. Curr Opin Cell Biol 2010;22(5):697–706.

[134] McSherry EA, Donatello S, Hopkins AM, et al. Molecular basis of invasion in breast cancer. Cell Mol Life Sci 2007;64(24):3201–18.

[135] Funasaka T, Raz A. The role of autocrine motility factor in tumor and tumor microenvironment. Cancer Metastasis Rev 2007;26(3-4):725–35.

[136] van Nes J, Chan A, van Groningen T, et al. A NOTCH3 transcriptional module induces cell motility in neuroblastoma. Clin Cancer Res 2013;19(13):3485–94.

[137] Crapoulet N, O'Brien P, Ouellette RJ, et al. Coordinated expression of Pax-5 and FAK1 in metastasis. Anticancer Agents Med Chem 2011;11(7):643–9.

[138] Kulbe H, Levinson NR, Balkwill F, et al. The chemokine network in cancer—much more than directing cell movement. Int J Dev Biol 2004;48(5-6):489–96.

[139] Madsen CD, Sahai E. Cancer dissemination—lessons from leukocytes. Dev Cell 2010;19(1):13–26.

[140] Lowell CA, Mayadas TN. Overview: studying integrins in vivo. Methods Mol Biol 2012;757:369–97.

[141] Miller DD. Integrins Lackey K, editor. Gene family targeted molecular design. Hoboken, NJ: John Wiley & Sons; 2009. pp. 85–118.

[142] Paschos KA, Majeed AW, Bird NC. Natural history of hepatic metastases from colorectal cancer—pathobiological pathways with clinical significance. World J Gastroenterol 2014;20(14):3719–37.

[143] Williams K, Motiani K, Giridhar PV, et al. CD44 integrates signaling in normal stem cell, cancer stem cell and (pre)metastatic niches. Exp Biol Med (Maywood) 2013;238(3):324–38.

[144] Van Roy FM, editor. The molecular biology of cadherins. Oxford, UK: Elsevier/Academic Press; 2013.

[145] Jeanes A, Gottardi CJ, Yap AS. Cadherins and cancer: how does cadherin dysfunction promote tumor progression? Oncogene 2008;27:6920–9.

[146] Talmadge JE, Fidler IJ. The biology of cancer metastasis: historical perspective. Cancer Res 2010;70(14):5649–69.

[147] Willis RA. The spread of tumours in the human body. London: Butterworth & Co; 1952.

[148] In: Ref. [128]. pp. 167–78.

[149] Loeb L. The biological basis of individuality. Springfield, IL: C.C. Thomas; 1945.

[150] Langdon SP. Animal modeling of cancer pathology and studying tumor response to therapy. Curr Drug Targets 2012;13(12):1535–47.

[151] Singh M, Darcy KM, Brady WE, et al. Cadherins, catenins and cell cycle regulators: impact on survival in a Gynecologic Oncology Group phase II endometrial cancer trial. Gynecol Oncol 2011;123(2):320–8.

[152] Rosai J, Carcangiu ML, DeLellis RA, Tumors of the thyroid gland. AFIP Atlas; 1990. p. 56.

[153] Figg W, Folkmann J. Angiogenesis: an integrative approach: from science to medicine. New York, NY: Springer; 2009.

[154] Bockhorn M, Jain RK, Munn LL. Active versus passive mechanisms in metastasis: do cancer cells crawl into vessels, or are they pushed? Lancet Oncol 2007;8(5):444–8.

[155] Weiss L, Orr FW. The microvascular phases of metastasis Simionescu N, Simionescu N, editors. Endothelial cell dysfunctions. New York, NY: Plenum Press; 1992. pp. 455–76.

[156] Burnier JV, Burnier Jr MN, editors. Introduction. Experimental and clinical metastasis: a comprehensive review. New York, NY: Springer; 2013. p. 5.

[157] Valastyan S, Weinberg RA. Tumor metastasis: molecular insights and evolving paradigms. Cell 2011;147(2):275–92.

[158] van den Berg YW, Osanto S, Reitsma PH, et al. The relationship between tissue factor and cancer progression: insights from bench and bedside. Blood 2012;119(4):924–32.

[159] Jain S, Harris J, Ware J. Platelets: linking hemostasis and cancer. Arterioscler Thromb Vasc Biol 2010;30(12):2362–7.

[160] Fidler IJ. The pathogenesis of cancer metastasis: the "seed and soil" hypothesis revisited. Nat Rev Cancer 2003;3:453–8.

[161] McGowan PM, Kirstein JM, Chambers AF. Micrometastatic disease and metastatic outgrowth: clinical issues and experimental approaches. Future Oncol 2009;5(7):1083–98.

[162] Fatatis A, editor. Signalling pathways and molecular mediators in Metastasis. New York, NY: Springer; 2012.

[163] Lyden D, Welch DR, Psaila B, editors. Cancer metastasis: biologic basis and therapeutics. New York, NY: Cambridge University Press; 2011.

[164] Wakefield L, Hunter K, editors. Metastasis. Amsterdam: IOS Press; 2006.

[165] Esteller M, editor. DNA Methylation, epigenetics and metastasis. New York, NY: Springer; 2005.

Etio-Pathogenesis IV: Heredity

L.P. Bignold: Principles of Tumors.
DOI: http://dx.doi.org/10.1016/B978-0-12-801565-0.00007-X

185

Heredity has been considered a factor in the development of tumors since the earliest medical writings. At the present time, the focus of this area of study of tumors is the 200 or so inherited cancer-susceptibility syndromes which have been discovered. The majority of these appear to be autosomal dominant with incomplete penetrance [1].

Together, the predisposition syndromes appear to contribute to the causation of less than 1% of all tumors. In addition to their significance in relation to families, as well as the individual patients, these inherited predispositions have the additional importance of potentially providing insights into the mechanisms of "familial" and "spontaneous"/sporadic tumors.

High-penetrance predispositions to individual types of tumors can be developed in animals through in-breeding. Also, it is possible to genetically engineer strains of animals which are similarly highly susceptible to tumor formation.

This chapter is concerned with aspects of these predispositions. Section 7.1 describes general aspects of high-penetrance predispositions to tumors. Sections 7.2–7.6 give details of some of the main inherited predispositions, emphasizing the variable complexities in phenotype–genotype associations. Section 7.7 addresses the question of the specificity of effects of the inherited predispositions for particular kinds of parent cells. Section 7.8 discusses low-penetrance human predispositions. Section 7.9 gives accounts of aspects of hereditary predispositions in experimental animals.

The principles of tumor-suppressor genes and the details of the functions of their protein products are discussed in Section 6.5.

7.1 GENERAL ASPECTS OF HIGH-PENETRANCE HEREDITARY PREDISPOSITIONS TO TUMORS IN HUMANS

7.1.1 Background

Up until the early twentieth century, factors in the causation of many diseases were thought to be intrinsic tendencies ("diatheses") of the individual to the specific disorder [2]. Diatheses were conceived as being either hereditary or acquired—usually by unknown mechanisms—*in utero* or in infancy or childhood. The concept was recorded by Hippocrates, but is currently little used because it is vague. If used in its original sense, it could include one disease complicating another. For example, diabetes mellitus could be considered a "diathesis" for susceptibility to bacterial infections.

In the eighteenth and nineteenth centuries, supposed "cancer diatheses" were a popular explanation of tumors [3]. Hereditable predispositions/diatheses to cancers generally (i.e.,

all types of tumors being more common than average in the same family) were widely credited. However, with the work of Warthin in 1913 [4], this idea was shown to be unlikely.

By this time also, it was known that cases of a few rare tumor types occur with increased frequency in certain families. In the first half of the twentieth century, after careful studies of family trees, the Mendelian inheritance of predispositions to some tumor types was established. Among these were familial polyposis [5], multiple exostoses [6], and xeroderma pigmentosum [7].

7.1.2 Specificity of Each of the Hereditary Predispositions to Particular Kinds of Parent Cells

In Mendelian genetics, many mutation-affected traits are related to a specific kind of cell. For example, in embryonic development, germ-line genomic events affect only one kind of cell, or one organ. Thus loss of function events in *wingless* genes affects only a limited number of kinds of cells in the *Drosophila* embryo [8]. In the human genome, sickle cell anemia is an obvious example. In a few kinds of inherited predisposition to tumors, the germ-line mutation is in a gene which is specifically active in the parent kind of cell. Tuberous sclerosis is an example, see Section 7.5.4. However, for most of the other hereditary predispositions, the protein product of the relevant mutant gene is unclear. Further, the gene product is expressed in many kinds of cells which do not suffer the tumorigenic effect of the mutation.

This issue is discussed further in Section 7.7.

7.1.3 Definitions and Characteristics of Inherited, Familial, and Spontaneous Predispositions

"Inherited" in the context of inherited predispositions means that the gonads of one of the parents are mutant, such that all gametes are mutant, and all offspring may be affected. The nongonadal cells (e.g., lymphocytes) of the parent are usually affected. Siblings of the patient may be affected (Table 7.1).

"Familial" is defined as the occurrence of a tumor type more frequently in blood relatives than might occur by chance.

TABLE 7.1 Characteristics of Hereditary, Familial, and Spontaneous Tumors

---

1. *"Hereditary cancer-type" characteristics*
   - Apparently autosomal dominant *or recessive* (added by present author) transmission of specific cancer type(s)
   - Earlier age of onset of cancers than is typical
   - Multiple primary cancers in an individual
   - Clustering of rare cancers of different types
   - Bilateral or multifocal cancers
   - First-degree relatives of mutation carriers are at 50% risk to have the same mutation
   - Incomplete penetrance and variable expressivity, such that obligate carriers of the family mutation may be cancer-free and the age of diagnosis of cancer among relatives will vary
   - Those who do not have the familial mutation have the general population risk for cancer
2. *"Familial cancer-type" characteristics*
   - More cases of a specific type(s) of cancer within a family than statistically expected, but no specific pattern of inheritance
   - Age of onset variable
   - May result from chance clustering of sporadic cases
   - May result from common genetic background, similar environment, and/or lifestyle factors
   - Does not usually exhibit classical features of hereditary cancer syndromes
3. *"Sporadic[a] cancers-type" characteristics*
   - Cancers in the family are likely due to nonhereditary causes
   - Typical age of onset
   - Even if there is more than one case in the family, there is no particular pattern of inheritance
   - Very low likelihood that genetic-susceptibility testing will reveal a mutation; testing with available technology/knowledge level will likely not provide additional information about cancer risk.

Source: Adapted from Ref. [8].
[a]*See also Figure 4.1.*

"Spontaneous" usually means "noninherited." In these, there is no mutation in the somatic cells, of either parent. Siblings of the patient are not affected.

The characteristics of these alternative aspects of all cancers can be adapted from descriptions originally made of breast cancer [9] as given in Table 7.1.

It is possible that some cases of apparently spontaneous tumors might have a genetic basis. These are difficult to detect, but include [10]:

i. A *de novo* germ-line mutation occurring during gametogenesis in the parental gonad. In these cases, no family members including siblings of the patient are affected, but all children of the patient may be affected.

ii. A mutation in embryonic life before organogenesis (Section A1.1.1) in the patient. There is no family history, but all cells of the individual will show the mutation, and the children of the patient are likely to be affected.

iii. A somatic mutation after the beginning of organogenesis in the embryonic life of the patient. Cells in some organs but not others will be affected (mosaicism in the individual with respect to the mutation). A child of the patient may be affected if the mosaicism affects both the organ in which the tumor forms and the gonads (see above). These kinds of events are probably rare [11].

iv. Incomplete family records.

7.1.4 Variabilities in Penetrances, Expressivities, and Timing of Inheritable Tumors

There are many in-principle aspects of inherited predispositions which are variable. These will be mentioned here, with examples also given in Section 7.2.

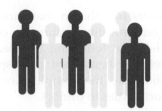

A **Penetrance**	B **Expressivity**
In conditions in which the trait is "all-or-nothing"/ discontinuous/qualitative, only some individuals carrying the germ-line mutation may show phenotypic changes (here, 50%). Unaffected individuals are assumed to have "modifying" genes, through the genetic diversity mechanism in meiosis known as "crossing-over" (see Section A2.1.3)	In some conditions, the trait is variable/continuous /quantitative. As with penetrance, this is thought to be due to the presence of modifier genes, perhaps having variable effects on the trait.

FIGURE 7.1   Penetrance and expressivity.

(a) In Penetrance and Expressivity Case-to-Case

"Penetrance" is defined as "the proportion of individuals with a mutation causing a particular disorder who exhibit clinical symptoms of that disorder" [12] (Figure 7.1). Many tumor predispositions also show variability in "expressivity," which is defined as "the relative capacity of a gene to affect the phenotype of the organism of which it is a part" [13] (Figure 7.1). Penetrance of hereditary traits can be affected by multiple complex genetic factors [14] as well as environmental conditions.

These aspects range from less than 5% to 100% in different tumor predispositions. Details for some of the predispositions are described in Sections 7.2–7.6. These variabilities are not always associated with different mutations within the affected gene in different families.

(b) Different Penetrances of the Different Components of Various Syndromes

In almost all human inherited tumor predispositions, the different manifestations of

a syndrome may show different penetrances. Thus in germ-line *RB* mutation, the penetrance of retinoblastoma is 80–90% at 5 years of age. The penetrance of osteosarcoma, however, is approximately 7% at 20 years [15,16]. (This percentage includes those—perhaps as many as half—which occur in the fields of radiation therapies given to the retinoblastoma.)

In the Carney Complex [17], there are pigmented lesions of the skin in all cases, but lower penetrance for pigmented lesions of the conjunctiva and adrenal glands, as well as for myxomas of the heart and skin. The penetrance of adrenocortical tumors in this condition is lower still. In the separate entity of Carney Triad [18], gastrointestinal stromal tumor occurs in most cases. There is lower penetrance of pulmonary chondroma, and lower still for extra-adrenal paraganglioma.

(c) Earlier Appearance of Tumors in Comparison with Spontaneous Cases

In general, cases of tumors in individuals with relevant hereditary predispositions occur at a younger age than tumors of the same type in individuals without the predisposition.

Examples are given in the next sections.

(d) Some Predispositions Are Mendelian Recessive

There are several conditions of predisposition to tumors, in which both germ-line alleles must be mutant for the individual to be predisposed. These conditions do not conform to the Knudson model (see Figure 6.6). A third mutation in a parent cell in the individual is presumably required for a tumor to form. This is discussed further in Section 7.7.5.

7.1.5 Nontumorous Lesions Occurring in Some Human Hereditary-Predisposition Syndromes

It is also relevant here that some tumor-predisposition syndromes are associated with non-tumorous lesions. The Carney Complex and Carney Triad have already been mentioned. As another example, the Peutz–Jeghers syndrome is characterized first of all by polyps of the small intestine which show only minor cyto-structural abnormalities but nevertheless have an increased risk of malignancies [19]. However, these patients also have focal oral mucosal hyperpigmentation, which is not related to any abnormal cell proliferation. This clearly implies that tumors are not the only biological effect of the mutation in the "tumorigenic when mutant" region of the genome.

7.1.6 Variabilities in Phenotype–Genotype Relationships

This is one of the most important issues in relation to understanding the genomic bases of tumors. Before the era of DNA sequencing of the genome, ideas of predisposition to a disease were generally similar to those of a disease itself—i.e., one particular genomic event in one gene. Information accumulated over the last

30 years has shown that most predispositions are genomically much more complex than this model. The following five sections describe a variety of hereditary predisposition, selected to illustrate the complexities of the genomic abnormalities in tumors.

7.2 PHENOTYPE–GENOTYPE RELATIONSHIP I: A SINGLE TUMOR TYPE DERIVING FROM ONE PARENT-CELL TYPE WITH ONE GENE OR MORE INVOLVED

This is in fact an unusual of the relationship and shown mainly by familial melanoma.

7.2.1 Familial Melanoma (CDK4 *and* CDKN2A)

Familial melanoma is a rare condition and has no other clinical manifestations (Figure 7.2). The definition of hereditary melanoma is now

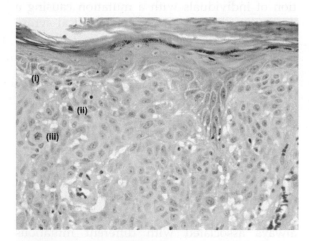

FIGURE 7.2   Nodular melanoma ×20. There are no histological differences between tumors arising in predisposed individuals and those arising spontaneously. This rule also applies to most of the other tumors described here. Note: This section shows (i) invasion of epidermis by melanoma cells, (ii) spontaneous necrosis of a single melanoma cell (Chapter 10), and (iii) an atypical mitotic figure (Chapter 8).

described as a family in which either two first-degree relatives are diagnosed with melanoma or families with three melanoma patients (irrespective of degree of relationship) [20].

The commonest genomic event responsible in all cases so far reported is a substitution at codon 24 in exon 2 of the CDK4 gene (12q13) [20]. The protein product (cyclin-dependent kinase 4) is a member of the serine/threonine protein kinase family (see Section 6.3.3).

At the time of writing, the pathogenetic mutations in CDKN2A (9p21) are not well established [21–23]. Certain other familial melanoma conditions have been described, and some nonmelanocytic tumors have been reported in association with these conditions.

7.2.2 MUTYH Polyposis Coli

This condition is autosomal recessive because mutations in both germ-line alleles are required for the phenotypic change. In comparison with familial adenomatous polyposis (FAP) (see above), the adenomas are less numerous and appear only in midlife. The risk of carcinoma is 28 times that of normal [24].

In 25% of cases, there are also duodenal adenomas, and a few cases develop carcinoma of the duodenum [25]. Carriers (single allele mutants) have no excess risk of carcinoma than normal.

The gene encodes a glycosidase which is important for DNA repair.

7.3 PHENOTYPE–GENOTYPE RELATIONSHIP II: SEVERAL TUMOR TYPES DERIVING FROM ONE (OR CLOSELY RELATED) PARENT-CELL TYPE WITH ONE GENE INVOLVED

There is only one relatively common example in this category.

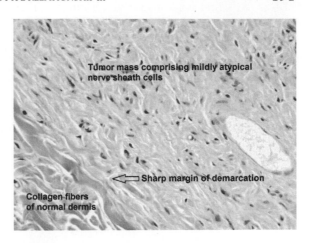

FIGURE 7.3   Neurofibroma in dermis ×20.

7.3.1 Neurofibromatosis 2

The incidence of neurofibromatosis 2 (NF2) (Figure 7.3) is approximately 1 in 30–60,000 people [26,27]. There is no family history in approximately half of the cases. Nearly 100% of cases manifest with Schwannomas on the acoustic nerves, with lesser penetrances for meningiomas, spinal nerve Schwannomas, and cutaneous neurofibromas. Severe/early onset and mild/late onset forms are recognized.

NF2 (on 22q) is the only gene known to be associated with the condition. Pathogenetic mutations appear to be associated with truncation of the protein product [26–28]. There are no mutation hotspots. Diagnosis is usually achieved on clinical manifestations.

7.4 PHENOTYPE–GENOTYPE RELATIONSHIP III: MULTIPLE TUMOR TYPES AND EVEN NONTUMOR LESIONS ARISING IN DIFFERENT KINDS OF PARENT CELLS IN THE ONE INDIVIDUAL: SINGLE GENE

7.4.1 Retinoblastoma

This is among the commonest of tumors in childhood (Figure 7.4). The incidence is

FIGURE 7.4   Retinoblastoma in retina ×10. The circular arrangements of cells with fibers in the middle (arrowed) are called 'rosettes'.

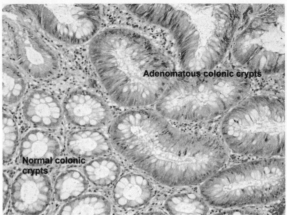

FIGURE 7.5   Colonic adenoma in FAP ×10. These are almost always typical tubular or tubulo-villous adenomas. The large, flat, 'carpet-like' villous adenoma is not common in FAP patients.

approximately 1 in 20,000 live births. Almost all cases manifest in the first 5 years of life. Between 10–15% of unilaterally occurring tumors are associated with a germ-line genomic event and almost all cases with bilateral tumors are associated with the germ-line mutation [29]. The tumors in this disorder are composed of retinoblast-like cells. There are no major histopathological differences between tumors arising spontaneously and those arising in association with the inherited predisposition.

This disorder occurs through mutations in the *rb* gene (on 13q) only. Penetrance is close to 100% and there is little variability in expressivity.

De novo germ-line mutations are suggested for approximately 60% of cases [30]. Germ-line lesions in the *RB* gene are point mutations, deletions, and insertions. The somatic (second "hit") events include point mutations, deletions, insertions, epigenetic silencing (somatic), and chromosomal rearrangements [13,31].

As mentioned in Section 7.1.3b, 7% of retinoblastoma patients develop soft tissue, cartilage, or bone sarcomas [12]. Forty percent of these arise in the field of radiation for treatment of the eye tumor. Abnormalities in the RB gene in these tumors, as well as in tumors of the same types not associated with retinoblastoma, have been found [32].

7.4.2   Familial Adenomatous Polyposis

This condition has an incidence of approximately 1 in 15,000 live births (Figure 7.5). It manifests mainly as very numerous adenomatous polyps which begin to appear in the colon and rectum in the early teens. Ultimately hundreds to thousands of adenomas appear, and carcinomas develop in almost all patients, with an average age at diagnosis of 39 years [33]. The adenomas and carcinomas are not significantly histologically different from spontaneously occurring adenomas and carcinomas [34]. The predisposition does not usually include other benign proliferative lesions of the colonic epithelium (e.g. hyperplastic and juvenile polyps).

The disorder occurs through mutations only in the *APC* gene. Penetrance is close to 100% and there is little variability in expressivity. Most mutations truncate the functional APC protein. *De novo* germ-line mutations are reportedly 20–25% [35].

The so-called attenuated FAP is due to different mutations in the APC gene compared to those responsible for typical FAP [35].

FAP cases may suffer associated tumors of the bones, soft tissues, and thyroid gland [33]. In addition, there are two main named associated syndromes.

i. *Turcot's syndrome*

This condition is rare. The original article [36] described families with classic FAP in which cerebral gliomas were more common. Other cases of FAP have suffered medulloblastomas. In these cases, germline mutations are found in the *APC* gene (on 5q), although a specific genomic event has not been identified [37].

It may be noted here that more recently, brain tumors in association with Lynch syndrome (in which there are colorectal carcinomas but not polyps; Section 7.4.3) have been referred to as Turcot's syndrome. These cases have mutations in mismatch repair genes of hereditary nonpolyposis colonic cancer [37], thus suggesting a separate condition. The situation is not clear because specific germ-line mutations for Turcot's syndrome versus typical cases of FAP have not been established [38].

ii. *Gardner's syndrome*

This refers to FAP families in which multiple connective tissue lesions including fibromas, osteomas, and lipomas, as well as occasional carcinomas are common [38]. As for Turcot's syndrome, genomic differences between families have not been established.

MUTYH-associated polyposis is a separate condition involving a different gene (see Section 7.6.2).

7.4.3 Neurofibromatosis 1

Neurofibromatosis was originally described in 1882 by von Recklinghausen [39] as one entity but are now known to comprise two distinct syndromes [40]. NF2 is described above.

The incidence of NF1 is approximately 1 in 3,000 live births [41,42]. Thirty to fifty percent of cases have no family history. There is nearly 100% penetrance, but expressivity is variable. The clinical manifestations include mildly pigmented patches on the skin (*café au lait* spots) and tumors on the peripheral nerves, including in the spinal canal and the autonomic nerves. Additional kinds of neural lesions are plexiform neurofibromas and neurofibrosarcomas (malignant peripheral nerve sheath tumors). Most of the malignant tumors are thought to arise by "malignant degeneration" of a previously benign tumor. There are also increased incidences of adrenal medullary cell tumors (pheochromocytomas), gliomas of the brain, and of the so-called juvenile onset chronic myeloid leukemia [41]. Diagnosis is usually achieved on the basis of clinical features.

7.4.4 Von Hippel–Lindau Disease

This disorder occurs in approximately 1 in 36,000 live births, with 90% penetrance by the age of 65 years [43]. Almost all the clinical manifestations are of tumors. Germ-line mutations mainly result in truncations of protein product. All mutations which cause marked or complete loss of function of the protein are associated with brain and retinal hemangioblastomas and renal cell carcinomas, but not pheochromocytomas. Partial loss of function of the protein is associated with pheochromocytomas and sometimes associated the hemangioblastomas and renal carcinomas. Other lesions include papillary adenomas of the epithelium of the endolymphatic sac (endolymphatic tumor), paragangliomas, and neuroendocrine tumors of the adrenal medulla. Nontumorous lesions are mainly benign cysts of visceral organs and the epididymis [44].

The tumors arise according to Knudson's hypothesis (Section 6.5.1): both alleles of the

VHL gene must be nonfunctional in a cell for a tumor to form [45].

7.4.5 BCC Syndrome (Syn Gorlin Syndrome; Basal Cell Nevus Syndrome) (PTCH Gene)

This disorder affects approximately 1 in 30,000 individuals [46]. The main manifestations—basal cell carcinomas—often develop these tumors in adolescence or early adulthood. Several benign lesions, such as keratocysts of the jaws and fibromas in the heart and ovaries, are common. Various developmental abnormalities can occur in these patients. Another malignant tumor type—medulloblastoma—affects only a few percent of cases [46,47].

All cases are due to mutations in the *Patched-1* gene (on 9q), which encodes the Patched receptor protein (Section 6.3.5b).

7.4.6 Ataxia-Telangiectasia

This rare autosomal recessive condition occurs when an individual inherits the mutant alleles of the *ATM* gene (on 11q) from both carrier parents. No other gene is known to be involved. Patients suffer a variety of neurological, dermatological, immunodeficient, and other disorders. Approximately 25% of patients develop leukemia or lymphoma.

The gene product (ATM protein) is important for repairs to DNA double strand breaks, so that chromosomal instability is a feature of the disorder [48].

7.4.7 Bloom's Syndrome

This rare autosomal recessive condition occurs in mainly Ashkenazi Jewish and Japanese ethnic groups. There are severe developmental disabilities, dermatological abnormalities, and chromosomal instability, but with considerable variability in expressivity. Fifty percent of patients have at least one tumor: 50% leukemias and lymphomas; 30% carcinomas, 10% benign lesions, and 10% other [81]. The mutations responsible are in the *BLM* gene (15q). The gene codes for a DNA helicase, which associates with several other proteins including *TP53*, *53BP1*, *WRN*, *MLH1*, *RAD51* in repairs of DNA strand breaks. This accounts for the chromosomal instability [49,50].

7.5 PHENOTYPE–GENOTYPE RELATIONSHIP IV: MULTIPLE TUMOR TYPES AND EVEN NONTUMOR LESIONS ARISING IN DIFFERENT KINDS OF PARENT CELLS: MULTIPLE GENES

7.5.1 Hereditary Nonpolyposis Coli (Lynch Syndrome)

There are various definitions of this syndrome, but all accounts generally indicate "a predisposition to various carcinomas, especially of the large bowel, associated with an autosomal dominant mutation in a gene for mismatch repair of DNA."

There are no absolutely diagnostic clinical features, or pathological abnormalities of the tumors in patients with this syndrome [51]. In contrast to FAP, there are no large numbers of associated adenomatous polyps in the colon.

The predisposition manifests as:

i. Carcinomas occurring, on average, at earlier ages than in noninherited colorectal carcinoma cases.
ii. Colorectal carcinomas often in the cecum and ascending colon, and being poorly differentiated.
iii. Tumors (mainly carcinomas) in more than one organ. Other organs may include endometrium, stomach, ovaries, small bowel, liver, biliary tree, bladder, and brain.
iv. Carcinomas in susceptible organs in family members.

The prevalence of the disease in a population is affected by the longevity of the individuals. Mutations in mismatch repair genes are found in 1–3% of all colorectal carcinomas. Extrapolating from this, it has been suggested that the carrier rate of the germ-line mutations is at least 1:2000 of the general population [51].

The known genetic basis of the condition(s) involves mutations in seven different genes, the main ones being *MSH2*, *MLH1*, *MSH6*, and *PMS6* (see Section 6.5.5). The penetrance (80% of cases develop the tumor in their life spans) appears to be the same regardless of which gene is involved in the particular case. The molecular lesion caused by all the mutations is failure to repair errors in synthesis of tandem repeat sequences of DNA.

Testing of individuals suspected of having the syndrome is complex because of the number of genes which may cause the syndrome. For patients, costs and other factors may be relevant [52,53].

7.5.2 BRCA1 and -2

Sometimes called the "hereditary breast and ovarian cancer syndrome," this condition is characterized by increased risks of breast and ovarian carcinoma, in association with mutations in these genes [54]. Similar to the same aspect of Lynch syndrome, there are no clinical features of the individual or pathological abnormalities in the tumors which are absolutely specific for carriers of the mutations. Suspicion of the genetic errors is indicated if the breast tumor:

i. Occurs at a young age.
ii. Is poorly differentiated, especially in relation to specific membrane receptors for estrogens, progesterone, and HER-2 (see Section 6.3.2).
iii. Additional tumors either in the breasts or other organs, mainly ovary, prostate, and pancreas.

iv. Tumors in the male breast.
v. Family members affected by any of the relevant tumors.

For incidence rates in the general population, and patient subgroups, the following summary of data are published [55]:

General population (excluding Ashkenazim): about 1 in 400 (approximately 0.25%).

- Women with breast cancer (any age): 1 in 50 (2%).
- Women with breast cancer (younger than 40 years): 1 in 10 (10%).
- Men with breast cancer (any age): 1 in 20 (5%).
- Women with ovarian cancer (any age): 1 in 8 to 1 in 10 (10–15%).

Mutations in the BRCA1 gene are twice as common as in the BRCA2 gene [55].

The penetrance of BRCA1 mutations collectively is a 70% risk of breast cancer, and a 60% risk of ovarian cancer by age 70 years. For BRCA2 mutations to the same age, the collective penetrances are approximately 75% and 30% respectively [56]. However, many mutations in these genes appear to be nonpathogenic.

Extensive testing in recent years has revealed large number of mutations in the population. They comprise a variety of sequence variants and deletions [54]. Up to the year 2014, the Universal Mutation Database [57] had collected 1,690 mutations with 886 variants in BRCA1 protein, and 2,223 mutations with 1,343 variants in BRCA2 protein. Deleterious mutations are taken to include all protein-ablating or -truncating events [56]. Other mutations can be considered deleterious if they are known to be associated with affected families. New mutations found by testing are often considered as of unknown significance [58].

7.5.3 Wilms' Tumor

In Western societies, this tumor occurs in approximately 7 per million children younger than 15 years [59–62]. One to two percent of cases are familial [61]. Most cases occur before the fifth year. Overall, in less than 5% of cases, the tumors are bilateral, although in the Dendy–Drash syndrome, both kidneys are almost always affected. Ten percent of Wilms' tumors are associated with nonrenal congenital abnormalities. Only a small proportion of these abnormalities occur in clusters which characterize named syndromes. The most prominent of the named syndromes is the Beckwith–Weidemann Syndrome [59].

Approximately 13 genes are associated with Wilms' tumor in the various syndromes [59,63]. The most commonly involved gene is WT1. The germ-line genomic event has only 50% penetrance, i.e., only 50% of individuals carrying the germ-line event develop the tumor [60].

7.5.4 Tuberous Sclerosis Complex

This condition occurs in approximately 1 in 6,000 individuals [64] (Figure 7.6). Patients suffer very slow-growing glial tumors ("tubers"/hamartomas) in the brain [65,66] together with cutaneous lesions and a particular type of benign renal tumor (angiomyolipoma). Sixty to seventy percent of cases appear to be sporadic. Penetrance and expressivity of the various clinical manifestations is considerable [67].

Two genes TSC1 and TSC2 encode distinct proteins—hamartin and tuberin respectively [64,66,67]. The germ-line mutation results in functional loss of the allele. Tumors form through somatic loss of the other allele, in accordance with Knudson's hypothesis for tumor-suppressor genes (see Section 7.2.1). There appear to be no definite clinical differences between the effects of TSC1 and TSC2 mutations [68].

7.5.5 Carney Complex

Carney complex [14,69] (see also Section 7.1.4) consists of myxomas of the heart and skin, spotty hyperpigmentation of the skin, and various

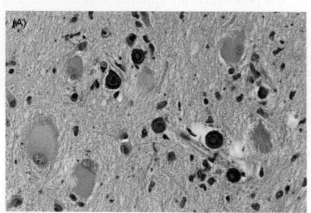
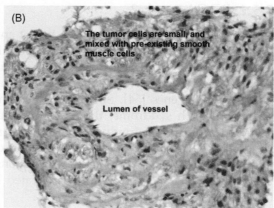

FIGURE 7.6 (A) Tuber in tuberous sclerosis. There is gliotic tissue containing irregular giant cells, which exhibit some features of astrocytes, and others of neurons. Calcification is common in larger lesions. According to the analytical approach in Chapter 8, these may be considered hyper-specializing very low grade astrocytic tumors, with either or both reactive changes in entrapped normal neurons, or a partial metaplasia in the tumorous giant astrocytes. Tubers do not occur in individuals without tuberous sclerosis. *Reproduced with permission from the American Registry of Pathology/ARP Press.*
(B) Angiomyolipoma in tuberous sclerosis. These lesions occur in both tuberous sclerosis patients, and in individuals without that disease. The characteristic feature is that the tumor cells appear to derive from the smooth muscle cells of vessels. The admixture of adipocytes is not explained. Almost all cases of these tumors exhibit staining with HMB 45, a feature which is otherwise mainly restricted to melanocytic tumors.

endocrine overactivities. It is distinct from Carney's triad. There is considerable phenotypic diversity. Ductal adenomas and osteochrondromyxomas may occur in some cases. Some manifestations appear to cluster as "subtypes" and many of the manifestations overlap with other syndromes such as those of McCune–Albright and multiple endocrine neoplasias (MEN).

Carney complex is most commonly associated with one of two mutations. The first is in the PRKAR1A gene on chromosome 17q23–q24. Approximately 80 different genomic events have been found in this gene, which may function as a tumor-suppressor gene. The encoded protein is a type 1A regulatory subunit of protein kinase A. Inactivating germ-line mutations of this gene are found in 70% of people with Carney complex.

The second genomic association is an event on chromosome 2 (2p16). The pathogenetic mechanisms of this germ-line mutation are unclear.

7.5.6 Li–Fraumeni Syndrome

This extremely rare disorder is characterized by up to 50-fold increases in incidences of soft tissue sarcomas, central nervous system tumors, as well as carcinomas of the breast and adrenal glands by the 30th year [70]. Patients frequently develop additional tumors later in life. A minority of cases appear to be new germ-line mutations.

In many references, Li–Fraumeni syndrome is said to occur in three types. LFS1, the "classic" form, is due to mutations in the TP53 gene [71]. Most of the germ-line mutations occur in nucleotide-site hotspots in the DNA binding domain of the protein. In LFS2, the CHEK2 gene—one of many regulators of TP53—is affected [72]. A family with similar clinical features was reported to have a mutation in an unnamed gene at 1q23 [73]. This was called LFS3, but the variant is controversial [74].

Diagnoses are based on clinical criteria. In most of the tumors in this condition, both the germ-line event of the syndrome and somatic loss of the other allele have been demonstrated.

7.5.7 Xeroderma Pigmentosum

This condition occurs in approximately 1 in 250,000 live births in Western countries, but has higher incidences in some other countries [75]. It was the first condition to be shown to be based in a defect in DNA repair, in particular, nucleotide excision repair [76]. There are seven complementation types and one variant (see Section 6.5.5). Because the epidermis exposed to sunlight suffers the greatest numbers of nucleotide-damaging events (see Section 4.3.2), the disorder manifests most prominently in atrophy, followed by tumor formation in those cells.

7.6 PHENOTYPE–GENOTYPE RELATIONSHIP V: PREDISPOSITIONS TO DIFFERENT SYNDROMES ACCORDING TO POSITION OF GERM-LINE EVENT IN A SINGLE GENE

7.6.1 RET Gene: Different Germ-Line Mutations Cause MEN Type 2 Versus Hirschsprung's Disease

All cases of MEN Type 2 have germ-line mutations in the RET (REarranged during Transformation) gene (on 10q). All cases of the three subtypes of MEN Type 2 [77] share a predisposition to medullary carcinoma of thyroid (MCT) (Figure 7.7):

i. Cases of MEN Type 2A develop this tumor in early adulthood and may also suffer pheochromocytoma.

ii. Cases of MEN Type 2B tend to develop MCT in childhood, suffer pheochromocytomas, and also exhibit various other abnormalities of development

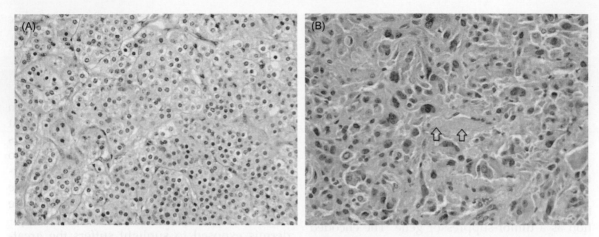

FIGURE 7.7   (A) Medullary carcinoma of thyroid; (B) pheochromocytoma.
These are two of several tumor types which are associated in multiple endocrine neoplasia syndromes. Both tumor types may exhibit many different histological patterns. The arrows in B indicates granularity of cytoplasm of a pheochromocytoma cell.

of the facial and skeletal bones as well as localized overgrowths of neural cells.

iii. Cases of the third subtype of MEN Type 2 (FMCT) suffer only MCT, which tend to appear only in middle age.

RET has many other names [78]. The gene product is a transmembrane signaling protein (Section 6.5.1) with the cytoplasmic tail being a tyrosine kinase. It is considered to belong to the superfamily of cadherin proteins [78].

In individuals with MEN Type 2A, a variety of genomic events have been detected.

In approximately 95% of individuals with the MEN Type 2B phenotype, there is a substitution at codon 918 in exon 16, which changes a threonine to a methionine (p.Met918Thr) in tyrosine kinase domain of *RET* [78].

In MEN Type FMCT, there seem to be various gene rearrangements. No specific point mutation has been found.

Deletions in 10q including *RET* are associated with Hirschsprung's disease (congenital aganglionosis of the colon) [77]. This disorder is not associated with any tumor. However, various deletions on other chromosomes and several named syndromes are associated with Hirschsprung's disease [75].

7.6.2 PTEN Gene: Cowden's and Related Syndromes

These syndromes were described separately but have clinical overlaps. The prevalence is about 1 in 200,000 individuals [79]. Penetrance is over 90% by middle age [78]. The main manifestations are sweat gland tumors and fibroepithelial polyps of the skin and mucous membranes [80]. There is a high incidence of tumors of the thyroid (but not MCT, see above), breast, and endometrium. Macrocephaly is a major diagnostic criterion of the syndrome [81]. The brains appear to exhibit megalencephaly, and there seems to be no definite neuropathology or cognitive defects in these patients.

The Bannayan–Riley–Ruvalcaba syndrome has similar features, but with intestinal polyps, lipomas, and pigmented macules on the glans penis [80].

In the Proteus syndrome (derivation: as for "protean"), there are slow-growing lesions of soft tissues, epithelium, and bone.

TABLE 7.2 Inherited Predisposition Syndromes Classified by Complexity of Their Phenotype–Genotype Relationships

1. A single tumor type deriving from one parent-cell type with one gene involved
  (a) Familial melanoma (CDK4) (Note deletions in *CDKN2A* (p16^{INK4A}) may play roles)
  (b) MUTYH polyposis coli (autosomal recessive)

2. Several tumor types deriving from one (or closely related) parent-cell type with one gene involved
  NF2: Schwannomas, meningiomas, and ependymomas

3. Multiple tumor types and even nontumor lesions arising in different kinds of parent cells; one gene involved
  (a) Retinoblastoma: Hard and soft connective tissue sarcomas (many are in the fields of irradiation)
  (b) FAP: Gliomas (Turcot's syndrome); soft tissue benign tumors (Gardner's syndrome)
  (c) NF1: Hamartomas of iris, nontumorous hyperpigmented patches in the skin
  (d) Von Hipple–Lindau disease
     Angiomas of retina and cerebellum: renal cell carcinomas, tumors of endocrine glands
  (e) BCC syndrome (*PTCH* gene)
     Multiple basal cell carcinomas: Epithelial cysts of jaws, multiple skeletal abnormalities.
  (f) Ataxia-telangiectasia (autosomal recessive)
     Mainly neurodegenerative and other nontumorous abnormalities. Tumors occur through genomic instability
  (g) Bloom's syndrome (autosomal recessive)
     Multiple nontumorous abnormalities. Tumors occur through genomic instability.

4. Multiple tumor types and even nontumor lesions arising in different kinds of parent cells; multiple genes involved
  (a) HNPPC/Lynch syndrome Carcinoma of colorectum without polyps | 6 genes
  (b) *BRCA1* and -2: Carcinoma of breast | 2 genes
  (c) Tuberous sclerosis complex: Slowly growing glial tumors | 2 genes
  (d) Carney complex | 2 genes
  (e) Li–Fraumeni syndrome | 2 (possibly 4) genes
  (f) Xeroderma pigmentosa (autosomal recessive) epidermal and melanotic malignancies | 6 (possibly 7) genes

5. Predispositions to different syndromes according to position of germ-line event in the one gene
  (a) *RET* gene: One germ-line mutations causes MEN Type 2, another causes Hirschprung's disease.
  (b) *PTEN* gene: Probably different germ-line mutations cause Cowden's, Bannayan–Riley–Ruvalcaba and Proteus-like syndromes

Some cases do not fit the criteria for these syndromes and may be called "Proteus-like" disorders [80].

Almost all cases of these syndromes carry a germ-line mutation in the *PTEN* (Phosphatase and TENsin homologue deleted on chromosome Ten) gene, which codes for the PTEN protein of cell signaling (Section 6.5.1). No other gene has been associated with this condition. Hence the collective term "*PTEN* Hamartoma tumor syndrome" may be used [80].

The precise mutations which cause the syndromes are not established [80].

7.7 GENOMIC MODELS FOR THE INHERITED PREDISPOSITIONS

The question of the specificity of the phenotypic effects of germ-line predispositions to tumors was raised in Section 7.1.2. Sections 7.2–7.7 have illustrated some of the many complexities of phenotype–genotype associations in inherited predispositions (Table 7.2). On the basis of current knowledge of the genome, some of these may have explanations, but the whole etiological sequences of most of the inherited predispositions are yet to be clarified.

7.7.1 Mutations in Different Genes Causing the Same Syndrome

An explanation of this is evident from knowledge of the tuberous sclerosis complex—*TSC1* and *TSC2*. The proteins—hamartin and tuberin respectively—coded by these genes participate in a complex having a specific function [82]. The function fails if either component is defective. This occurs when both alleles are lost—the Knudson model. Hence the outcome—loss of a glial cell-specific tumor-suppressor protein complex—is the cause of tumor formation.

7.7.2 Different Mutations in the Same Gene Causing Different Clinical Features

This phenomenon is shown by the *RET* gene (Section 7.6.1). The explanation lies in the multidomain nature of relevant protein. Thus in Hirschsprung's disease, the disease seems to occur with complete loss of *RET* protein signaling. However, for the MEN Type 2 syndromes, small genomic lesions in *RET* presumably lead to gain of function in the signaling protein product. The basis of the parent-cell specificity for the tumor formation, however, remains unclear.

7.7.3 Different Penetrances of Different Tumors Associated with the Same Germ-Line Event

This phenomenon is common (see Section 7.1.4). There are various possible explanations as follows.

a. *Modifier genes*
This is the commonest suggestion, e.g., Refs. [83–85] for the phenomenon. In principle, it implies that third, fourth, etc. genes *which are only expressed in the susceptible parent kind of cell* may play permissive or enhancing roles with respect to the effects of mutations in the main genes on the phenotype of the cell.

b. *Somatic mutations occur less commonly in one somatic cell compared to another*
This idea suggests the possibility that the occurrence of second genomic events may be dependent on the kind of parent cell of origin [86]. This could be based on different relative exposures to carcinogens. However, it seems an unlikely explanation because the patterns of different penetrances are different for each predisposition.

c. *The second somatic event occurs during embryonic development, when the parent cells of the tumors are, for some reason, similarly vulnerable to somatic mutation*
This possibility would explain the occurrence of multiple tumor types from neural cells (NF1 and NF2), or the association of large bowel and endometrial carcinomas in DNA mismatch repair syndromes.

7.7.4 Models for the Parent Cell–Type Specificities of Dominant Inherited Predispositions

This is perhaps the outstanding question concerning the inherited predispositions (Figure 7.8). The question is why the genomic events—part germ-line and part somatic—alter the phenotypes in only certain kinds of cell (see Section 7.1.2). As shown in the previous section, almost all of the genes so far identified as being associated with inherited predispositions are active in more kinds of cells than those in which the tumors develop.

Some possible explanations are:

a. *A parent cell–specific gene product is a necessary enhancer modifier of the tumorigenic effect of the loss of the tumor suppressor*
This might especially apply to transcription factors, many of which are known to affect multiple genes. Also,

Scheme 1, based on Knudson option 1 (second hit on same allele).

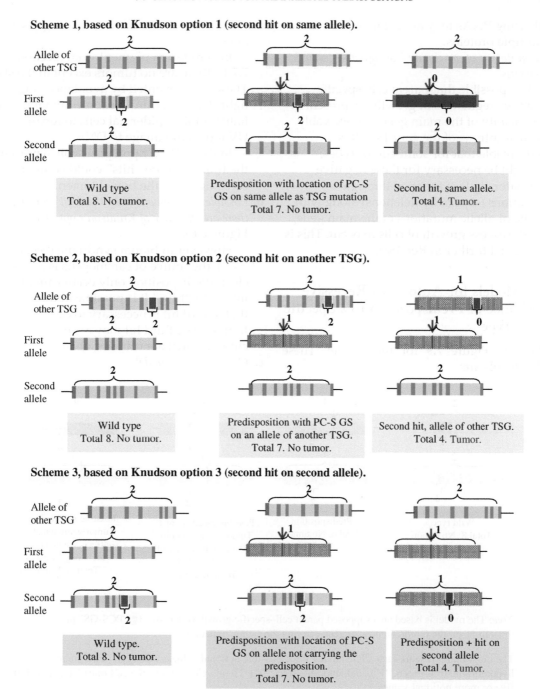

FIGURE 7.8 Possible genomic model for one germ-line genomic event predisposing to one tumor type of one (or closely related) kinds of cell based on supposed heterozygous intronic parent cell–specific growth suppressors (PC-S GS). The locations of these may be in three locations as in Knudson's three options for second hits on tumor suppressor genes (see Figure 6.6).

silencing RNAs may affect the translation of multiple proteins.

b. *Co-mutation of proximate (contiguous) genomic elements*

It is possible that parent cell–specific negative regulators of growth are present in the vicinity of the main gene—conceivably even as introns of the gene [87]. It is conceivable that for some tumor types, it might be necessary for these negative regulators to lose function through mutations (especially deletions) along with as the bi-allelic mutations of the main gene, for the excess growth of cells to occur. This is discussed further in Ref. [88].

7.7.5 Models for Autosomal Recessive Inheritance of Predisposition to a Specific Tumor Type

Refer to Figure 7.9 for this model. These model details are:

a. *The autosomal state allows greater ingress of carcinogen*

In xeroderma pigmentosum (Section 7.7.1), there are no tumors except in the skin. These are prevented by avoidance of UV light. Thus the third hit comes about through failure of the epidermal cells to repair UV-induced damage to DNA.

In the *MUTYH* polyposis coli syndrome, the third or more "hits" could come from carcinogens in the bowel lumen.

b. *Possible requirement of a "third hit" for which there are the three of Knudson's options (see Figure 6.6).*

However in Bloom syndrome (Section 7.7.4), the source of carcinogens is not as clear. Predisposition only occurs after two inherited "hits." It is possible in those cases that a third hit is necessary to cause the tumors. The third hit may arise though chromosomal aberrations alone.

c. *Other possible model*

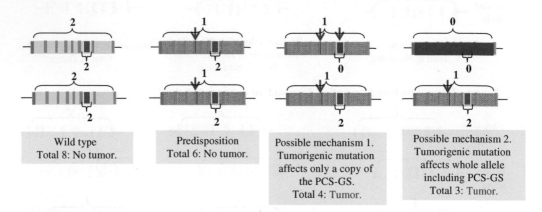

| Wild type Total 8: No tumor. | Predisposition Total 6: No tumor. | Possible mechanism 1. Tumorigenic mutation affects only a copy of the PCS-GS. Total 4: Tumor. | Possible mechanism 2. Tumorigenic mutation affects whole allele including PCS-GS Total 3: Tumor. |

Note: The model is based on a supposed parent cell–specific growth suppressor (■)"PCS-GS" in the introns of the tumor-suppressor gene carrying the exonic mutation.

The diagram represents each allele having three possible notional values: wild type (dark green; "2"), hypomorphic (red-green, "1"), and amorphic (red, "0"). To prevent a tumor, the cell must have a total of five or more notional values.

FIGURE 7.9 Two possible genomic schemes for parent-cell specificity of tumor formation in autosomal recessive conditions which do not involve carcinogens (see Section 7.7.5).

By suggesting the possible involvement of (i) parent cell–specific tumor-suppressor genomic elements and (ii) reposition effect (Section A2.3.2), models can be hypothesized for all the clinical aspects of inherited predispositions, such as different tumors in different parent cells, and different families with the same germ-line mutation suffering different syndromes of tumor susceptibility [88].

7.8 LOW-PENETRANCE INHERITED SUSCEPTIBILITY SYNDROMES IN HUMANS

7.8.1 Background

Low-penetrance predispositions are defined as genetic predispositions which cause only slightly more cases than occur spontaneously [89]. As a result, only a small proportion of individuals carrying the genomic lesion will develop the disease. They cannot be identified by studies of individual family trees, but only by statistical analyses of incidences in large numbers of families. Because of this, they are difficult to detect [90–91]. Methods for identifying them are generally [92]:

i. via candidate genes: studying particular genes for mutations in cancer patients versus noncancer controls;
ii. via genome-wide association studies. This usually means of the "exome"—all the exonic DNA—which is 1–2% of the genome (see Section A2.1.1).

7.8.2 In Relation to Carcinoma of the Breast

Large numbers of studies have been reported, and genes involved in DNA repair, such as *CHEK2, ATM, BRIP1 (FANCJ), PALB2 (FANCN),* and *RAD51C (FANCO),* appear to be associated with moderate risk of this tumor type [93,94].

7.8.3 In Relation to Other Tumors

For colonic carcinoma, *APC 1130 K* appears to double the risk of colorectal carcinoma. Other low-penetrance genes are those associated with hamartomatous and other polyposis syndromes [95,96]. For prostatic carcinoma, various genes including *RNaseL, ElaC2, MSR1* have been implicated in a hereditary component to this disease [97].

7.8.4 Studies Involving Genome-Wide Association Studies in Human Populations

The question here is whether or not any specific combinations of these germ-line predispositions might contribute to a greater likelihood of an individual developing a tumor of a specific tumor type. Conventional studies of family trees and gene mapping are not able to answer this question. Theoretically, possible answers to these questions might come through:

i. Genomic lesion data through high-throughput DNA sequencing
ii. Gene transcription data through high-throughput RNA sequencing
iii. Gene expression data through microarray studies.

However, at the present time, the problems essentially comprise the following:

i. What is "normal" in the genome?
ii. What is "normal" in RNA expression?
iii. What abnormalities, if found, are experimental errors, present but spurious, or pathogenetic?

Results of sequencing of DNA of individuals require considerable skill in interpretation and discussion [98–100].

7.9 HEREDITARY PREDISPOSITIONS TO TUMORS IN EXPERIMENTAL ANIMALS

7.9.1 Incrementally Increasing Susceptibility with In-Breeding

Although these tumors are of limited direct relevance to human medicine, this group of hereditary predispositions do support concepts relevant to principles of pathogenesis for human tumors.

Studies of this type began in the early twentieth century using mainly mice. The goal was to establish whether the "spontaneous" tumors in these animals arise through the simple Mendelian model of dominant or recessive individual genes [101,102]. By the 1930s, however, it was accepted that the tumors increased in incidence through "multiple small increments"—a non-Mendelian pattern of inheritance—due to polygenism (see Section A2.2.3) [103]. This explanation appears to be confirmed by recent work in many different experimental models [104–107].

7.9.2 Spontaneous Large Impact Genomic Events Predisposing to Tumors in Animals

These predispositions in animals have similar genetic and genomic features to those of human large impact predispositions. The known animal predispositions comprise a small number of disorders, e.g., melanomas and lymphomas in pigs [108] and melanomas, lymphomas, gastrointestinal stromal tumors, and renal cell carcinomas in mice [109] An example of an inherited predisposition in animals to more than one type of lesion is hereditary renal cell carcinoma in dogs which is associated with nodular dermatofibrosis [110]. Intestinal lymphomas in cats have been found to be associated with mutations in mismatch repair genes in these animals [111]. All these predispositions are either Mendelian "dominant" or "recessive." Mechanisms of these tumors are generally thought to be via "somatic second hits" according to Knudson's model for human predispositions.

7.9.3 Genetically Engineered Large Impact Inherited Predispositions to Tumors in Animals

The fact that DNA prepared from one organism can enter another and result in transfers of traits was noted by Avery et al. [112]. In 1952, Hershey and Chase [113] reported that bacteriophage viruses could carry genes into bacteria. In the 1970s, "restriction" DNases—of the type which cut DNA at specific nucleotide sequences—were discovered [114]. This meant that so that a particular short piece of DNA (rather than random fragments) could be prepared. Genes could be either isolated or synthesized, and injected into pluripotential embryonic cells, which then developed into genetically modified mature animals [115]. This would increase the copy numbers of genes in the recipient animals' genomes. With numerous technical developments [116], genes can be inserted at particular sites, wholly or partially deleted, or modified in other ways.

Using these technologies, it has been possible to create models for almost every human cancer [117]. These enable testing of hypotheses concerning the possible roles of oncogenes and tumor-suppressor genes, as well as various other hypotheses concerning phenotype–genotype relationships in human tumors [116–120]. Large numbers of new models, including ones which closely mimic the histopathological features of human tumors [121], are being developed at the present time.

References

[1] Nagy R, Sweet K, Eng C. Highly penetrant hereditary cancer syndromes. Oncogene 2004;23(38):6445–70.
[2] Ackerknecht EH. Diathesis: the word and the concept in medical history. Bull Hist Med 1982;56(3):317–25.

[3] Wolff J. Die Lehre von der Krebskrankheit von den Aeltesten Zeiten bis zur Gegenwart /The science of cancerous disease from earliest times to the present (1907). [Ayoub B, Trans.]. Introduced by P. Sarco; Canton, MA: Science History Publications USA; 1989. esp ps 78ff, 144, and 260ff.

[4] Warthin AD. Heredity with reference to carcinoma: as shown by the study of the cases examined in the pathological laboratory of the University of Michigan, 1895–1913. Arch Intern Med 1913;12:546–55.

[5] Lockhart-Mummery A. Cancer and heredity. Lancet 1925:427–9.

[6] Bauer KF. Zentralblatt. f. Chirurg 1927;No 15:943ff.

[7] Cockayne EA. Inherited abnormalities of the skin and its appendages. Oxford: Oxford University Press; 1933.

[8] Society for Developmental Biology Wingless gene. Available at: http://www.sdbonline.org/sites/fly/segment/wingles1.htm.

[9] Berliner JL, Fay AM. Risk assessment and genetic counseling for hereditary breast and ovarian cancer: recommendations of the National Society of Genetic Counselors. J Genet Couns 2007;16(3):241–60.

[10] Poduri A, Evrony GD, Cai X, et al. Somatic mutation, genomic variation, and neurological disease. Science 2013;341(6141):1237758.

[11] Hafner C, Groesser L. Mosaic RASopathies. Cell Cycle 2013;12(1):43–50.

[12] Genetics Home Reference. Penetrance. Available at: http://ghr.nlm.nih.gov/glossary=penetrance.

[13] Genetics Home Reference. Expressivity. Available at: http://ghr.nlm.nih.gov/glossary=expressivity.

[14] Miko I. Phenotype variability: penetrance and expressivity. Nat Educ 2008;1(1):137.

[15] Kleinerman RA, Schonfeld SJ, Tucker MA. Sarcomas in hereditary retinoblastoma. Clin Sarcoma Res 2012;2:15.

[16] Chauveinc L, Mosseri V, Quintana E, et al. Osteosarcoma following retinoblastoma: age at onset and latency period. Ophthalmic Genet 2001;22:77–88.

[17] Carney JA, Hruska LS, Beauchamp GD, et al. Dominant inheritance of the complex of myxomas, spotty pigmentation, and endocrine overactivity. Mayo Clin Proc 1986;61(3):165–72.

[18] Carney JA. Gastric stromal sarcoma, pulmonary chondroma, and extra-adrenal paraganglioma (Carney Triad): natural history, adrenocortical component, and possible familial occurrence. Mayo Clin Proc 1999;74(6):543–52.

[19] van Lier MG, Wagner A, Mathus-Vliegen EM, et al. High cancer risk in Peutz-Jeghers syndrome: a systematic review and surveillance recommendations. Am J Gastroenterol 2010;105(6):1258–64.

[20] Atlas of Genetics and Cytogenetics in Oncology and Haematology. Familial melanoma. Available at: http://atlasgeneticsoncology.org/Kprones/Familial MelanomID10088.html.

[21] Online Mendelian Inheritance in Man. Melanoma, Cutaneous Malignant, Susceptibility 1; CMM1. Available at: http://www.omim.org/entry/155600.

[22] Marini F, Falchetti A, Luzi E, et al. Multiple Endocrine Neoplasia Type 1 (MEN1) Syndrome. Cancer Syndromes (Internet). Available at: http://www.ncbi.nlm.nih.gov/books/NBK7029/.

[23] Sheppard KE, McArthur GA. The cell-cycle regulator CDK4: an emerging therapeutic target in melanoma. Clin Cancer Res 2013;19:5320.

[24] Online Mendelian Inheritance in Man. Ataxia-Telangiectasia mutated gene; ATM. Available at: http://omim.org/entry/607585.

[25] Theodoratou E, Campbell H, Tenesa A, et al. A large-scale meta-analysis to refine colorectal cancer risk estimates associated with MUTYH variants. Br J Cancer 2010;103:1875–84.

[26] Genetics Home Reference. What is neurofibromatosis type 2? Available at: http://ghr.nlm.nih.gov/condition/neurofibromatosis-type-2.

[27] Evans DG. Neurofibromatosis 2. In: GeneReviews (Internet). Available at: http://www.ncbi.nlm.nih.gov/books/NBK1201/.

[28] Genetics Home Reference. Home page. Available at: http://ghr.nlm.nih.gov/.

[29] Lohmann D. Retinoblastoma. Adv Exp Med Biol 2010;685:220–7.

[30] Lohmann D, Gallie B, Dommering C, et al. Clinical utility gene card for: retinoblastoma. Eur J Hum Genet 2011;19(3).

[31] Lohmann D, Gallie B. Retinoblastoma. GeneReviews (Internet). Available at: http://www.ncbi.nlm.nih.gov/books/NBK1452/.

[32] Hansen MF, Koufos A, Gallie BL, et al. Osteosarcoma and retinoblastoma: a shared chromosomal mechanism revealing recessive predisposition. Proc Natl Acad Sci USA 1985;82(18):6216–20.

[33] Rodriguez-Bigas MA. Genotype phenotype correlation in familial adenomatous polyposis. In: Rodriguez-Bigas MA, Cutait R, Lynch PM, editors. Hereditary colorectal Cancer. New York, NY: Springer; 2010. pp. 197–202.

[34] Jass JR. The pathologist and the phenotype of hereditary colorectal cancer. In: (33). pp. 175–94.

[35] Carvajal-Carmona LG, Silver A, Tomlinson IP. Molecular genetics of familial adenomatous polyposis. In: (33). pp. 45–66.

[36] Turcot J, Despres JP, St Pierre F. Malignant tumors of the central nervous system associated with familial polyposis of the colon: report of two cases. Dis Colon Rectum 1959;2:465–8.

[37] Digilio MC. Turcot syndrome. Atlas Genet Cytogenet Oncol Haematol 2006;10(1):38–9.

[38] Online Mendelian Inheritance in Man. Familial adenomatous polyposis 1; FAP1. Available at: http://omim.org/entry/175100.

[39] von Recklinghausen FD. Über die multiplen Fibrome der Haut und ihre Beziehung zu den multiplen Neuromen. Festschrift für Rudolf Virchow. Berlin; 1882.

[40] Antônio JR, Goloni-Bertollo EM, Trídico LA. Neurofibromatosis: chronological history and current issues. An Bras Dermatol 2013;88(3):329–43.

[41] Genetics Home Reference. Neurofibromatosis type 1. Available at: http://ghr.nlm.nih.gov/condition/neurofibromatosis-type-1.

[42] Friedman JM. Neurofibromatosis 1. GeneReviews (Internet). Available at: http://www.ncbi.nlm.nih.gov/books/NBK1109/.

[43] Chou A, Toon C, Pickett J, et al. von Hippel–Lindau syndrome. Front Horm Res 2013;41:30–49.

[44] Frantzen C, Links TP, Giles RH. Von Hippel-Lindau Disease. GeneReviews (Internet). Available at: http://www.ncbi.nlm.nih.gov/books/NBK1463/.

[45] http://ghr.nlm.nih.gov/condition/von-hippel-lindau-syndrome.

[46] Genetics Home Reference. Von-Hippel-Lindau syndrome. Available at: http://ghr.nlm.nih.gov/condition/gorlin-syndrome.

[47] Evans DG. Nevoid basal cell carcinoma syndrome. GeneReviews (Internet). Available at: http://www.ncbi.nlm.nih.gov/books/NBK1151/.

[48] Atlas of Genetics and Cytogenetics in Oncology and Haematology. Ataxia-Telangiectasia and variants. Available at: http://atlasgeneticsoncology.org/Deep/ATMID20006.html.

[49] Atlas of Genetics and Cytogenetics in Oncology and Haematology. Bloom syndrome. Available at: http://atlasgeneticsoncology.org/Kprones/BLO10002.html.

[50] Online Mendelian Inheritance in Man. Bloom syndrome; BLM. Available at: http://omim.org/entry/210900.

[51] de la Chapelle A. The incidence of Lynch syndrome. Fam Cancer 2005;4(3):233–7.

[52] Kohlmann W, Gruber SB. Lynch syndrome. GeneReviews (Internet). Available at: http://www.ncbi.nlm.nih.gov/books/NBK1211/.

[53] Vasen HFA.Hardwick JCH. An overview of the Lynch syndrome (hereditary non-polyposis colorectal cancer). In: Rodriguez-Bigas MA, Cutait R, Lynch PM, editors. Hereditary colorectal Cancer. New York, NY: Springer; 2010. pp. 271–99.

[54] Petrucelli N, Daly MB, Feldman GL. BRCA1 and BRCA2 hereditary breast and ovarian cancer. GeneReviews (Internet). Available at: http://www.ncbi.nlm.nih.gov/books/NBK1247/.

[55] National Cancer Institute. Genetics of breast and gynecologic cancers–for health professionals. Available at: http://www.cancer.gov/cancertopics/pdq/genetics/breast-and-ovarian/HealthProfessional/page2.

[56] Evans DG, Shenton A, Woodward E, et al. Penetrance estimates for BRCA1 and BRCA2 based on genetic testing in a clinical cancer genetics service setting: risks of breast/ovarian cancer quoted should reflect the cancer burden in the family. BMC Cancer 2008;8:155.

[57] Caputo S, Benboudjema L, Sinilnikova O, et al. French BRCA GGC Consortium. Description and analysis of genetic variants in French hereditary breast and ovarian cancer families recorded in the UMD-BRCA1/BRCA2 databases. Nucleic Acids Res 2012;40(Database issue):D992–1002.

[58] Spurdle AB, Healey S, Devereau A, et al. ENIGMA. ENIGMA—evidence-based network for the interpretation of germline mutant alleles: an international initiative to evaluate risk and clinical significance associated with sequence variation in BRCA1 and BRCA2 genes. Hum Mutat 2012;33(1):2–7.

[59] National Cancer Institute. Late effects of treatment for childhood cancer–for health professionals. Available at: http://www.cancer.gov/cancertopics/pdq/treatment/wilms/HealthProfessional/page2.

[60] Davidoff AM. Wilms' tumor. Curr Opin Pediatr 2009;21(3):357–64.

[61] Erson AE, Petty EM. Kidney: nephroblastoma (Wilms tumor). Atlas Genet Cytogenet Oncol Haematol 2007;11(1):50–3.

[62] Dome JS, Huff V. Wilms tumor overview. GeneReviews (Internet). Available at: http://www.ncbi.nlm.nih.gov/books/NBK1294/.

[63] Choufani S, Shuman C, Weksberg R. Molecular findings in Beckwith–Wiedemann syndrome. Am J Med Genet C Semin Med Genet 2013;163C(2):131–40.

[64] Genetics Home Reference. Tuberous sclerosis complex. Available at: http://ghr.nlm.nih.gov/condition/tuberous-sclerosis-complex.

[65] Burger PC, Scheithauer BW. Tumors of the central nervous system. AFIP Atlas of Tumor Pathology, Ser 4, vol. 7; 2007. p. 512.

[66] Sosunov AA, Wu X, Weiner HL, et al. Tuberous sclerosis: a primary pathology of astrocytes? Epilepsia 2008;49(Suppl. 2):53–62.

[67] Northrup H, Koenig MK, Au KS. Tuberous sclerosis complex. GeneReviews (Internet). Available at: http://www.ncbi.nlm.nih.gov/books/NBK1220/.

[68] Sancak O, Nellist M, Goedbloed M, et al. Mutational analysis of the TSC1 and TSC2 genes in a diagnostic setting: genotype—phenotype correlations and

comparison of diagnostic DNA techniques in Tuberous Sclerosis Complex. Eur J Hum Genet 2005;13:731–41.

[69] Courcoutsakis NA, Tatsi C, Patronas NJ, et al. The complex of myxomas, spotty skin pigmentation and endocrine overactivity (Carney complex): imaging findings with clinical and pathological correlation. Insights Imaging 2013;4:119–33.

[70] Sorrell AD, Espenschied CR, Culver JO, et al. Tumor protein p53 (TP53) testing and Li–Fraumeni syndrome: current status of clinical applications and future directions. Mol Diagn Ther 2013;17(1):31–47.

[71] Ognjanovic S, Olivier M, Bergemann TL, et al. Sarcomas in TP53 germline mutation carriers: a review of the IARC TP53 database. Cancer 2012;118(5):1387–96.

[72] Genetics Home Reference. Li-Fraumeni syndrome. Available at: http://ghr.nlm.nih.gov/condition/li-fraumeni-syndrome.

[73] Bachinski LL, Olufemi SE, Zhou X, et al. Genetic mapping of a third Li–Fraumeni syndrome predisposition locus to human chromosome 1q23. Cancer Res 2005;65(2):427–31.

[74] Malkin D. Li–Fraumeni syndrome. Genes Cancer 2011;2(4):475–84.

[75] DiGiovanna JJ, Kraemer KH. Shining a light on xeroderma pigmentosum. J Invest Dermatol 2012;132(3 Pt 2):785–96.

[76] Cleaver JE, Lam ET, Revet I. Disorders of nucleotide excision repair: the genetic and molecular basis of heterogeneity. Nat Rev Genet 2009;10(11):756–68.

[77] Moline J, Eng C. Multiple endocrine neoplasia Type 2. GeneReviews (Internet). Available at: http://www.ncbi.nlm.nih.gov/books/NBK1257/.

[78] Genetics Home Reference. RET. Available at: http://ghr.nlm.nih.gov/gene/RET.

[79] Genetics Home Reference. RET. Available at: http://ghr.nlm.nih.gov/condition/cowden-syndrome.

[80] Eng C. PTEN hamartoma tumor syndrome. Available at: http://www.ncbi.nlm.nih.gov/books/NBK1488/.

[81] Mester JL, Tilot AK, Rybicki LA, et al. Analysis of prevalence and degree of macrocephaly in patients with germline PTEN mutations and of brain weight in Pten knock-in murine model. Eur J Hum Genet 2011;19(7):763–8.

[82] Steffann J, Munnich A, Bonnefont JP. Tuberous sclerosis (TSC). Atlas Genet Cytogenet Oncol Haematol 2002;6(4):314–5.

[83] Bignon Y-J. Biological basis of cancer predisposition. In: Eng R, Easton D, Ponder B, editors. Genetic predisposition to cancer (2nd ed.). Boca Raton, FL: CRC Press; 2004. pp. 16–19.

[84] Ripperger T, Gadzicki D, Meindl A, et al. Breast cancer susceptibility: current knowledge and implications for genetic counselling. Eur J Hum Genet 2009;17(6):722–31.

[85] de la Chapelle A. Genetic predisposition to colorectal cancer. Nat Revs Cancer 2004;4:769–80.

[86] Sigal A, Rotter V. The oncogenic activity of p53 mutants. In: Zambetti GP, editor. The p53 tumor suppressor pathway and Cancer. New York, NY: Springer; 2007. p. 207.

[87] Lutter D, Marr C, Krumsiek J, et al. Intronic microRNAs support their host genes by mediating synergistic and antagonistic regulatory effects. BMC Genomics 2010;11:224.

[88] Bignold LP. The cell-type-specificity of inherited predispositions to tumours: review and hypothesis. Cancer Lett 2004;216(2):127–46.

[89] Croager E. Locating low-penetrance genes. Nat Revs Cancer 2003;3(8) 553.

[90] Turnbull C, Rahman N. Genetic predisposition to breast cancer: past, present, and future. Annu Rev Genomics Hum Genet 2008;9:321–45.

[91] Lalloo F, Evans DG. Familial breast cancer. Clin Genet 2012;82(2):105–14.

[92] National Cancer Institute. Genetics of breast and gynecologic cancers–for health professionals. Available at: http://www.cancer.gov/cancertopics/pdq/genetics/breast-and-ovarian/HealthProfessional/page3.

[93] Filippini SE, Vega A. Breast cancer genes: beyond BRCA1 and BRCA2. Front Biosci (Landmark Ed) 2013;18:1358–72.

[94] Shuen AY, Foulkes WD. Inherited mutations in breast cancer genes—risk and response. J Mammary Gland Biol Neoplasia 2011;16(1):3–15.

[95] Jeter JM, Kohlmann W, Gruber SB. Genetics of colorectal cancer. Oncology (Williston Park) 2006;20(3):269–76.

[96] Whiffin N, Houlston RS. Architecture of inherited susceptibility to colorectal cancer: a voyage of discovery. Genes (Basel) 2014;5(2):270–84.

[97] Alberti C. Hereditary/familial versus sporadic prostate cancer: few indisputable genetic differences and many similar clinicopathological features. Eur Rev Med Pharmacol Sci 2010;14(1):31–41.

[98] Krepischi AC, Pearson PL, Rosenberg C. Germline copy number variations and cancer predisposition. Future Oncol 2012;8(4):441–50.

[99] Hu H, Roach JC, Coon H, et al. A unified test of linkage analysis and rare-variant association for analysis of pedigree sequence data. Nat Biotechnol 2014;32(7):663–9.

[100] Thrift AP, Whiteman DC. Can we really predict risk of cancer? Cancer Epidemiol 2013;37(4):349–52.

[101] Cuddihy J. Maud slye. Cancer Cells 1991;3(9):366–8.

[102] Crow JF. C. C. Little, cancer and inbred mice. Genetics 2002;161:1357–61.

[103] Gaudilliere J-P. Circulating mice and viruses. In: Fortun M, Mendelsohn E, editors. The practices of human genetics. Dordrecht: Kluwer; 1999. p. 119.

[104] Dragani TA, Canzian F, Pierotti MA. A polygenic model of inherited predisposition to cancer. FASEB J 1996;10(8):865–70.

[105] Szpirer C. Cancer research in rat models. Methods Mol Biol 2010;597:445–58.

[106] Heaney JD, Nadeau JH. Testicular germ cell tumors in mice: new ways to study a genetically complex trait. Methods Mol Biol 2008;450:211–31.

[107] De Miglio MR, Pascale RM, Simile MM, et al. Polygenic control of hepatocarcinogenesis in Copenhagen x F344 rats. Int J Cancer 2004;111(1):9–16.

[108] Misdorp W. Congenital and hereditary tumours in domestic animals. 2. Pigs. A review. Vet Q 2003;25(1):17–30.

[109] Hino O, Kobayashi T, Okimoto K. Genetic and environmental factors in hereditary predisposition to tumors: a conceptual overview. EXS 2006;vol. 96:269–92.

[110] Lingaas F, Comstock KE, Kirkness EF, et al. A mutation in the canine BHD gene is associated with hereditary multifocal renal cystadenocarcinoma and nodular dermatofibrosis in the German Shepherd dog. Hum Mol Genet 2003;12(23):3043–53.

[111] Aberdein D, Munday JS, Howe L, et al. Widespread mismatch repair expression in feline small intestinal lymphomas. J Comp Pathol 2012;147(1):24–30.

[112] Avery OT, MacLeod CM, McCarty M. Studies on the chemical nature of the substance inducing transformation of pneumococcal types: induction of transformation by a desoxyribonucleic acid fraction isolated from *Pneumococcus* Type III. J Exp Med 1944;79(2):137–58.

[113] Hershey A, Chase M. Independent functions of viral protein and nucleic acid in growth of bacteriophage. J Gen Physiol 1952;36(1):39–56.

[114] Smith HO, Wilcox KW. A restriction enzyme from *Haemophilus influenzae*. I. Purification and general properties. J Mol Biol 1970;51(2):379–91.

[115] Jaenisch R, Mintz B. Simian virus 40 DNA sequences in DNA of healthy adult mice derived from preimplantation blastocysts injected with viral DNA. Proc Natl Acad Sci USA 1974;71(4):1250–4.

[116] Primrose SB, Twyman RM. Principles of gene manipulation and genomics (7th ed.). Malden, MA: Blackwell; 2006.

[117] Vignjevic D, Fre S, Louvard D, et al. Conditional mouse models of cancer. Handb Exp Pharmacol 2007;178:263–87.

[118] Genetically engineered mice for cancer research. In: Green JE, Reid T, editors. Design, analysis, pathways, validation and pre-clinical testing. New York, NY: Springer; 2012.

[119] Walrath JC, Hawes JJ, Van Dyke T, et al. Genetically engineered mouse models in cancer research. Adv Cancer Res 2010;106:113–64.

[120] Taneja P, Zhu S, Maglic D, Fry EA, et al. Transgenic and knockout mice models to reveal the functions of tumor suppressor genes. Clin Med Insights Oncol 2011;5:235–57.

[121] Saborowski M, Saborowski A, Morris 4th JP, et al. A modular and flexible ESC-based mouse model of pancreatic cancer. Genes Dev 2014;28(1):85–97.

Morphology, Type Characteristics, and Related Features of Tumors

L.P. Bignold: Principles of Tumors.
DOI: http://dx.doi.org/10.1016/B978-0-12-801565-0.00008-1

Almost every tumor exhibits morphological abnormalities, which include various deviations from the morphology of their parent cell. On the basis of these abnormalities, often in conjunction with other features, tumors are classified into a thousand or so different types. All the abnormalities can be seen as variable components of "broad complex phenotypes," which theories of tumors—and particularly genetic theories—might be expected to explain (Section 2.5). Issues in the "broad phenotypes" include:

i. The phenomenology of the individual traits
ii. The phenomenology of the combinations of traits which characterize each tumor type.

Historically, the writings of David Paul Hansemann [1] included this kind of analysis. Currently however this topic is not frequently discussed. Much of the difficulty in approaching the morphological and related features of tumors according to basic genetic principles may lie in the fact that the abnormalities in features are extraordinarily diverse and variable—both individually and in their combinations (see Section 2.5.1). This diversity is similar in principle to the diversity in the expressions of proteins and other antigens in tumor cell populations (Section 9.3.5), as well as to the clinical behaviors of the tumor types (Chapter 1). The diversity is also similar to that in the susceptibilities to treatments (see Chapters 11–13).

This chapter introduces the morphological features and the characteristics of their combinations in the various tumor types.

Section 8.1 discusses general aspects of the features in tumor cells. Section 8.2 describes tumor types in terms of combinations of traits. Section 8.3 is concerned with variabilities in tumors and Section 8.4 with progression in tumor cell populations. Section 8.5 mentions some possible genomic mechanisms of the combinations of traits exhibited by the various tumor types. Section 8.6 deals with whether or not leukemias and related hematopoietic cell tumors are fundamentally similar to "solid" tumors.

8.1 MORPHOLOGICAL AND OTHER FEATURES OF TUMOR CELLS

8.1.1 Classification of Features of Tumors According to Possible Origins

The "parent" cells of tumors are almost entirely in the categories of "labile"—i.e., those which proliferate either as part of their normal physiological role—and "stable," i.e., those which proliferate only in response to appropriate stimuli (Section A1.3.3). "Permanent" cells in adults (Section A1.3.3) do not give rise to tumors. There are no "generic" or completely unspecialized cells from which multiple types of completely different tumor types originate. That is to say, e.g., there is no cell which sometimes gives rise to carcinomas and sometimes to sarcomas. Certain cases of tumor occur in which carcinoma and sarcoma appear to be present together. These are considered due to focal ectopic specializations in descendants of the original tumor cell (see Section 8.1.2d).

The traits in tumors, in relation to the features of their normal cells, can be seen to fall into the following categories (Table 8.1).

(a) Quantitative Gains in Features in the Parent Cell, Including Those of Its Reactive Repertoire

These features of tumor cells include the following:

i. Increases in a feature which is present in the parent cell in normal physiological circumstances. This is an explanation of excess growth (see Section 6.1) in "labile" populations of cells. Another example is the occasional increases in mucus production seen in some cases of tumors, such as carcinoma of the colon (Section 8.1.3).

TABLE 8.1   Classification of Traits of Tumors by Possible Origin

1. *Gains in feature of parent cell*
   Increase in normal parent-cell rate of cell production
      (labile parent cells)
   Instigation of cell production without normal stimulus
      (stable parent cells)
   Increased specialization product per cell (uncommon)
   Increased biomarker production per cell (uncommon,
      see Section 9.7)
2. *Gains in feature of a nonparent cell*
   Greater rate of cell production (possibly as an
      embryonic trait)
   Invasion and metastasis (possibly as a leukocytic trait)
   –all embryonic traits and all metaplasias
3. *Losses of features of parent cell*
   Reduced specialization product (extremely common)
   Reduced cyto-structural regularity (extremely common)
   Reduced chromosomal and mitotic regularity
      (extremely common)
   Reduced biomarker production per cell (very common,
      see Section 9.7)

Note: For most types of tumors, the traits in individual cases may be individually focal and quantitatively variable.

---

ii. Those which the kind of cell of origin does not show in normal circumstances, but which the cell of origin may exhibit given the right stimulus. This is one possible mechanism of excess growth (see Section 6.1). An example could be in the reactive change of fibrocytes into fibroblasts (Section A1.2.7).

(b) Gains in Features Exhibited Only by Other Kinds of Cells

This refers to features which occur in a tumor type, but which are never seen in the parent kind of cell of that tumor type. These features do, however, occur in other kinds of cells in the body, including embryonic cells.

Thus in particular, excess growth could occur through gains in activity of embryonic gene activities. As mentioned in Sections 6.2 and 6.3, many oncogenes are normally expressed in particular embryonic cells. Invasion and metastasis are further phenomena

which have counterparts in normal hemato-lymphoid cells (see Sections 6.1.1 and A1.5).

A morphological example of features in tumor cells deriving from other kinds of cells is particularly demonstrated in metaplasias, as discussed in Sections 8.2 and A1.4.4. It can be noted here that this idea of metaplasia in the features of tumors was described in the nineteenth century by Virchow [2]. He stressed that almost all individual features of tumors can be seen in one or other normal human cell. The view was also supported by Foulds [3].

(c) Quantitative Reductions in the Features of the Parent Cells

These reductions are most obvious in degrees of specialization (Section 8.1.2). Others are also reductions in cyto-structural regularity (see Section A1.2.2). This loss of regularity of tumors is often referred to as "cytological atypia."

8.1.2 Abnormalities in Specialization

Overall, tumor cell populations usually show the full range of specialization in comparison with the parent cell from which the tumor arose (i.e., of lineage-faithful specialization). These degrees in specialization are as follows:

(a) Normo-Specialization in Tumors

Normal specialization of the main cell types is described in Section A1.2.2. The cells of a few tumor types show normal degrees of specialization. Examples include lipomas and hemangiomas [4], and some adenomas of the adrenal gland [5].

(b) Hypo-Specialization

As mentioned in Section 1.3, this is the commonest alteration in specialization among tumor cell populations. There is less specialized product in the tumor cell than in the corresponding cell type from which the tumor arose (Figure 8.1A). In a small proportion of tumors, the cells have no residual morphological

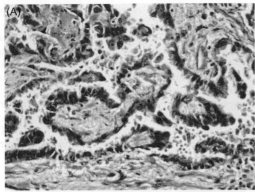

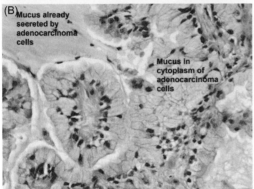

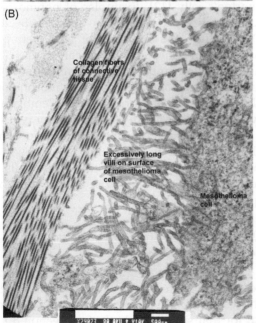

◀ **FIGURE 8.1** Reduced and increased specialization in tumors. (A) Hypospecialization indicated by absence of mucus production in a papillary carcinoma of lung. (B) Hyperspecialization indicated by excessive mucus production in a carcinoma of lung. (C) Hyperspecialisization indicated by elongation of microvilli in malignant mesothelioma (EM ×29,000). Courtesy Dr P. Sutton-Smith.

evidence of specialization and are called "undifferentiated" or "anaplastic." In many of these cases, some evidence of the kind of cell of origin can be found by the kinds of antigens which they express (see in Section 9.1). Reduced specialization, however, also occurs in nontumorous pathological processes, being associated with regeneration and wound healing (see Section A1.4.3). This indicates that specialization and viability are not always related traits of the cells, but that the specialization process can be inhibited or suppressed while the cell lives and grows.

(c) Hyper-Specialization

This feature is indicated by excess specialization product in or adjacent to tumor cells. The simplest example is in the "blue" nevus, in which all cases show marked overproduction of melanin per cell. As another example, in "mucinous" (syn. "colloid") adenocarcinoma cells, there is excessive accumulation of mucus (Figure 8.1B). This change is seen in many adenocarcinomas, e.g., of the colon, breast, and also in the mucinous variant of "broncho-alveolar" carcinomas of the lung [6]. Hyper-specializations may be seen in certain reactive pathological conditions, such as the keratosis seen in chronic mechanical trauma to the epidermis (e.g., lichen simplex chronicus or clavus) [7].

The concept of hyper-specialization may be applied also in mesotheliomas. In this condition, electron microscopy can be used to show a diagnostic feature in this tumor type: elongation rather than shortening of microvilli on mesothelioma cells [8] (Figure 8.1C).

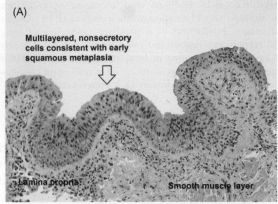

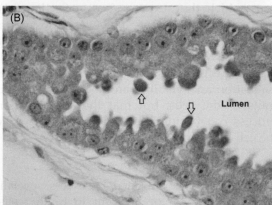

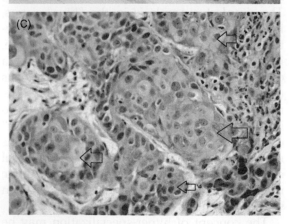

FIGURE 8.2 Example of ectopic specialization (metaplasia) in tumor cells. (A) Early squamous metaplasia in bronchial epithelium. This may be a pre-malignant condition. (B) Apocrine metaplasia in glandular breast tissue. This metaplasia is very commonly found in benign mammary dysplasia, and is not thought to be a pre-malignant condition. The phenomenon of apocrine secretion by breaking-off of the upper parts of the cells (arrowed) into the lumen of the gland is clearly seen. (C) Squamous metaplasia in transitional cell carcinoma of the bladder. The squamous cells are mainly in clusters (arrowed). It is not thought to have any prognostic significance.

(d) Embryonic Reversions

These abnormal specializations refer to the situation in which a descendant of an adult stem cell acquires certain morphological appearances or biochemical properties of an embryonic cell (see Section 2.3.3 and [9–11]). Historically, examples of the former were thought to include myxomas, because of their resemblance to embryonic connective tissue and in particular the myxoid kind known as Wharton's jelly [12]. Examples of nonmorphological embryonic reversions (see also Section 9.3.2) include the secretion of alpha-feto-proteins by some hepatocellular carcinomas [13] and the carcino-embryonic antigen produced by some colonic carcinomas [14]. Like morphological reversions, nonmorphological "embryonic reversions" occur in only specific tumor types.

Tumor cells with morphological or biochemical resemblances to embryonic cells, however, do not become functional embryonic cells: if they did, they would give rise to new adult cell types in the tumor masses.

(e) Ectopic Specializations, Including Metaplasias in Tumors

This feature (see Figure 8.2) consists of the presence at a site in the body of tumors cells which specialize in a direction which is not normal for that site in the body (see Section 8.1.1c and [9]).

Ectopic specializations are seen in tumors in three circumstances:

i. All the cells of the tumor manifest only the specialization of a type of cell not normally found in the site—i.e., a metaplasia. For example in the bronchus, carcinomas of keratin-producing squamous type are common, although no squamous cells normally exist in the bronchus [15]. The term "metaplastic carcinoma" is commonly used for primary squamous cell carcinoma of the breast.

ii. Foci are present in some cases of a tumor type which is always made up mainly of another cell type. Examples are numerous and include occasional foci of squamous cell carcinoma in otherwise typical transitional cell carcinomas of the bladder [16] and Figure 8.2C.

iii. "Nonmorphological ectopic specializations" as can be identified in some tumor types through endocrinological and immunological investigations. The most obvious are the "inappropriate hormone secretion" syndromes, such as of adrenocorticotrophic hormone by some cases of carcinoma of the lung [17]. Yet another example may be of "neuroendocrine" antigen expression in morphologically typical other types of carcinomas (see Section 8.1.3) [18].

(f) Mixed Specializations

This refers to tumor types in which all cases contain cells of more than one kind of specialization from their inception. A common example is the fibro-adenoma of the breast (Figure 8.3A). Both components of normal breast are increased.

Other examples are "pleomorphic adenoma" ("mixed tumor") of salivary gland [19] (Figure 8.3B) and the muco-epidermoid tumor of salivary gland [20] and less commonly of the bronchus (Figure 8.3C). These tumors show both epithelial-type and connective tissue-type specializations. In these tumors, at least one of the kinds of specializations in the tumors cells has no normal counterpart at the site of the tumor.

In the literature of histopathology, tumors comprising more than one kind of cell have been named in various ways depending on the theoretical concepts of tumor cell formation (see Section 2.3). For example, where a tumor is named according to "lineage specificity" ideas, those tumors which have both epithelial and mesenchymal components may be called "biphasic" as in synoviosarcomata [21] or "mixed" as "mixed mesotheliomas" [22].

However, when tumors are named according to ideas of "metaplasia," they are called "metaplastic" tumors, e.g., in the breast ([23] and see above).

Other terms do not indicate any theory and are descriptive only. For example "carcinosarcomas" arising in the adult lung ([24] and Figure 8.3D) or in the uterus [25] merely indicate the kinds of specialization present.

We can note here (see Sections 8.1.3 and 9.3.2) that when a tumor exhibits both lineage-faithful and -unfaithful antigen expressions (usually a subgroup of a previously recognized tumor type), the situation is often described in terms of "co-expressions" of antigens.

8.1.3 Abnormal Spatial Arrangements of Tumor Cells

Like almost all other features of tumors, these are complex. They occur in various categories as follows:

(a) Disturbances in Spatial Arrangements of Tumor Cells One to Another ("Architecture")

The regular spatial relationships between normal parent cells (see Section A1.2.2) are almost always at least partly lost between adjacent tumor cells (Figure 8.4). This is almost universal in the various types of tumors, including

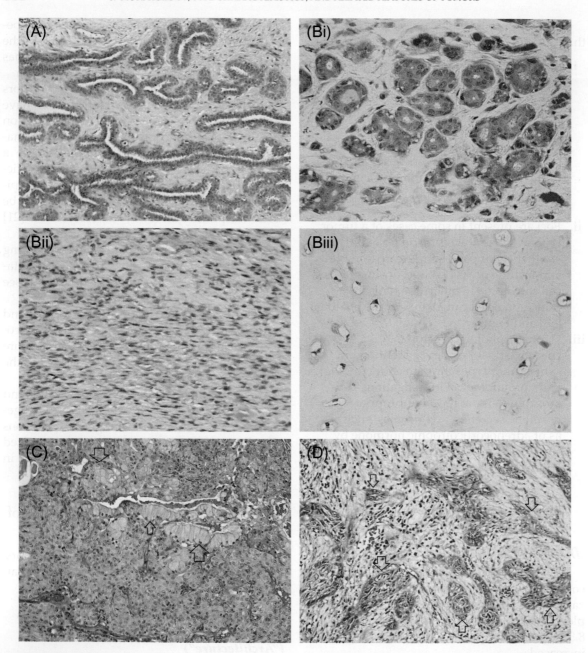

FIGURE 8.3   Examples of mixed specialization in tumors. (A) Epithelial and connective tissue overgrowth in fibroad-enoma of breast. (B) Tubular epithelial (Bi), myoepithelial (Bii) and cartilaginous (Biii) specialization in a typical case of pleomorphic adenoma of salivary gland. (C) Mucous (arrowed) and squamous specialization in muco-epidermoid tumor of salivary gland. (D) Epithelial (arrowed) and connective tissue overgrowth in 'carcinosarcoma' of lung. Note: Some authors interpret this appearance as 'mesenchymal transformation' of epithelial cells.

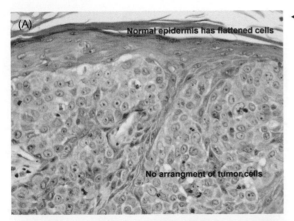

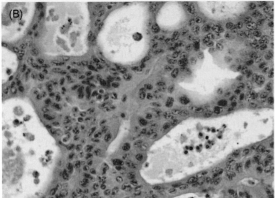

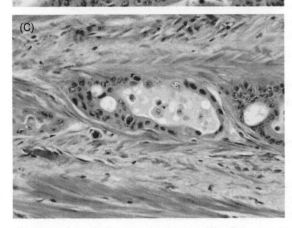

FIGURE 8.4 Abnormalities in spatial arrangements of tumor cells. (A) The cells have no arrangement to each other, and in particular, no parallel flattening with the surface. Squamous cell carcinoma of skin. (B) Cells have no orientation to the lumens in the mass. Note: Cells in the lumina appear necrotic and can be interpreted as 'apoptotic bodies (see Chapter 10). Case of carcinoma of the colon. (C) Information as for (B). Some of the detached cells in the lumen appear viable.

"benign" tumors. It is most obvious in tumors arising in cells of tissues which specialize in a morphologically contiguous way.

Common "architectural patterns" include "papillary" (Figure 8.1A), "tubular" (Figure 8.3Bi), "organoid," and "sheet-like."

(b) Disturbances in Spatial Arrangements of Tumor Cells to Supportive Cells

Abnormal spatial relationships between tumor cells and adjacent normal cells are the characteristic of invasion and metastasis, as are described in Sections 6.6 and 6.7.

8.1.4 Other General Cyto-Structural Abnormalities

The common morphological features of the individual cells are loss of polarity of nuclei within the cell, "multilayering," and loss of cohesiveness (Figure 8.4). The changes in size and shape of cells and of the nuclear:cytoplasmic ratios are variable (Figure 8.4, and see also Section 8.3). The cell biological basis of these cellular changes is unknown, but implies a loss in function of some component of the extracellular/transmembrane/intracellular cytoskeletal arrangements (see Section 2.9.3).

(a) Central Repositioning of the Nucleus

In tumors which form parent cells which have nuclei toward one end of the cell (Section A1.2.4), the tumor cell nuclei tend to locate in the middle of the cell.

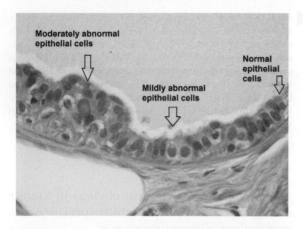

FIGURE 8.5   Adjacent cyto-structural irregularity in the same kind of parent cell. Case of atypia of ductal cells in the vicinity of carcinoma of the breast. ×40.

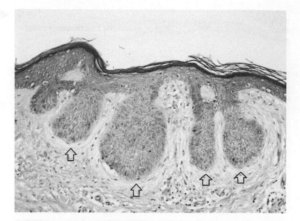

FIGURE 8.6   Cyto-structural abnormalities discontinuously in a tumor (superficial multicentric variant of basal cell carcinoma). This phenomenon is common in superficial multi-centric variant of basal cell carcinoma (shown here, arrowed) and in ductal carcinoma of the breast, but less common in other tumor types.

The phenomenon implies a possible defect in the cytoskeletal positioning of the nucleus in the cell. Parent cells with central nuclei (such as fibrocytes) give rise to tumors in which the nuclei are always central.

(b) Adjacent Cyto-Structural Irregularity in Continuity

This phrase describes diminishing nuclear regularity in cells of the same type as the cell of origin of the tumor, beginning immediately adjacent to the tumor cells (Figure 8.5). These cells are considered potentially tumorous, because if not removed at the time of operation (see Section 1.4.3) they may give rise to additional tumors. However, the lesion which constitutes "atypia" is not completely understood, because what the histopathologist sees as "atypia" is in fact probably karyo-instability combined with other intranuclear metabolic and physiological disturbances (see Sections A1.2.6 and A1.2.7). The phenomenon may depend on intercellular communication mechanisms, as discussed in Section A1.2.4.

(c) Adjacent Cyto-Structural Irregularity in Discontinuous Foci

In certain tissues, the atypia is discontinuous—that is to say, abnormal cells form near the mass

but with histologically normal cells in between. The most obvious example is the superficial multicentric version of basal cell carcinoma (Figure 8.6) [26] in which, at intervals usually of about 0.5 mm, new foci of tumor appear in the basal epidermis. Often a focus of solid tumor is found in the middle. Whether such foci are formed from the superficial tumor, or vice versa, cannot be established in the clinical situation.

A similar phenomenon is found in approximately 10% of cases of invasive ductal carcinoma of the breast [27].

8.1.5 Specific Nuclear Abnormalities

In addition to spatial irregularities of cells, tumors differ from normal tissues—and tumor types from each other—in the abnormalities of the internal structure of their constituent cells (Figure 8.7). The main intracellular abnormalities relate to nuclei and are as follows.

(a) Size of Nuclei

The populations of most tumor types include cells with nuclei which are larger than either

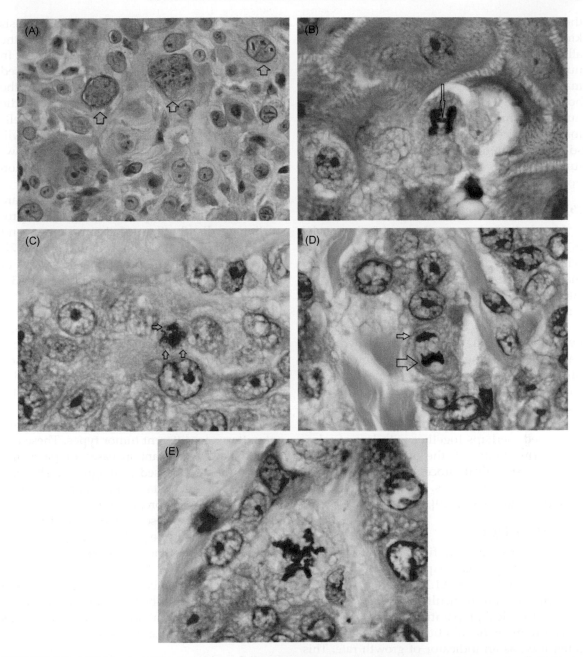

FIGURE 8.7 Abnormal nuclei and mitoses in tumor cells. All images are from one case of squamous cell carcinoma of the skin. (A) Nuclear pleomorphism and multinucleated giant cells (arrowed), ×40. (B) Ana(sub)phase bridge (arrow) in mitotic figure, ×100. (C) Fragmentation of chromosomes, ×100. (D) Apparent asymmetric distribution of chromosomes. (The possibility that one ana(sub)phase group of chromosomes has been 'cut through' by sectioning is not excluded), ×100. (E) Mitotic figure with five asters (ends of spindles), ×100.

the normal or reactive versions of their parent kind of cell. The enlargement of nuclei is usually attributed to increases in chromosome numbers (Section 5.3). However, other factors may be involved especially in increased metabolic activity. It should also be remembered that nuclei are shrunken by the steps in processing for histological studies (see (c) and (e) below) so that measurements of nuclear sizes in cells of tumors (e.g., as reviewed by Rather [28]) may not be entirely accurate.

(b) Nuclear Shape

Morphological appearances of the nuclei in tumor cells do not necessarily correlate with cell biological phenomena. This probably mainly derives from the fact that steps in tissue processing affect the appearances of cell structures.

Tumor nuclei in the unfixed state are usually ovoid or spherical. After fixation, dehydration, paraffin embedding, and staining however, the nuclei can assume irregular shapes. Irregular shapes are therefore presumably an artifact of fixation. It is possible that in tumor cells, the nuclear "matrix" including the envelope is weakened, perhaps together with loss of selective permeability of the nuclear membrane so that histological processing methods have greater or altered effects. These issues are discussed further in Ref. [29].

(c) Mitotic Figures

The phrase "mitotic figures" refers to cells seen in metaphase, anaphase, and telophase (Sections A1.3.4 and A1.3.5). Other phases of the cell cycle are difficult to visualize in ordinary histological preparations.

The number of mitotic figures in tumors is often used as an indicator of growth rate. This is subject to two limitations. First, it does not take into account cell death and hence cannot be an indicator of the net rate of cell accumulation (see Section 6.1). Second, these numbers may not be a perfect indicator of production

rate [30]. This is because in a tumor, mitosis may be prolonged, along with the rest of the division period of the cell (Section A1.3.4). In this situation, mitosis will become an increased proportion of the cycle, without change in the cell production rate. Moreover, cells can die during mitosis. At the moment when the cells are fixed for histological processing, there will be more than the normal number of cells showing "mitotic figures," but the cell production rate may be normal.

The abnormalities in the mitotic figures are well known in terms of multipolarity, disproportionate distributions of chromosomes, and losses of individual chromosomes (Figure 8.7C and see Ref. [31]). The possible mechanisms of these disturbances are described in Section 5.3.

(d) Chromatin Patterns

Unlike in histopathology, in cytopathology and hematology, there is not usually embedding of cells in paraffin. Also, the fixatives used in histopathologic preparations of cells are different in cytopathology and hematology. After cytological preparative methods, clumping of the chromatin (see above) can occur in different patterns in different tumor types. These can be sufficiently constant in cases of particular tumor types to be used in diagnosis. The patterns are not so constant, and so are less used, in histopathological preparations.

The details of these patterns are described in texts of cytology (e.g., Ref. [32]). Chromatin clumping can be "fine" or "coarse," and "regularly" (evenly) or "irregularly" (unevenly) distributed in the nucleus. Failure to clump in preparations is described as "homogenization." This change is seen almost exclusively in cells infected by intranuclear viruses.

(e) Other Abnormalities of Nuclei in Tumor Cells

Among tumors of different types, certain characteristic changes of nucleoli have been identified. For example in anaplastic small celled carcinoma of the lung, nucleoli are inconspicuous while in the Reed–Sternberg cells of Hodgkin's disease, they are characteristically large.

Silver-stainable nucleolar organizing regions "Ag-NORs" (Section A1.2.6f) show no tumor-type-specific abnormalities and hence are of limited diagnostic use [33]. The number of Ag-NORs in the tumor cells may have no more prognostic significance than nuclear pleomorphism.

8.1.6 Miscellaneous Cytoplasmic Abnormalities

The ratios of the volumes of cytoplasm to nuclei are usually reduced in tumor cells, although there are exceptions, such as "mucoid" adenocarcinoma cells (see Section 8.1.2). Most cytoplasmic abnormalities are related to abnormalities in specializations or are artifacts of tissue processing (see above sections).

A few cytoplasmic abnormalities are much more common in some tumor types than others. For example, tumors of the salivary gland and renal tubular epithelium occasionally show "oncocytic change" [34,35], but this change is never seen in carcinomas of the stomach or colon.

8.1.7 Induction of Abnormalities in Adjacent Other Kinds of Cells

(a) Desmoplasia

This term refers to the formation of excess fibrous tissue immediately adjacent to individual tumor cells (see also Figure 8.8). The classic example is scirrhous carcinoma of the breast [36], in which tumor cells are seen to lie in "Indian files" between the collagen (Figures 3.9C and 8.11A). The fibrocytes which produce the collagen are not considered themselves tumorous. Collagen-producing cells in metastases are thought to be local fibrocytes recruited by the metastatic carcinoma cells.

These phenomena are little studied, but presumably depend on some factor of intercellular communication, such as a juxtacrine or paracrine event (see Figure 6.1). It is possible that the Wnt signaling pathway (see Section 6.3) has a particular role [37]. Myelofibrosis may be a special case of desmoplasia [38].

(b) Lymphoid Associations

In Warthin's tumor of the salivary gland, there is a dense layer of lymphoid tissue surrounding the epithelial cells of the tumor [39]. This may represent some "tonsil-like" embryonic

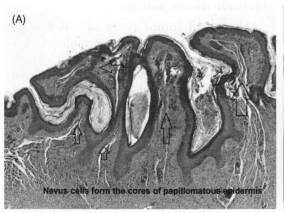

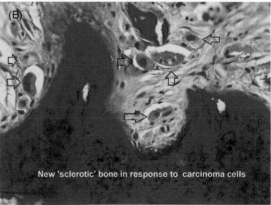

FIGURE 8.8   Induced abnormalities in adjacent cells. (A) Nevus of skin, in which nevus cells form the cores (arrows) of papillomatous epidermal overgrowths ×4. (B) Metastatic lung carcinoma cells (arrowed) causing excess growth of bone, ×20.

developmental phenomenon. The lymphocytes are not, and never develop into, lymphoma cells.

(c) Vessels

A few types of tumors are associated with markedly increased numbers of blood vessels throughout their substance. Examples include telangiectatic osteogenic sarcomas [40] and angio-lipomas [41]. This may be related to vascular endothelial growth factors (see Section 2.3.3c).

8.2 CHARACTERIZATION OF THE TUMOR TYPES IN TERMS OF COMBINATIONS OF BEHAVIORAL, MORPHOLOGICAL, AND OTHER FEATURES

8.2.1 Each Tumor Type Is Characterized by a Specific Combination of Features

An essential step in approaching the possible genomic bases of tumors is to recognize that each tumor type is characterized by a combination of variably expressed individual trait-like changes. Some of the combinations are similar except for features which relate to those of the different kinds of cell of origin. However, some combinations of features are virtually unique.

The specific combinations (Table 8.2) variably differ according to:

i. The kind of parent cell
ii. The number of the different component abnormal features
iii. The degrees of alteration in particular features of the parent-cell populations
iv. The intensities of the acquired (i.e., nonparent) abnormalities.

8.2.2 The Different Numbers of Features in the Combinations of Different Tumor Types

The tumor types can be categorized according to the number of traits in the combination as follows.

(a) Tumors with Excess Growth and No Other Abnormalities

A few types of tumors exhibit only excess numbers of morphologically normal cells. One example is the simple lipoma [4]. Another example is simple fibro-epithelial polyps of the skin [42] which consist of excess dermal tissue with overlying normal epidermis. Simple polyps of the endometrium and endocervix [43] probably also qualify for this category.

(b) Tumors with Excess Growth with Abnormalities in Specialization but No Cytological Atypia

This is a relatively uncommon combination of features. To be included in the category, the nuclei must be normal. Increased specialization with normal nuclei is seen in some cases of lesions known as "hyperplastic" polyp of the colon [44]. These tumors show growth with excess retention of mucus without any other feature. Reduction in specialization in cytologically normal cells may be seen in the "mucin-poor" variant of the hyperplastic polyp [44].

(c) Tumors with Excess Growth and Multiple Abnormalities

Almost all tumor types are in this group, because specialization abnormalities, loss of cyto-structural regularity, and nuclear abnormalities occur so often together. The cells of adenomas of the colon show linked growth, reduction of specialization, loss of cyto-structural regularity, and reduction of spatial arrangements to each other [45].

Cells of carcinomas of the colon show linked excess growth, reduced specialization, and reduced spatial arrangements between each other and to adjacent tissues, as well as potential for metastatic growth. They may also show a feature of embryonic reversion: production of carcino-embryonic antigen (see Section 8.1.2d).

TABLE 8.2 Diversity in Combinations of Traits Among the Types of Tumors

	Rate of Cell Accumulation	Invasion	Metastasis	Abnormal Specialization	Cyto-Structural Irregularity	Mitotic and Chromosomal Irregularity	Other
1. Common benign tumors							
Lipoma	+	−	−	−	−	−	−
Colon adenoma	+	−	−	↓	+/++	−	−
Colon hyperplastic polyp	+	−	−	↑	−/+	−	−
2. Lung carcinoma							
Large cell anaplastic	+++	+++	++	↓↓↓	+++	+++	−
Adenocarcinoma	+++	+++	+++	↓↓/↑	+++	+++	−
Squamous	+++	+++	+++	Metaplastic	+++	+++	−
Small cell anaplastic	+++	+++	+++	↓↓↓	+++	+++	Often neuroendocrine biomarkers
3. Colorectal carcinoma	++	++[a]	++	↓↓/↑	++/+++	−/++	Embryonic antigen expression common; Mucinous pattern common
4. Breast carcinoma	++	++[a]	++	↓/↑	+/++	−/+	Mucinous pattern occasional
5. Prostate carcinoma	++	++[a]	++	↓/↓↓	+/++	−/+	Mucinous pattern rare
6. Epidermal carcinoma							
Squamous cell carcinoma	+	+	+	↓/↑	+/+++	+/+++	Few architectural variants
Basal cell carcinoma	+	−/+	−	↓↓↓	+	−	Several architectural variants

[a]Some differences in diagnostic criteria may be relevant (see Chapter 3).

8.2.3 There Is No Universal Association of Nongrowth Features in the Different Tumor Types

When all the details of all the types of tumors are taken into account, it is apparent that there is no particular combination of nongrowth features which is present in all tumor types ([46] and see Table 8.2).

Every possible combination of features can be seen in one or more tumor types. From this lack of any particular association of features can be inferred:

i. Each of the combinations of traits may be a unique grouping.
ii. Hence the genomic event underlying the same trait may be different in different tumor types (see Section 8.5).

8.2.4 Different Numbers and Ratios of Tumor Types Arise from Different Kinds of Parent Cells

There are further complexities in the relationships between the parent cells and the tumor types which derive from them.

This is evident from the fact that there are over a thousand different tumor types and only 200 or so kinds of parent cells. It follows that some kinds of parent cells must give rise to more than two types of tumors. The "two" here refers to the common misconception that each kind of normal cell gives rise to only one benign and only one malignant type of tumor.

The fact is that, quite strikingly, some kinds of parent cells give rise to only a few tumor types while other kinds of parent cells give rise to many tumor types (Table 8.3).

Further examples of the range in numbers of tumor types arising from individual normal kinds of cells are as follows:

i. In the esophagus, the squamous cells give rise to only one type of benign tumor (fibro-epithelial polyp) but more frequently

TABLE 8.3 Parent-Cells and the Tumor Types Which Arise From Them

Lung, bronchial epithelium	B: Papillomas (uncommon)
	M: Adenomas, squamous cell, and anaplastic large cell carcinomas (all common)
Breast, ductal epithelium	B: Several types including fibroadenomas, papillomas, and lactating adenomas
	M: Carcinomas with many variant patterns (see Section 3.4)
Colon, epithelium	B: Many types of adenomas, including tubular, villous, hyperplastic, juvenile, and serrated.
	M: Adenocarcinoma. A few variant patterns
Prostate, epithelium	B: None recognized
	M: Carcinoma. Variant patterns are uncommon
Skin, epidermis	B: Many types, including solar keratosis and seborrheic keratosis
	M: Basal cell carcinomas (several variants) and squamous cell carcinomas (few variant patterns)
Skin, melanocytes	B: Nevi: many types, including common, congenital, blue
	M: Melanoma: several types including nodular, superficial spreading, acral lentiginous, and nevoid

B, benign; M, malignant.*Note*: It can be seen from this that all kinds of parent cells do not give rise to equal ratios of benign to malignant tumors, nor do they all give rise to all the possible types of either benign or malignant tumors.

to a malignant tumor type (keratinizing squamous cell type) [47].
ii. The epidermis of the skin gives rise to many different benign lesions, e.g., fibro-epithelial polyps, seborrheic keratoses, clear cell acanthomas and inverted follicular keratosis, and to solar keratoses [48].

The epidermis also gives rise to two "malignant" clinicopathological tumor types:

- The squamous cell carcinoma
- The basal cell carcinomas.

The latter tumor type occurs in several subtypes/variants: nodular, insular, infiltrative, pigmented, and superficial multicentric.

iii. Breast duct epithelium gives rise to multiple types of benign types of tumors which include "papillary adenomas" and "fibroadenomas." Likewise, carcinomas of the breast duct epithelium are of many types and subtypes (see Table 8.3 and [49]).

A related point to the one in the previous paragraph is that different normal cell types give rise to different ratios of benign: malignant tumor types. For example, in the colon, benign epithelial lesions are more common than are malignant ones. On the other hand, in the bronchi, the surface epithelium gives rise to almost no benign tumors, but frequently to a malignant tumor (squamous celled carcinoma). (Other types of bronchial malignancies—adenocarcinomas, anaplastic carcinomas, etc.—may arise from different cell populations in the bronchial mucosa.)

In both colon and bronchus, the major carcinogens are believed to arrive at the surface of the epithelium because the carcinogens are assumed to be in the lumen of the colon and of the bronchus respectively.

There is no known explanation of these differences in ratios of numbers of benign: malignant tumor types.

8.2.5 Some Potential Tumor Types (i.e., Combinations of Traits) Do Not Occur in Many Kinds of Normal Cells

Mention must be made of a frequently overlooked fact of tumor phenomenology. It is that certain particular combinations of traits (i.e.,

tumor types) which occur in some kinds of normal cells do not occur in others. Thus "diffuse carcinoma" ("linitis plastica") occurs commonly in the stomach but is rare in the small intestine and the colon. As another example, "clustered multifocality" of tumors occurs in the epidermis as superficial multicentric basal cell carcinoma. Multifocality is also common in carcinomas of the breast. However this pattern of multifocality does not occur in most other epithelia.

8.2.6 Tumor Types Which Exhibit Continuous Spectra from Benign to Malignant

Another aspect of some tumor types is that cases constitute a continuous spectrum from benign to malignant, manifest in spectra both of cytological abnormality and clinical aggressiveness (Figure 8.9). This means that there is no sharp distinction between benign and malignant at all. This phenomenon is shown by the urothelium of the urinary tract [50]. Cases of tumors of this kind of parent cell form a complete spectrum from benign to malignant. At the benign end of the spectrum are noninvasive, cytologically normal papillomas. At the malignant end of the spectrum are undifferentiated invasive tumors with marked cytological atypia (Figures 8.9). Most cases, however, are intermediate in aggressiveness and exhibit all degrees of these architectural and cytological abnormalities.

8.2.7 Exceptions to the General Rule That Benign Tumors Are Less Morphologically Abnormal than Malignant Tumors

There are exceptions to the general rule that benign tumors are less abnormal than malignant tumors (Section 1.1.2). As one example, the cells in almost all cases of carcinoid tumors

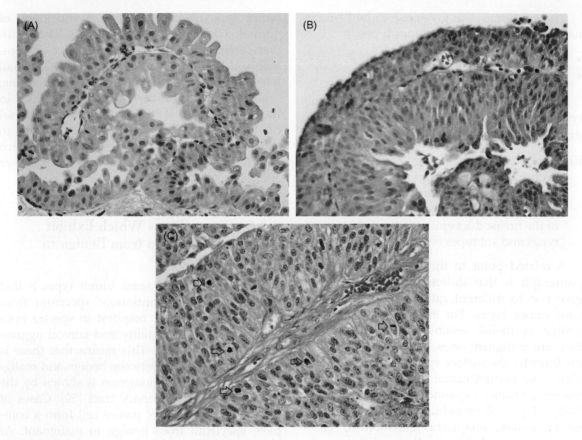

FIGURE 8.9   Tumor type showing a continuous spectrum from benign to malignant. Worsening of the histological and biological abnormalities in transitional cell carcinoma is indicated by excessive thickness of the layers of tumorous transitional cells, as well as cyto-structural irregularities of the tumor cells. (A) Bladder papilloma ×20. (B) Bladder papillary transitional cell carcinoma, low grade ×20. (C) Bladder, invasive papillary transitional cell carcinoma, ×20. Mitotic figures are arrowed.

(one of the "neuroendocrine" types of tumors) of the small bowel are very like their parent cell—i.e., they appear benign. However, carcinoid tumor cells show invasion and often metastasize [51].

The reverse relationship is seen in atypical fibroxanthomas of the skin [52]. In this tumor, the cells are bizarre, with frequent mitotic figures—i.e., appear "malignant." This tumor, however, does not invade significantly and never metastasizes.

8.2.8 Tumor Types with Unusual Behavioral Features

Many types of tumors have unusual if not unique features. Two examples can be given here.

(a) Melanocytic Tumors

Melanocytes are well known to give rise to benign nevi and to malignant melanomas. Nevi, however, are peculiar lesions. They form in the normal location of the melanocytes in the "junctional zone" of the skin. They then invade the dermis so that they are located both in the junctional zone and the dermal zone forming what are known as "compound nevi." Then for unknown reasons, the proliferation and invasion stop, the junctional component disappears, and all that is left is a nodule in the dermis—the "intradermal nevus." Nevi therefore exhibit invasion, which is a "malignant" characteristic of tumors, but are self-limiting and do not significantly harm the patient in most cases.

Spitz nevi are melanocytic lesions mainly of childhood, which often have the same marked morphological changes as do malignant melanomas. However, these tumors are benign (see Ref. [53]).

(b) Keratoacanthoma

This lesion arises from the squamous cells of the epidermis. This tumor—as it is generally accepted to be—grows rapidly and invades the dermis and shows abnormalities which are very similar to squamous cell carcinoma. However, this lesion actually spontaneously involutes after about 6 weeks, ultimately leaving a scar, but no other trace of its existence [54].

8.2.9 In Situ Tumors

An in situ tumor is a lesion comprising cells having features of malignancy, in the normal site, but without having invaded the stroma of the tissue. The phenomenon is seen where ever a micro-anatomical boundary exits between one kind of cell and the supporting cells, as shown particularly be epithelium and stroma. The malignant-like cells almost always appear surrounded by less atypical cells. These "less than malignant-like" cells are referred to as dysplastic, and often graded "mild," "moderate," and "severe."

The whole phenomenon is seen particularly in epidermis [55], the uterine cervix [56], the ducts of the breast [57], and the nipple of the breast (Paget's disease of the nipple) [58] and to a lesser extent in the bronchi [59] and bladder [60]. The same features are often called "severe dysplasia" when seen in the colon and rectum [61] or in the stomach [62].

In the uterine cervix, the lesions are associated with infections by human papilloma virus (see Section 4.5). In experimental skin carcinogenesis, these lesions often precede the appearance of invasive tumor [63]. Causative factors for in situ lesions in other sites are largely unclear.

The in situ phenomenon is not seen in some organs, such as the renal tubules, the follicles of the thyroid, the mesothelium of the pleura, or the ependymal lining of the ventricles of the brain.

8.3 DETAILS OF VARIABILITIES IN TUMORS

The phenomenon of variabilities in the cell populations of tumors was identified in the earliest microscopic studies of tumors (see Sections 1.3 and 2.5.1). Through the whole twentieth century, most authors described these variabilities using the words "anaplasia" and "dedifferentiation," deriving from of Hansemann's theory, or used descriptive words such as "pleomorphism" which do not relate to any theory of tumors. Only in the last decade or so has there been sufficient knowledge of somatic cell genomics to suggest possible mechanisms of these phenomena. This section classifies the levels at which variabilities occur and then gives examples of the main features of tumors which show variability.

As a point in terminology, "variation" is the term used in species taxonomy for the presence or absence of traits in members of a species or strain. In germ-line genetics, it is often used for quantitative differences in expressions of traits. Here, it would be confusing to use the word, because tumor variabilities have many more complex aspects than species variation, as the following shows.

8.3.1 Tumors Are Variable Between the Types, Between Cases of the Same Type, and Within Individual Cases

These are readily appreciated as follows:

a. *As emphasized in* Section 8.2, *the tumor types are markedly variable in respect of each other (i.e., inter-type variability).*

Most obviously, they vary by being "benign" or "malignant" (see Section 1.1.2). Even among invasive tumors, there are marked variabilities according to type. For example, lung carcinomas almost always metastasize, while basal cell carcinomas of the skin rarely do so.

b. *Between cases of the same tumor type.*

Degrees of specialization are commonly variable between cases of the same kind of tumor. As an example among tumors of fibrocytes, cellular dermatofibromas show little deposition of collagen per cell, while in the sclerotic form, there is much more collagen production.

Another example is in carcinoma of the colon, in which the mucinous form produces more mucus than the commoner, less well-differentiated form.

For most common carcinomas, the whole range between complete loss and complete preservation of specialization can be seen among different cases of the same tumor type.

c. *Different parts of the same case of a tumor may show different degrees of changes in morphological and antigen expression (intra-case variability).*

In most of the common carcinomas, the whole range of loss of specialization (see Figure 2.4) is often manifested in the one case of tumor.

8.3.2 In Growth Rates and Life Cycles/Cytokinetics

The different life cycles/cytokinetics of normal cells are described in Section A1.3.3 (see also Figure 8.10). Here, some points concerning tumor cell populations are noted.

(a) General

The fundamental feature of tumors is the accumulations of cells. Accumulation depends on an excess of production over losses of cells.

(b) Increased Production

Increased production of cells essentially involves reduced inhibition of entry into the cell cycle (see Section A1.3). Reduction in length of the normal period of suppression of cell division means shortening of the durations of time which cells spend in G0.

This has different implications according to the kind of cell involved.

    i. For the stem cells of "labile" cell populations (see Section A1.3.3)—those which normally continuously replenish their numbers—this means premature entry into G1.

   ii. For stable cells—those which only divide in response to stimuli (Section A1.3.3), it means inappropriate entry into G1. The importance of this distinction is emphasized in the discussions of stem cells, as in Section A1.3.2 and Chapter 13.

There is no evidence that any growth factors shorten the division period (i.e., G1 to cytokinesis, see Figure A1.1). On the contrary, the division period is almost certainly always longer in tumor cells than in normal cells (see Section A1.3.6b).

(c) Reductions in Cell Losses

This is possible mainly in "labile" kinds of cells because they normally mature to terminal

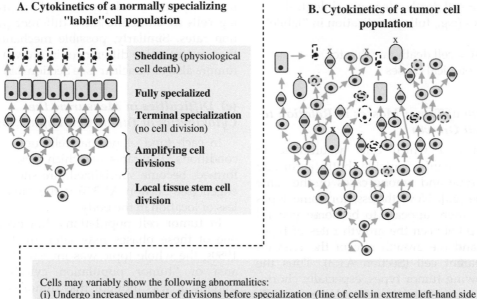

A. Cytokinetics of a normally specializing "labile" cell population

Shedding (physiological cell death)

Fully specialized

Terminal specialization (no cell division)

Amplifying cell divisions

Local tissue stem cell division

B. Cytokinetics of a tumor cell population

Cells may variably show the following abnormalities:
(i) Undergo increased number of divisions before specialization (line of cells in extreme left-hand side of population).
(ii) Cease dividing before beginning specialization (unphysiological vegetative state –).
(iii) Failing to proceed to full specialization (and).
(iv) Die through losing genomic material during an abnormal cell division or other reason (cells with dashed outlines).

N.b. The original tumorigenic event is shown in a stem cell. This may not necessarily occur in all tumors (see Section 2.9.1).

FIGURE 8.10 Tumor cell factors which can be variable in the cytokinetics of tumor masses.

specialization, and then die and are lost. Reductions in these losses of cells may occur through failure of completion of specialization in tumor cell. This means failure to progress to permanent G0, or failure to shed or lyse or be scavenged (see Section 10.1.1). In this situation, the transit-amplifying phase may continue with greater numbers of transit-amplifying divisions and hence continue to add cells to the tumor mass (see also Section 2.9.1).

In the kinds of cells which only proliferate in response to stimuli ("facultative" cells—Section A1.3.3) specialization-related losses probably do not occur to great degrees at all. Thus when stimulated by a skin wound, fibrocytes

change to their corresponding blast form and then slowly return to their inactive forms (see Section A1.2.7) when the wound heals.

In some tumor types, it appears that inappropriate expression of the embryonic cell death proteins of the Bcl2 family may be involved in reducing the deaths of cells [64]. This is discussed further in Chapter 10.

The concepts above can be expressed as the equation:

$$CAR = CPR - (P - CDR + UP - CDR)$$

where

CAR = cell accumulation rate
CPR = cell production rate

P-CDR = cell death rate in physiological processes (e.g., full specialization in "labile" cells)

UP-CDR = cell death rate due to unphysiological causes

(d) Growth of Tumor Masses in Relation to Parent-Cell Growth

The rates at which cells accumulate in tumors are highly variable both according to tumor type and among cases of the same tumor type [65]. Nevertheless, for some types of tumors, there appears to be some general relationship between the growth rates of types of tumor and the growth rate of the relevant kind of parent cell (Section A1.3). Thus the fastest-growing tumor types, especially chorio-carcinoma, arise from the testicular germ cells (in males) and from the placenta (in females). These normal parent cells are among the most rapidly proliferating normal cells in the body.

Lymphomas derive from lymphoid cells which—with appropriate stimulation—can accumulate at medium rates. Lymphomas exhibit a range of growth rates which on average could be considered medium among tumors.

Melanomas have a wide range of rates of cell accumulation. The parent cell normally has a low proliferation rate, and some melanomas have phases of very slow growth and can be "dormant" for years if not decades (see also Sections 4.1.4 and 8.4). In other cases, especially in late phases, growth may be at medium rates.

A wide range of growth rates is also seen in carcinomas of the breast. The breast ductal cells from which most of these tumors arise are almost inactive in the nonpregnant state, but can grow at medium rates in pregnancy. Tumors of these ductal cells can be slow growing, such that the disease can have a natural history of years if not decades. Other cases of tumors of the breast duct cells grow at medium rates.

As noted in Section A1.3, little is known of the mechanisms by which different normal cells

in embryonic development and in proliferating cells in adult life have different proliferation rates. Similarly, possible mechanisms for these quantitative differences in growth rates in tumors also are unclear (see Section 8.5).

(e) Difficulties in Measuring the Cytokinetics of Tumor Cell Populations

In each kind of normal cells under normal conditions, the rates at which new cells are formed, become specialized, are shed or scavenged (see Section A1.3.3) is the same regardless of location in the body.

In tumor cell populations, however, every one of these phases may alter. In the1970s—1980s, the whole topic was investigated as the area of "tumor population cytodynamics." Initially, it was thought that tumor cell populations would consist simply of "proliferative" and "vegetative" subpopulations [66–68].

However, enormous degrees of heterogeneity were found between different cell lines from the same individual case of tumors, between different cases of the same type of tumor, and between different types of tumors. As a result, the concept of tumor cell kinetics is less studied, except in relation to radiotherapy [69,70].

In detail, the variabilities were found to be in:

i. The frequencies of division of cells.
ii. The periods of time taken for the whole process of cell division. Division time often lengthens and probably never shortens in tumor cells.
iii. The degrees of specialization achieved by cells. Thus the unspecialized dividing cells may have an increased replication rate, but may, to various degrees, fail to progress to fully functional "mature" cells—i.e., remain fully or partially unspecialized.
iv. At any stage of cell division, the cells may enter inappropriately into G0 (i.e., spontaneously become nondividing, or "vegetative"—see Section 10.6.2).

v. The rates at which daughter cells die (Section 10.3) and disappear by lysis from the population.

8.3.3 In Biomarker Expressions (as Used in Diagnosis and Therapy)

General aspects of biomarkers are discussed in Section 9.7. Here it can be noted that the variabilities in biomarker expressions are seen both within the same case of tumor (intra-case variation) and between different cases of the same tumor type (inter-case variability). Thus as one example in widely used diagnostic staining for epithelia tumors: cytokeratin7 is expressed by 13% of colonic carcinomas, 57% of gastric carcinomas, 89% of pancreatic carcinomas, and 99% of breast carcinomas, while the corresponding percentages of cases expressing cytokeratin20 are 90%, 66%, 55%, and 12% respectively [71].

However, some whole tumor cell populations show supra-normal biomarker expressions, while other tumor cell populations can show scattered cells with supra-normal expressions (see Section 9.7.1).

All of these changes could be manifestations of one or more forms of genomic instabilities, as discussed in Section 8.5.

8.3.4 For Malignant Tumors, in Degrees of Malignancy

The rates at which tumors are liable to destroy adjacent tissues and metastasize are referred to as their "clinical aggressiveness." Different malignant tumor types have different degrees of aggressiveness. However, this is not related in any way to the general functional category of parent cell (Section A1.2.1). Thus tumors arising from the same general category of cell, but in different organs and tissues, may show quite different morphological abnormalities and clinical behaviors. Among malignant tumors of squamous epithelial cells, carcinomas of the epidermis are only mildly aggressive, carcinomas of the esophagus and uterine cervix are moderately aggressive, and carcinomas of the lung are highly aggressive. The same phenomenon is seen in adenocarcinomas. Adenocarcinomas of the sweat glands have a low proportion of cases exhibiting aggressive behavior. Adenocarcinomas of endometrium are generally low grade, while those of the breast, colon, prostate, and stomach show intermediate malignancy. Adenocarcinomas of the lung are almost all highly aggressive tumors [72].

There is no obvious reason why the different kinds of parent cells should give rise to tumors with different features in this way.

Finally here, the aggressiveness of tumors has no relationship to the location of the tumor. Thus malignancies arising in the same site, but from different kinds of cells, often exhibit different degrees of malignancy. For example, carcinomas of the epidermal cells are almost always less aggressive than malignant tumors of the melanocytes of the skin (which are located side by side with the basal epithelial cells of the epidermis).

Similarly chondrosarcomas and osteogenic sarcomas arise from different connective tissue parent cells in the same organs in the body. Chondrosarcomas, however, are less aggressive than osteogenic sarcomas. Another example is the tumors of the hematopoietic system. They all arise in the same organ (the bone marrow) but comprise types with different degrees of clinical aggressiveness [73].

8.3.5 Variably Abnormal Proportions of Tumor and Supporting Cells, Especially Relating to Microenvironments

In different normal tissues, there are different rations of supporting cells to parenchymal cells (Section A1.1.2) (Figure 8.11).

In tumor tissues the relative proportions of tumor cells to supporting cells may be highly variable, not just between tumor types, but between cases of the same tumor type, and

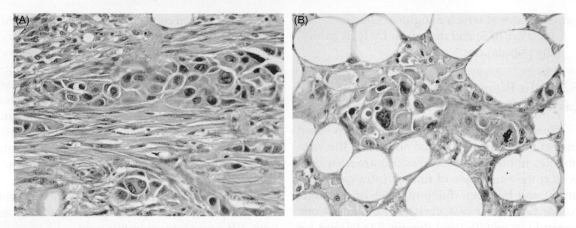

FIGURE 8.11 Variable microenvironmental features. (A) Focus of breast carcinoma cells with surrounding fibro-sclerosis, ×20. (B) Focus of cells in the same case of breast carcinoma, showing no fibro-sclerosis, ×20.

even in different foci of the same case of tumor. The commonest change in tumors is that there are fewer supportive cell types in relation to tumor cell numbers, and that vascularity in the middles of tumor masses may be reduced [74]. As mentioned in Section 8.1.7, some kinds of tumors are associated with relative excesses of supportive cells, as in "desmoplasia" (fibrous tissue) or even blood vessels (telangiectatic osteogenic sarcomas). Some tumors secrete angiogenic factors which both act to supply more nutrients to tumor cells and act as increased avenues for tumor cells to begin the metastatic process (see Section 6.7.3).

These variabilities may be relevant to cytotoxic therapy of tumors, because they determine the microenvironment of the tumor cells and have many other effects on defenses of tumor cells to noxious agents (see Section A3.1.2).

8.3.6 Use of "Heterogeneity" in Reference to Tumor Cell Populations

"Heterogeneity" would obviously be appropriate to describe the variabilities in tumors. Although it has never been part of the histopathologists' vocabulary, discussion of it is appropriate here.

In fact, the term was used in the 1950s [75] but came into wide use in the late 1970s, experimental investigations of invasion and metastasis were being undertaken [76] (see Sections 6.6 and 6.7). Widely variable degrees of invasiveness and metastatic capabilities were found among different cell lines from the same individual cases of tumor. From this it was concluded that the original tumor cell populations were "heterogeneous" for these features [77–80]. Subsequently, when almost all other features of tumor cells were investigated, variability was found to be a general phenomenon in tumors (Table 8.4).

The possible genetic bases of heterogeneities are generally thought to be in genomic instabilities (see Section 2.8 and Chapter 5).

8.4 "PROGRESSION" IN TUMOR CELL POPULATIONS

8.4.1 General

This term is used for different situations. It is usually used for a phase of accelerated growth in a malignant tumor, as was described in the nineteenth century [81]. In

TABLE 8.4 Main Evidence on Which the Concept of Tumor Heterogeneity Is Based

1. Observation of human tumors

a. Different patterns of histology in the one tumor
b. Histology of metastases "worse" than primary
c. Different patterns of metastasis among similar tumors of the same organ
d. Different outcomes: age, stage, grade, treatment
e. Observable karyo-instability in tumors

2. Experimental evidence

a. Variable behavior of tumor explants hetero-transplanted into normal laboratory animals (see Section 6.7.2)
b. Primary tumors yield different cell lines in culture
c. Primary tumors yield different cell lines in athymic mice

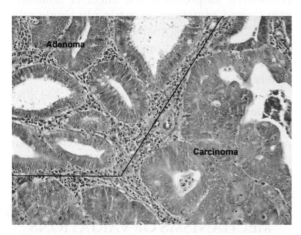

FIGURE 8.12 "Progression" as in carcinoma arising in adenoma. The line separates adenoma (above) from carcinoma (below).

the 1920s, a similar phenomenon was recognized in the almost invariable terminal acute phase of cases of chronic myelogenous leukemia [82]. Progression of less aggressive forms of Hodgkin's disease to more aggressive forms was documented in the 1960s [83].

The term is also applied to malignancy developing in preexisting benign solid tumors (Figure 8.12). In particular, Foulds [84] thought that all tumorous phenomena can be viewed as the result of a single "neoplastic" process which "develops" such that all malignant tumors appear through an initial benign phase of tumor growth.

This form of progression, however, occurs in different frequencies among the various "benign" tumor types. Solar keratoses of the epidermis, nevi of the skin, and villous adenomas of the colon are probably among the commonest benign solid tumors to give rise to malignancies. Carcinomas appearing in pleomorphic adenomas are less common (see also (ii) below).

8.4.2 Progression Occurs Only in Some Types of Tumors

Further to the previous section, "progression" is not a feature of all types of tumors (Table 8.5). The reasons are:

i. Many malignant tumors develop entirely *de novo* and without any precursor benign lesion. For example, anaplastic small celled carcinomas of the bronchus do not arise from any known benign precursor tumor [85].

ii. The different timings of malignancy arising in benign tumors. In chronic myelogenous leukemic, "blast transformation" almost always occurs within 2–5 years after diagnosis [86]. Malignant melanomas usually arise in nevi usually only after several years [87]. Similar delays occur with squamous cell carcinomas arising in solar keratoses. Carcinomas in pleomorphic adenomas of the parotid gland, occur only after decades [88].

It may be pointed out here that no data are available for durations of time associated with the appearance of carcinomas in adenomas of the colon. This is because colonic adenomas are not usually observed repeatedly over long periods (see also Section A1.2.4).

iii. Most benign tumors in fact do not undergo malignant progression at all. Seborrheic

TABLE 8.5 Differences in Tendencies of Benign Tumors to Give Rise to Malignancies

1. Some benign tumor types with greater than 1% rates of malignant progression

Colon	Villous adenoma
Breast	Papillary adenoma
Skin, epidermis	Solar keratosis
Skin, melanocytes	Hutchinson's pre-melanotic freckle

2. Some benign tumor types with less than 1% rates of malignant progression

Colon	Tubular adenoma
Breast	Fibroadenoma
Parotid gland	Pleomorphic adenoma
Uterus	Leiomyoma
Skin, melanocytes	Common nevi
Bone	Enchondroma
Nerves	Neurofibromas
Ovary	Brenner tumor

3. Some benign tumor types which virtually never give rise to malignancies

Lung	Adenochondroma
Breast	Lactating adenoma
Colon	Peutz–Jegher's polyp
Skin, epidermis	Seborrheic keratosis
Salivary gland	Warthin's tumor
Bone	Osteochrondroma
Kidney	Angiomyolipoma
Testis	Sertoli cell tumor
Adrenal medulla	Pheochromocytoma
Anterior pituitary gland	All adenomas

Note: Chronic lymphocytic and myeloid leukemias are considered by some authors as "benign" conditions, which undergo progression to acute leukemias in most cases.

keratoses, dermatofibromas, and lobular hemangiomas of the skin, solitary lipomas of adipose tissue, and Warthin's tumor of the salivary glands are all benign tumors in which development of malignancy is extremely rare.

8.4.3 Genomic Basis of Tumor Progression

This is a controversial issue. Some authors see tumors as originating in single cells as results of relatively rare single mutations, producing homogenous clones of cells (Section 2.7). According to this view, progression can be considered as probably the effect of relatively rare second mutations occurring in single cells in this "clone." Progression therefore is the formation of a new homogenous clone with more aggressive features in the original, less aggressive homogenous clone.

On the other hand, other authors see progression as the effect of the development of genomic instabilities, either *ab initio* in the original tumor cell, or over time, in the homogenous clone developing from that cell (Section 2.8).

8.5 POSSIBLE GENOMIC MECHANISMS OF VARIATIONS COMBINED WITH ASSOCIATIONS OF FEATURES, AS CHARACTERIZED BY THE TUMOR TYPES

Despite the evidence of the variabilities in tumors, there remains the fact that to a large degree and variably, a general association exists between excess growth, alteration in a parent-cell-specific activity (specialization) and then, to degrees, invasiveness, and metastasis.

As emphasized at the beginning of this chapter, the phenotypic features of tumors are:

i. Combinations of features
ii. Continuous/quantitative variation in all aspects

iii. Often associated with particular kinds of parent cells of origin.

In classical genetics, combinations of phenotypic changes are often accounted for by co-mutation of "linked"/proximate relevant genes.

Also in classical genetics, continuous variation is explained by polygeneism and "morphisms" of genes (see Section A2.1.2).

It may be that growth and altered parent-cell-specific activity are one linkage, and that the supervening invasion and metastasis is another process caused—probably differently in different tumor types—*with* or *by* the first process (Figure 8.13).

8.6 ARE HEMATOPOIETIC TUMORS SIMILAR IN PRINCIPLE TO "SOLID" TUMORS?

In the second half of the nineteenth century, accumulations of abnormal leukocytes in the blood and bone marrow (leukemias) and of lymphocytes in lymph nodes (lymphomas) were first recognized [89,90]. It was assumed that these conditions are fundamentally the same as "true tumors" because:

i. They are characterized by proliferating cells.
ii. These cells damage tissues in which they grow. In cases of leukemias, the damaged other kinds of cells are mainly in the bone marrow. In cases of lymphomas, damage occurs in the cells of any organ in which the lymphoma cells grow.
iii. Untreated, leukemias and lymphomas usually cause death of the patient.

With specialization in medical practice in the twentieth century, leukemias and lymphomas became more or less the exclusive province of hematologists, who both examined the tissues of the patients (especially the bone marrow) and treated the patients.

Certain traditions in terminology developed as follows.

8.6.1 Specialization Morphology in Hematopoietic Tumors

In leukemias, the cells in all cases appear morphologically like one or other kinds of cell in the specialization sequence in the bone marrow. There is little cyto-structural variability in the leukemic cell populations [85]. Hence, there is no such thing as an "anaplastic" leukemia, in the sense of being composed of cells which show the great degrees of nuclear pleomorphism and irregularity as are seen in solid tumors.

Because all the tumorous leukocytes morphologically resemble precursor cells in the bone marrow, the leukemias are described according to the specialization which the largest numbers of cells exhibit. Thus if the cells lack any specialization, the leukemia is called "blastic," or "stem cell." Leukemias with slightly specialized cells are called pro-myelocytic; leukemias with moderately specialized cells are called "myelocytic," and leukemias with fully specialized cells are called "myeloid."

8.6.2 Lack of Particular Specialization Morphology in Lymphomas

In lymphomas, the specialization issue is complicated. In the immune system, lymphocytes begin as "naïve" cells [91]. After exposure first to a suitably presented antigen, they undergo "blastic" change and become "committed" lymphocytes. As the antigen presentation ceases, they subside to "memory" lymphocytes, which morphologically are indistinguishable from naïve cells. On second and subsequent exposures to specific antigen, the memory lymphocytes undergo blastic change, which is morphologically similar to the first blastic change. After that, the lymphocytes become fully specialized, producing antigen-specific B-cells (including plasma cells) and effector T-cells. All lymphoid cells which do not appear active/"blastic" tend to be described as "mature."

A. Parent cell specialization gene (∗) and suppressor (∗∗) of growth factor-oncogene (GFO ∗∗∗)

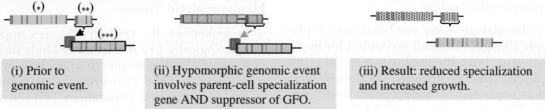

(i) Prior to genomic event.

(ii) Hypomorphic genomic event involves parent-cell specialization gene AND suppressor of GFO.

(iii) Result: reduced specialization and increased growth.

B. Suppressor of GFO (∗∗) is intronic in the parent cell specialization gene (∗)

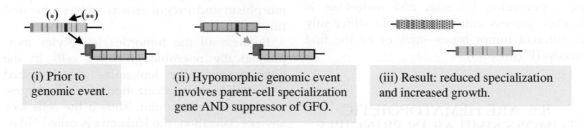

(i) Prior to genomic event.

(ii) Hypomorphic genomic event involves parent-cell specialization gene AND suppressor of GFO.

(iii) Result: reduced specialization and increased growth.

C. Damaged insulator model. Parent cell specialization gene (∗) is upstream of an insulator (∗∗) and an inactive GFO (∗∗∗).

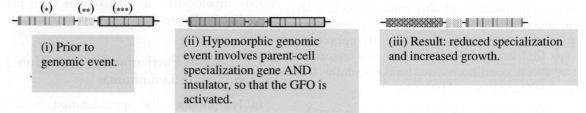

(i) Prior to genomic event.

(ii) Hypomorphic genomic event involves parent-cell specialization gene AND insulator, so that the GFO is activated.

(iii) Result: reduced specialization and increased growth.

D. Model of parent cell specialization gene (∗) and tumor suppressor gene (∗∗). No third allele involvement (see Figure 7.8b scheme 3)

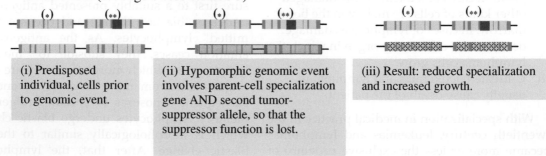

(i) Predisposed individual, cells prior to genomic event.

(ii) Hypomorphic genomic event involves parent-cell specialization gene AND second tumor-suppressor allele, so that the suppressor function is lost.

(iii) Result: reduced specialization and increased growth.

FIGURE 8.13 Possible genomic models for tumor types without subsequent change. Note: The original tumorigenic event is shown in a stem cell. This may not necessarily occur in all tumors (see Section 2.9.1).

Thus concepts such as "well-differentiated" and "poorly differentiated" lymphocytes [92] belong to the era before the functions of lymphocytes were recognized—all that is being seen are phases in physiological processes.

8.6.3 Clinical Features of Leukemias

In many texts of hematology, acute leukemias are mentioned before the chronic ones. However, the chronic leukemias tend to "transform"/undergo "blastic" change, which is to an acute leukemia. On this pattern, the chronic leukemia could be considered as the "benign" tumor and the acute leukemia as the "malignantly transformed" variant.

8.6.4 Clinical Features of Lymphomas

Clinically, most lymphomas appear as swellings of lymph glands and progress to involve other organs over months to years. They are thus chronic disorders. "Acute" lymphomas are perhaps mainly those with many cells in the blood circulation, e.g., the natural killer lymphoblastic leukemia/lymphoma [93].

8.6.5 Aspects Other Kinds of Hemato-Lymphoid Tumors

There are many other kinds of hematopoietic disorders which may be considered "benign," "of low-grade malignancy," or "of high-grade malignancy" [94]. The erythroid cells of the marrow give rise to a slow-proliferative condition known as "polycythemia rubra vera." Leukemias of erythroid precursors are rare. Cases of polycythemia rubra may terminate in acute myeloid leukemia. Myelofibrosis is a sclerosing condition of the marrow, spleen, and liver which tends to terminate in acute myeloid leukemia. Tumors composed of plasma cells are usually malignant ("multiple myeloma"). Benign tumors of plasma cells are known as "plasmacytomas," but many cases of this condition progress to multiple myeloma.

On these general points, there seems to be enough evidence to consider that hemato-lymphoid tumors are fundamentally the same as other kinds of tumors. Their differences seem to be mainly related to the particular normal features of their parent cells of origin (Section 8.1.1).

References

[1] Bignold LP, Coghlan BL, Jersmann HP. David Paul Hansemann: contributions to oncology: context, comments and translations. Basel: Birkhäuser, ; 2007.
[2] Virchow R. Ueber metaplasia/On metaplasia. See in Bignold LP, Coghlan BL, Jersmann HP. Virchow's "cellular pathology" 150 years later. Semin Diagn Pathol. 2008;25(3):140–46.
[3] Foulds L. Neoplastic development, 2 vols. London: Academic Press; 1969, 1976.
[4] Weiss SW, Goldblum JR. Enzinger and Weiss' soft tissue tumors, 5th ed. Philadelphia, PA: Mosby/Elsevier; 2008. pp. 433, 637.
[5] Lack EL. Tumours of the adrenal glands and extraadrenal paraganglia. AFIP Atlas, Ser 4, Fasc 8, 2007. p. 101.
[6] Colby TV, Koss MN, Travers WD. Tumors of the lower respiratory tract. AFIP Atlas, Ser 3, Fasc 13, 1995, pp. 213–19.
[7] Weedon D. Skin pathology. New York, NY: Churchill-Livingstone; 1997. pp. 74–5.
[8] In: Ref. [22]. p. 64.
[9] Bignold LP. Embryonic reversions and lineage infidelities in tumour cells: genome-based models and role of genetic instability. Int J Exp Pathol 2005;86(2):67–79.
[10] Guarino M, Rubino B, Ballabio G. The role of epithelial–mesenchymal transition in cancer pathology. Pathology 2007;39(3):305–18.
[11] Polyak K, Weinberg RA. Transitions between epithelial and mesenchymal states: acquisition of malignant and stem cell traits. Nat Rev Cancer. 2009;9(4):265–73.
[12] In: Ref. [1]. p. 61.
[13] Flores A, Marrero JA. Emerging trends in hepatocellular carcinoma: focus on diagnosis and therapeutics. Clin Med Insights Oncol 2014;8:71–6.
[14] Beachy SH, Repasky EA. Using extracellular biomarkers for monitoring efficacy of therapeutics in cancer patients: an update. Cancer Immunol Immunother 2008;57(6):759–75.
[15] Kurman RJ, Ronnett BM, Sherman ME, et al. Tumors of the cervix, vagina, and vulva. AFIP Atlas. Ser 3, Fascicle 13. pp. 157–78.
[16] Murphy WM, Grignon DJ, Perlman EJ. Tumors of the kidney, bladder and related urinary structures. 2004. pp. 270–71.

[17] Gandhi L, Johnson BE. Paraneoplastic syndromes associated with small cell lung cancer. J Natl Compr Canc Netw 2006;4(6):631–8.

[18] In: Ref. [15]. p. 288.

[19] Ellis GL, Auclair PL. Tumors of the salivary glands. AFIP Atlas, Ser 4, Fasc 9, 2008. pp. 49–70.

[20] In: Ref. [19]. pp. 173–92.

[21] In: Ref. [4]. pp. 1164–78.

[22] Churg A, Cagle PT, Roggli VL. Tumors of the serosal membranes. AFIP Atlas, Ser 4, Fasc 3, 2006. p. 42.

[23] Tavassoli FA, Eusebi V. Tumors of the mammary gland. AFIP Atlas. Ser 4, Fascicle 10. pp. 217–26.

[24] In: Ref. [15]. pp. 411–17.

[25] Silverberg SG, Kurman RJ. Tumors of the uterine corpus and gestational trophoblastic disease. AFIP Atlas. Ser 3, Fasc. 3, 1991. pp. 166–157.

[26] In: Ref. [7]. p. 649.

[27] In: Ref. [23]. p. 157.

[28] Rather LJ. The significance of nuclear size in physiological and pathological processes. Ergeb Allg Pathol u Pathol Anat 1958;38:127–99.

[29] Bignold LP. Hypothesis for the influence of fixatives on the chromatin patterns of interphase nuclei, based on shrinkage and retraction of nuclear and perinuclear structures. Br J Biomed Sci 2002;59(2): 105–13.

[30] Cooper EH, Bedford AJ, Kenny TE. Cell death in normal and malignant tissues. Adv Cancer Res 1975;21:59–120.

[31] Koller P. The role of chromosomes in cancer biology. New York, NY: Springer-Verlag; 1972.

[32] DeMay RM. The art and science of cytopathology. Chicago, IL: ASCP Press; 1996. pp. 40–42.

[33] Derenzini M, Ceccarelli C, Santini D, et al. The prognostic value of the AgNOR parameter in human breast cancer depends on the pRb and p53 status. J Clin Pathol 2004;57(7):755–61.

[34] In: Ref. [19]. p. 109.

[35] In: Ref. [16]. pp. 164–75.

[36] Walker RA. The complexities of breast cancer desmoplasia. Breast Cancer Res 2001;3(3):143–5.

[37] Fonar Y, Frank D. FAK and WNT signaling: the meeting of two pathways in cancer and development. Anticancer Agents Med Chem 2011;11(7):600–6.

[38] Tefferi A. Pathogenesis of myelofibrosis with myeloid metaplasia. J Clin Oncol 2005;23(33):8520–30.

[39] In: Ref. [34]. pp. 85–99.

[40] Unni KK, Inwards CY, Bridge JA et al. Tumors of the bones and joints. AFIP Atlas. Ser 4, Fascicle 2. pp. 155–58.

[41] Weiss SW, Goldblum JR. Enzinger and Weiss's soft tissue tumors, 5th ed. Philadelphia, PA: Mosby/Elsevier; 2008. pp. 437–40.

[42] Patterson JW, Wick MR. Nonmelanocytic tumors of the skin. AFIP Atlas. Ser 4, Fascicle 4. pp. 215–16.

[43] Robboy SJ, Mutter GL, Pratt J, editors. Robboy's pathology of the female reproductive tract (2nd ed.). Philadelphia, PA: Churchill-Livingstone/Elsevier; 2009. pp. 169, 361.

[44] Bosman FT, Carneiro F, Hruban RH, et al. WHO classification of tumours of the digestive system. Lyon: IARC; 2010. p. 162.

[45] Shepherd NA, Warren BF, Williams GT, editors. Morson and Dawson's gastrointestinal pathology (5th ed.). Chichester: Wiley Blackwell; 2013.

[46] Bignold LP. The mutator phenotype theory of carcinogenesis and the complex histopathology of tumours: support for the theory from the independent occurrence of nuclear abnormality, loss of specialisation and invasiveness among occasional neoplastic lesions. Cell Mol Life Sci 2003;60(5):883–91.

[47] Lewin KJ, Appleman HD. Tumours of the esophagus and stomach. AFIP Atlas, Ser 3, Fasc 18, 1995. pp. 31–98.

[48] In: Ref. [42]. pp. 14–70.

[49] In: Ref. [23]. pp. 31–248.

[50] In: Ref. [16]. pp. 249–94.

[51] Riddle RH, Petras RE, Williams GT, Sobin LH. Tumors of the intestines. AFIP Atlas Ser, Fasc 32, 2003. pp. 279–15.

[52] In: Ref. [42]. pp. 253–55.

[53] Elder DE, Murphy GF. Melanocytic tumors of the skin. AFIP Atlas, Ser 4. Fasc. 12. pp. 15–114.

[54] In: Ref. [42]. pp. 42–44.

[55] In: Ref. [54]. pp. 27–37.

[56] In: Ref. [15]. pp. 75–104.

[57] In: Ref. [23]. pp. 67–100.

[58] In: Ref. [57]. pp. 350–54.

[59] In: Ref. [6]. pp. 145–56.

[60] In: Ref. [16]. p. 9.

[61] In: Ref. [51]. pp. 86–89.

[62] In: Ref. [47]. pp. 257–66.

[63] Klein-Szanto AJ. Pathology of human and experimental skin tumors. Carcinog Compr Surv 1989;11:19–53.

[64] Kelly PN, Strasser A. The role of Bcl-2 and its pro-survival relatives in tumourigenesis and cancer therapy. Cell Death Differ 2011;18(9):1414–24.

[65] Friberg S, Mattson S. On the growth rates of human malignant tumors: implications for medical decision making. J Surg Oncol 1997;65(4):284–97.

[66] Baguley BC, Marshall ES, Finlay GJ. Short-term cultures of clinical tumor material: potential contributions to oncology research. Oncol Res 1999;11(3):115–24.

[67] Meyer JS, McDivitt RW, Stone KR, et al. Practical breast carcinoma cell kinetics: review and update. Breast Cancer Res Treat 1984;4(2):79–88.

[68] Perry S. Cell kinetics and cancer therapy: history, present status, and challenges. Cancer Treat Rep 1976;60(12):1699–704.

[69] Norton L, Gilewski TA. Cytokinetics Hong WK, Bast RC, Hait WN, editors. Holland–Frei cancer medicine (8th ed.). Shelton, CT: PMPH; 2010. pp. 550–7.

[70] Wilson GD. Cell kinetics. Clin Oncol (R Coll Radiol) 2007;19(6):370–84.

[71] Tot T. Cytokeratins 20 and 7 as biomarkers: usefulness in discriminating primary from metastatic adenocarcinoma. Eur J Cancer 2002;38:758–63.

[72] In: Ref. [6]. pp. 179–201.

[73] Swerdlow SH, Campo E, Harris NL, editors. WHO tumours of the haematopoeitic and lymphoid tissues (4th ed.). Lyon: IARC; 2008.

[74] Vaupel P, Mayer A. Availability, not respiratory capacity governs oxygen consumption of solid tumors. Int J Biochem Cell Biol 2012;44(9):1477–81.

[75] Potter VR. Biochemical uniformity and heterogeneity in cancer tissue (further discussion). Cancer Res 1956;16(7):658–70.

[76] Fidler IJ. Tumor heterogeneity and the biology of cancer invasion and metastasis. Cancer Res 1978;38(9):2651–60.

[77] Owens AH, Coffey DS, Baylin S, editors. Tumour cell heterogeneity: origins and implications. New York, NY: Academic Press; 1982.

[78] Heppner GH. Tumor heterogeneity. Cancer Res 1984;44:2259–65.

[79] Dexter DL, Leith JT. Mammalian tumor cell heterogeneity. Boca Raton, FL: CRC Press; 1986.

[80] Mantovani M, D'Incalci A. Heterogeneity of cancer cells. New York, NY: Raven Press; 1994.

[81] In: Ref. [1]. p. 84.

[82] Anon.: Acute leukemia as a terminal event in chronic leukemia. BMJ 1920;i:231.

[83] Rappaport H. Tumors of the hematopoietic system. AFIP Atlas. Ser 1, Fascicle 8. pp. 161–206.

[84] See Ref. [3], vol 2. pp. 6-13.

[85] In: Ref. [6]. pp. 235–50.

[86] Shet AS, Jahagirdar BN, Verfaillie CM. Chronic myelogenous leukemia: mechanisms underlying disease progression. Leukemia 2002;16:1402–11.

[87] In: Ref. [54]. pp. 235–50.

[88] In: Ref. [19]. pp. 259–68.

[89] Kampen KR. The discovery and early understanding of leukemia. Leuk Res 2012;36(1):6–13.

[90] Aisenberg AC. Historical review of lymphomas. Br J Haematol 2000;109(3):466–76.

[91] Cavaillon JM. The historical milestones in the understanding of leukocyte biology initiated by Elie Metchnikoff. J Leukoc Biol 2011;90(3):413–24.

[92] In: Ref. [83]. pp. 101–56.

[93] In: Ref. [73]. pp. 155–56.

[94] In: Ref. [73]. pp. 31–108.

Molecular Abnormalities in Tumors: "Molecular Pathology" and "Biomarkers"

L.P. Bignold: Principles of Tumors.
DOI: http://dx.doi.org/10.1016/B978-0-12-801565-0.00009-3

Chapters 4–7 of this book deal with etio-pathogenesis of tumors with emphasis on genomic events known to occur in tumors, and the genes which may be involved in certain types of tumor. Chapter 8 describes the complex aspects of the morphology and behavior of the different types of tumors, as well as the complexities in the combinations of features which individually characterize the various types of tumors. It is noted that there are so far no definite genomic explanations of either the variabilities in intensities of the individual traits of tumors, or of the complexities in the combinations of traits.

This chapter follows on from Chapter 8, being mainly concerned with the molecular abnormalities that have been identified in cases of human tumor types. Particular attention is given to those genomic abnormalities that have been applied to the diagnosis and/or treatment of the tumor types.

Section 9.1 gives an introduction to the kinds of studies of these abnormalities, indicating the various uses of the terms "molecular pathology" and "biomarkers." Section 9.2 provides a background to the identification of substances in tissues. Section 9.3 describes aspects of protein antigens in tumor tissues. Section 9.4 gives an overview of abnormalities in RNA and DNA in human tumor tissues used in diagnosis. Section 9.5 deals with chromosomal abnormalities in tumors. Section 9.6 discusses aspects in correlating the findings in the different studies with the clinical behavior of individual cases of tumor. Section 9.7 gives an account of medical aspects of the field of biomarkers in body fluids.

In each section, observations, which bear on theories of tumors (Chapter 2), and knowledge of possible genomic lesions (Chapter 5) are mentioned.

9.1 GENERAL TERMS USED IN THE STUDY OF MOLECULAR ABNORMALITIES IN TUMORS

Molecular abnormalities of tumors are studied as "molecular pathology," and as "biomarkers." The general points can be summarized as follows:

a. *In medicine*

The term "molecular pathology" is applied to additional data for pathologic diagnosis of specimens of tumor tissue (Chapter 8). In any particular study, the target molecule in the tissue may be called "the antigen," "the transcript," or "the gene," according to whether the "target"/"marker" molecule is protein, RNA, or DNA. The term "molecular pathology" may be used for all these tests collectively. A synonymous term is "molecular diagnostics" [1].

Occasionally, e.g. [2], "molecular pathology" is reserved for the tests on genes. In these tests, it is common to use "genotyping" in reference to the detection of known mutations. "Mutation scanning" is the detection of any lesion in a target region of the genome (usually, exons of a gene).

b. *In genetics*

In this discipline, "molecular pathology" may refer to an aspect in the assessment of the effects of variant DNA sequences for function-altering changes which might lead to disease [3]. That is to say, which changes in the sequence of nucleotides result in what degrees of loss of function/"hypomorphism" (see A2.1.2).

c. *Definitions of "biomarkers"*

Currently, different definitions of biomarkers describe quite variable circumstances. For example:

(Any) anatomic, physiologic, biochemical, or molecular parameters associated with the presence and severity of specific disease states. Biomarkers are detectable and measureable by a variety of methods including physical examination, laboratory assays and medical imaging. [4]

Using this definition, biomarkers may include morphological changes used for diagnosis, i.e., the type-specific histopathological features seen by microscopy. And in another definition:

A biomarker can be a substance that is introduced into an organism as a means to examine organ function or other aspects of health. [5]

d. *Biomarkers for diagnosis*

"Biomarker" can be used for a tumor type-specific molecule that can be measured in the blood or other fluid. In these situations, a high level of the molecule allows diagnosis of the presence of tumor without microscopic examination of a biopsy. The molecule may then be called a "biomarker" [6].

e. *Biomarkers for assessing effects of therapies*

Similarly, during treatment of a case of a type of tumor which releases measurable amounts of a specific molecule, improvements, and relapses of the case can be followed by measuring the levels of the molecule/"biomarker" in the blood or fluid [7,8].

f. *"Biomarkers" in epidemiology*

In this discipline, a "biomarker" is any abnormal molecule which may be used as a measure of exposure to a carcinogen. This is discussed in Section 14.3.2.

As a broad generalization, "molecular pathology" is probably most often used for studies in which molecular technology is directed at localizing the molecule in tissues and cells. "Biomarker" is probably most often used for substances which are measurable in blood or tissue fluid, and hence without any implication of localizing the substance in a region of the body, or in a tissue, or in cells.

9.2 BACKGROUND TO THE IDENTIFICATION AND LOCALIZATION OF SPECIFIC CHEMICALS IN TISSUES

This section only deals with the identification of substances in sections of tissues (Figure 9.1). Identification of substances in body fluids is mentioned in Section 9.7 and identification of specific nucleic acid abnormalities in extracts of tissues in Section 9.4.

9.2.1 Chemical Basis of Histological Staining

Since the nineteenth century, advances in histology have generally been dependent on discoveries in chemistry, biochemistry, and related fields (see Table 9.1). The discovery of iodine in 1813 [9] led to its use as a biological stain for cell membranes and nuclei, upon which histology and histopathology were dependent for many

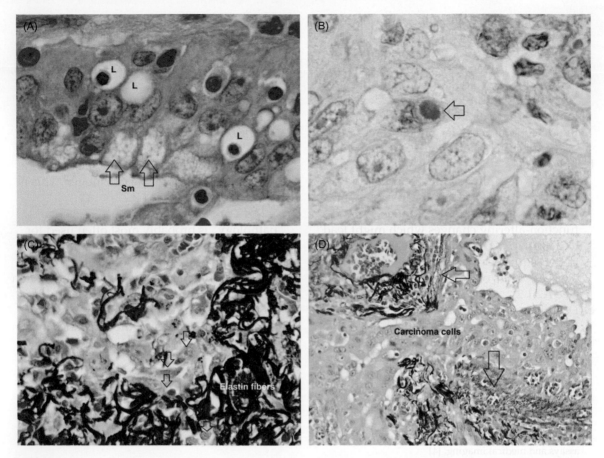

FIGURE 9.1   Some histochemical stains used for the diagnosis of tumors. (A) Vacuoles in lung carcinoma cells. L = large, Sm = small vacuoles, ×100. (B) Vacuole in lung carcinoma cells containing mucus (arrowed), Periodic acid - Schiff with diastase stain, ×100. (C) Elastin of pleura showing invasion by lung cancer cells (arrowed) ×40. (D) Elastin stain of the wall of a vein (arrowed) showing invasion by lung carcinoma cells, ×20.

decades [10]. In the second half of the nineteenth century, stains from natural sources, especially hematoxylin, became available. Later, synthetic dyes, especially derivatives of aniline, were developed [10]. In parallel with these studies were investigations of the particular substances in tissues to which the stains bind.

9.2.2 Histochemistry

However, only in the twentieth century, with understanding of the natures of carbohydrates,

proteins, and lipids, could the specific field of histochemistry develop [11]. Only a relatively small number of substances can be detected in these ways. An example of this is the periodic acid - Schiff ("PAS") stain for carbohydrates, which is widely used in histology and histopathology today.

9.2.3 Fluorescence-Labeled Antibodies

In the 1960s, techniques depending on specific antibodies began to be applied to localizing

TABLE 9.1 Chronology of Technological Advances for Visualizing Cells

1800s	Better single lens microscopes with improvements in lens making
1820s	Iodine solutions for staining biological materials
1830s	Achromatic lenses used in tandem ("compound microscopes"). This development allowed reliable magnifications up to approximately 300×.
1860s	Paraffin embedding for better section-cutting.
1870s	Apoachromatic lenses, substage condensers, hematoxylin, and other stains
	By the 1880s, light microscopes achieved maximum possible resolution (approximately 1000×).
1920s and 1930s	New tissue stains, including those using silver compounds.
1950s	Beginnings of modern histochemistry, including application of Fuelgen reaction to estimating DNA in nuclei.
	Phase-contrast microscopy
	Electron microscopy
1960s	Fluorescence microscopy
1980s	Immunohistochemistry
1990s	In situ hybridizations for nucleic acids

Decades are those in which methods came into widespread use.

substances in tissues. In "direct" immunofluorescence techniques, the antibodies are labeled with fluorescent dyes and the complex is applied to histological sections of tissues. The antibody with attached dye localizes at sites of the relevant antigen. The sites of the antigen are then visualized in the sections with wavelengths of UV light (mainly UVA), which excites the dyes to fluoresce. In "indirect" immunofluorescence techniques, first, an unlabeled antibody is applied to the sections. After washing, a labeled antibody to the first antibody (e.g., goat anti-rabbit IgG) is applied. The indirect method has certain advantages over the direct methods. These methods are still used in histopathological examinations of renal biopsies [12].

9.2.4 Immunohistochemistry

Subsequently, antibody-based methods were made easier by the use of immuno-peroxidase and similar methods, collectively referred to as "immunohistochemistry" [13].

The method is an indirect type (see Figure 9.2). The first antibody binds to the protein in the tissue. The second protein, which has an enzyme attached (either to itself or via a linker molecule such as biotin or protein A), binds to the first protein. Once bound, the second antibody is localized by adding a particular substrate for enzyme. The product of the activity of the enzyme must be insoluble, so that a microscopically visible precipitate forms at the site of the protein.

All of these methods were improved as the quality of antibodies was enhanced by monoclonal techniques [14].

These methods have become a major extension of "routine" staining methods used by pathologists to achieve more accurate typing of cases of tumor (see Figure 9.3). However, there are many technical issues involved in these methods [15]. These may contribute to the differences in findings between different laboratories.

9.2.5 Total Nuclear DNA

The deoxyribose in DNA was identified in the 1920s, and a reliable histochemical method specific for it was described by Feulgen in the same decade [16]. The method remains in use in several applications [17], and has been used to confirm the polyploidy of various tumor types. However, in routine histopathological practice, it does not offer significant advantages over visual assessments of chromatin content of nuclei

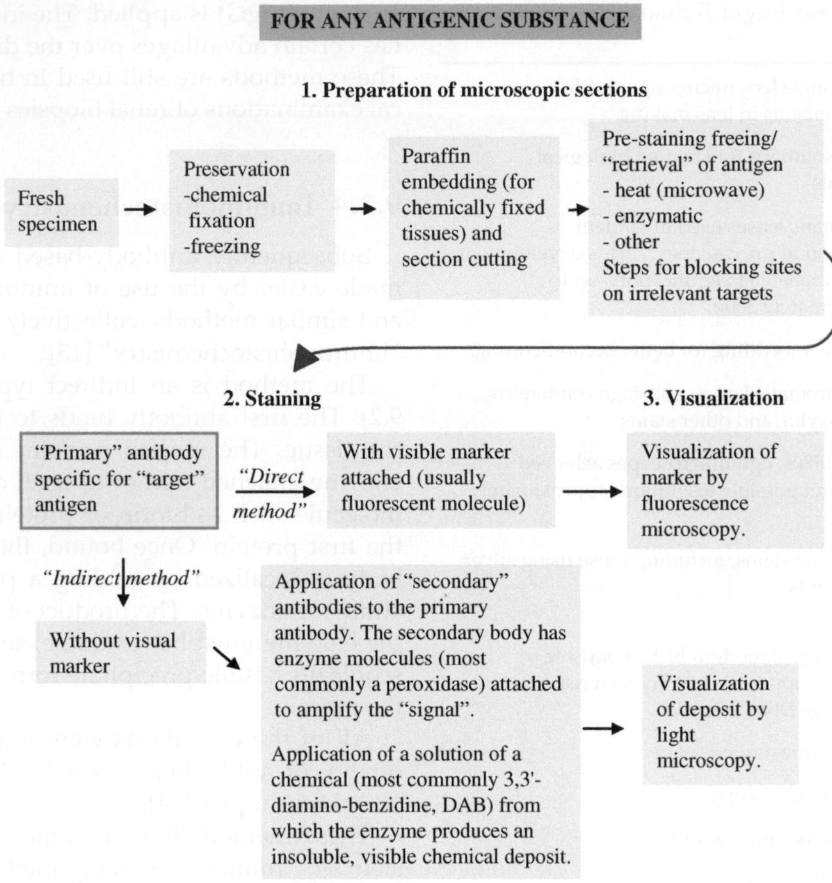

FOR ANY ANTIGENIC SUBSTANCE

1. Preparation of microscopic sections

| Fresh specimen | → | Preservation -chemical fixation -freezing | → | Paraffin embedding (for chemically fixed tissues) and section cutting | → | Pre-staining freeing/ "retrieval" of antigen - heat (microwave) - enzymatic - other Steps for blocking sites on irrelevant targets |

2. Staining   **3. Visualization**

"Primary" antibody specific for "target" antigen

"Direct method" → With visible marker attached (usually fluorescent molecule) → Visualization of marker by fluorescence microscopy.

"Indirect method" ↓

Without visual marker →

Application of "secondary" antibodies to the primary antibody. The secondary body has enzyme molecules (most commonly a peroxidase) attached to amplify the "signal".

Application of a solution of a chemical (most commonly 3,3'-diamino-benzidine, DAB) from which the enzyme produces an insoluble, visible chemical deposit.

→ Visualization of deposit by light microscopy.

FIGURE 9.2   Principles of immunohistochemical techniques.

according to intensity of hematoxylin staining. It is probably most accurate when used on monolayers or thinly smeared cells. It may be less accurate when used to study nuclei in sections of paraffin-embedded tissues. This is because

i. The process of sectioning leaves many "cut-through" nuclei in the sections. These will not contain all the DNA of the original nucleus, and

ii. DNA may be washed out of nuclei during preparation of histologic sections, as discussed in Section 9.7.1.

9.2.6 Identification of Specific RNA and DNA in Tissues

In the 1990s, another whole field of study of tumor tissues developed from technology for localizing particular DNA sequences in cell nuclei. In principle, this is done by using complementary DNA or RNA labeled with fluorescent or other dyes, and hybridizing this nucleic acid to the target RNA or DNA in the sections of tissue [18–20] (see Figure 9.4).

In some situations, more complex studies may be accomplished. For example in "dual SISH" for

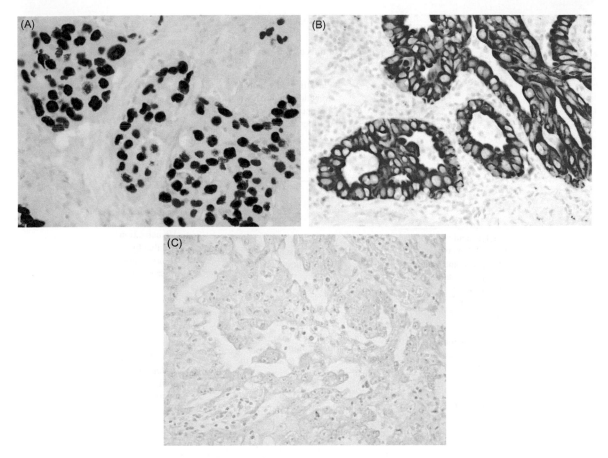

FIGURE 9.3 Tumor cells stained for specific antigens by immunohistochemical methods. All images are from the same case. (A) Lung carcinoma cells showing strong staining of nuclei for TTF1, ×20. (B) Lung carcinoma cells showing strong staining of cytoplasm for CK7, ×20. (C) Lung carcinoma cells staining very weakly for CK20, ×20.

counting *HER2* gene copies in cells, one probe (typically black) is for part of the *HER2* gene, while the other probe (typically red-colored) is for a site elsewhere on the chromosome [17] which carries the *HER2* gene (Figure 9.5). If both the gene and chromosome 17 probes stain excess numbers of sites in a cell, the *HER2* gene is probably amplified through amplifications of the whole chromosome 17, as in hyperploidy (see 5.3). However if the sites identified by the *HER2* gene probe greatly exceed the chromosome 17 sites, the amplification is more likely by sub-chromosomal mechanisms (see Figures 5.8 and 5.9).

9.3 ANTIGENS AND RELATED NON-NUCLEIC ACID MATERIALS IN HUMAN TUMOR CELLS

9.3.1 Antigens Specific to Parent Cells in Tumors

(a) General

The antigens that are most easily identified in tumors are those of the kind of parent cell from which any particular tumor arose [21]. In diagnostic histopathology, these antibodies are used especially for cases of tumors which are so

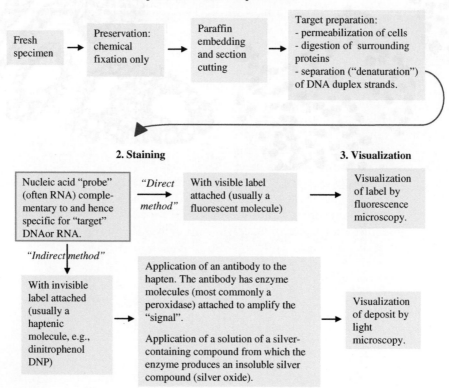

FOR RNA OR DNA

1. Preparations of microscopic sections

Fresh specimen → Preservation: chemical fixation only → Paraffin embedding and section cutting → Target preparation:
- permeabilization of cells
- digestion of surrounding proteins
- separation ("denaturation") of DNA duplex strands.

2. Staining

Nucleic acid "probe" (often RNA) complementary to and hence specific for "target" DNA or RNA.

"Direct method" → With visible label attached (usually a fluorescent molecule)

3. Visualization

Visualization of label by fluorescence microscopy.

"Indirect method" ↓

With invisible label attached (usually a haptenic molecule, e.g., dinitrophenol DNP) → Application of an antibody to the hapten. The antibody has enzyme molecules (most commonly a peroxidase) attached to amplify the "signal".

Application of a solution of a silver-containing compound from which the enzyme produces an insoluble silver compound (silver oxide). → Visualization of deposit by light microscopy.

FIGURE 9.4   In situ hybridization methods.

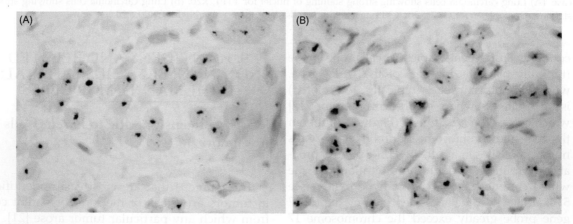

FIGURE 9.5   *In situ* hybridization preparations in assessment of carcinoma of the breast. (A) Nuclear staining for HER-2. The black dots are the site of the gene in the nuclei (pale blue). The cytoplasm of the cells is not stained, ×60. (B) Dual nuclear staining for copies of *HER-2* (black dots) and for chromosome 17 (red dots) in nuclei (pale blue). Cytoplasm of cells is not stained, ×60.

"undifferentiated"/lacking specialized features (see Section 1.3.2 and Table 9.2) that the kind of cell of origin cannot be determined by standard histological features. For example, tumors of epithelial cells can be indistinguishable from those of soft tissue cells and may be called "sarcomatoid" [22]. Another situation is when the histological appearances show no evidence of any kind of specialization, e.g., [23], and may rarely show prominent giant cell formation, e.g., [24]. In many of these cases, the tumor cells express only proteins of the kind of cell of origin, and not proteins of other kinds of cells. For example, an undifferentiated tumor of squamous cells tends to express cytokeratins (as in all epithelial cells) and not express the actins of smooth muscle cells (as in most fibrocytic cells). Once the kind of cell of origin is established, the case of tumor can be classified according to the criteria mentioned in Section 1.1.2.

TABLE 9.2 Some Common Antigens Identified by Immunohistochemical Studies of Tumors

Purpose	Antigens
1. Identification of tumor type	
Identifying general type of parent cell	
Epithelial	Combinations of cytokeratins
Hematolymphoid	CD45
Neural	S100
Melanocytic	S100, HMB45, melan-A
Sarcomas in general	None
Mesothelioma	Calretinin, mesothelin, HMBE-1
Identifying specific carcinomas	
Lung (adeno)	TTF1
Prostatic carcinoma	Prostate-specific antigen, prostatic alkaline phosphatase
Colorectal carcinoma	CK7 and CK20
For diagnosis of Lynch Syndrome	MLH1, MSH2, MSH6, PMS2 proteins
Breast	Estrogen receptor, progesterone receptor proteins
Malignant lymphoma subtypes	CD ("cluster of differentiation") 3, 5, 10, 20, 30, 45
Granulocyte tumors	Myeloperoxidase
Identifying specific sarcomas	
Gastrointestinal stromal tumor ("GIST")	CD117, DOG1
Angiosarcomas	CD31, CD34, Fli-1
Leiomyosarcoma	Desmin
Fibroblastic tumors	Factor XIIIa, various actins
2. Guidance of therapy and prognostic evaluation of tumors	
Therapy and prognosis of breast carcinoma	HER-2 protein, BRCA1 protein
Prognosis in various tumors	Cyclin D1, Ki67, P16

(b) Supra-Normal Expressions of Parent Cell Antigens

While reductions from normal in bio-marker expressions have received most attention in tumor research (see above), the opposite phenomenon—supra-normal expressions—has rarely been considered in detail (see Figure 9.6). Supra-normal specialization, in the sense of excess specialization-associated substance visible by ordinary microscopy, is well known (Section 8.1.2c). However, supra-normal expressions of cancer-related proteins, as detected by immune-histochemical methods are less frequently noted. Examples include supra-normal gp100 (also known as Pmel 17, detected with HMB45 antibody) [25] and tyrosinase [26] in malignant melanoma.

9.3.2 Lineage Infidelities, Embryonic Reversions, and Neo-antigen Expressions

(a) Lineage Infidelities

While most tumors express only antigens of their parent kind of cell, a proportion of cases

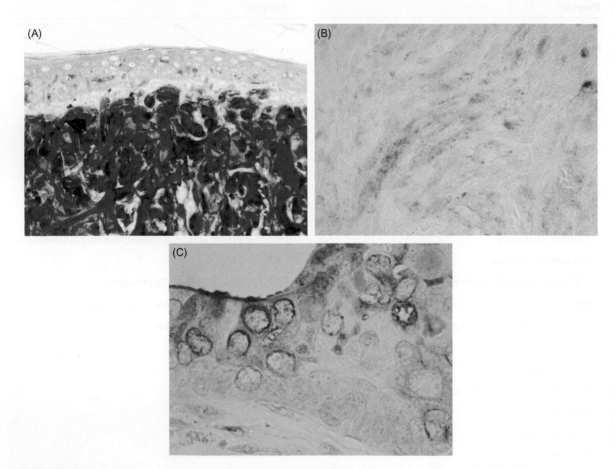

FIGURE 9.6 Supra-normal, embryonic, and ectopic expressions of antigens in tumors. (A) Xs S100 in melanoma. The cytoplasmic stain is so heavy that the nuclei are difficult to discern. ×40. (B) HMB45 in angiomyolipoma. The staining is the brown dots in the cytoplasm. The nuclei are pale blue. The darker blue dots in the nuclei are nucleoli, ×40. (C) Carcinoembryonic antigen (CEA) stained in the cytoplasm of cells of carcinoma of colon, ×60.

express antigens of other kinds of cells. This appears to be a parallel phenomenon with the lineage-unfaithful/metaplastic morphological changes which are known to occur in a few tumor types (Section 8.1.2). However, lineage infidelities in tumor cell antigen expressions appear to be much commoner than morphological infidelities.

Patterns of lineage infidelities appear to be divisible into:

1. Those between different cells of the same general kind (e.g., epithelia). For example, prostate-specific antigen is normally found only in prostatic epithelium, and prostatic epithelial tumors. However, it is also found in approximately 50% of carcinomas of the female breast and approximately 20% of salivary gland tumors [27].

    Another example is CDX-2, which is normally expressed only in the epithelium of small and large bowel epithelium, and organs derived from them—pancreas and appendix. However, it is often expressed in carcinomas of stomach, bladder, and ovary [28].

2. Those which are between quite different general kinds of cells (e.g., epithelial and neural cells). For example prostate-specific antigen has been reported in approximately 50% of malignant melanomas [29]. As another example, CD34 is normally expressed on myeloid and lymphoid progenitor cells, as well as on the surface of endothelial cells. However, it is also expressed by 80% of cases of gastro-intestinal stromal tumors [30]. Further, Melan A is a specialized protein in melanocytic cells. However, reactivity with this protein is present in nearly the quarter of the mesenchymal tumor called angiomyolipoma, see Figure 9.6B and [31].

(b) Embryonic Reversions

In certain tumor types, expressions of embryonic antigens are common. This applies particularly in germ-cell tumors, e.g., placental alkaline phosphatase in seminomas and human chorionic gonadotrophin in non-seminomatous tumors [32]. However embryonic antigens are also expressed in non-gonadal tumors. For example, carcino-embryonic antigen is expressed in many adenocarcinomas [33] and alpha-fetoprotein in approximately one-third of hepatocellular carcinomas [34] (see Figure 9.6C Section 8.1.2). Another example is the family of cancer/testis antigens, which are expressed in a variety of tumors of cells in adults [35].

These instances sometimes reduce the diagnostic value of the antigen for particular tumor types because the value of the antibody in discriminating between tumor types is compromised. In relation to theories of tumors, ectopic expressions appear to support the concept of de-repression of genes occurring in specific tumor types, as indicated in Section 5.2.5 and mentioned in A2.1.3. In considering this issue, the question of specificity of the antibodies in use is always raised. Just because an unknown substance reacts with an apparently specific antibody—say to "substance A"—the reactivity does not prove that the unknown substance is "substance A". The possibility may exist that the antibody is cross-reacting with the unknown substance.

(c) Neo-Antigens

Historically, there was little evidence that many tumors produce antigens against which cytotoxic immune responses either occur or can be induced (Section 2.9.5). However, with increasing knowledge of genomic events in tumor cell populations (Section 5.1.6), many neo-antigens—reflecting these genomic alterations—are being discovered.

The first main example was the BCR/ABL fusion gene in chronic myeloid leukemia (see Section 9.5.1). Another example is the BRAF-^{V600E} gene mentioned below.

At the present time, fusion genes are being searched for in many types of tumors, because they provide a basis for personalized therapy of individual patients (Section 12.4) [36].

9.3.3 Antigens Relevant to Growth

The presence or absence of these antigens is of little value for diagnosing particular tumor types. This is because they are expressed—at least in a proportion of cases—in a wide variety of tumor types. These antigens are studied, however, for their prognostic value and as possible targets of specific therapies.

(a) Growth and Cell Signaling

There have been extensive studies of expressions of proto-oncogenes and proteins of cell signaling using immunohistochemical methods. These methods can detect increases and reductions in concentrations, as well as certain specific mutant variants, of these proteins. Depending on their function, the change in expression or structure of the protein may have prognostic value.

Some examples are:

i. HER2/neu receptor protein (one of the epithelial growth factor receptors, see Section 6.3.2) is overexpressed in approximately 30% of cases of carcinoma of the breast [37], as well as in many other malignancies, including those of ovary, bladder, salivary gland, endometrium, pancreas, and lung (non–small cell type) [38].

ii. *PTEN* (a tumor suppressor gene by virtue of its phosphatase activity, see Section 6.5.5) is normally expressed in prostatic epithelium, as well as other cells. Reduction in the quantity of PTEN protein in prostatic carcinoma can be demonstrated by histochemical methods [39].

iii. BRAF is an upstream component of the MAP/ERK signaling pathway (Section 6.3.2). BRAF ^{V600e} is the protein product of the commonest kind of allele mutation found in cancer cells. The mutant protein can be detected by immunohistochemical methods [40,41].

(b) Those Associated with the Cell Division Process

Essentially, these are the downstream effector molecules of the cell signaling pathways. Like upstream signaling molecules, they are commonly expressed in a wide variety of tumor types.

i. Cyclin D1 is synthesized during cell division and, in complexes with other proteins, promotes G1→S phase progress of the cell cycle (see A1.3.6). Increased amounts of cyclin D1 are seen in many different types of cancers [42]. It is increased in amounts in cells of mantle cell lymphoma, but not other lymphomas [43].

ii. Ki-67 is a nuclear protein which is present in cells only during cell division (i.e., phases except G0). The presence of the protein is a marker of proliferation. Overexpression of the Ki-67 is associated with pre-malignant change in some conditions such as Barrett's esophagus. The increased amount of the protein may be a reflection of aneuploidy [44].

9.3.4 Antigens Related to Invasion or Metastasis

It is generally accepted that tumors are due genomic events. It is also widely thought that malignant behavior is a single biological process and is due to a small number of sequential events in tumor cell populations (A2.2). According to these views, the products of the "invasion and metastasis" genes should be present in the tumor cell populations. However, no single antigen for all malignant tumors versus all benign tumors has been found.

Invasion-specific clusters of gene activities have been described [45,46].

9.3.5 Variabilities in Expressions of Almost All Antigens

Considerable variabilities in antigen expressions are often seen in the diagnostic

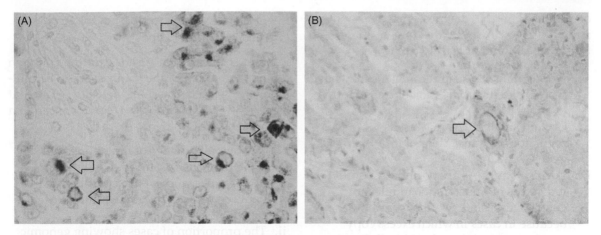

FIGURE 9.7 Variabilities in expressions of antigens in tumors. Heavily stained cells are arrowed. In all normal parent cells, the staining is uniform. (A) Variable staining with HMB45 in melanoma, ×40. (B) Variable staining for thyroglobulin in papillary carcinoma of thyroid, ×20.

immunohistochemistry (see Figure 9.7). Staining is usually expressed as weak/strong, or on a scale 0– + + + [47]. Additionally the focal or diffuse pattern of staining may be recorded.

In general, tumor cell populations over time often lose biomarker expression per cell in parallel with diminution in specialization (i.e., "progression"; change to a higher "grade," see Sections 1.1.3 and 8.4). These losses are assumed to be due to the accumulation of loss-of-function genomic events in the population through genomic instability.

Another general finding of molecular pathological studies is that the patterns of variability of antigen expressions—inter-type, inter-case, intra-case—are the same as for loss of specialization (see Section 8.3.3). The principle applies as much to lineage-faithful expressions as to lineage-unfaithful expressions. These variabilities may be explained by technical factors, such as variable fixation both between different cases, or between different areas of the same tumor. However, no techniques have been developed which eliminate these variabilities in antibody staining.

9.3.6 Different Degrees of Expressions between Primary Tumors and Their Metastases

Until the last few years, it had been assumed that tumor cell populations are monoclonal, and that metastases are simply a colony of the same clone growing in another place in the body from the primary tumor. According to this, metastatic tumor cell populations were thought to be likely to be genomically the same as the primary tumor. However, these several studies have shown discordant results between primary tumors and paired metastases [48]. In some cases, it is possible that the genomic lesions in the metastases may be a better indicator of prognosis than those in the primary tumor [49].

9.4 NUCLEOTIDE SEQUENCES IN RNA AND DNA IN HUMAN TUMOR CELLS

9.4.1 RNA

At the present time, studies of these molecules in tumor tissues or body fluids are not often used in the diagnosis or treatment of patients [50]. However, they are of interest in tumor research for at least two reasons:

i. Non-coding RNAs in the nucleus may have major roles in suppression of transcription of genes (A2.2.8).
ii. Certain small RNAs may act as regulators of translations of mRNAs of genes (A2.2.6).
iii. Levels of mRNAs are potentially useful indicators of transcription of relevant genes. This is an important consideration, because in cases in which excess copy numbers of genes are found in the tumor cells, there may be no apparent increase in protein product [51,52]. Messenger RNA levels could allow distinction between deficient transcription and deficient translation as the cause of the lack of protein product (see also Section 9.6). This issue is an area of active research which may lead to diagnostic or treatment applications in the future.

9.4.2 DNA

The known kinds of nucleotide errors are described in 5.1. The detection of particular forms of these has been mainly limited to studies of exons of known genes [53] especially in testing for germline predispositions to tumors (Chapter 7). Genes being most studied have been oncogenes (Section 6.3), signaling proteins (Section 6.4), and tumor suppressor genes (Section 6.5). To the time of writing, these studies of nucleotide errors in particular genes are not widely used in assessments of prognosis and as guides to therapy.

However, with the high-throughput nucleotide sequencing of DNA, large databases of genomic abnormalities in tumors of different types are being established [54–56], and may soon be applied more widely to assessments of individual cases of tumor [57].

9.4.3 Nucleotide Error Spectra in Different Tumor Types

From studies so far on most of the common tumors, the errors in the exons seem to be multiple and largely random in many different genes [53,54,58,59]. However, some nonrandom phenomena are apparent.

i. The overall number of mutations varies more than a thousand fold between tumor types, and even a thousand fold between cases of the same tumor type [58].
ii. The proportion of cases showing genomic events of all types in a particular gene differ according to tumor type. For *P53* the proportions ranges from 56% in lung cancers to 0% in Wilm's tumors, testicular tumors, and pheochromocytoma [60].
iii. The relative percentages of kinds of event (sites of base substitutions, etc., see above) vary. For example, of all mutations in *P53*, G:C in CpG (cytosine-pair-guanine) events vary from 47% in colonic carcinomas to 9% in hepatocellular carcinomas [61].
iv. The sites of "mutational hot spots" vary according to the tumor type [60,61].
v. The numbers of clusters ("hot-spots") vary in different genes and different tumor types. In addition, the proportions of total mutations in the hotspots vary between different tumor types. This is exemplified by the detailed studies of the *P53* gene [61] and the *CDNK* gene [62].

9.4.4 Complexities of Findings in Tumor Genomes

While nucleotide sequence data is being accumulated, it must be noted that there is considerable "noise" in these results. To quote the website of one of the databases:

Cancer genomes can be a very noisy source of data. It is estimated that an individual's tumor is caused by 5–10 driver mutations, but genome re-sequencing regularly reveals over 10,000 somatic mutations per tumor, with much larger numbers not unusual in hypermutated samples (we've seen samples with over 100,000 mutations each, the greatest being 178,763). Across studies from different groups using different techniques, it is unclear whether these huge numbers reflect true hypermutation, substantial germline variation or technical artefacts. To try to improve the value of these data, we are beginning to define a cancer genome noise reduction strategy. Initially, we will exclude any sample with over 15,000 mutations, as this immediately introduces huge noise into COSMIC; these can be reintroduced at a later date. In addition, we are removing all known SNPs from new genome uploads (initially, these are defined by the 1000 genomes project and a panel of normal (non-cancer) samples from Sanger CGP sequencing). We will be assessing how to enhance these filters and best apply them to our curated genes over the next release. Ultimately, we aim to identify the most significant high-value data within cancer genomes, making it much easier to identify actionable biomarkers. [53]

9.5 CHROMOSOMAL ("CYTOGENETIC") ABNORMALITIES DETECTED IN TUMORS

As noted in Sections 2.4 and 2.9.7, chromosomal abnormalities in tumor cells were considered at length by many authors beginning with Hansemann in the 1890s. In the 1950s, the development of the technique of "metaphase squash" preparations led to the discoveries of the chromosomal abnormalities in patients with Down syndrome as well as in the leukemic cells of patients with chronic myeloid leukemia (Figure 9.8A). Studies of chromosomal structure have then proceeded with new techniques allowing the identification of smaller and smaller chromosomal aberrations (Figure 9.8, see also in Chapter 5).

This section deals with general aspects of chromosomal abnormalities in tumors, mentioning those which are associated with particular tumor types (Figure 9.8B).

9.5.1 General

Overall, it has been established that there is no known individual kind of chromosomal abnormality present in all tumors, nor in all cases of the common carcinomas, for example of the lung, colon, breast, and prostate. No cell can exist without one of each chromosome pair. Most malignant tumor cell populations exhibit karyo-instability and are hyperploid as discussed in Section 5.3.4.

Chromosomal abnormalities are potentially pathogenetic via the following:

a. *Deletions and inversions*

These are well recognized in relation to loss-of function genomic events in tumor suppressor genes, as described in Section 6.5 and sections of Chapter 7.

b. *Amplifications (i.e., increased copy numbers of genes)*

Alleles of genes can be detected in metaphase preparations or in any non-mitotic phase by *in situ* hybridization (A2.1.1). The different kinds of amplifications of genes in tumor cell populations are noted in Section 5.3.1. They are essentially the same in general kinds as those found as variations in the genomes of normal human populations (2.1.3). Amplifications of growth factor-oncogene copy numbers are a mechanism of tumor formation, as discussed in Section 6.4.

c. *Creation of fusion genes*

These do not occur in normal cells, but are created by translocations of parts of genes. The main example is *BCR/ABL*, which is associated with chronic myeloid leukemia. The gene comprises the breakpoint cluster gene (*BCR*) on chromosome 22q, and the Abelson (*ABL*) tyrosine kinase gene on chromosome 9q. The gene product is a tyrosine kinase which is not affected by any physiological inhibitors. As a result, the gene has high constitutive activity, which causes excessive growth of

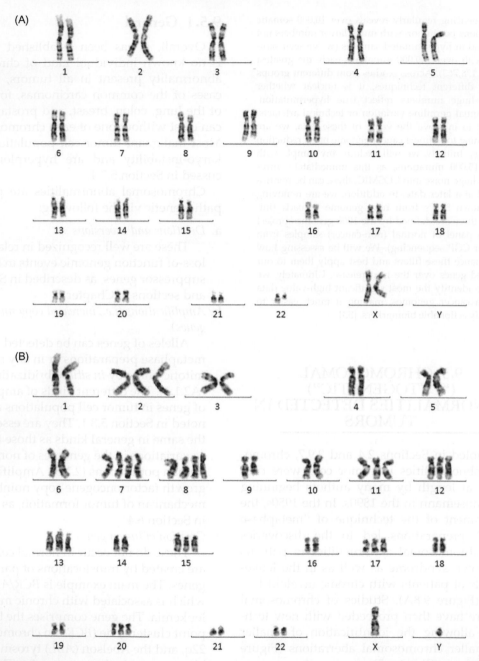

FIGURE 9.8   Karyograms of Philadelphia chromosome in chronic myeloid leukemia cells (A) and abnormal chromosome in embryonal rhabdomyosarcoma cells (B).

(A) The Philadelphia rearrangement is identified in chronic myeloid leukemia, acute lymphoblastic leukemia and rare cases of acute myeloid leukemia. The arrowed chromosomes show the breakpoints on the abnormal chromosomes resulting in translocation of ABL1 (chromosome 9) and BCR (chromosome 22) sequences.

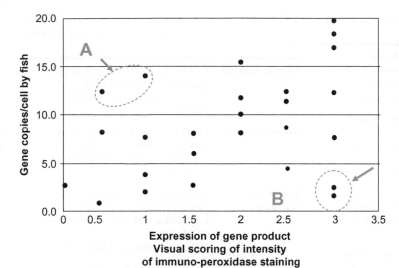

FIGURE 9.9 Gene copy numbers and expressions or protein products.

Cells in rectangle "A" have the high copy numbers but low expression. These may have suffered 'loss-of-function' genomic events in many copies, with full detection (retention of probe-binding sites).

Cells in circle "B" have the highest expression per copy. This may be because a 'gain-of-function' genomic event has occurred in one or more copies, or that functional copies have not been detected (due to e. g. loss of probe-binding site).

the myeloid cells. The hyper-proliferative cells are liable to further mutations due to reductions in DNA repair and genomic instability [63].

It may be noted, however, that half of cases of acute myelogenous leukemia have normal karyotypes, while the other half show a variety of chromosomal lesions, of which trisomy 8 is the commonest [64].

There are currently over 900 documented fusion genes in the Genome Atlas [65], many of which are seen in epithelial cancers [66]. Many of these fusion genes have no established activity or role in tumor formation (see Figure 9.9).

d. *Others*

Breakpoints in chromosomes can potentially deactivate a gene by disruption. Transpositions and inversions can potentially activate or deactivate genes by re-position effects of transposed DNA (see A2.2.3). Many breaks with or without re-joinings and also transpositions may have no effect on any gene, because at least 50% of the genome is made up of repetitive DNA, which so far as is known at present, is of no certain function.

9.5.2 Examples of Specific Chromosomal Abnormalities with Specific Tumor Types

These are given in Table 9.3.

◀ **FIGURE 9.8** (Continued) The BCR-ABL1 fusion gene lies on the derivative chromosome 22 and is pathogenic. ABL1-BCR (derivative chromosome 9) may, or may not be expressed. The BCR-ABL1 fusion gene encodes a chimeric transcript that is translated into a hybrid protein with novel properties. It demonstrates a constitutive and increased tyrosine kinase activity (as compared to wild type ABL1) that is responsible for downstream activation of many pathways including the STAT, RAS and PI3-K signaling pathways. (B) Cytogenetically, the Rhabdomyosarcomas are divided into two main groups – those characterized by translocations that result in fusion gene formation i.e. the t(2;13) PAX3-FOXO1 and t(1;13) PAX7-FOXO1 of alveolar rhabdomyosarcoma (ARMS) and those characterized by chromosomal gain as is seen in the above example of embryonal rhabdomyosarcoma (ERMS). The most commonly gained chromosomes are 2, 8, 11, 13 and 20. ERMS may show chromosomal rearrangements, but these are not recurrent.

Images and legends for both karyograms courtesy of Sarah Moore, Head of Cytogenetics, Genetic Pathology, SA Pathology, Adelaide, South Australia.

TABLE 9.3 Common Chromosomal Abnormalities in Common Tumor Types

1. Lung carcinoma	3p deletions; Many relatively rare abnormalities, e.g., translocations (2;3) (p23;q21) and t(2;10)(p23.2;p11.22) producing *ALK* fusion genes. Ref [67]
2. Breast carcinoma	Few reported studies of chromosome lesions in primary tumors. Because of usual low mitotic rates in breast cancers. X chromosome duplication in basal-like variant. Ref [68]
3. Colorectal carcinoma	Deletions in chromosomes 4, 8, and 18. Duplications on chromosomes 7, 8, 13, 20, and X. Ref [69]
4. Prostate carcinoma	Losses in chromosomes 5q, 6q, 8p, 10q, 13q, 16q, 17p, and 18q, Gains in 7p/q, 8q, 9p, and Xq. Rearrangement in 21q. Ref [70]

TABLE 9.4 Some Comparisons of Amounts of Protein Product and Respective Gene Copy Numbers in Various Tumors

Lung carcinoma	EGRF expression and gene copy number correlated [71,72] Not consistently associated for EGFR [73]
Breast carcinoma	HER2 expression and gene copy number correlated [74,75] Correlations differ with different methods [76]
Colorectal carcinoma	Poor correlation between EGFR protein and gene copy number [77] Correlation better for HER2 than for EGFR [78] Poor correlation for EGFR [79]
Prostate carcinoma	Poor correlation between HER2 expression and gene amplification [80] HER2 amplification not associated with detectable increase in HER2 protein expression [81].

9.6 CORRELATIONS BETWEEN THE MOLECULAR PATHOLOGICAL DATA AND THE CLINICAL BEHAVIOR OF THE TUMOR MASSES

9.6.1 General

Numerous studies have been undertaken comparing expressions of genes in individual cells in a tumor cell population with the gene copy number in the same individual cells. In these studies, gene copies are detected in interphase nuclei by FISH, and biomarker production of the cells individually is estimated by immunohistochemical staining for the relevant gene product.

Initially, it was assumed that all copies found would probably be equally active, and hence that expressions of biomarkers would correlate with copy numbers. However, direct studies have almost uniformly shown that the numbers of copies of particular genes in tumor cells do not correlate with biomarker expression in that cell (see Figure 9.9 and Table 9.4).

9.6.2 Lung Carcinoma

There have been few studies of this type. One study found correlation between copy numbers of *MET* gene and protein expression. However, neither result correlated with clinical outcome [82]. The topic is reviewed in [83,84].

9.6.3 Breast Carcinoma

There have been many studies in this area especially in relation to estrogen and progesterone receptors as well as *HER-2*. The results appear to be inconclusive [52,85–88].

9.6.4 Colorectal Carcinoma

The topic has been studied in this type of cancer in relation to genes on particular chromosomes [89], as well as in relation to epithelial growth factor receptors [90,91].

9.6.5 Prostate Carcinoma

Androgen receptor gene amplification has been correlated with receptor protein expression [92]. However, many additional factors appear to be involved in the progress of these tumors to androgen independence [93]. It has been shown that TOP2A protein expression in these tumors is not associated with altered copy numbers of the alleles of the relevant gene [94].

9.6.6 Sources of Error

a. *Sampling error*

Because many tumors are heterogeneous (see Section 8.3), different foci in the same case of tumor may be different in all characteristics which are not essential for cell viability [95].

Even at the level of histopathological diagnosis, sampling error may account for up to 50% of all errors [96]. At the level of molecular pathology, sampling errors are known to occur [97,98].

b. *Unassessed additional molecular biological factors in the sequence: regulators → allele of the gene → mRNA → peptide chain → functional protein product*

The tests available provide information on nuclear DNA which is recognized by appropriate complementary DNA probes, and proteins which possess epitopes recognized by antibodies to known substances. As noted in the previous subsections, the numbers of copies of the DNA many not relate to the amounts of biological activity of the corresponding protein products.

Certain unassessed molecular biological factors (A2.2) may be involved in the lack of this correlation.

i. *No increase in detected protein product, but allele copy numbers are increased*
  - Some of the alleles are not transcribed, due to transposition with inversion, or other event deactivating transcription.
  - Splicing errors delete the mRNA for the antibody-binding epitope on the protein.

ii. *Increased protein product and increased activity of protein product without change of gene allele copy number* [99]
  - Increased transcription from an activating transcription factor for the gene, or presence of a neo-transcription factor due to a genomic event in another gene.
  - Loss-of-function genomic alterations in the nucleotide sequences of inhibitor sites in the promoter regions of oncogenes so that excessive transcription of the gene occurs.
  - Activation of alleles which are normally inactive. An example is loss of an insulator (A2.2.3) between an active gene and an inactive downstream allele of gene may activate the latter, without affecting the other allele-copy(ies) of the second gene.
  - The gene has sustained a gain of function genomic events in the exons of the gene. This can occur when the enzymatic site of a protein product becomes fixed in the active state. The protein, although not increased in amount is permanently active, rather than only intermittently active. The exons are genomically altered.

iii. *Protein product and alleles are increased, but there is no apparent biological effect*
  - Some alleles are mutant so that the functional domain is lost, but the epitope for the antibody is preserved.
  - Splicing errors in the mRNA cause the active site on the protein protein product to be non-functional.
  - Appropriately spliced mRNA may not be translated due to silencing RNAs.
  - Defects in posttranslational modifications occur in protein product.
  - The mutant protein from one allele has a dominant negative effect on the functional produce of the other.

c. *Technical issues in laboratory tests*
- Errors in test performance as are recorded in most instruction documents dealing with each test.

d. *Assessments of clinical outcomes*
These are discussed in Chapter 14.

9.7 "BIOMARKERS"

9.7.1 General

As mentioned in Section 9.1, molecular methods have been used to study cancer-related substances in blood and other specimens. In an individual patient being investigated for symptoms, a substance may be present in the bloodstream which indicates the presence of a particular type of tumor.

These biomarkers include the specialization product of the normal cell in the bloodstream. Thus for adrenal medullary tumors (pheochromocytomas), the biomarkers are serum levels of catecholamines [100], and for adrenal cortical tumors, the biomarkers are corticosteroid metabolites [101]. Another example is human chorionic gonadotropin (HCG) for choriocarcinoma [102].

Other biomarkers are specific to the parent cell of the tumor type, but not specialization products. Examples are the prostate-specific antigen (PSA) for prostatic carcinoma [103].

After initial treatment, success of therapy and possible relapses of the tumor in the body may be detected by falls and rises in the levels of the same chemicals in the blood. These substances were then being used as "biomarkers" of the presence, and also extent, of disease in the individual patient.

9.7.2 Validation of Biomarkers

Because biomarkers may be widely used in epidemiology for causation, for diagnosis (see Section 9.3) as well as for prognosis and therapy (see in Section 12.4.2), drug design,

and therapy-monitoring situations, it is important that each biomarker accurately reflects the biological process for which it is being used as a proxy/surrogate measurement. The development of each of these potentially cost-saving biomarkers is an expensive process in itself. The term "validation of a biomarker" essentially refers to the process by which reliable data is accumulated in respect of the efficiency or otherwise of the marker in relation to the target disease process [104,105].

The components of the validation vary according to the particular disease process being studied, and /or the particular outcome which the biomarker usage is designed to cause or avoid. The same common issues of sensitivity and specificity apply to biomarkers as to any laboratory test of a disease. Additional criteria of "truth," "discrimination," and "feasibility" may be applied [106].

9.7.3 Biomarkers and Clinical Outcomes

As for the molecular pathological studies involving samples of tumor tissue, there are sometimes imperfect correlations of biomarker levels with the outcomes for the patients. For example, serum prostate-specific antigen levels are not perfect indicators of presence or recurrence of prostate cancer [107,108]. Similarly serum markers such as CA 15-3 are imperfect indicators of breast cancer recurrence, and are not currently recommended for routine follow-up of patients with this disease [109].

9.7.4 "Circulating" Solid Tumor Cells

Leukemic cells are usually readily identifiable in the bloodstream of the patient. Occasionally, lymphoma cells enter the bloodstream in large numbers in a so-called leukemic phase [110]. In both disorders, they commonly come to populate the reticuloendothelial organs, especially the liver and spleen. Cells of solid tumors which metastasize to sites beyond the lungs (e.g., colonic carcinoma metastasizing to the brain) are usually accepted to have entered the bloodstream from the primary tumor, and passed through the lung vascular bed to impact at the distal site. The alternative possibility is that cells from the primary tumors have metastasized to the lung, and cells of this metastatic tumor have then metastasized to the distant site. In autopsies however, primary tumors have been shown to give rise to the distant metastases without any lung metastasis being present.

Until the last few decades, it was thought unlikely that, on a continuous basis, solid tumor cells could behave like leukemic cells, i.e., passing through the lungs, and also then passing through the capillary beds of the distant organs again and again (i.e., circulating). This was essentially because such cells could not be found by ordinary microscopy of blood [111].

In the 1980s, techniques were developed for sorting subpopulations of cells in the blood using specific "sorter antibodies" to the cell type-specific antigens on the surfaces of the cells. The antibodies might be bound to filters or a magnetic particle ("flow cytometry" [112]). When anti-epithelial antibodies are used as the "sorter molecule," carcinoma cells are liable to adhere to the filter and then be detectable by ordinary staining methods.

There are technical difficulties associated with the methods currently in use [113]. One of these is that in any malignant tumor cell population, a proportion of cells are likely to be dying spontaneously (see Section 8.3.2). These cells may release antigens into the bloodstream. If these antigens bind to the surfaces of normal blood cells, especially monocytes, then the normal cell may bind to the filter or be influenced by the equivalent sorting mechanism, and be recorded as tumor cells.

The clinical usefulness for the individual patient of data on tumor cells circulating in his/her bloodstream has not yet been fully established, although many new technical developments are being assessed [114].

References

[1] Hennessy BT, Bast RC, Mills GB. Molecular diagnostics in cancer. In: Hong WK, Bast Jr RC, Hait WN, et al, editors. Cancer medicine. Shelton, CT: PMPH. pp. 335–46.

[2] Nakamura S, Muller-Hermelink HK, Delabie J, et al. Lymphoma of the stomach. In: Bosman FT, Carbeiro F, Hruban RH, editors. WHO classification of tumours of the digestive system (4th ed.). Lyon: IARC; 2010. pp. 69–73.

[3] Strachan T, Read A. Human molecular genetics (4th ed.). New York, NY: Garland Science; 2011. pp. 428–35.

[4] Massachusetts General Hospital Center for Biomarkers in Imaging. Homepage. Available at: http://www.mass-general.org/.

[5] The Biomarkers Consortium. Homepage. Available at: http://www.biomarkersconsortium.org/.

[6] National Cancer Institute. Biomarker. Available at: http://www.cancer.gov/dictionary?cdrid=45618.

[7] Strimbu K, Tavel JA. What are biomarkers? Curr Opin HIV AIDS 2010;5(6):463–6.

[8] Nass SJ, Moses HL, editors. Cancer biomarkers. Washington DC: National Academies Press; 2007.

[9] Swaine PA. Bernarc Courtois (1777–1838), famed for discovering iodine (1811) and for his life in Paris from 1798. Bull Hist Chem 2005;30:103–11.

[10] Clark G, Kasten FH. History of staining (3rd ed.). Baltimore: Williams & Wilkins; 1983.

[11] Pearse AGE, Stoward PJ. Histochemistry, theoretical and applied, 4th ed. Edinburgh, UK: Churchill-Livingstone; 1991.

[12] Ghiran IC. Introduction to fluorescence microscopy. Methods Mol Biol 2011;689:93–136.

[13] Matos LL, Trufelli DC, de Matos MG, et al. Immunohistochemistry as an important tool in biomarkers detection and clinical practice. Biomark Insights 2010;5:9–20.

[14] Köhler G, Milstein C. Continuous cultures of fused cells secreting antibody of predefined specificity. Nature 1975;256:495–7.

[15] Taylor CR, Shi S-R, Barr NJ, et al. Techniques in immunohistochemistry: principles, pitfalls and standardization. In: Dabbs DL, editor. Diagnostic immunohistochemistry. Philadelphia, PA: Churchill-Livingstone / Elsevier; 2009. pp. 1–42.

[16] Lessler MA. The nature and specificity of the Feulgen nucleal reaction. Int Rev Cytol 1953;3:231–47.

[17] Chieco P, Derenzini M. The Feulgen reaction 75 years on. Histochem Cell Biol 1999;111:345–58.

[18] Cassidy A, Jones J. Developments in in situ hybridisation. Methods 2014 pii: S1046-2023(14)00150-9.

[19] Bonin S, Stanta G. Nucleic acid extraction methods from fixed and paraffin-embedded tissues in cancer diagnostics. Exp Rev Mol Diagn 2013;13(3):271–82.

[20] Cody NA, Iampietro C, Lécuyer E. The many functions of mRNA localization during normal development and disease: from pillar to post. Wiley Interdiscip Rev Dev Biol 2013;2(6):781–96.

[21] Chu PG, Weiss LM. Modern immunohistochemistry. New York, NY: Cambridge University Press; 2009. pp. 8–9.

[22] Bosman FT, Carbeiro F, Hruban RH, editors. WHO classification of tumours of the digestive system (4th ed.). Lyon: IARC; 2010.

[23] In: Ref. [22]. p. 52.

[24] In: Ref. [22]. p. 90.

[25] Rothberg BE, Moeder CB, Kluger H, et al. Nuclear to non-nuclear Pmel17/gp100 expression (HMB45 staining) as a discriminator between benign and malignant melanocytic lesions. Mod Pathol 2008;21(9):1121–9.

[26] Boyle JL, Haupt HM, Stern JB, et al. Tyrosinase expression in malignant melanoma, desmoplastic melanoma, and peripheral nerve tumors. Arch Pathol Lab Med 2002;126(7):816–22.

[27] In: Ref. [21]. p. 271.

[28] In: Ref. [21]. pp. 211–12.

[29] Bodey B, Bodey Jr B, Kaiser HE. Immunocytochemical detection of prostate specific antigen expression in human primary and metastatic melanomas. Anticancer Res 1997;17(3C):2343–6.

[30] In: Ref. [21]. p. 206.

[31] In: Ref. [21]. pp. 17; 147–8.

[32] In: Ref. [21]. p. 323.

[33] In: Ref. [21]. p. 124.

[34] In: Ref. [21]. p. 254.

[35] Scanlan MJ, Simpson AJ, Old LJ. The cancer/testis genes: review, standardization, and commentary. Cancer Immun 2004;4:1.

[36] Brown SD, Warren RL, Gibb EA, et al. Neo-antigens predicted by tumor genome meta-analysis correlate with increased patient survival. Genome Res 2014;24(5):743–50.

[37] In: Ref. [21]. p. 358.

[38] Scholl S, Beuzeboc P, Pouillart P. Targeting HER2 in other tumor types. Ann Oncol 2001;12(Suppl. 1):S81–7.

[39] Lotan TL, Gurel B, Sutcliffe S. PTEN protein loss by immunostaining: analytic validation and prognostic indicator for a high risk surgical cohort of prostate cancer patients. Clin Cancer Res 2011;17(20):6563–73.

[40] Cantwell-Dorris ER, O'Leary JJ, Sheils OM. BRAFV600E: implications for carcinogenesis and molecular therapy. Mol Cancer Ther 2011;10(3):385–94.

[41] Long GV, Wilmott JS, Capper D. Immunohistochemistry is highly sensitive and specific for the detection of V600E BRAF mutation in melanoma. Am J Surg Pathol 2013;37(1):61–5.

[42] Rao RN. Targets for cancer therapy in the cell cycle pathway. Curr Opin Oncol 1996;8(6):516–24.

[43] In: Ref. [21]. p. 523.

[44] Sikkema M, Kerkhof M, Steyerberg EW, et al. Aneuploidy and overexpression of Ki67 and p53 as markers for neoplastic progression in Barrett's esophagus: a case-control study. Am J Gastroenterol 2009;104(11):2673–80.

[45] Ryu B, Jones J, Hollingsworth MA, et al. Invasion-specific genes in malignancy: serial analysis of gene expression comparisons of primary and passaged cancers. Cancer Res 2001;61(5):1833–8.

[46] Walter-Yohrling J, Cao X, Callahan M, et al. Identification of genes expressed in malignant cells that promote invasion. Cancer Res 2003;63(24):8939–47.

[47] In: Ref. [15]. p. 14.

[48] Aitken SJ, Thomas JS, Langdon SP, et al. Quantitative analysis of changes in ER, PR and HER2 expression in primary breast cancer and paired nodal metastases. Ann Oncol 2010;21(6):1254–61.

[49] Chia S. Testing for discordance at metastatic relapse: does it matter? J Clin Oncol 2012;30(6):575–6.

[50] Denkert C, Loibl S, Kronenwett R, et al. RNA-based determination of ESR1 and HER2 expression and response to neoadjuvant chemotherapy. Ann Oncol 2013;24(3):632–9.

[51] Hyman E, Kauraniemi P, Hautaniemi S, et al. Impact of DNA amplification on gene expression patterns in breast cancer. Cancer Res 2002;62(21):6240–5.

[52] Geiger T, Cox J, Mann M. Proteomic changes resulting from gene copy number variations in cancer cells. PLoS Genet 2010;6(9):e1001090.

[53] Wellcome Trust Sanger Institute. Catalogue of somatic mutations in cancer. Available at: http://cancer.sanger.ac.uk/cancergenome/projects/cosmic/.

[54] Wellcome Trust Sanger Institute. Catalogue of somatic mutations in cancer - Gene census. Available at: http://cancer.sanger.ac.uk/cancergenome/projects/census/.

[55] Soon WW, Hariharan M, Snyder MP. High-throughput sequencing for biology and medicine. Mol Syst Biol 2013;9:640.

[56] Loewe RP. Combinational usage of next generation sequencing and qPCR for the analysis of tumor samples. Methods 2013;59(1):126–31.

[57] Kanagal-Shamanna R, Portier BP, Singh RR, et al. Next-generation sequencing-based multi-gene mutation profiling of solid tumors using fine needle aspiration samples: promises and challenges for routine clinical diagnostics. Mod Pathol 2014;27(2):314–27.

[58] Lawrence MS, Stojanov P, Polak P, et al. Mutational heterogeneity in cancer and the search for new cancer-associated genes. Nature 2013;499(7457):214–8.

[59] Watson IR, Takahashi K, Futreal PA, et al. Emerging patterns of somatic mutations in cancer. Nat Rev Genet 2013;14(10):703–18.

Sublethal Injuries and Deaths of Cells and Tissues

L.P. Bignold: Principles of Tumors.
DOI: http://dx.doi.org/10.1016/B978-0-12-801565-0.00010-X

The previous chapters of this book have dealt with the incidences, causation, as well as the morphological and related manifestations of the different tumor types.

The next general area for discussion is therapy. All therapies are concerned with both injuring—hence reducing proliferation—and causing the death of tumor cells. In many cases of malignant tumor, however, some tumor cells survive and regrow.

In addition, many forms of therapy injure normal cells and provoke reactions—especially different kinds of inflammation—in tissues.

This chapter is concerned with all the known forms of cell injury, cell death, survival, as well as reactions of normal tissues to therapies. The major perspective is how these phenomena may be relevant to the behavior of malignant tumor masses and to the effects of therapies.

Sections 10.1 and 10.2 describe sublethal injuries to cells. Section 10.3 discusses physiological deaths of cells in normal cell populations.

Sections 10.4 describes necrosis; Section 10.5 gives an account of apoptosis; and Section 10.6 mentions other forms of cell death.

Sections 10.7 mentions aspects of inflammation which are relevant to the treatment of tumors.

10.1 SUBLETHAL NONGENOPATHIC EFFECTS IN CELLS: "DEGENERATIONS," "CELL STRESS," AND "CELL STRESS RESPONSES"

10.1.1 Terminology

Sublethal injuries are disturbances from which the cell may

1. recover completely or
2. suffer permanent alteration in some cellular function but nevertheless remain alive.

These effects are distinguished from lethal injuries, from which the cell does not recover.

Nongenopathic effects are defined as all injurious effects which do not primarily or directly affect the genome, i.e., there is no damage to nuclear membrane, chromosomes, or DNA. Genopathic effects primarily affect the genome of the cell. The two kinds of injury are mentioned also in Sections 4.1.1 and 10.2.

10.1.2 Different Effects with Different Doses of Injurious Agent

There is no necessary difference in mechanism between sublethal and lethal injuries.

Almost all kinds of sublethal injuries may be lethal to cells if the dose of harmful agent/noxin is increased sufficiently.

The period of time over which the dose of noxin is given is also important, because many cells have capacities—within limits—for countering, or adapting to, injuries. Thus the total dose, and rate of application (dose rate), are both important to the actual effects of injurious agents.

Many harmful agents produce no morphological change at low doses, but cause cytoplasmic changes in high doses, and often cell death if even higher doses are administered. A good example of the principle is poisoning of liver cells by ethanol. At low doses, the liver cells fail to produce normal amounts of serum albumin, so the level of this protein in the blood falls [1]. At this stage, no structural change is seen microscopically. With higher doses of ethanol, lipid processing in the liver cells is damaged, and triglycerides accumulate in vesicles in the cytoplasm of the cells. This is detectable microscopically and is referred to as "degeneration." The lesion is also seen as a side effect of methotrexate (see in 12.2.1) [2]. Higher doses still of ethanol cause death of cells, which takes a particular form ("eosinophilic" death [3,4]).

In the case of poisoning with lead, the toxin is cumulative, and effects increase with total accumulated dose [5].

10.1.3 Morphological Manifestations: "Degenerations"

Traditionally, sublethal effects in cells which produce morphological changes are termed "degenerations" [1] (Figure 10.1).

The main ones are as follows:

1. Watery swelling
   a. When the swollen structures are seen by EM to be mitochondria, the term "cloudy swelling" may be used (Figure 10.1A).
   b. If larger dilated structures such as endoplasmic reticulum are found on EM, the term "hydropic degeneration" or "vacuolar degeneration" may be used (Figure 10.1B).
   c. If the cytoplasm is swollen with water but without dilatation of any organelle, the term "cytoplasmic edema" is used [6].
2. "Fatty change" (accumulations of triglycerides in vacuoles) (Figure 10.1C and D).
3. "Hyaline degeneration"
   This is characterized by accumulation of proteinaceous masses in the cytoplasm. Amyloid is an extracellular hyaline lesion, in which the proteins accumulate by forming insoluble crystals of a specific kind [7].
4. "Zeiosis" and "blebbing"
   This is a collection of acute morphological changes which can be effects of a variety of toxins, such as cytochalasin B [8] and others [9]. These lesions are difficult to appreciate in most human pathological lesions, but are easily demonstrated in experimental conditions using special light microscopic techniques (Figure 10.1E and F). Blebs are defined as rounded, usually multiple, projections from the cell surface. Their sizes range from approximately 3 μm down to less than 1 μm in diameter. The term "zeiosis" was introduced in the 1960s for larger protrusions and is now little used. "Blebbing" is now very commonly used, especially for smaller protrusions. The blebs may contain only fluid, or contain cytoplasm, and even autosomes and fragments of nuclei. They may be associated with other toxic changes [10] which together may be a predeath state for the cell. The mechanisms of their formation appear to involve toxic effects on the binding between components of the membrane and one or more structures in the cytoplasm [10,11].
5. "Granularity" of cytoplasm is seen in tumor cells of certain kinds. This change is not one of cell injury, but a morphological manifestation of the particular tumor type.

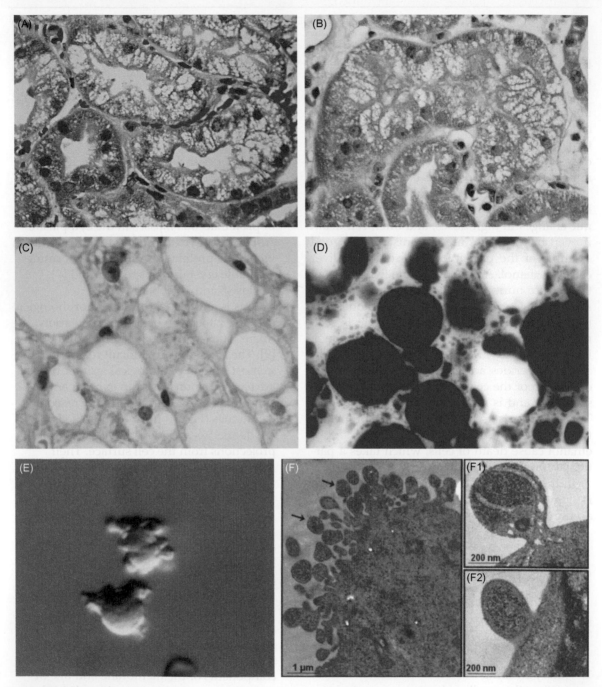

FIGURE 10.1 Images of sublethal nongenopathic injuries. (A) Watery small vacuolar change (mitochondrial) in renal tubular epithelium, ×40. (B) Watery large vacuolar change in renal tubular epithelium, ×40. (C) Fatty change (H&E) ×60. The large round spaces contained the lipid before processing dissolved the lipid out of the section. (D) Fatty change (Oil red O), ×60. In this preparation, no lipid solvents are used, so that the lipid remains and is stainable by oil-based stains. (E) Zeiosis. Neutrophil leukocytes treated with cytochalazin B and N-formyl peptide, ×40. Courtesy A/Professor D Harkin. (F) Blebbing of cultured mammalian cells induced by plasmin as shown by transmission EM. This change occurs before any sign of cell death. However, longer exposures to plasmin caused the cells to undergo apoptosis [10]. Courtesy The Biochemical Journal.

10.1.4 Lysosomes: Autophagy

Lysosomes are specific cytoplasmic organelles (Figure 10.2). They are packed with enzymes held in an inactive state at neutral pH. Their function was first studied extensively in relation to killing by phagocytes of microorganisms which they ingest [12]. Ingested organisms are initially held in a vesicle known as a phagosome. With changes in the surface of the lysosomal membrane, the lysosome fuses with the phagosome which then secretes hydrogen ions into the lumen. The lysosomal enzymes are activated by this acidification and digest the organism in the combined lysosome–phagosome structure ("phagolysosome"). Lysosomal enzymes are mainly hydrolases for the major macromolecules: nucleic acids, proteins, carbohydrates, and lipids [13]. The enzymes involved appear to act in supramolecular complexes known as "proteas(e-)omes" [14]. Undigested material may be ejected from the cell (exocytosis) as "residual bodies."

In inflammatory conditions, it is thought that lysosomal enzymes can be activated and released by cells, causing local inflammation [15].

It is also known that damaged cytoplasmic proteins and other structures in injured cells are accumulated in vacuoles (sometimes called "autosomes") as part of normal cellular metabolism. These may be ejected undigested from the cell (exocytosis). However, they may alternatively fuse with lysosomes and be digested within the cells. This is known as autophagy [12,16].

1. Lysosomes must be activated to function

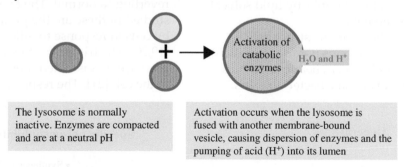

The lysosome is normally inactive. Enzymes are compacted and are at a neutral pH

Activation occurs when the lysosome is fused with another membrane-bound vesicle, causing dispersion of enzymes and the pumping of acid (H^+) into its lumen

2. Several different vesicles can activate lysosomes (L)

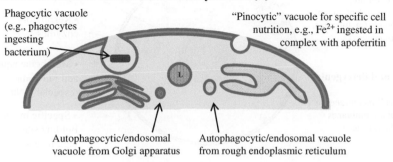

Phagocytic vacuole (e.g., phagocytes ingesting bacterium)

"Pinocytic" vacuole for specific cell nutrition, e.g., Fe^{2+} ingested in complex with apoferritin

Autophagocytic/endosomal vacuole from Golgi apparatus

Autophagocytic/endosomal vacuole from rough endoplasmic reticulum

FIGURE 10.2 Lysosomal destruction of material: autophagy.

10.1.5 Cell Stress

There are many biochemical and other injuries in cells which do not cause any morphological changes. The term "cell stress" has been introduced both for circumstances in which metabolic and morphological manifestations are induced, as well as for circumstance in which there are metabolic effects only (Figure 10.3).

The general categories of damage include

1. Inhibition of specific enzymes of biochemical pathways, e.g., by cyanide, which inhibits the enzyme cytochrome c oxidase [17], and diisopropylfluorophosphate (DFP) which reacts with acetylcholine esterase in the synapses of neurons, and thus has a powerful neurotoxic effect [18]
2. Deprivation of essential nutrients, especially oxygen, i.e., hypoxia
3. Damage to lipid-containing structures, such as membranes, for example, by lipid solvents including detergents
4. General disturbance in the structural conformation of proteins (i.e., denaturation), e.g., by heat, radiations, chemicals such as ethanol (above) or some bacterial toxins.

This occurs with all agents that have nonspecific effects on macromolecules, such as disturbance in noncovalent bonds, which sometimes is associated with coagulation (see Section 4.3.1).

The results are overall general metabolic inhibition, as well as suppression of cell division. However, it should be noted that any inhibition of DNA and RNA synthesis is a genopathic effect, because the slowed synthesis may be associated with more nucleotide errors in the new nucleic acid molecules. Also, mitosis can be slowed and possibly deranged (see Section 10.3.1).

10.1.6 Cell Stress Responses

The term "cell stress response" can refer to

1. Evoked metabolic effects in stressed cells which promote the likelihood of those cells reverting to normal. The most extensively studied of these are the proteins produced by cells in response to mild thermal injury [19,20]. The trigger for the responses is the presence of misfolded/denatured proteins in the cell [21]. The response is mainly the

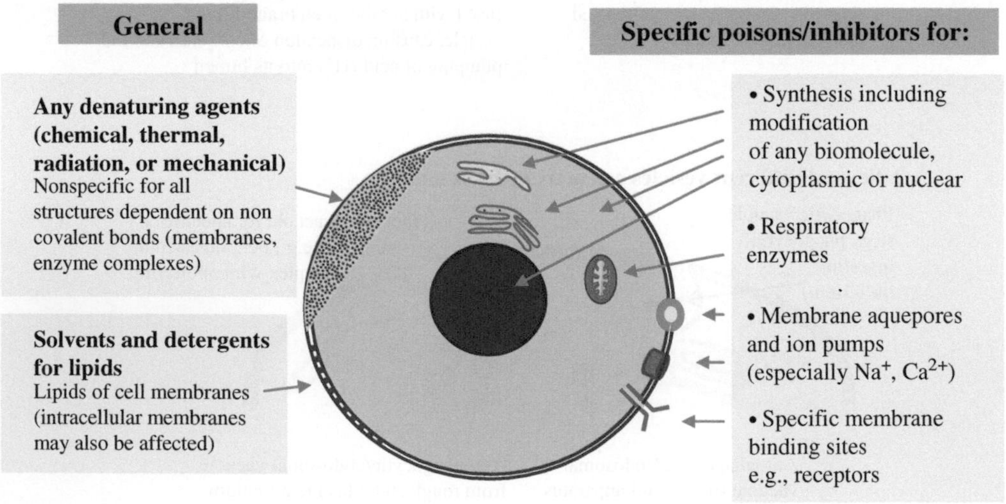

General

Any denaturing agents (chemical, thermal, radiation, or mechanical)
Nonspecific for all structures dependent on non covalent bonds (membranes, enzyme complexes)

Solvents and detergents for lipids
Lipids of cell membranes (intracellular membranes may also be affected)

Specific poisons/inhibitors for:

• Synthesis including modification of any biomolecule, cytoplasmic or nuclear

• Respiratory enzymes

• Membrane aquepores and ion pumps (especially Na^+, Ca^{2+})

• Specific membrane binding sites e.g., receptors

FIGURE 10.3   Origins of cell "stress."

production of specific proteins called "heat shock proteins." The presence of these proteins protects a cell from subsequent higher intensities of the injury (usually, a temperature which previously, would have killed the cell). The main function of heat shock proteins is to assist in refolding partially denatured proteins [21].

2. Evoked metabolic effects which are associated with the destruction of damaged cellular materials [22,23]. This is mainly the autodigestive processes involving activation of enzymes in lysosomes (see Section 10.1.4). If the injury has been sufficient to cause lethal damage to the cell, these activations of enzymes occur concurrently with the dying process of the cell. As is discussed in Section 10.3.2, these are related to mechanisms of autolysis.

10.2 SUBLETHAL GENOPATHIC INJURIES TO CELLS

Severe damage to the genome is not consistent with life. However, certain toxic agents—usually in low dose and administered over long periods of time—induce a variety of biochemical and morphological changes in the genomic compartment of the cell. Examples of this possibility have been recognized since the earliest studies of chromosomal aberrations (see Section 5.3).

10.2.1 Transient Reductions in DNA and RNA Syntheses

Generally, these are part of the general metabolic inhibition in cells caused by cell stress. Certain anticancer drugs affect mainly these processes. Radiation-induced inhibition of DNA synthesis is a well-established phenomenon [24,25]. It is the basis of the so-called S phase "checkpoint" in the cell cycle [26].

Like synthesis of DNA, total synthesis of RNA (i.e., of all kinds of RNA, see Section A2.2.7) is inhibited by radiations [25].

In a few studies, differential effects of alkylating agents [27] and of arsenic compounds [28] on synthesis of DNA and total RNA have been reported.

10.2.2 Limited Damage to Genes

As is described in Section A2.1.1, the germline genomes of normal individuals consist of approximately 50% repetitive DNA of uncertain function. In the remainder, genomes of individuals vary considerably one to another. It is therefore not surprising that the genomes of normal cells might survive limited amounts of damage [29] and see A2.1.5.

10.2.3 Limited Chromosomal Aberrations

As with nucleotide errors, minor chromosomal differences occur between the karyotypes of normal individuals (see Section A2.1.3). Thus small numbers of additional aberrations are consistent with continued functions of cells (see Section 5.2).

10.2.4 Formation of "Micronuclei"

The term "micronucleus" is used in different ways.

1. In biology, it refers to the smaller of the two nuclei in certain ciliate organisms, such as *Paramecium*. The smaller nucleus is diploid and participates fully in mitosis and meiosis of the organism [30].
2. In hematology, "micronucleus" is used for residual nuclear fragments—called "Howell–Jolly bodies"—when they are still present in occasional erythrocytes when they enter the blood circulation [31].
3. In a variety of toxic circumstances, small pieces of chromatin material may be found in the cytoplasm of mammalian cells during interphase [32]. They are thought to represent

condensed chromosomes which have separated from the spindle during metaphase–anaphase of a prior nuclear division. The cell is not necessarily unviable in this situation, because other chromosomal maldistributions in the same division may have given the nucleus of the cell the second chromosome of the pair.

To be a "micronucleus" in the biological sense, these toxin-induced structures should have a nuclear membrane, as seen in [33]. Even in EM studies, it is often difficult to establish whether a membrane, if present, is a normal nuclear membrane, or is the membrane of an autosome into which condensed extranuclear chromosomal material may be taken to be digested by lysosomal mechanisms (see above) [34].

Assays involving micronuclei test two aspects of the genome processes: tendency of chromosomes to break and to separate from the spindle in meta(sub)phase–ana(sub)phase (see Sections A1.3.4 and 1.3.5). The presence of membranes would imply another phenomena: that the separation has occurred at telophase-cytokinesis.

10.2.5 Other

The following abnormal processes can be seen as sublethal genopathic effects, because the cell lives and divides, but does not function normally.

(a) Carcinogenesis

This is thought to be due to a permanent genomic change—including liability to further change—in populations of cells (see Sections 2.5–2.8).

(b) Transformation In Vitro

This tumor-like change in cultured cells is inducible by viruses (see Section 4.5) and is understood to have a genomic basis (see Section 14.2.3).

(c) Transient Proliferative Effects

A variety of infectious agents produce proliferative lesions in organs. Examples are certain viruses such as human papilloma viruses, produce proliferative lesions such as warts. Another example is the proliferation of morphologically abnormal lymphocytes in infectious mononucleosis. The relationship between the presence of the virus in cells and their proliferation is complex and not fully understood [35] and see Section 5.7.

(d) Nuclear "Inclusions"

These constitute a general category of change. They are most commonly seen in virally induced proliferative lesions, and in certain "degenerative" conditions of the brain [36].

(e) Teratogenesis

This phenomenon (the induction of congenital malformations) usually involves inappropriate loss of proliferation of a subpopulation of cells in embryos [37]. It is therefore a genopathic effect and does not necessarily involve any other sublethal kind of injury to cells.

10.3 CELL DEATHS IN NORMAL CELL POPULATIONS IN VIVO

10.3.1 General

Before describing deaths of cells as effects of noxious agents, it is necessary to discuss the normal physiological circumstances in which cell deaths occur.

In all kinds of cells which produce mature, fully specialized cells from local tissue stem cells (see Section A1.3.3), the mature functional cells die immediately prior to being deleted from the body.

Examples of these cell deaths include as follows (Figure 10.4):

1. The scavenged erythrocytes which have circulated in the blood stream for approximately 3 months. The removal of these aged cells is performed by the reticuloendothelial cells in the lymph nodes, bone marrow, spleen, and liver (see Section A1.1.2j).

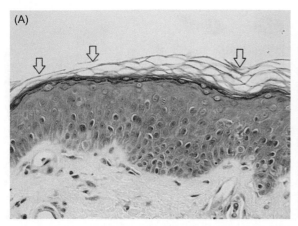

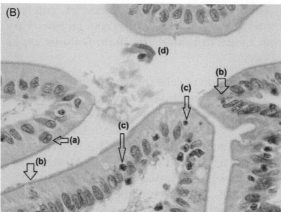

FIGURE 10.4 Some physiological cell deaths. (A) Shedding of obsolescent keratinocytes from the surface of the skin, ×20. (B) Apparently different methods of shedding of obsolescent epithelial cells (arrowed) from the tips of villi in the small intestine, ×40. (a) Apparently discharged whole; (b) Apparently undergoing lysis in the epithelial layer prior to discharge; (c) Apparently undergoing shrinkage necrosis. These 'apoptotic bodies' are difficult to distinguish from lymphocytes. Lymphocytes are shed through the intestinal mucosa as a defense against infections; (d) Some cells may be artifactually displaced into the lumen by handling of the specimen.

2. The fully keratinized cells which are shed from surface of the epidermis.
3. The epithelial cells which are shed from the surface of the gastrointestinal tract, including all organs from the oral cavity to the lower rectum.
4. The neutrophils which migrate through the mucosa and die in the oral cavity itself.
5. The lymphocytes which migrate through the epithelium into the lumen of the intestine followed by their death.
6. All eggs and spermatozoa which are not involved in fertilization.

The terminology for these apparently simple biological phenomena has never been fully developed, except for phrases such as "normal shedding" [38] and "scavenged" [39]. Some of the examples mentioned above would be adequately described as "shrinkage necrosis" and others are often included in "apoptosis" (see Section 10.5). "Necrobiosis" is appropriate to these physiological deaths of cells [40], but in pathology, this word is mainly used for death of collagenous and adipose tissues in specific inflammatory

conditions [41]. "Senescence" is another word which has been used for aged cells, but it also for other situations (see Section 10.6.4).

The word "obsolescence" may be the best word for physiological losses of fully mature and functional cells in labile cell populations. This is mainly because the word has not been widely used for any other circumstance.

10.3.2 Autolysis

This term refers to disintegration of cells or tissues due to actions of enzymes within themselves [42]. It is typically seen after tissues have been surgically resected, but not placed in sufficient fixative (enzyme-inactivating) fluid, such as formaldehyde. The enzymes responsible include (i) enzymes of serum and/or plasma in the blood vessels of the tissues and (ii) in the cytoplasm of the cells themselves, especially in their lysosomes (see Section 10.1.4).

In some circumstances, autolysis may be accelerated as a result of bacterial enzymes, associated with growth of these organisms in the tissues. This is hence early putrefaction [43].

The histological appearances of autolytic tissues in the absence of microorganisms vary from tissue to tissue. The main features are as follows:

1. Dissociation of cells from each other and from the connective tissue macromolecules. This is probably mainly due to serum enzymes.
2. Loss of cellular macromolecules, especially nucleic acids and proteins. This is probably mainly associated with release of lysosomal enzymes into cytoplasm (see Section 10.1.4). In this, it may be remembered that hypoxic tissues will tend to become acidic due to accumulation of lactic acid and other acidic tissue breakdown products [44].

10.4 NECROSIS

10.4.1 Macroscopic and Microscopic Features

At the time of forming the basic concepts in histology and histopathology (mid-nineteenth century, see Section 2.1), microscopical techniques could demonstrate the corpses of once-living cells. These were seen in various pathological conditions and were referred to as "necrotic" cells [45,46]. The term was also applied to deaths of tissues, organs, or parts of the body. Early on, the mechanisms of necrosis were recognized to include deprivation of a vital nutrient to the cell or toxin-induced damage to the structures of metabolism of the cell (see Sections 10.1.5 and 10.4.2). In immunology, both antibody and cytotoxic T-cell-induced cell deaths are thought to occur entirely via attacks on the cell membrane of the target cell [47].

Macroscopically, necroses of tissues are divided into [2]:

1. "Coagulative": i.e., being like tissue which has been coagulated by heat or chemicals
2. "Liquefactive": i.e., tissue has turned into liquid

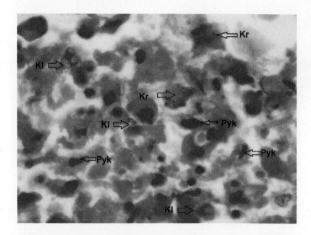

FIGURE 10.5 "Classical" nuclear changes of necrosis. Nuclear changes found in necrosis. Pyk = pyknosis, Kr = karyorrhexis, Kl = karyolysis
Some of the rounded dark nuclear fragments may be lobes of neutrophil nuclei. ×60.

3. "Caseous": i.e., being like cheese—pale and tending to crumble. This is typically seen in tuberculous lesions.
4. Hemorrhagic: any necrosis involving blood vessels so that bleeding from the lesion occurs.

By light microscopy, necrosis of cells is definitively associated with abnormalities in the nuclei of the relevant cells. If the nucleus is destroyed, the cell is recognized to be dead. Several different kinds of nuclear abnormality indicate early necrosis as follows (Figure 10.5):

1. Pyknosis: Shrinkage to a densely staining, small angulated body in the cell.
2. Karyolysis: The nuclear membrane and histological chromatin (Section A1.2.6) lose their definition and staining, and appear to dissolve away.
3. Karyorrhexis: Fragmentation of nucleus, in which the nucleus appears to break up with parts of the chromatin remaining attached to nuclear membrane.

It should be noted that these changes occur almost always in different cells in the same pathological lesion. In any early necrotic tissue,

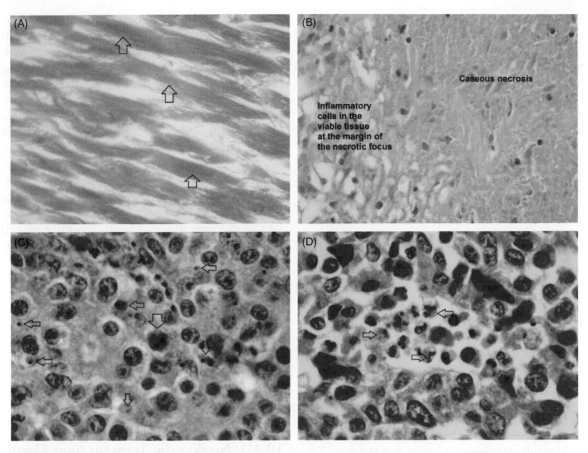

FIGURE 10.6 Some microscopic forms of tissue necrosis. (A) Coagulative necrosis of myocardium, ×40. The nuclei have disappeared, but the cross-striations (arrows) in the cytoplasm remain. (B) Caseous necrosis, ×40. Lung, tuberculosis. A few nuclear fragments are seen in the amorphous material. (C) Fragmentation necrosis of tumor cells, arrows, ×60. (D) Focus of neutrophil leukocytes in tumor mass. The leukocytes are probably attracted by break-down (lytic) products of necrotic tumor cells. Some of the neutrophil leukocytes are undergoing apoptosis/ fragmentation necrosis (arrows). ×60. This phenomenon has been described in [64].

different individual cells appear to undergo these changes individually. The reason for this is not known. With time, however, ultimately all nuclei are completely lysed.

Microscopy also reveals different patterns of appearances for each of the macroscopic forms of necrosis (Figure 10.6). In coagulative necrosis caused by ischemia, the histologic changes are similar to those of autolysis in the absence of microorganisms (see above). The cells become separated, but the outlines of the cells and connective tissue structures of the tissue are preserved. All the nuclear changes (above) may be seen. In liquefactive necrosis, individual cells are seen in a protein-rich fluid. In caseous necrosis, there is only pale material showing no structure of any kind.

Additional kinds of necrosis are demonstrated microscopically. In "eosinophilic necrosis" the cytoplasm stains heavily with eosin and the nucleus is shrunken (pyknotic). Also, the cytoplasm may be shrunken to greater or lesser degrees. "Fibrinoid necrosis" refers to the accumulation of fibrin in the vicinity of dead cells. This occurs in the arterioles in malignant hypertension and some other conditions. It is probably not a specific kind of cell death, but simply the occurrence of two processes—cell death and deposition of serum proteins—in the one focus.

10.4.2 "Necrosis": Electron Microscopic Appearances and Biochemical Changes

Electron microscopic studies have contributed some additional observations of dying cells. In addition to the nuclear changes, many authors [2,48] consider a cell dead if there is any one of the following ultrastructural features:

1. The plasma membrane is ruptured. This is often associated with generalized swelling of cytoplasmic organelles and has been called "oncosis" [49]. It is to be understood as an early phase in the morphological changes of cell death. Irreversible damage to the nucleus is assumed following such membrane disruptions.
2. The cell membrane is intact, but there is marked disruption to cytoplasmic organelles and the nucleus. The assumption is that the DNA is no longer available for transcription of RNA, or translation to proteins, and hence the cell must eventually die.
3. A whole cell is present in the cytoplasm of another cell. This phenomenon, known as "emperipolesis," is uncommon, but was originally identified in relation to lymphocytes [50] and is currently used in relation to a macrophage-rich disorder of lymphoid tissue known as Rosai–Dorfmann disease [51]. Whether the phenomenon is one of the lymphocytes being ingested by the surrounding cell or is a kind of possibly hypermotility of the one cell in respect of the other is unclear [52].

Finally, the biochemical alterations underlying necroses have been investigated. Different noxious agents act in different ways. Heat, acids, and alkalis denature proteins, including enzymes; detergents dissolve lipids especially of membranes; aldehydes cross-link macromolecules, especially proteins; and alcohols dissolve lipids and precipitate proteins. Cold injury inhibits enzymes, and ice forming in cells disrupts cellular structures. Loss of blood supply causes primarily loss of oxygen utilization followed by loss of energy molecules (e.g., ATP). This leads to loss of the sodium pump, loss of calcium ion exclusion, integrity of cell membranes, and dissolution of intracellular structures, including nuclei [53,54].

Certain poisons affect particular structures or enzymes in cells, such as membranes or organelles, or the enzymes of respiration. For example, cyanide specifically inhibits the enzymes of respiration. High doses of radiations are assumed to be due to ionization-induced damage to vital cell structures such as membranes and enzymes (see Sections A1.2.4 and A1.2.5).

10.4.3 No Specific Term for Slow Death of Cells in Pathological Conditions

In some pathological conditions such as chronically reduced blood supply to the kidney, the cells die and their corpses in part become incorporated into proteinaceous masses. There is no specific term for this kind of cell death. "Necrosis" is not regularly applied to this phenomenon, and general words such as "hyalinization" or "sclerosis" are most often used.

10.5 APOPTOSIS

10.5.1 Original Description

The discovery of apoptosis was achieved by the use of electron microscopy. The original descriptions were based on observations of acute experimental insults to adrenal cortical cells as well as other kinds of cells [55–57]. The morphological changes characterizing apoptosis were described [55] as follows.

The structural changes take place in two discrete stages. The first comprises nuclear and cytoplasmic condensation and breaking up of the cell into a number of membrane-bound, ultrastructurally well-preserved fragments. In the second stage, these apoptotic bodies are shed from epithelial-lined surfaces or are taken up by other cells, where they undergo a series of changes resembling *in vitro* autolysis within phagosomes, and are rapidly degraded by lysosomal enzymes derived from the ingesting cells.

There were therefore actually three consecutive phases in the process (Figure 10.7):

1. Shrinkage
2. Fragmentation of the cell into membrane-bound droplets containing parts of nucleus bound by nuclear membrane
3. Lysis of the contents of the droplets.

Later in the text, it was indicated that lysis of the droplets could occur either

- in adjacent cells through the phagocytic lysosomal mechanisms (see Section 10.1.4) or
- outside cells by autodigestion of the droplets by lysosomes within them or
- outside cells by ambient serum proteases.

Of these three phases, the droplet fragmentation with intact membranes was the only apparently unique phenomenon. As the article stated [55], it was already known that both shrinkage necrosis as well as phagocytosis of nonmembrane bound part cells by other cells is followed by digestion in the phagolysosomes of the other cells.

The process was called "apoptosis" because it was seen as analogous to the normal dying of leaves on deciduous plants in autumn [55]. (Leaves dry out and shrink, and then fall off the tree as whole leaves, after which they fragment on the ground and are degraded in the soil.)

In the text of the original article, "apoptosis" was associated with very particular acute tissue insults such as (i) the abrupt withdrawal growth-maintaining stimuli (in embryos), as well as (ii) abrupt withdrawal of activity-maintaining hormones, and (iii) with exposure to particular toxins (both in adult organisms).

10.5.2 Various Uses of the Term

Kerrs' original findings have been confirmed in many circumstances, e.g., rapid cell death is seen when lymphoid tissue is exposed to high doses of corticosteroid hormones [58].

However, since then, "apoptosis" has been applied to

1. All cell deaths manifesting as fragmentation and/or shrinkage cell deaths (above).
2. Toxin-induced cell deaths in which cells burst their cell membranes without prior droplet fragmentation. In studies of human pathological lesions, these are features of "necrosis" (see Section 10.4).
3. Although the involvement of lysosomal enzymes was not an essential part of the original descriptions, "apoptosis" has been applied to any condition in which lysosomal enzymes are activated in cells [49,59–63].

Apoptosis/death via fragmentation of neutrophil leukocytes has been observed in tumors [64].

Although rarely used in this way, the term "apoptotic bodies" could be applied to the larger micronuclei seen in *in vitro* genopathic tests (see Section 14.2.4).

10.5.3 Biochemical "Surrogates" for Apoptosis

Many biochemical studies of cell deaths via fragmentation have shown that the process can be particularly associated with activation

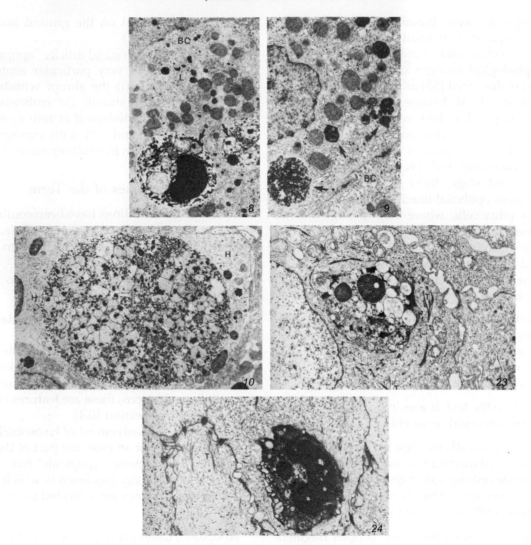

FIGURE 10.7 Images of apoptosis from original paper by Kerr et al. [55]. Atrophying rat liver tissue 3 days after obstruction of its portal blood supply. The two apoptotic bodies indicated by arrows in *Fig. 8* lie within phagosomes in the cytoplasm of a hepatocyte; their ergastoplasm is degenerate and their mitochondria are swollen and show focal matrix densities (autolytic changes). Note that normal numbers of secondary lysosomes (L) are still present in the cytoplasm bordering the bile canaliculus (BC). In *Fig. 9*, residues of degraded apoptotic bodies (arrows) are seen in the paracanalicular cytoplasm, and secondary lysosomes of the type found in normal hepatocytes have disappeared, suggesting that they have previously fused with phagosomes. *Fig. 10* shows a large apoptotic body without a nuclear remnant in the cytoplasm of a "histiocytic" (H): note the autolytic changes. Fig. 8: x 4600; Fig. 9: x 9200; Fig. 10: x 6600. *Figs 23 and 24* - Untreated squamous cell carcinoma of the human cervix uteri. Apoptotic bodies with nuclear remnants are seen within carcinoma cells, which can readily be identified by the presence of tonofibrils (T). Autolytic changes are evident in the body illustrated in Fig. 23: more advanced degradation is shown in Fig. 24. Fig. 23: x 14,200; Fig. 24: x 16,000. *Reproduced with permission from Nature Publishing Group.*

Present author's note: the processes of mitochondrial swelling, autolysis, nuclear fragmentation and the further degradation of nuclear fragments by lysosomes are identified. The phagocytic cells are considered to be mainly of the mono-nuclear/ macrophage system. The distinction between all apoptotic carcinoma cells and fragments of other cells is established by the tonofilaments seen by electron microscopy. In the original descriptions, the features of apoptosis were almost entirely morphological.

of the caspase group of proteases [65–68]. These enzymes lyse cysteine–aspartate bonds and are activated by various other enzymes and some receptors for substances such as tumor necrosis factor. However, the enzymes are also involved in nondeath cellular phenomena, such as inflammation and specialization. Thus caspase activity is not essential or exclusive to apoptosis [62].

Fragmentation of DNA has been another suggested biochemical marker of apoptosis, on the basis that caspases activate DNAses which break the DNA strands between nucleosomes [68]. This also currently is not fully accepted to be specific for apoptosis [62].

An alternative view is that these enzymes and effects are those of cytoplasmic autolysis which is one aspect of the general phenomenon of autolysis of tissues after withdrawal of vital nutrients (see Section 10.3.2).

10.5.4 Apoptosis in Nontumorous Human Pathological Conditions

Apoptosis as it was defined by Kerr et al. [55] is common in a few particular conditions such as graft versus host disease [69] and some viral diseases [70]. However, it is a rare finding in most other human nontumor pathological conditions. Using less strict criteria, apoptosis has been recorded in a wide variety of circumstances [71–73].

10.6 OTHER FORMS OF CELL DEATH

10.6.1 "Area" Coagulative Necrosis

Certain types of tumors exhibit "area" coagulative necrosis more than others (Figure 10.8A). For example, carcinomas of the kidney almost invariably show it. Area necrosis is probably always due to poor vascularization of the tumor masses or thrombosis of tumor vessels. It is also seen strikingly after therapeutic embolization of arteries to tumors.

10.6.2 Inappropriate Vegetative State ("Reproductive Death") in Tumor Cell Populations

This is the most characteristic cytokinetic abnormality in tumor cell populations and

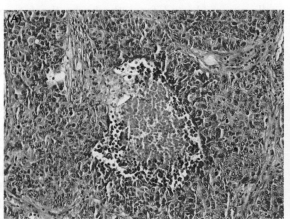 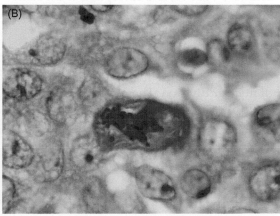

FIGURE 10.8 Miscellaneous cell deaths. (A) Area necrosis in carcinoma of the lung, ×10. (B) Death during mitosis. In this cell, condensation of cytoplasm is occurring while a multipolar ana(sub) phase is in progress. This may be called 'mitotic catastrophe', but it is not certain that the abnormal mitosis has caused the necrosis. The two processes could have occurred coincidentally at the same time, ×100.

refers to spontaneous cell cycle arrest [74]. The cells remain alive in the tumor mass but are incapable of dividing. The phenomenon is also termed "senescence" [75,76].

They include the following categories:

1. Cells which progress through only some amplifying divisions and remain partly specialized in the tumor mass. These cells remain in place as a viable but not expanding population in the mass (Figure 10.8) and may form significant components of tumor deposits which clinically remain unchanging in size over long periods of time. Presumably, they may be susceptible to antimetabolites (see Section 12.2), but not to antimitotic drugs. The term "mitotic arrest" is sometimes used for G0→G1 block (checkpoint) [77,78].
2. Cells may lose the capacity for mitosis, but still retain the capacity for DNA synthesis. This phenomenon is also known as "endoreduplication" of DNA (see Section 5.4.2). The process causes these vegetative (G0) cells to develop enlarged "hyperchromatic" nuclei, containing more than the normal diploid amounts of DNA. They can be considered nondividing, permanently viable cells (see 1 above) and susceptible to the same types of drugs as in (1).
3. Less commonly in the same masses, some cells may lose the capacity to divide at the level of the whole cell (cytokinesis) but retain the capacities for all the components of nuclear division. These cells then become multinucleated "tumor giant cells."

In treated tumors, these changes may be more common than are seen in untreated experimental populations (see Section 10.6.3).

10.6.3 Mitotic Catastrophe

A particular form of inappropriate vegetative state is the condition in which cells can be seen to have entered mitosis to the extent of formation of chromosomes, (meta(sub)phase, ana(sub)phase, and telo(sub)phase) but to have not completed the process. These are referred to as being in "mitotic catastrophe" [63,79] (Figure 10.8B). The term "metaphase–anaphase block (checkpoint)" is sometimes used for the same event [80].

Whether or not damage to the centrosome has a role in reproductive inactivation of cells is unclear. In the sequence of events in cell division, DNA synthesis begins before duplication of the centrosome, so that such damage would not necessarily lead to inhibition of DNA synthesis, but could lead to failed centriolar separation and then cessation of the cell cycle at G2 (G2/M checkpoint [81]). The resulting cells would be tetraploid without having gone through a phase of chromosome assembly. Several biochemical types of damage may be involved [82].

10.6.4 "Senescence" of *In Vitro* Cultures

If the problems of cell death were not already difficult enough, they are further complicated by that fact that the term "death" has been applied to cessations of growth by populations of cells in *in vitro* cultures. When normal cells are explanted from the body, the fate of the culture is usually predictable. First, there is a phase of continuous growth. The duration of this depends mainly in the kind of cells being cultured. Ultimately, however, all cultures of normal cells stop growing and the populations in these "ageing" cultures have been called "senescent" [83–85].

The senescent phase of the cells is usually heralded by chromosomal instability and hence has been seen as a surrogate for neoplastic change [86]. This is a conceptual contrast with transformation (see Section 14.2.3), in which cell division increases, and chromosomal abnormalities are less prominent. It is distinct from apoptosis, because the cells do not actually die, and chromosomal abnormalities are not seen in apoptosis. Here, senescence is seen as a form of "mitotic catastrophe" (see Section 10.6.3).

It should be noted that there is no normal "senescence" of entire cell populations in the body. The intestinal mucosa remains the same in old age as in early life. No normal cells in the body undergoes premortem mitotic catastrophe.

10.6.5 "Autophagic" Cell Death

A similar issue is "autophagic" cell death. Autophagic vacuoles are normal cellular organelles and may be increased as a sublethal injury (see Section 10.1.4). Cells dying with prominent autophagic vacuoles should probably be considered as dying *with* autophagy, not *because of* autophagy [63].

10.7 INFLAMMATION AND OTHER TISSUE EFFECTS OF NOXINS

In addition to necrosis and atrophy (discussed above), noxins may induce inflammation and scarring (Figure 10.9). A dictionary definition of inflammation is "a protective tissue response to injury or destruction of tissues, which serves to destroy, dilute, or wall off both the injurious agent and the injured tissues" [87]. This is generally appropriate. The only significant exception is allergic/hypersensitivity inflammation. Allergic inflammation, while it may assist in the protection of some tissues, is harmful and may be lethal to the patient. Thus allergic inflammation is defined as "one in which the body mounts an exaggerated or inappropriate immune response to a substance perceived as foreign, resulting in local or general tissue damage" [88].

10.7.1 "Acute" and "Chronic" Inflammation

Broadly, inflammations are considered acute if they begin within hours to days after the injury and last less than 3–6 months even if untreated. Chronic inflammations have more insidious onsets and tend to persist if untreated for longer than 6 months. Many forms of acute inflammation can last for more than 6 months, and then the patient is arbitrarily described as suffering a chronic inflammation.

Additional terms are "hyperacute" inflammation, for onsets of seconds or a few minutes of exposure to the injurious agent. This is especially characteristic of some forms of hypersensitivity inflammation. "Subacute" may be used for "an inflammation that is intermediate in duration between that of an acute inflammation and that of a chronic inflammation, usually persisting longer than 3 or 4 weeks" [89].

Most texts include in their descriptions of inflammation the Ancient observations of "heat, swelling, redness, and painfulness" [90]. This mainly applies to acute inflammations which are now known to be due to bacteria, such as staphylococci and streptococci. However, some acute kinds of inflammation, and some kinds of chronic inflammation, may not feel hot. Also, particular kinds may not show swelling, may not be reddened, and may not be painful.

10.7.2 Morphological Forms of Inflammation

The fundamental anatomical change in most kinds of inflammation is the emigration of leukocytes from the blood stream into the tissues [91]. Only a very few kinds of inflammation, such as certain forms of urticaria (hives), show no emigrated leukocytes or other abnormality. Hives exhibit swelling (local edema) and sometime pain, but no redness or heat [92].

Many kinds of acute inflammations are associated with other morphological features. For example, those associated with pus are called "purulent"; those associated with bleeding "hemorrhagic"; and those associated with necrosis of tissues (see Section 12.4) "necrotizing" [93].

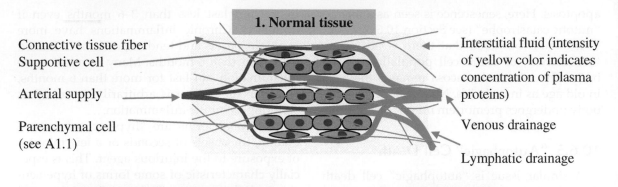

1. Normal tissue

Connective tissue fiber

Supportive cell

Arterial supply

Parenchymal cell
(see A1.1)

Interstitial fluid (intensity
of yellow color indicates
concentration of plasma
proteins)

Venous drainage

Lymphatic drainage

2. Necrosis of tissue (death of all cells in the part)

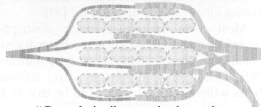

"Coagulative" necrosis shown here.

3. Atrophy

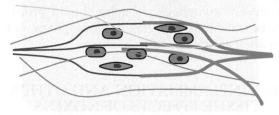

- slow loss of all constituent cells
- relative devascularization

4. Acute inflammation

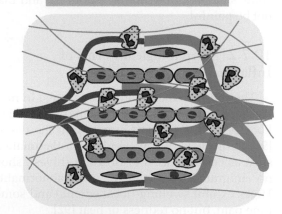

- vasodilatation,
- protein-rich oedema,
- emigration of leukocytes ( )

5. Scarring (= "repair")

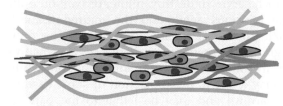

- as for atrophy, but with
overgrowth of fibroblasts and laying
down of excess collagen

FIGURE 10.9   Tissue effects of noxins.

Almost all chronic inflammations show fibrosis at the margins of the lesion. A few particular kinds of chronic inflammation, such as in response to tuberculous mycobacteria, show granulomas, which consists of microscopic clusters of macrophages. In most situations, these are not associated with edema.

10.7.3 Healing by Scarring

Any defect in a tissue is likely to be followed by scarring [93]. In this process, local fibrocytes are activated to their "-blast" form (see Section A1.2.7) and begin to lay down new collagen across the gap. When the gap is filled, collagen production ceases and the fibroblasts revert to fibrocytes. Changes in local blood vessels accompany and support this reaction. After months to years, the excess local blood vessels disappear, and the scar remains as a permanent, pale, mechanically strong structure joining normal tissue on either side.

References

[1] Kumar V, Abbas AK, Fausto F, editors. Robbins and cotran pathologic basis of disease (8th ed.). Philadelphia, PA: Elsevier; 2009. p. 3–42.

[2] US National Library of Medicine. Methotrexate. Available at: http://livertox.nih.gov/Methotrexate.htm.

[3] Johns Hopkins Medicine. Alcohol liver disease. Available at: http://www.hopkinsmedicine.org/gastroenterology_hepatology/diseases_conditions/faqs/alcoholic_liver_disease.html.

[4] Angulo P. Nonalcoholic fatty liver disease. N Engl J Med 2002;346(16):1221–31.

[5] Skerfving S, Bergdahl IA. Lead. In: Nordberg GF, Fowler BA, Nordberg M, editors. Handbook on the toxicology of metals (3rd ed.). Philadelphia, PA: Elsevier; 2007. pp. 599–643.

[6] Cheville NF. Ultrastructural pathology: the comparative cellular basis of disease. Hoboken, NJ: John Wiley & Sons; 2007. p. 7.

[7] Sipe JD, editor. Amyloid proteins: the beta sheet conformation and disease. Weinheim: Wiley-VCH; 2008.

[8] Godman GC, Miranda AF. Cellular contractility and the visible effects of cytochalasin. Front Biol 1978;46:277–429.

[9] Charras GT, Coughlin M, Mitchison TJ, et al. Life and times of a cellular bleb. Biophys J 2008;94:1836–53.

[10] Doeuvre L, Plawinski L, Goux D, et al. Plasmin on adherent cells: from microvesiculation to apoptosis. Biochem J 2010;432(2):365–73.

[11] Fackler OT, Grosse R. Cell motility through plasma membrane blebbing. J Cell Biol 2008;181(6):879–84.

[12] Alberts B, Johnson A, Lewis J, et al. Molecular biology of the cell, 5th ed. New York, NY: Garland Science; 2008. pp. 779–99.

[13] Lüllmann-Rauch R. History and morphology of the lysosome. In: Saftig P, editor. Lysosomes. New York, NY: Springer; 2005. pp. 1–16.

[14] Weissmann G, Serhan C, Korchak HM, et al. Neutrophils: release of mediators of inflammation with special reference to rheumatoid arthritis. Ann N Y Acad Sci 1982;389:11–24.

[15] Voorhees PM, Dees EC, O'Neil B, et al. The proteas(e-) ome as a target for cancer therapy. Clin Cancer Res 2003;9(17):6316–25.

[16] Ryter SW, Cloonan SM, Choi AM. Autophagy: a critical regulator of cellular metabolism and homeostasis. Mol Cells 2013;36(1):7–16.

[17] Agency for Toxic Substances and Disease Registry. Cyanide. Available at: http://www.atsdr.cdc.gov/toxprofiles/tp8-c2.pdf.

[18] Abou-Donia MB, Lapadula DM. Mechanisms of organophosphorus ester-induced delayed neurotoxicity: type I and type II. Annu Rev Pharmacol Toxicol 1990;30:405–40.

[19] Fulda S, Gorman AM, Hori O, et al. Cellular stress responses: cell survival and cell death. Int J Cell Biol 2010;2010:214074.

[20] Simmons SO, Fan CY, Ramabhadran R. Cellular stress response pathway system as a sentinel ensemble in toxicological screening. Toxicol Sci 2009;111(2):202–25.

[21] Shiber A, Ravid T. Chaperoning proteins for destruction: diverse roles of Hsp70 chaperones and their co-chaperones in targeting misfolded proteins to the proteasome. Biomolecules 2014;4(3):704–24.

[22] Benbrook DM, Long A. Integration of autophagy, proteasomal degradation, unfolded protein response and apoptosis. Exp Oncol 2012;34(3):286–97.

[23] Chen F, Evans A, Pham J. et al., editors. Cellular stress responses: a balancing act. Mol Cell 2010;40(2) (special issue).

[24] Lajtha LG, Oliver R, Kumatori T, et al. On the mechanism of radiation effect on DNA synthesis. Radiation Res 1958;8:1–16.

[25] Prasad KN. Handbook of radiobiology, 2nd ed. Boca Raton, FL: CRC Press; 1995. pp. 106–22.

[26] Jaspers NG, Zdzienicka MZ. Inhibition of DNA synthesis by ionizing radiation: a marker for an S-phase checkpoint. Methods Mol Biol 2006;314:51–9.

[27] Skipper HE, Bennett Jr. LL. Biochemistry of cancer. Annu Rev Biochem 1958;27(3):137–66.

[28] Abdullaev FI, Rivera-Luna R, García-Carrancá A, et al. Cytotoxic effect of three arsenic compounds in HeLa human tumor and bacterial cells. Mutat Res/Genetic Toxicol and Environ Mutagen 2001;493(1–2):31–8.

[29] Friedberg EC. Suffering in silence: the tolerance of DNA damage. Nat Rev Mol Cell Biol 2005;6(12):943–53.

[30] Beale G, Preer JR. Paramecium: genetics and epigenetics. Boca Raton, FL: CRC Press; 2008. pp. 139–45.

[31] Hilberg RW, Ringle RD, Balcerzak SP. Howell–Jolly bodies. Intracellular or extracellular? Arch Intern Med 1973;131(2):236–7.

[32] Fenech M. The micronucleus assay determination of chromosomal level DNA damage. In: Martin CC, editor. Environmental genomics. New York, NY: Springer; 2008. pp. 185–216.

[33] Schiffmann D, De Boni U. Dislocation of chromatin elements in prophase induced by diethylstilbestrol: a novel mechanism by which micronuclei can arise. Mutat Res 1991;246(1):113–22.

[34] Rello-Varona S, Lissa D, Shen S, et al. Autophagic removal of micronuclei. Cell Cycle 2012;11(1):170–6.

[35] Klein G, Klein E, Kashuba E. Interaction of Epstein–Barr virus (EBV) with human B-lymphocytes. Biochem Biophys Res Commun 2010;396(1):67–73.

[36] Mayo MC, Bordelon Y. Dementia with Lewy bodies. Semin Neurol 2014;34(2):182–8.

[37] Johnson EM, Kochhar DM, editors. Teratogenesis and reproductive toxicology. Berlin: Springer Verlag; 2011.

[38] Creamer B, Shorter RG, Bamforth J. The turnover and shedding of epithelial cells. I. The turnover in the gastro-intestinal tract. Gut 1961;2:110–8.

[39] Newcombe DS. The immune system, an appropriate and specific target for toxic chemicals. In: Newcombe DS, Rose NR, Bloom JC, editors. Clinical immunotoxicology. New York, NY: Raven; 1992. p. 33.

[40] The American Heritage® Dictionary of the English Language, Fourth Edition copyright ©2000 by Houghton Mifflin Company. Updated in 2009.

[41] McKee PH, Calonje E, Granter SR. Pathology of the skin, 3rd ed. Philadelphia, PA: Mosby/Elsevier; 2005. pp. 305–10.

[42] Dorland's Medical Dictionary for Health Consumers. © 2007 by Saunders, an imprint of Elsevier, Inc.

[43] The American Heritage® Dictionary of the English Language, Fourth Edition copyright © 2000 by Houghton Mifflin Company. Updated in 2009.

[44] Herdson PB, Kaltenbach JP, Jennings RB. Fine structural and biochemical changes in dog myocardium during autolysis. Am J Pathol 1969;57(3):539–57.

[45] Kölliker A. Handbuch der Gewebelehre des Menschen. Translated and edited Busk G, Huxley T. Manual of human histology. London: New Sydenham Soc. pp. 1853–4.

[46] Virchow R (1858). Die Cellularpathologie in Ihrer Begründung Auf Physiologische Und Pathologische Gewebelehre, 2nd German ed. Trans by Chance F, as Cellular Pathology as Based upon Physiological and Pathological Histology. With a new introductory essay by L. J. Rather. Dover, New York, NY; 1971.

[47] Golstein P, Ojcius DM, Young JD. Cell death mechanisms and the immune system. Immunol Rev 1991;121:29–65.

[48] Zong W-X, Thompson CB. Necrotic death as a cell fate. Genes & Dev 2006;20:1–15.

[49] Majno G, Joris I. Apoptosis, oncosis, and necrosis. An overview of cell death. Am J Pathol 1995;146(1):3–15.

[50] Humble JG, Jayne WHW, Pulvertaft RJV. Biological interaction between lymphocytes and other cells. Brit J Haematol 1956;2:283–94.

[51] Carbone A, Passannante A, Gloghini A, et al. Review of sinus histiocytosis with massive lymphadenopathy (Rosai–Dorfman disease) of head and neck. Ann Otol Rhinol Laryngol 1999;108(11 Pt 1):1095–104.

[52] Yang YQ, Li JC. Progress of research in cell-in-cell phenomena. Anat Rec (Hoboken) 2012;295(3):372–7.

[53] Trump BF, Berezesky IK, Smith MW, et al. The relationship between cellular ion deregulation and acute and chronic toxicity. Toxicol Appl Pharmacol 1989;97(1):6–22.

[54] Trump BF, Berezesky IK. Calcium-mediated cell injury and cell death. FASEB J 1995;9(2):219–28.

[55] Kerr JF, Wyllie AH, Currie AR. Apoptosis: a basic biological phenomenon with wide-ranging implications in tissue kinetics. Br J Cancer 1972;26(4):239–57.

[56] Kerr JF, Searle J. The digestion of cellular fragments within phagolysosomes in carcinoma cells. J Pathol 1972;108(1):55–8.

[57] Kerr JF. Shrinkage necrosis of adrenal cortical cells. J Pathol 1972;107(3):217–9.

[58] Planey SL, Litwack G. Glucocorticoid-induced apoptosis in lymphocytes. Biochem Biophys Res Commun 2000;279(2):307–12.

[59] Trump BF, Berezesky IK, Chang SH, et al. The pathways of cell death: oncosis, apoptosis, and necrosis. Toxicol Pathol 1997;25(1):82–8.

[60] Searle J, Kerr JF, Bishop CJ. Necrosis and apoptosis: distinct modes of cell death with fundamentally different significance. Pathol Annu 1982;17(Pt 2):229–59.

[61] Walker NI, Harmon BV, Gobé GC, Kerr JF. Patterns of cell death. Methods Achiev Exp Pathol 1988;13:18–54.

[62] Kanduc D, Mittelman A, Serpico R, et al. Cell death: apoptosis versus necrosis. Int J Oncol 2002;21(1):165–70.

[63] Galluzzi L, Vitale I, Abrams JM, et al. Molecular definitions of cell death subroutines: recommendations of the nomenclature committee on cell death 2012. Cell Death Differ 2012;19(1):107–20.

[64] Searle J, Lawson TA, Abbott PJ, et al. An electron-microscope study of the mode of cell death induced by cancer-chemotherapeutic agents in populations of proliferating normal and neoplastic cells. J Pathol 1975;116(3):129–38.

[65] Abu-Qare AW, Abou-Donia MB. Biomarkers of apoptosis: release of cytochrome c, activation of caspase-3, induction of 8-hydroxy-2'-deoxyguanosine, increased 3-nitrotyrosine, and alteration of p53 gene. J Toxicol Environ Health B Crit Rev 2001;4(3):313–32.

[66] Krysko DV, Vanden Berghe T, D'Herde K, et al. Apoptosis and necrosis: detection, discrimination and phagocytosis. Methods 2008;44(3):205–21.

[67] Aschoff A, Jirikowski GF. Apoptosis: correlation of cytological changes with biochemical markers in hormone-dependent tissues. Horm Metab Res 1997;29(11):535–43.

[68] Saraste A, Pulkki K. Morphologic and biochemical hallmarks of apoptosis. Cardiovasc Res 2000;45(3):528–37.

[69] Washington K, Jagasia M. Pathology of graft-versus-host disease in the gastrointestinal tract. Hum Pathol 2009;40(7):909–17.

[70] Upton JW, Chan FK. Staying alive: cell death in antiviral immunity. Mol Cell 2014;24:54.

[71] Green DR, editor. Means to an end: apoptosis and other cell death mechanisms. Cold Spring Harbor, NY: CSHL Press; 2011.

[72] Erhardt P, Toth A, editors. Apoptosis methods and protocols (2nd ed.). New York, NY: Humana Press/Springer; 2009.

[73] Lavrik IN, editor. Systems biology of apoptosis. New York, NY: Springer; 2013.

[74] Balcer-Kubiczek EK. Apoptosis in radiation therapy: a double-edged sword. Exp Oncol 2012;34(3):277–85.

[75] Vargas J, Feltes BC, Poloni J, de F, et al. Senescence; an endogenous anticancer mechanism. Front Biosci (Landmark Ed) 2012;17:2616–43.

[76] Hwang ES. Replicative senescence and senescence-like state induced in cancer-derived cells. Mech Ageing Dev 2002;123(12):1681–94.

[77] Dash BC, El-Deiry WS. Cell cycle checkpoint control mechanisms that can be disrupted in cancer. Methods Mol Biol 2004;280:99–161.

[78] Aguirre-Ghiso JA. Models, mechanisms and clinical evidence for cancer dormancy. Nat Rev Cancer 2007;7(11):834–46.

[79] Vakifahmetoglu H, Olsson M, Zhivotovsky B. Death through a tragedy: mitotic catastrophe. Cell Death Differ 2008;15(7):1153–62.

[80] Jordan MA. Mechanism of action of antitumor drugs that interact with microtubules and tubulin. Curr Med Chem Anticancer Agents 2002;2(1):1–17.

[81] Rieder CL. Mitosis in vertebrates: the G2/M and M/A transitions and their associated checkpoints. Chromosome Res 2011;19(3):291–306.

[82] Surova O, Zhivotovsky B. Various modes of cell death induced by DNA damage. Oncogene 2013;32(33):3789–97.

[83] Kuilman T, Michaloglou C, Mooi WJ, et al. The essence of senescence. Genes Dev 2010;24(22):2463–79.

[84] de Magalhães JP. Cellular senescence. Available at: http://www.senescence.info/cell_aging.html.

[85] Francis Rodier F, Campisi L. Four faces of cellular senescence. J Cell Biol 2011;192:547–56.

[86] Artandi SE, Attardi LD. Pathways connecting telomeres and p53 in senescence, apoptosis, and cancer. Biochem Biophys Res Commun 2005;331(3):881–90.

[87] The American Heritage® Medical Dictionary. Acute inflammation. Available at: http://medical-dictionary.thefreedictionary.com/acute+inflammation.

[88] The American Heritage® Medical Dictionary. Allergic inflammation. Available at: http://medical-dictionary.thefreedictionary.com/Allergic+inflammation.

[89] Medilexicon. Subacute inflammation. Available at: http://www.medilexicon.com/medicaldictionary.php?t=44465.

[90] Punchard NA, Whelan CJ, Ian Adcock I. The journal of inflammation. J Inflamm (Lond) 2004;1:1.

[91] In: Ref. [1]. pp. 66–8.

[92] Grattan CEH, Black AK. Urticaria and mastocytosis Burns DA, Breathnach SM, Cox NH, editors. Rook's textbook of dermatology (8th ed.). Chichester, UK: Blackwell Publishing; 2010. pp. 22.1–22.36.

[93] In: Ref. [1]. pp. 79–110.

Therapies I: General Principles

L.P. Bignold: Principles of Tumors.
DOI: http://dx.doi.org/10.1016/B978-0-12-801565-0.00011-1

287

As is mentioned in previous chapters, all therapies are directed at killing tumor cells or at least limiting their growth in the body. Surgery is the main mode of treatment for most "solid" (nonhematological) malignancies, especially because surgical excision cures cases of tumor if all the cells of the tumor are removed. Even if the surgical removal of a primary tumor mass does not cure the patient, the operation can prevent debilitating, if not fatal, local complications of the primary tumor.

Nonsurgical therapies are used for tumor cells which, for one reason or another, cannot be removed surgically. The means for killing tumor cells in the living body are radiations, drugs, and other methods.

This chapter is concerned with some of the general principles in these therapies. The topics are arranged according to when they may arise in the course of a typical case of a malignant tumor.

Section 11.1 deals with the objectives of therapies. Section 11.2 discusses the principles of surgical therapies. Section 11.3 deals with the principles of nonsurgical therapies. Section 11.4 describes resistance of tumors to nonsurgical anticancer agents, while Section 11.5 is concerned with other aspects of nonsurgical therapies.

11.1 OBJECTIVES

At the time that a patient is diagnosed with a tumor, various possible treatment options are usually considered. In some cases, it may be questioned whether or not the tumor should be specifically treated at all. Thus if a patient is very elderly, and the tumor is benign or very slow growing, no treatment may be indicated. Also if patient is too frail due to comorbidities for treatment, then it may be appropriate that only supportive care (see Section 12.7) is offered. However, in most cases, some specific therapy for the tumor is appropriate. In these cases, treatment may proceed with one or more objectives.

1. Cure

   Cure is defined as lifelong freedom from any recurrence of the disease without any further treatment. The curabilities of the different types of tumors are quite different to one another (see Section 11.3.2).

2. Prolongation of life through limitation of growth of tumor cells in the body

   This is the usual aim in patients who have a tumor cell population which cannot be removed surgically. The objective is sometimes referred to as "increasing the survival time" of the patient. All the common tumor types are now considered treatable by one or more forms of therapy.

   For example, a case of carcinoma of the breast may be treated by surgery, radiotherapy, hormonal modification and various chemotherapies.

3. Amelioration of specific side effects

   These treatments are designed to relieve specific short-term disorders induced by the nonsurgical anticancer therapies. The commonest may be antinausea drugs for patients undergoing radio- and/or chemotherapy. Another common group of treatments is transfusions of whole blood or separated components of blood. These may be given for hematological complications of therapy, e.g., erythrocytes for anemia, leukocytes for immune deficiency, and platelets for deficiency causing excessive bleeding.

Bone marrow transplantation is probably the most complex form of therapy in this category, as is discussed in Section 11.5.3.

4. Palliation

   These treatments are designed to relieve intractable long-term effects of the disease, such as pain, ulceration, or anemia.

5. Supportive care

   This treatment is in principle, relief of symptoms or prevention of complications by methods not requiring surgery or specific anticancer drugs.

11.2 PRINCIPLES OF SURGICAL THERAPY

11.2.1 Background

In the last decades of the nineteenth century, there were dramatic improvements in surgical techniques for removing tumors from the body. This occurred in association with the development of better surgical techniques in general [1]. The improvements were principally the discovery of anesthesia and the invention of antiseptic methods. Most of the details of the most commonly performed surgical operations—the precise parts of the body which are cut and those parts which are removed—are based on gross anatomy and macroscopic pathology and were established by the mid-twentieth century.

Numerous ancillary technical developments have made surgical operations more anatomically accurate, safer, and less liable to complications, as well as to degrees, easier to perform. These advances include those in imaging, for example, computerized axial tomography—CAT scans and magnetic resonance imaging—MRI scans; and technical developments such as electrocautery, laser scalpels, and stents.

11.2.2 Classification of Operations

(a) General

Surgical operations are classified first according to the organ or structure involved, and then according to whether it is removed ("-ectomy"), incised ("-otomy"), or otherwise anatomically altered (e.g., "-plasty").

For certain organs, several different operations may be possible.

The lung can be removed completely (pneumonectomy), a lobe only cut out (lobectomy), or only a smaller part ("segment" or "wedge") resected.

The colon is usually removed in part only: e.g., in cecectomy, right-sided hemicolectomy, extended right-sided hemicolectomy and sigmoidectomy.

The breast can be operated on to remove a mass (lumpectomy); the breast tissue alone (simple mastectomy, with or without overlying skin); or the breast tissue, skin, and adjacent lymph nodes (extended mastectomy). More extensive operations, for example, in which the breast, the axillary nodes, and the pectoralis minor muscle are removed ("radical mastectomy"—Halsted); and more extensive *en bloc* resections (e.g., "supraradical mastectomies") are rarely carried out [2] (Table 11.1).

(b) By purpose

There are several possible reasons for surgical operations for tumors.

1. *For cure*

    Surgery alone can cure many tumors, e.g., almost all benign tumors; most epidermal malignancies; many cases of melanomas of the skin; as well as many cases of carcinomas of the colon and rectum, the kidney, the bladder, the endometrium, and the thyroid and adrenal glands. In these cases, there must be no tumor outside the surgical margin of excision (Figures 1.11 and 11.1).

2. *For cure in combination with another mode of treatment*

    For some tumors, especially of the breast, surgery alone appears to cure only a proportion of cases. In these, additional therapy, such as radiotherapy or chemotherapy, may be given as part of initial treatment.

TABLE 11.1   Common Cancers, Surgical Operations, and Possible Complications

Lung carcinoma	
Lobectomy or pneumonectomy	Breakdown of bronchial closure causes pneumothorax
Breast carcinoma	
Lumpectomy, "simple" mastectomy	Usually none
Radical mastectomy	Lymphedema of the arm
Colorectal carcinoma	
Partial colectomy	Breakdown of anastomosis
Prostate carcinoma	
Trans-urethral resection (TURP)	Usually none significant
Suprapubic prostatectomy	Urinary incontinence, impotence

These are in addition to the general complications of hemorrhage and infection.

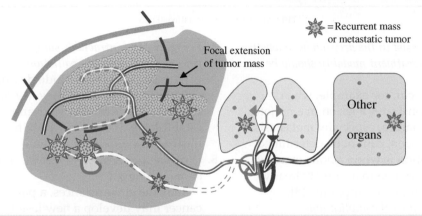

1. After this operation, tumor can potentially recur from focal extension of tumor mass, from tumor cells in transit to lymph nodes, in lymph nodes, as well as in the blood vessels, in the lungs and in other organs

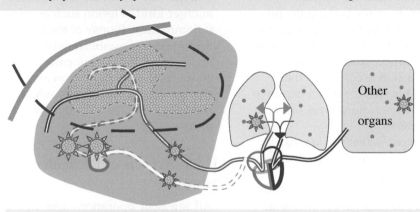

2. After this operation, tumor can potentially only recur from tumor cells in transit in lymph or venules, in lymph nodes, in lungs or in other organs

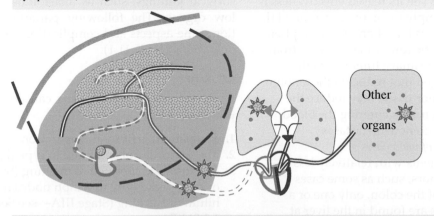

3. In this operation, tumor can potentially only recur from tumor cells in transit in lymph beyond local lymph nodes or in venules, in lungs or in other organs

NB: Probably only a small and unpredictable proportion of micrometastases, also at an unpredictable rate, will grow into clinically appreciable metastases.

FIGURE 11.1 Sites of potential recurrence of malignant tumors depend on extent of surgical resection of primary tumor (see also Figure 1.11).

3. *For staging to assist in the decision as to whether or not another treatment modality should be given*

This may occur, for example, when imaging demonstrates a lesion in a remote organ, but its nature, especially metastasis versus other lesion, cannot be determined. Another situation is when a bone marrow biopsy is undertaken in a case of already diagnosed malignant lymphoma [3].

4. *For prevention of local complications*

As an example, this may occur when a carcinoma of the breast has already metastasized widely in the body at the time of diagnosis. Removal of the primary tumor does not affect the growth of the metastases, but prevents ulceration, infection, and bleeding of breast skin involved by the primary tumor. This was the major objective when mastectomy operations were first introduced (see above).

5. *For debulking ("cytoreduction") of the amount of tumor in the body prior to use of another therapeutic modality*

In certain cases, it may be appropriate to surgically remove as much tumor tissue as possible so that adjuvant therapies will be more effective. The increased effectiveness is on the basis simply of agent: target ratio [4]. Another reason is based on the assumption that all metastatic tumor masses arise from the cells of the primary tumors. On this basis, it is alleged that reducing the amount of primary tumor tissue may reduce the numbers of such cells entering the blood, and growing into new metastases.

6. *For the removal of metastases*

In some patients with relatively less aggressive tumors, such as some cases of carcinoma of the colon, only one or a few metastases are found in the liver at diagnosis and assessment, and none are found elsewhere in the body. It is reported that in these cases, removal of the hepatic metastases prolongs life [5].

7. *For relief of an intractable symptom without possibility of cure (i.e., palliation)*

These operations include destruction of nerves to reduce pain. They also include relief of obstruction as for example by establishing a colostomy proximal to an obstructing colonic carcinoma.

8. *For follow-up biopsies for possible further chemo- or radiotherapy*

In some circumstances, a patient with cancer may develop a new lesion on imaging. However, it may be unclear whether the lesion is a metastasis or another kind of lesion such as a focus of inflammation. Biopsy of the lesion may be undertaken to establish nature of the lesion, and hence allow an informed decision about further therapy. However, if no change in therapy is likely whatever the result of the biopsy, the biopsy may not be necessary [6].

11.2.3 Aspects of Particular Cancer Operations

All surgical operations can be complicated by bleeding, wound infection, and failure of surgical joinings, such as anastomoses of hollow organs. The following paragraphs mention some aspects and complications of specific operations (Table 11.1).

(a) Lung

Operations to remove lung cancers are only undertaken if

1. The patient is fit for the operation
2. The tumor is not small cell in type [7]
3. The tumor is confined to the lung (Stage I) or has spread only to lymph nodes near the hilum of the lung (Stage IIIA—see Section 1.3.3) [7].

Complications include reduced lung capacity and bronchopleural fistula leading to pneumothorax, and recurrences in the other lung.

(b) Breast

The principles of the main operations are mentioned above. The main complication of breast surgery is the cosmetic effect. Operations which involve removal of the axillary lymph nodes can be complicated by edema of the arm (lymphedema), which is disfiguring, uncomfortable, and often difficult to treat.

Local recurrence of stage I/II carcinoma in the breast or skin occurs in 5–10% of cases overall, with higher percentages occurring in premenopausal women [8].

(c) Colon and Rectum

All operations on the colon may be complicated by a period of loss of colonic movements and hence failure to pass feces. The condition is almost always transient.

All anastomoses of bowel may breakdown with leakage of bowel contents and local peritonitis. Removal of the lower rectum and anus usually necessitates a surgical operation to permanently allow drainage of the large bowel through the abdominal wall (colostomy).

Significant local recurrences of colonic carcinomas are not common, because a wide margin of mesocolon is usually available. However, approximately 10% of all resectable carcinomas of the rectum recur. Recurrences are more common with carcinomas of the rectum than colon because lymphatic drainage is more diffuse, and the possible width of the margin of excision of tumors at this site is often limited to a few centimeters [9].

(d) Prostate

Prostatic carcinomas may be first identified in pieces removed at trans-urethral prostatectomy for urinary obstruction. This operation, however, does not include all prostatic tissue. The usual definitive operation is complete prostatectomy with seminal vesicles. This may be done by retropubic or suprapubic approaches. The common complications of prostatectomy are loss of urinary control (continence) and of erectile (sexual) function [10].

(e) Other (see in Ref. [11])

Parotidectomy. More than 90% of these operations are for benign tumors (adenomas of various types) of this salivary gland. The surgical procedure can damage the facial nerve.

Gastrectomy. This operation can be partial or total. Possible chronic complications include persistent vomiting, anemia, deficient vitamin absorption, and a disorder of fluids and electrolytes known as the "dumping syndrome."

Pancreatectomy. This operation can be partial or total. Particular complications of partial pancreatectomy are leakage of pancreatic enzymes into the retroperitoneal tissue, followed by peritonitis. Total pancreatectomy includes the Islets of Langerhans, and so is almost always followed by insulin deficiency and hence diabetes mellitus.

Thyroidectomy. During any thyroidectomy, the recurrent laryngeal nerve may be damaged, causing chronic hoarseness. In total thyroidectomy, the parathyroid glands may be inadvertently removed, resulting in hypoparathyroidism. This is now a rare complication. Total thyroidectomy is always followed by loss of production of normal thyroid hormones.

Part of cerebrum or cerebellum. These operations are almost always followed by neurological impairments of kinds depending on the part of the brain involved. Epileptic seizures are another complication. These can usually be prevented by antiepileptic drugs.

Oophorectomy, orchidectomy, adrenalectomy. These usually have few specific complications beyond loss of function of the organ removed.

11.3 PRINCIPLES OF NONSURGICAL THERAPIES

11.3.1 Background

Attempts have been made to treat tumors by diet and toxins since time immemorial. At the time when tumors were identified as cellular masses (see Section 2.1), rapid developments were being made in chemistry. Numerous new compounds were identified and tested for biological effects. Long before that, arsenic compounds had been known to be toxic to humans. In the 1870s, arsenic trioxide was used as an anticancer agent [12]. This preceded the work of Paul Ehrlich (1854–1915), who studied arsenic compounds for their antibacterial activities, and in 1909, discovered the successful antisyphilis drug "Salvarsan" [13].

Radiotherapy began in the first decades of the twentieth century, with use of X-rays to treat superficial tumors, and implants of radioactive "seeds," especially of radon, into deeply placed tumors [14].

Selenium is adjacent to arsenic in the periodic table, and some selenium compounds such as hydrogen selenide have toxic properties similar to arsenic compounds [15]. Selenium compounds were tested in the early twentieth century for antitumor effects [16], but found to be too toxic for clinical use.

Modern anticancer chemotherapy began with the development of nitrogen analogues of the poison gas bis-dichloroethyl sulfide ("mustard gas," the weapon first used in WWI) [17] and the separate group of substances known as antimetabolites such as methotrexate (see Section 12.2.1) [18].

The different drugs in current use derive from

1. Naturally occurring compounds
2. Synthetic compounds fortuitously found to have cell-killing potencies
3. Synthetic compounds designed to "target" a biochemical process or enzyme (see Section 12.3.3)

4. An antibody directed against a particular cell surface molecule (see Section 12.5.1).

11.3.2 Variabilities in Sensitivities of Tumor Types to Individual Agents

Over the decades, various regimens of radiotherapies and hundreds of thousands of potential drugs have been tested for antitumor effects against the various types of malignant tumors. Several hundreds of drugs have been sufficiently effective and relatively nontoxic to be used in medical practice.

The general results of this testing of radiations and drugs against the various tumor types have been that all agents are useful against some tumor types, but none are effective against all tumor types.

For example, radiotherapy is effective against most lymphomas and squamous cell carcinomas, but of less use against adenocarcinomas and melanomas [19]. And among anticancer drugs, antimetabolite agents such as methotrexate can be effective against leukemias and lymphomas [20] but less effective against solid tumors.

These differences in sensitivities to radiations according to tumor type are not explained by DNA damage since all tumor types share the same DNA. Similarly, these differences are not explained by rates of cell division, since rapidly growing lymphomas are usually sensitive, but equally rapidly growing carcinomas and melanomas are often insensitive.

Similarly, the differences in sensitivities to anticancer drugs of the different tumor types are not fully explained. The mechanisms of these differences may lie in pharmacokinetic factors at the subcellular level, as are discussed in more detail in Section A3.1 (see also Figure 11.2).

It is possible that susceptibilities of tumor cells to anticancer agents may be at least partly a characteristic inherited from parent cell. Examples of sensitive cells are hematopoietic cells and epidermal cells. Both the normal cells and the tumor cell populations which derive from them are

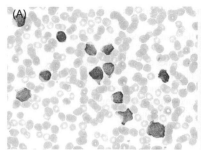 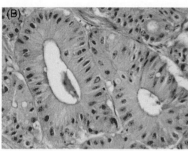 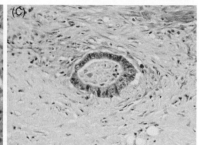

FIGURE 11.2   Microscopic features of kinds of sensitive versus resistant cells. (A) Leukemic cells in blood, ×60. These cells are fully exposed to blood-borne anti-cancer drugs. (B and C) Cells of colonic carcinoma, ×20. Many of these are likely to be protected to a degree from blood-borne cytotoxic agents by other cancer cells.

sensitive to radiations. Examples of resistant cells are fibrocytes and melanocytes, which are little affected by anticancer agents, as are the tumors which derive from them.

As well as the differences in resistances according to tumor type, different cases of the same tumor type may have quite different resistances. For example, significant proportions of indolent malignant lymphomas are resistant to radiotherapy, while the parent cells (lymphocytes) are sensitive to radiations [21].

11.3.3 The Basis of Tumor Population Sensitivity to Anticancer Agents and the Therapeutic "Window"

So far as is known, tumors have no specific mechanism of resistance to cytotoxic agents (Figure 11.3). In general, it is believed that no kind of mechanism of defense against noxious agents occurs only in tumors and not in normal cells. This is because no "neoplasia-specific" defense against toxic agents has been found.

Thus the differences between the resistances of the tumor types probably are based on quantitative differences in the cells' defensive factors—pretarget, target, and posttarget, as described in Appendix 3—to each agent. The "therapeutic window" is the range of dosage of the anticancer agent between lack of effect and intolerable toxicity (Figure 11.4).

The basis of these quantitative differences in resistances in tumor cell populations may not lie simply in the numbers of genomic events in tumor cell populations compared with normal cells. This is because the most genomically unstable populations of cells are often the most resistant to therapies.

11.3.4 Combination Therapies

For almost every type of tumor, multiple agents given concurrently are more effective than single agents or multiple agents given consecutively (Figure 11.3). The mechanism is probably that tumor cells

1. may have reductions in the efficiencies of only some of the their many possible defensive mechanisms (see Appendix 3)
2. may have increased resistance mechanisms (see Section 11.4.5) to only some agents.

Thus if a cell sustains damage to its cell membrane, its cytoplasmic metabolism, and its genomic mechanisms, it will be less likely to survive than if only one function is damaged.

Because of this, more tumor cells in any given tumor are likely to be destroyed by combinations of agents when the targets—or range of targets—of the agents are different to each other and their effects are additive.

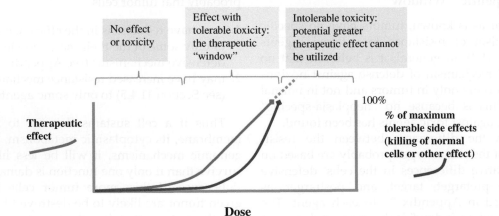

Membrane targets
- receptors
- aquepores
- ion pumps
- other exporter mechanisms

Cytoplasmic targets
- metabolic enzymes
- synthetic enzymes
- signaling enzymes

Nucleus-related targets could include:
- the centrosome
- DNA
- enzymes of DNA processes
- histones
- nuclear membrane (which reconstitutes as spindle fibers during mitosis)

General
    Radiations, but few drugs, are general noxins (see Figure 10.3). No therapeutic agent acts on lipid solubility. Cytotoxic drugs probably affect mainly enzymes and other cytoplasmic and nuclear contents. Selective drugs may act on surface receptors or on cytoplasmic signaling molecules (chapter 6).

Some bases of greater sensitivities of some cells in tumor cell populations to cytotoxic drugs
    Potentially, tumor cells could be more susceptible to anticancer agents because of
- poorer extracellular defences (not shown)
- poorer cellular defences, for example, exporter mechanisms for drugs
- lower quantities of the targets (i.e., lower target:agent ratios)
- poorer capacities for repairs and regenerations

Rationale for multiple agents concurrently
If all at the same time, a cell sustains damage to its cell membrane, its cytoplasmic metabolism and its genomic mechanisms (see also Figure 10.3), it will be less likely to survive than if only one site is damaged.

FIGURE 11.3   Targets and issues relating to anticancer agents.

| No effect or toxicity | Effect with tolerable toxicity: the therapeutic "window" | Intolerable toxicity: potential greater therapeutic effect cannot be utilized |

Therapeutic effect

100%

% of maximum tolerable side effects (killing of normal cells or other effect)

Dose
(Single dose or cumulative dose depending on agent and tissue)

FIGURE 11.4   The therapeutic "window" of dosage.

11.3.5 Adjuvant Regimens

The term "adjuvant" (Latin, *adjuvare*: to aid) in cancer therapy refers to any additional modality of treatment given in addition to the primary one [22]. There are special regimes for treating cancers of particular types. The commonest situation is when, after a cancer has been resected, the tissues around the primary site are subjected to radiotherapy, and/or chemotherapy, hormonal, or other therapy is administered in the same period. The rationale for this therapy is that the micrometastases (see Section 1.3.4) will be killed. Anticancer drugs may kill micrometastases in the body generally, while radiotherapy will only kill cells in the irradiated field.

11.3.6 Neoadjuvant Regimens

This term refers to giving chemo- and/or radiotherapy to a primary tumor before surgery [23]. Various rationales are appropriate, including

1. Inoperable tumors can be reduced to an operable size
2. Numbers of metastases appearing after surgery will be reduced
3. A response in the primary tumor will assist in choosing postsurgical therapeutic modalities.

Whether or not neoadjuvant therapies have any benefits over ordinary adjuvant therapeutic regimens, especially in breast cancer, has been controversial [24]. Nevertheless, for this type of malignancy, neoadjuvant therapy is widely used at the present time [25,26].

11.3.7 "First-Line," "Second-Line," and "Third-Line" Therapies and Regimens

There are many published regimens for each type of tumor. For most tumor types, and for each stage or grade of the particular case, there are recognized "best current practices" which have been established by clinical trials (see Section 11.11). To these the term "first-line therapy" is often applied. If the patient's particular circumstances render that regimen inappropriate, other regimens—termed "second-line therapies"—may be used. Also if a patient's tumor has been resistant to a first-line therapy, a second, or even a "third-line" regimen may be used.

11.4 PARTIAL RESPONSES OF TUMORS TO NONSURGICAL ANTICANCER AGENTS

The expectations for a patient undergoing nonsurgical therapies for metastatic malignant tumors are highly variable according to tumor type (see Section 11.3.2).

For certain types of tumors, complete responses (i.e., cure) are frequently obtained. For the remainder, however, partial responses are the commonest clinical pattern (see Figure 1.12) [27–32].

This section is mainly concerned with the causes of partial responses.

11.4.1 The Tumor Cell Population Has a Component Which Is Not Reached by the Agent

This is also called resistance by way of microenvironmental 'niche' [33,34]. It implies that the survival and regrowth of cells in a mass are because those cells were not exposed to the agent which killed other cells in the mass.

Factors within the tumor mass allowing for tumor cell survival may include

(a) Hypovascularization and Hypoxia

It is well established that the blood vessels at the margin of a tumor may be increased in number, while those in the center of the mass are usually diminished. Thus drugs may be able to kill cells at the periphery of a mass, but not reach and affect cells in the middle of the mass [35].

For radiation therapy, such pharmacodynamic factors do not apply. However, the hypovascular middle parts of solid tumor masses are likely to be relatively hypoxic. These zones could be the origins of the cells surviving radiation therapy [36].

(b) Desmoplasia

This phenomenon is described in Section 8.3.5. If it protects tumor cells from blood-borne anticancer agents (mainly drugs), it may be due to the associated reduced vascularity rather than simply the presence of excess collagen.

11.4.2 The Tumor Cell Population Is Heterogeneous for Viability (Cytokinetic Explanation)

In this concept, therapy hastens the process of death in already dying tumor cells (Figure 11.5). It means that resistance may be variable between foci in the same individual tumor.

11.4.3 The Tumor Cell Population Is Heterogeneous in Some Other Way, with a Resistant Component ("Cancer Stem Cells")

This is a frequently invoked phenomenon, but raises the question of the precise nature of the properties of the resistant cells (Figure 11.6). There are various theories for this.

(a) The Resistant Cells Are Present Before Therapy Because of Genomic Instability

In Section 9.3.5, it is pointed out that many tumor cell populations show variable numbers of cells which express antigens to supranormal degrees. The question then arises: if the population of tumor cells is heterogeneous, are *any* of the cells in the tumor cell population resistant to the point of having supranormal resistance to these agents? There seems to be no reason why hyperresistance should not spontaneously emerge in parallel with other forms of hyperexpressions of biochemical activity. These cells would then only become apparent after therapy has destroyed the susceptible cells. This is discussed further in Section 11.4.6.

(b) The Resistant Cells Are Those in the Interdivision Period

This concept is that the cells in dividing populations (see Section A1.3.4) which are most sensitive to cytotoxic agents are those in the division period. Death of the susceptible component of the population gives the apparent effect of "partial response" of the tumor masses.

Some support for the ideas comes from results of laboratory studies of irradiation of cultured cells using selected doses of X-rays. In these studies, it has been discovered that a constant proportion of the population of cells survive [37]. However, this observation could be explained on the basis that it represents the bell-shaped "normal distribution" of sensitivity of cells in the population without any influence of cell cycle factors.

In experimental studies, it has been found that different agents affect different proportions of cells according to phase of cell cycle [38].

(c) The Resistant Cells Are Descendants of Local Tissue Stem Cells

Normal "labile" tissues are maintained by unspecialized, immortal local tissue stem cells, which, when they divide, produce one stem cell and one line of mortal, specializing, "transit-amplifying" cells (see Section A1.3.2) leading to the fully specialized "mature" cells.

Because these kinds of tissues usually regenerate (see Section A1.4.3) after cessation of the drugs, it is usually assumed that the relevant local stem cells are relatively therapy resistant. Cytotoxic therapies, especially antimitotic

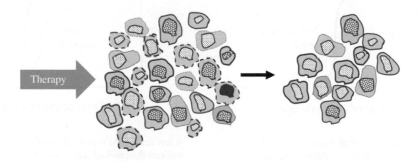

Tumor cell populations comprise: (i) proliferating cells, (ii) nonproliferating cells which are viable ("inappropriate vegetative state"), and (iii) cells which are subviable and already in the course of dying.

Therapy to a tumor cell population may cause a substantial reduction in size of mass, due to deletion of subviable cells. i.e., therapy has hastened the destruction of cells which were already in the process of dying.

Therapy often also converts previously proliferative cells to vegetative cells.

The rate of regrowth of the mass will depend mainly on the proliferation capacities of the remaining proliferative cells. These are one kind of cells which are referred to as "cancer stem cells" (see Section 2.9.1).

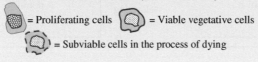

= Proliferating cells   = Viable vegetative cells

= Subviable cells in the process of dying

FIGURE 11.5   Cytokinetic factors in partial responses of tumor cell population.
Note: This discussion assumes uniform exposure of all tumor cells in the population to agent. (The assumption applies in radiotherapy, but not necessarily in anti-cancer drug therapy – see Section 11.4.1).

drugs, probably affect transit-amplifying cells as well as stem cells.

It has been thought that perhaps this resistance is a manifestation of their low rates of cell division on the basis that all cells may be more vulnerable when they are in G1, S, G2, and M phases than when they are in G0 [39–43]. However, it is not clear whether stem cells in fact are more resistant in GO than when in cell division for all agents, or whether local micro-environmental factors may be responsible [38].

(d) The Resistant Cells Are Descendants of Clones Formed at the Beginning of the Tumor

According to the polyclonal theory of tumors (see Section 2.7), tumors are made up of different clones of cells [43,44]. Thus partial responses may simply reflect destruction of the susceptible clones and survival of the resistant clones.

(e) The Resistant Cells Are Generated During Growth of the Tumor Because of Genomic Instability

The mutator phenotype theory (see Section 2.8) suggests that tumor cell populations are aclonal and heterogeneously heterogenizing. In this scenario, hyper-resistant cells may emerge *ab initio* at random in the tumor cell population.

The implications of this possibility include that even multiagent therapy may not kill

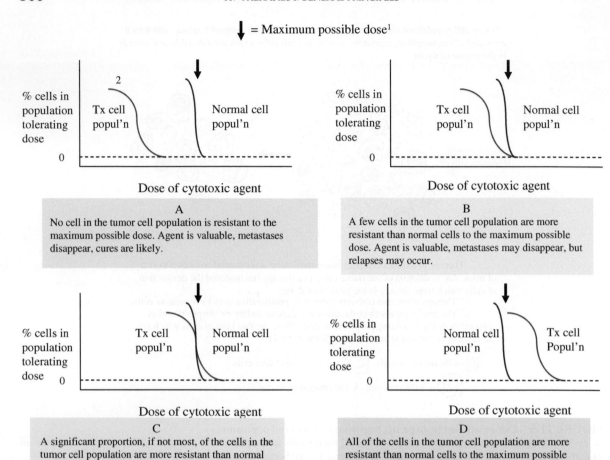

FIGURE 11.6 Relationship between heterogeneities in resistances of tumor cell populations compared to normal cells in the clinical responses of malignant tumors.

them. It further implies that the patient's disease may not be curable, because if higher doses are given toxicity to normal cells will itself threaten the patient's life.

11.4.4 "Acquired" Resistance

Conventionally, this phrase is used when recurrences of tumor masses appears to be more resistant to therapy (i.e., hyperresistant) than were the masses during the first course of therapy. This occurs both in radiotherapy and anticancer chemotherapy.

(a) In Radiotherapy

It is sometimes observed that recurrent masses of tumors grow more rapidly, and/or arc more resistant to further therapy, after radiotherapy or chemotherapy in comparison with the original tumor.

In radiotherapy, mechanisms of this apparent acquired radioresistance have been considered for many decades [45–48]. For example, to quote directly from [45]:

a. Radiation selection of spontaneous mutant resistant cells in the tumor population,
b. radiation-induced mutation in a tumor cell to resistance, followed by selection, and
c. a complex of radiation effects on the host that interacts with the tumor to give the appearance of resistance to the tumor.

(b) In Anticancer Chemotherapy

In anticancer chemotherapy, essentially the same observation has been made. After responding to a combination of drugs, many tumor masses regrow more rapidly than previously, and the regrowth does not respond to any further drug therapy [49,50].

In some cases, the sensitivity of the recurrent cells to other drugs may be preserved, so that the acquired resistance is specific to the first drug. In other cases, however, the resistance after exposure to one drug extends to a wide range of other anticancer drugs.

The mechanisms of these observations are presumed to be similar to that for recurrences after radiotherapy: either a resistant subpopulation of tumor cells already exists at the time of diagnosis or arises during the period of treatment.

11.4.5 Biochemical Aspects of Resistance of Tumors to Anticancer Agents

In general, the biochemical mechanisms of resistance of tumor cell populations include the normal cellular defenses against noxins, especially transporter and degradative enzymes described in Section A3.1.2. The roles of these mechanisms in the pharmacokinetic phenomena are often different for different drugs [28,32].

(a) Decreased Activation of the Prodrug

An example of this is reduction in the activity of the enzyme which activates 5-FU (see Section 11.3.3).

(b) Decreased Accumulation of the Active Drug in the Tumor Cells

This may occur through several mechanisms, including:

– decreased active influx into the cell. An example is the reduced polyglutamination of methotrexate
– increased active efflux of the drug (either before or after activation) out of the cell
– increased sequestration or destruction within the cell.

It may be noted that these transport and sequestration mechanisms can be specific for a kind of drug, or affect a wide range of drug (i.e. causing "multi-drug resistance"). The relevant proteins are called P-glycoproteins, and the genes encoding them are referred to as "MDR" (Multi Drug Resistance) genes.

(c) Increased Production by the Cell of "Target" Molecules in Response to the Agent

This is a form of "adaptation" (see Section A1.4.2). Examples are increased thymidylate synthetase production in response to 5-FU, and of increased dihydrofolate reductase production in cells treated with methotrexate [32].

11.4.6 Genomic Bases of Resistance

For resistance mechanisms within the tumor cells, function-altering genomic events in genes in the tumor cell are likely to be responsible for the alterations in pharmacokinetic factors underlying the resistance. A relevant gain-of-function or a loss-of-function effect may contribute to resistance to the drug.

For acquired resistance (see Section 11.4.4), it is possible that during therapy (and possibly accelerated by therapy), karyo-unstable and mutator phenotypes produce new populations of tumor cells, with new genomic events which increase the defenses of the cells against the drug, or made the target more resistant to the actions of the particular agent [51–54].

For this, the genopathic effects of inducing either chromosomal aberrations or nucleotide errors would be operative.

The alternative explanations might be that these cases represent a "progression" involving increasing resistance analogous to the acceleration in growth rates which occurs in untreated tumors (see Section 8.4).

In relation to resistances to drugs, it would seem that specific resistances imply that some additional defensive mechanism had developed in relation to the specific target of the particular drug. The simplest of these would be that the target resistance mechanism of amplification of amount of target (see Section A3.2).

In relation to non-specific acquired resistance, it would seem to imply amplification involving a non-specific cell defense, such as the detoxifying mechanism P450 (see Section A3.1).

It is also possible that in some cases, the micro-environment of the tumor cells (see previous subsection) are changed by the original drug therapy, possibly with the effect of increasing defensive conditions outside the tumor cells.

The whole issue cannot be fully investigated by clinical studies because the normal tissues at the site of the recurrent tumor cannot be further irradiated / treated, owing to "limitations on treatment" (see Section 11.5.2) in adjacent normal tissues.

11.5 OTHER ASPECTS OF NONSURGICAL THERAPIES

11.5.1 Greater Efficacies in Split Doses: Resistance Factors Combined with Recovery Factors

A major feature of radiotherapies and anticancer chemotherapies is that better clinical outcomes are achieved when these therapies are given in split doses (Figure 11.7). They aim to administer the greatest total doses of agents to be given, with less of the side effects which would occur if the total dose was given in one administration. The side effects can be the direct action of the agent on the normal tissues, and the secondary effects of tumor cell breakdown products on normal cells [55].

(a) Radiotherapy

Early in the history of radiotherapy, radiations were administered in large single doses [56]. This practice was supplanted by "fractionation" of doses, as pioneered in France in the 1910s [57]. Since then, different regimens have been developed for the different tumor types, especially as different technologies have become available and are being tested [58].

(b) Anticancer Chemotherapy

At the beginning of the widespread use of cytotoxic drugs for malignant tumors (the alkylating agents, in the 1940s), the usual regime was to give a large dose at the time of diagnosis, and perhaps additional treatments thereafter. This was found to be unsatisfactory, and for many years, treatment regimens consisted of continuous "maintenance" doses. Finally, from about the late 1960s, large doses of anticancer drugs were given in periods of intensive doses, interspersed with periods of no therapy [59]. A typical

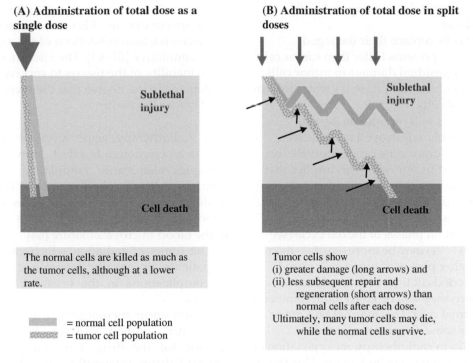

(A) Administration of total dose as a single dose

Sublethal injury

Cell death

The normal cells are killed as much as the tumor cells, although at a lower rate.

= normal cell population
= tumor cell population

(B) Administration of total dose in split doses

Sublethal injury

Cell death

Tumor cells show
(i) greater damage (long arrows) and
(ii) less subsequent repair and
   regeneration (short arrows) than
   normal cells after each dose.
Ultimately, many tumor cells may die,
   while the normal cells survive.

FIGURE 11.7 Greater sensitivity and poorer repairs and regeneration by tumor cell populations as the basis of the greater efficacy of administration of therapies in split doses.

regime became: three daily doses in 1 week followed by 3 weeks "recovery" without treatment.

(c) Principle of Effect

Split-dose regimes take advantage of the poorer repair and regenerative powers of the tumor cells (Figure 11.7). After each part of the total dose, the sublethally injured normal cells are able to recover sufficiently to resist the next or subsequent fractions (see Section A1.4.3). In contrast, the sublethally injured tumor cells will not be able to recover in time and will be killed by the next or subsequent part doses.

For any particular agent, there are no known ways to measure the relative capacities of tumor cell populations and normal populations to recover after each dose. The details of the regimens of the treatment for each agent and each tumor type have been established necessarily by clinical trials (see Section 11.11).

It can be noted, however, that at any time, the doses can begin to have no effect, and the tumor masses may begin to regrow despite therapy (the tumor is said to have acquired resistance, see above).

(d) Factors in Increased Efficacies of Split Doses

Factors responsible for the value of fractionation of doses have been explained by many authors. The following is adapted from [60], and the general phenomena of recovery from injury is described in Section A1.4.

1. Intrinsic sensitivity (of tumor cells versus normal cells) is important, but does not completely explain the greater

benefits of split doses compared with single doses.

2. Normal cells replace their damaged cytoplasmic proteins faster than tumor cells, so that the residual damage to tumor cells at the end of the rest phase is greater than in the normal cells.

3. Normal cells will repopulate (by cell division) foci of destroyed normal cells faster than tumor cells can repopulate foci of destroyed tumor cells. This implies that cell replication mechanisms either are less damaged, or are repaired faster, in normal cells compared to tumor cells.

4. Cells in certain phases of the cell cycle (see Section A1.3.4) may be more sensitive than cells in other phases. If tumor cells spend longer periods of time in those phases of sensitivity, they will be killed more commonly than normal cells. This may not be a major factor for some anticancer agents.

5. In relation to radiotherapy, reoxygenation of the tumor tissue may be a factor. This factor is based on the fact that oxygen increases the killing of cells (see Section 12.1.3). Generally, the centers of solid tumor masses are less oxygenated than normal cells. However, initial doses of radiation may create local inflammation, which increases perfusion of tumor tissues with arterial oxygen concentrations in the blood. This greater blood supply raises the intratumoral oxygen tension, so that later doses of anticancer therapy have a more powerful effect. This may not be a major factor for some types of tumors.

11.5.2 Limitations to Doses for the Individual Patient

Many side effects of anticancer agents are of two kinds.

1. Those which resolve completely when the therapy ceases. These do not usually prevent additional therapy at a later date.

2. Those which are permanent. In this situation, the severity of the side effect increases with successive administrations of the agent, i.e., is cumulative [61–63]. The effect manifests as an inability of the tissues to survive cytotoxic damage to the degree that they were capable of before the therapy.

In radiotherapy, acute reversible inflammation is a common side effect. However, radiotherapies also cause cumulative irreversible damage to tissues. An example of this is nonhealing ulceration of the skin. This side effect appears to be at least partly due to destruction of the blood microvasculature [64].

The limits of radiation which can be given to patients are usually determined by the risk of complications in the organ or site within 5 years.

In anticancer chemotherapy, irreversible side effects are common, and sometimes relatively specific for the drug. For example, nitrogen mustard drugs frequently cause bone marrow suppression, but most patients recover over time. Cisplatin may cause cumulative and irreversible neurotoxicity.

The usual situation is that the limit of cumulative dose for each drug is set by its major side effect [64].

The mechanisms of life-time dose limits are not clear. They may involve the same principles as are involved in the resistance of all tissues to noxins (See Appendix 3).

Thus the therapies reduce one or more of the normal cell properties as follows:

1. Permanent reduction in the supply of nutrients to the tissue
2. Permanent reduction in the defenses of normal cells to intercurrent noxins
3. Permanent reduction in the intracellular repair functions
4. Permanent reduction in the regenerative function of adjacent cells of the same kind.

11.5.3 "Rescue" from Therapy

This term applies to two situations. One is where an antimetabolite is given in high doses. If the patient suffers severe side effects, the situation is "rescued" by giving large doses of the relevant metabolite [65].

The other situation is when high doses of cytotoxic agents are given and the normal bone marrow is severely depleted. The situation can then be "rescued" by transplantation of hematopoietic cells. The patient's peripheral blood contains hematopoietic stem cells, which can be cultured *in vitro* to provide sufficient numbers to repopulate the patient's bone marrow [66]. It is used to "rescue" the patient from the otherwise lethal effects of certain treatments of hematological malignancies. For other malignancies, it has generally been found to be unrewarding [67].

11.5.4 Effects of Agents on Genomic Stability; Potential for Carcinogenic Effect

Certain anticancer agents are themselves carcinogenic. The main example is radiotherapy, because radiations in large doses are carcinogenic in many tissues (see Section 4.3). In general, tumors caused by radiations only appear many years after the rays were administered (i.e., a long latent period, see Section 4.1.4). In current practice, with better targeting of radiations to tumors, few tumors arise as a result of this therapy. The tumors which do arise are most prominent in children treated with radiotherapy [68]. This may be at least partly because of their greater life spans posttherapy compared to adults.

The majority of cytotoxic drugs either affect DNA directly by forming adducts or by interfering with the genome indirectly, for example, via disturbing the polymerization of tubulin (the protein subunit of spindle fibers; see Section 12.2.1). However, few of these drugs are carcinogenic in humans.

As exceptions, certain poisons of topoisomerase II (see Section A2.4.2c), especially etoposide, doxorubicin, and mitoxantrone, are associated with the development of specific types of leukemia [69]. Analogues of etoposide are cytotoxic, but do not bind to topoisomerases [70]. The mechanisms of action of etoposide, and the reasons why it is carcinogenic, are unclear.

References

[1] Laurence Jr. W. History of surgical oncology. In: Norton JA, Barie PS, Bollinger RR, editors. Surgery: basic science and clinical evidence. New York, NY: Springer; 2008. pp. 1889–900.

[2] Halsted CP, Benson JR, Jatoi I. A historical account of breast cancer surgery: beware of local recurrence but be not radical. Future Oncol 2014;10(9):1649–57.

[3] Boot H. Diagnosis and staging in gastrointestinal lymphoma. Best Pract Res Clin Gastroenterol 2010;24(1):3–12.

[4] Kelly KJ, Nash GM. Peritoneal debulking/intraperitoneal chemotherapy-non-sarcoma. J Surg Oncol 2014;109(1):14–22.

[5] Adam R, De Gramont A, Figueras J, et al. The oncosurgery approach to managing liver metastases from colorectal cancer: a multidisciplinary international consensus. Oncologist 2012;17(10):1225–39.

[6] Pollock RE, Doroshow JH, Khayat D, editors. UICC manual of clinical oncology (8th ed.). Hoboken NJ: John Wiley and Sons; 2004.

[7] McKenna Jr. R. Surgical management of primary lung cancer. In: Silberman H, Silberman AW, editors. Principles and practice of surgical oncology. Philadelphia, PA: LWW; 2010. pp. 921–6.

[8] Freedman GM, Fowble BL. Local recurrence after mastectomy or breast-conserving surgery and radiation. Oncology (Williston Park) 2000;14(11):1561–81.

[9] Yeo HL, Paty PB. Management of recurrent rectal cancer: practical insights in planning and surgical intervention. J Surg Oncol 2014;109(1):47–52.

[10] Reyblat P, Skinner DG, Stein JP. Urologic cancers. In: (6). pp. 945–74.

[11] Norton JA, Barie PS, Bollinger RR, et al. Surgery: basic science and clinical evidence. New York, NY: Springer; 2008.

[12] Samuel Waxman S, Anderson KC. History of the development of arsenic derivatives in cancer therapy. Oncologist 2001;6(Suppl. 2):3–10.

[13] Zaffiri L, Gardner J, Toledo-Pereyra LH. History of antibiotics. From salvarsan to cephalosporins. J Invest Surg 2012;25(2):67–77.

[14] Slater JM. From X-rays to ion beams: a short history of radiation therapy. In: Linz U, editor. Ion beam therapy. Berlin: Springer Verlag; 2012. pp. 3–16.

[15] WebElements. Selenium. Available at: http://www.webelements.com/selenium/.

[16] Bignold LP, Coghlan BL, Jersmann HP. David Paul Hansemann: contributions to oncology: context, comments and translations. Basel: Birkhäuser; 2007. p. 304.

[17] Bignold LP. Alkylating agents and DNA polymerases. Anticancer Res 2006;26(2B):1327–36.

[18] Chabner BA, Roberts Jr. TG. Timeline: chemotherapy and the war on cancer. Nat Rev Cancer 2005;5(1):65–72.

[19] Halperin EC, Schmidt-Ulrich RK, Perez CA, et al. The discipline of radiation oncology. In: Perez CA, Brady LW, Halperin EC, editors. Principles and practice of radiation oncology (4th ed.). Philadelphia, PA: LWW/Wolters Kluwer; 2004. pp. 16–17.

[20] de Beaumais TA, Jacqz-Aigrain E. Intracellular disposition of methotrexate in acute lymphoblastic leukemia in children. Curr Drug Metab 2012;13(6):822–34.

[21] Gustavsson A, Osterman B, Cavallin-Ståhl E. A systematic overview of radiation therapy effects in non-Hodgkin's lymphoma. Acta Oncol 2003;42(5–6):605–19.

[22] National Cancer Institute. Adjuvant therapy. Available at: http://www.cancer.gov/dictionary?cdrid=45587.

[23] National Cancer Institute. Neoadjuvant therapy. Available at: http://www.cancer.gov/dictionary?CdrID=45800.

[24] Mauri D, Pavlidis N, Ioannidis JP. Neoadjuvant versus adjuvant systemic treatment in breast cancer: a meta-analysis. J Natl Cancer Inst 2005;97:188–94.

[25] Colleoni M, Goldhirsch A. Neoadjuvant chemotherapy for breast cancer: any progress? Lancet Oncol 2014;15:131–2.

[26] Thompson AM, Moulder-Thompson SL. Neoadjuvant treatment of breast cancer. Ann Oncol 2012;23(Suppl. 10):x231–6.

[27] Moscow JA, Schneider E, Sikic BI, et al. Drug resistance and its clinical significance. In: Hong WK, Bast RC, Hait WN, editors. Holland–Frei Cancer medicine (8th ed.). Shelton, CT: PMPH; 2010. pp. 597–610.

[28] Siddik ZH, Mehta K, editors. Drug resistance in Cancer cells. New York, NY: Springer; 2009.

[29] Teicher BA, editor. Cancer drug resistance. New York, NY: Springer; 2006.

[30] Gladstone M, Su TT. Radiation responses and resistance. Internat Rev Cell Molec Biol 2012;299:235–53.

[31] Multi-drug resistance in CancerZhou J, editor. Methods in Molecular Biology, (Springer Protocols), vol. 596; 2010.

[32] Pinedo HK, Giaccone G, editors. Drug resistance in the treatment of cancer. Cambridge UK: Cambridge University Press; 1998.

[33] Minchinton AI, Kyle AH. Drug penetration and therapeutic resistance. In: Siemann DW, editor. Tumour microenvironment. Chichester, UK: John Wiley & Sons; 2011. pp. 329–52.

[34] Hodkinson PS, Sethi T. Extracellular matrix-mediated drug resistance. In: (27). pp. 115–35.

[35] Tomida A, Tsuruo T. Drug resistance pathways as targets. In: Baguley BC, Kerr DJ, editors. Anticancer drug development. San Diego, CA: Academic Press; 2001. p. 83.

[36] Nais AHW. An introduction to radiobiology (2nd ed.). Hoboken, NJ: Wiley. pp. 156–73.

[37] Hall EJ, Giaccia AJ. Radiobiology for the radiologist, 6th ed. Philadelphia, PA: LWW; 2006. p. 453.

[38] Gorczyca W, Gong J, Ardelt B, et al. The cell cycle related differences in susceptibility of HL-60 cells to apoptosis induced by various antitumor agents. Cancer Res 1993;53(13):3186–92.

[39] Cogle CR. Cancer stem cells: historical perspectives and lessons from Leukemia. In: Allan AL, editor. Cancer stem cells in solid tumors. New York, NY: Springer; 2011.

[40] Burkert J, Wright NA, Alison MR. Stem cells and cancer: an intimate relationship. J Pathol 2006;209(3):287–97.

[41] Antoniou A, Hébrant A, Dom G, et al. Cancer stem cells, a fuzzy evolving concept: a cell population or a cell property? Cell Cycle 2013;12(24):3743–8.

[42] Alison MR, Lim SML, Nicholson LJ. Cancer stem cells: problems for therapy? J Pathol 2011;223(2):147–61.

[43] Aapro MS, Eliason JF, Krauer F, et al. Colony formation in vitro as a prognostic indicator for primary breast cancer. J Clin Oncol 1987;5:890–6.

[44] Farrar WL. Cancer stem cells. Cambridge UK: Cambridge University Press; 2009.

[45] Shimura T. Acquired radioresistance of cancer and the AKT/GSK3β/cyclin D1 overexpression cycle. J Radiat Res 2011;52(5):539–44.

[46] Windholz F. Problems of acquired radioresistance of cancer; adaptation of tumor cells. Radiology 1947;48(4):398–404.

[47] Dacquisto MP. Acquired radioresistance; a review of the literature and report of a confirmatory experiment. Radiat Res 1959;10(2):118–29.

[48] Balmukhanov SB, Yefimov ML, Kleinbock TS. Acquired radioresistance of tumor cells. Nature 1967;216(5116):709–11.

[49] Göker E, Gorlick R, Bertino JR. Resistance mechanisms to antimetabolites. In: Siemann DW, editor. Tumour microenvironment. Chichester, UK: John Wiley & Sons; 2011. pp. 1–13.

[50] Panasci L, Paiement JP, Christodoulopoulos G, et al. Chlorambucil drug resistance in chronic lymphocytic leukemia: the emerging role of DNA repair. Clin Cancer Res 2001;7(3):454–61.

[51] Skov KA. Radioresponsiveness at low doses: hyper-radiosensitivity and increased radioresistance in mammalian cells. Mutat Res 1999;430(2):241–53.

[52] Joiner MC, Lambin P, Marples B. Adaptive response and induced resistance. C R Acad Sci III 1999;322(2-3):167–75.

[53] Wilson TR, Fridlyand J, Yan Y, et al. Widespread potential for growth-factor-driven resistance to anticancer kinase inhibitors. Nature 2012;487(7408):505–9.

[54] Ashworth A. Drug resistance caused by reversion mutation. Cancer Res 2008;68(24):10021–3.

[55] Cancer Research UK. Why plan chemotherapy? Available at: http://www.cancerresearchuk.org/cancer-help/about-cancer/treatment/chemotherapy/plan/why-plan-chemotherapy.

[56] Heilmann HP. Radiation oncology: historical development in Germany. Int J Radiat Oncol Biol Phys 1996;35(2):207–17.

[57] Fletcher GH. Regaud lecture perspectives on the history of radiotherapy. Radiother Oncol 1988;12(4):iii–v. 253-71.

[58] National Health Services (UK). Radiotherapy - Clinical trials. Available at: http://www.nhs.uk/Conditions/Radiotherapy/Pages/clinical-trial.aspx.

[59] Calvert AH. Introduction: a history of the progress of anticancer chemotherapy. Cancer Surv 1989;8(3):493–509.

[60] Nais AHW. An introduction to radiobiology (2nd ed.). Chichester, UK: Wiley & Sons; 1998. pp. 274–95.

[61] Gunderson LL, Tepper JE. Clinical radiation oncology, 3rd ed. New York, NY: Elsevier; 2011. p. 12.

[62] Atkinson Jr. AJ, Huang S-M, Lertora JJL, editors. Principles of clinical pharmacology (3rd ed.). San Diego, CA: Academic Press/Elsevier; 2012. p. 351.

[63] Dunton CJ. Management of treatment-related toxicity in advanced ovarian cancer. Oncologist 2002;7(Suppl. 5):11–19.

[64] Park HJ, Griffin RJ, Hui S, Levitt SH, Song CW. Radiation-induced vascular damage in tumors: implications of vascular damage in ablative hypofractionated radiotherapy (SBRT and SRS). Radiat Res 2012;177(3):311–27.

[65] Cohen IJ, Wolff JE. How long can folinic acid rescue be delayed after high-dose methotrexate without toxicity? Pediatr Blood Cancer 2014;61(1):7–10.

[66] Hopman RK, DiPersio JF. Advances in stem cell mobilization. Blood Rev 2014;28(1):31–40.

[67] Dillman RO, Seagren SL, Taetle R. Failure of low-dose, total-body irradiation to augment combination chemotherapy in extensive-stage small cell carcinoma of the lung. J Clin Oncol 1983;1(4):242–6.

[68] Yock TI, Caruso PA. Risk of second cancers after photon and proton radiotherapy: a review of the data. Health Phys 2012;103(5):577–85.

[69] Pendleton M, Lindsey Jr RH, Felix CA Topoisomerase II and leukemia. Ann N Y Acad Sci 2014;1310:98–110.

[70] Daley L, Guminski Y, Demerseman P, et al. Synthesis and antitumor activity of new glycosides of epipodophyllotoxin, analogues of etoposide, and NK 611. J Med Chem 1998;41(23):4475–85.

L.P. Bignold: Principles of Tumors.
DOI: http://dx.doi.org/10.1016/B978-0-12-801565-0.00012-3

309

The previous chapter indicated the general principles of the treatment of tumors. This chapter gives more specific detail about particular therapies.

Section 12.1 describes issues in radiation therapy, while Section 12.2 discusses particular aspects of cytotoxic anti-cancer chemotherapy. Section 12.3 is concerned with "target-selective" drugs, and Section 12.4 with aspects of personalized therapies. Section 12.5 discusses some other non-surgical therapies, while Sections 12.6 deals with assessing the effects of therapy in the patient. Section 12.7 describes palliative and support care and Section 12.8 discusses alternative therapies in cancer care.

12.1 ASPECTS OF RADIATION THERAPY

The general issues in radiation damage to tissues are described in Section 4.3.1 and in Section 11.5.2. Here, the emphasis is on clinical aspects of radiation therapies.

12.1.1 Units of Therapeutic Radiation and Absorption

In physics, energy can be indicated in terms of the electron volt (eV), being the energy absorbed or lost by the charge on an electron moved by one volt of electric potential difference. In medicine (see Figure 12.1), units of energy

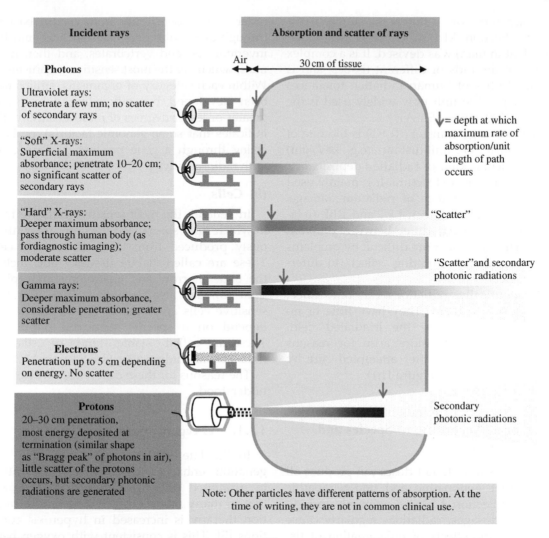

FIGURE 12.1 Absorption, penetration, and scatter for different particles and different energies (EVs) in water.

deposition have been derived in relation to tissue damage [1] reflecting that:

i. Only absorbed energy has any effect on the tissues.
ii. The proportion of the energy in the incident radiations which is absorbed by the tissue or part of the body depends on the kind of tissue as well as the energy carried by the particles of the radiation.

Incident radiation energy is quantitated in terms of the "Roentgen," which is the amount of radiation necessary to create by ionization of one *esu* (electrostatic unit) of electricity in $1\,cm^3$ of air under standard conditions.

Absorbed radiation energy in any structure can be referred to in units called "rads" (radiation absorbed dose), or as more often as "Gray" ($100\,rads = 1\,Gy$).

To account for differences between tissues (see Section A1.1.3), the "rem" (radiation equivalent in man) was devised. It is a complex average of the rads of different tissues, and is used in studies of human radiation tolerances and damage. The unit now widely used is the "Sievert" (100 rem = 1 Sievert).

"Linear energy transfer" (LET) is the rate of energy transfer per unit distance (e.g., keV/μm) that a particular kind of radiation deposits in a particular matter or structure. It is mainly used in experimental studies of radiation damage. The relationships between LET and RBE (relative biological effect) differ between different tissues. This is made more difficult by problems of defining and quantitating "effect" in different biological systems.

Therapeutically, radiations can only affect cells exposed to them. They have little or no effect on cells outside the irradiated field. Recurrences may therefore arise for reasons similar to those seen after attempted cure by surgical resection (see Figure 11.1).

12.1.2 Aspects of Kinds of Damage: Differences According to Species and Kinds of Cells

Radiations in different doses can produce all the different genopathic and non-genopathic effects (see Section 4.1.1b). Commonly in high short-term dosages, radiations regularly cause non-genopathic effects of inflammation of tissues and acute cell death (see Section 10.6.6a). At lower doses given over longer periods of time, radiations cause mild inflammation and tissue atrophy, as well as often the whole range of genopathic effects (see Section 10.2). Many of these effects are variable between the different species which are used experimentally.

(a) Species

There are enormous differences in the radioresistance of species according to evolutionary complexity. There is a well-recognized ascending order of radiosensitivity from certain bacteria, through unicellular organisms such as amoebae, invertebrates, and vertebrates, and then mammals, which are the most sensitive organisms [2]. Within each category of organisms, strains have been developed by in-breeding, which have greater or lesser degrees of radiosensitivity. This indicates that some genomic factor(s), probably acting through a gene product, are associated with radiosensitivity.

(b) Cells

In general, the most sensitive cells are those in which fully specialized versions are continuously produced from local tissue stem cells. These are called "labile tissues," and include the bone marrow and gastrointestinal epithelium [3] (see also Section A1.3.3). The least sensitive cells are those of tissues that do not depend on a specific stem cell population. These are called "stable" kinds of cells, e.g., fibrocytes and adipocytes (see Section A1.3.3).

The reasons for these differences are not well understood.

12.1.3 Oxygen Effects

In the late 1940s, it was found that oxygenation enhances chromosomal aberrations in irradiated cells [4]. Other studies showed that in many models, cell killing due to radiation therapy is increased in hyperoxic conditions [5]. This is consistent with oxygen being the most important atom from which reactive species can be generated by radiation therapy (see Section 4.3). It should be noted that oxygen effects are only seen with radiation therapies. There is no oxygen effect with any of the common cytotoxic drugs used in non-surgical anticancer therapy. This is because these agents do not use oxygen as an intermediary source of toxic products.

Attempts have been made to enhance tumor cell sensitivity by hyperbaric oxygen treatment to the patient. The results have generally been disappointing, although interest for these procedures continues [5].

Hypoxia induces a variety of responses in cells. One is a transcription factor (hypoxia-inducible factor 1, HIF-1) which enhances production of a number of other proteins. Inhibitors of HIF-1 are potentially useful enhancers of radiation-induced cell killing [6].

As another factor, hypoxic cells may lack energy due to deficiencies in the respiratory enzyme cycle. These cells, although not dead, may be sub-viable/under "cell stress" (see Section 10.1.5). For these cells, oxygen therapy may not assist in limiting tumor cell growth.

12.1.4 Hyperthermia as a Possible Adjunct in Radiation and Chemotherapy Distinct from Thermal Ablation

Temperatures in the range of 40°C to 45°C sensitize cells to radiations and chemotherapy [7,8]. For radiation therapy, possible mechanisms may be that

i. The tissue-heating effect of radiations raises the temperature in the tumor masses to the lethal range (above 45°C), and
ii. Increased ionization of atoms occurs because the electron-energizing effects of radiations are greater at higher temperatures.

In relation to drugs, the chemical reactions of these agents with cellular macromolecules may be increased by increased temperatures.

As a different form of treatment, raising tumor temperature over 45°C kills all cells directly, and is used in the procedure known as thermal ablation, to destroy tumors [9], including in situ tumors.

12.1.5 Specific Issues in Radiations Acting on Genomic Stability in Cells

It has long been known that radiations can induce genomic events, such as chromosomal aberrations and heritable phenotypic changes (mutations) [10]. However, the idea that radiation administered to parent cells could lead to progressive genomic instability in daughter cells is relatively recent [11–13]. Genomic instability might usefully increase tumor cell loss in the short term. However, it might later lead to radio-resistant strains in surviving cells, with radiation and chemical resistance, and even overall shorter patient survival [14].

12.1.6 Electron Beam Radiation Therapies

In beams, electrons have only slight penetrance of tissues, and are useful against some skin tumors [15].

Electrons are also used for isotope treatment of thyroid cancers. Iodine^{125} decays into tellurium with the release of eight low energy electrons, but not gamma rays. The same isotope is also used for brachytherapy (direct introduction into tumor masses of sealed sources of radiation) for carcinoma of the prostate.

The isotope is cytotoxic and causes chromosomal aberrations, but is not known to be carcinogenic. This paradox is not fully understood (see Section 4.3).

12.1.7 Nuclear Particle Beams

For a given amount of absorbed energy, nuclear particle radiations cause more tissue damage than photonic radiations. This has created interest in their possible superiority as therapeutic agents against tumors. Machines have been developed to emit protons, neutrons, helium nuclei (α rays), and carbon nuclei in beams. Of these beams, at the time of writing, only proton-beam irradiation is being used regularly for tumor therapy. In the United States, there are 14 operative proton therapy centers and a further 10 under construction [16]. There is a small number in the United Kingdom and elsewhere. Only approximately 1% of cancers are suitable to this mode of treatment. There is little convincing evidence that for many patients, the treatment is significantly more effective than conventional photonic radiation therapy [17].

(a) Protons in Comparison with High Energy Radiations

Protons deposit their energy in broader tracks, but have lower penetrations of tissues than photons [16–19] (see Figure 12.2). The benefit of beams of photons over beams of high-energy photons lies in the fact that in tissues, protons deliver energy with a peak at a particular depth (often referred to as a "Bragg peak"), while electron beams deposit energy in a slowly attenuating distribution with depth (see Figure 12.1). When used for treating tumors, the proton beams are adjusted so that peak energy deposition occurs in the tumor. The existence of the peak results in relatively little damage to the tissues deep to the tumor.

As mentioned above, photons cause greater damage to tissues compared with electrons for the same amount of deposited energy. This phenomenon can be understood in terms of the absorption of energy by the two forms of radiation. Because photons have low masses, the ejection of an electron from an outer shell of an atom probably requires approximately eight photons of sufficient energy moving in the same direction to strike within a short period of time (Figure 12.2). The energy of these photons is absorbed in the one ionization event. Other ionizing events are scattered at random in the tissue.

On the other hand, due to their relatively high mass, protons only lose part of their energy in an ionizing event. After one such event, a proton can continue in its path causing multiple lesions in a line. This line can include sites in the one macromolecule, so that this clustered damage causes greater effects (Figure 12.2). As a result, proton beams inflict more chromosomal damage and more lethality per unit energy deposited compared to photonic radiations [16].

All nuclear particle rays can displace nuclear particles from irradiated atoms, especially "secondary neutrons." The significance of this is at present unclear [20].

(b) Neutrons

These rays, especially with high energy levels ("fast neutrons"), have less penetration of tissues, but inflict more damage on tissues for the energy deposited than X-rays. In clinical trials, neutron beams produced more damage to normal tissues, without better effects on tumor tissues, than X-rays [21].

(c) Helium Nuclei (α Rays)

These rays have very low penetration of air and tissues, and are used in clinical practice via administered isotopes rather than as beams [22].

(d) "Heavy" Ions: Carbon Nuclei

These are not currently in wide use, and are undergoing trials in a few centers only. At the present time, their superiority over X-ray therapies is unclear [19].

12.1.8 Radiosensitizers and Protectors

A variety of physical agents (e.g., heat—see Sections 11.4.5 and 12.1.4) and chemicals have been suggested to assist cell killing by radiation therapy. They include conventional cytotoxic drugs, as well as those which act through hypoxia-related phenomena [5,23]. Another group comprises inhibitors such as misonidazole, of radical-scavenging chemicals, especially thiols. Recently, metal-containing nanoparticles have been studied in this role [24].

12.1.9 Aspects of Applications in the Clinic

(a) Anatomical Precision

The principle of radiation therapy is to deliver the maximum dose of radiation to the

Electron shell level

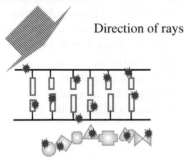

Outer electron energy level of atom

Additive effects of at least eight high energy photons (as in X-rays and gamma rays) are required to ionize an atom (in biology, mainly of oxygen)

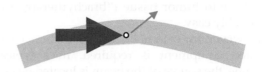

A single energized nuclear particle is sufficient to ionize an atom (relevant to biology, mainly of oxygen)

Molecular level

Direction of rays

Direction of rays

The energy of the photons is absorbed by the ionizing event, so that these photons cause little further damage.
 Low numbers of ionizing events in a molecule may not affect its structure, so that the function of the complex of which it is part may not be affected.

The energy of the proton is only slightly reduced by the ionizing collision, so that it continues in a path causing further ionizations.
 High numbers of ionizing events in a molecule are likely to affect its structure, so that the function of a complex of which it is part may be affected.

FIGURE 12.2   Basis of different biological effects of photonic and nuclear particle radiations.

tumor cells for the minimum damage to normal tissues.

 The precision of delivery of the radiations depends on accurate prior knowledge of the anatomical extent of the tumor. This involves understanding of the shape of the tumor mass in three dimensions. However, the demarcation between the tumor and the adjacent apparently normal tissue may be blurred because there may be micro-metastases and intravascular tumor deposits beyond the macroscopic (and imaging) margins of the main tumor mass (see Section 1.3.4 and Figure 12.3).

In practice, anatomical precision of radiation delivery from inserted pellets of radioisotopes directly into tumor tissue ("brachytherapy") is reasonably easy.

However, for external beam irradiation, complicated equipment is required. In a typical machine, the source of the beam is located on circular gantry, can be rotated around the patient's body, and as well, can be moved up and down (toward the head or feet). The beam source can also be angled in any direction within the "cylinder" of the head-to-foot movement of the gantry. Thus, potentially, every part of the body can be irradiated from any angle. The beam can be modified in various ways, and changed in shape (mainly made more oval) by alterations in the collimator of the beam source [25].

The most modern machines have a computerized tomographic ("CT") or magnetic resonance imaging ("MRI") function added to the radiation-source function. With these, it is possible to deliver radiations of the required dose according to the detailed three-dimensional images of the tumor in the patient.

(b) Treatment of Regular Side Effects

Regular local side effects occur because damage to untargeted adjacent tissue is a regular side effect of all radiation therapies. Damage to normal tissue can occur

i. Between the beam source and the tumor,
ii. Beyond the tumor, and also
iii. To the sides of the tumor, by way of "scatter" of radiations.

Regular systemic side effects occur because most forms of cytotoxic damage to parts of the body result in tissue breakdown products entering the circulation. These substances can cause mild malaise and nausea (see Section 10.4), for example in myocardial infarction. In radiation therapy, fatigue, nausea, vomiting and malaise occur in all patients, and often in severe degrees [26]. The mechanisms are not understood. Radiation-killed cells and tissues

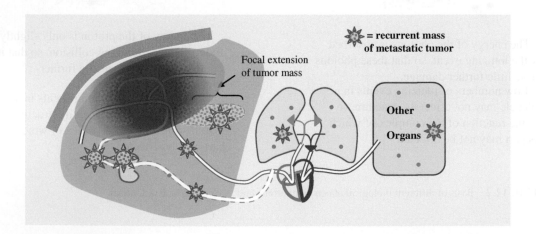

Focal extension of tumor mass

= recurrent mass of metastatic tumor

Other Organs

In principle, this "ablative" radiotherapy has similar limitations to those of surgery: tumor outside the field of irradiation is little affected.

FIGURE 12.3 Sources of recurrences after radiation therapy.

may release products of greater potency than from un-irradiated dead cells, due to the degrees of denaturation of the substances which may occur. Various inflammatory mediators, including 5-hydroxytryptamine [27], may play roles in these side effects.

Various anti-nausea drugs and corticosteroids are often given for relief of these symptoms. They do not affect the efficacy of the radiation treatment.

(c) Limitations to Total Doses of Radiation Therapy

These can be classified as local and systemic (see also Section 11.5.2). Local limiting factors to total dose of radiation therapy mainly relate to the site of the tumor. In cases in which the tumor is close to a vital organ or structure in the path of the proposed radiation beams, too much radiation may damage the vital structure. Any concurrent disease in those organs (e.g., previous glomerular disease in a kidney in or near the radiation field) must be taken into account.

Systemic limiting factors to total possible therapeutic dose include the presence of any other diseases, such as emphysema, heart disease, or renal failure, which would make the patient less able to tolerate radiation therapy.

(d) Protection of Radiation Therapy Staff

Staff members working in radiation therapy departments are often long-term employees. All staff members have protective shielding when in treatment areas, and wear exposure monitors when at work. The question of acceptable long-term exposure to radiations is an important, but to a degree controversial, issue (see e.g. [28]).

12.2 ASPECTS OF CYTOTOXIC ANTI-CANCER CHEMOTHERAPY

Historical and general issues including activation and resistance in anti-cancer drug therapy

were described in Sections 1.4.6 and 11.3. This section takes up some aspects in more detail.

12.2.1 Differences in Chemical Structures and Mechanisms of Effects of Cytotoxic Drugs

Cytotoxic anti-cancer drugs, like many carcinogens, react with many macromolecules in the cell and, as noted in Appendix 3, are subject to tissue and cellular resistance factors before they reach their target(s).

Anti-cancer drugs range from small molecules such as arsenic trioxide and *cis*-platin through chain structures such as "mustard" alkylating agents to complex multi-ring structures, such as taxanes. Most of these drugs appear to have "non-genopathic" and "genopathic effects" (Figure 12.4 and Sections 4.1.1, 10.1, and 10.2). Neither of these kinds of effects are predictable on the basis of their chemical natures or structures [29].

It seems difficult to understand how such chemical diversity in drugs can be consistent with cytotoxic reactions involving only DNA. Rather, the diversity may be indicated as a large number of different "targets" of the anti-cancer agents in the cell, which might better correspond to the large number of "sites of effect" on the various proteins which process DNA.

This subsection reviews the chemical and therapeutic effects of each of the major kinds of anti-cancer agent.

(a) Small-Molecule Drugs

i. *Arsenic trioxide*

This compound was used as an anti-cancer agent in the nineteenth century (see Section 11.3.1) and is now used in the treatment of promyelocytic leukemia. It acts on multiple enzymes, especially those metabolizing glutathione, in cells [30,31].

ii. *Hydroxyurea and nitrosoureas*

This drug has a hydroxyl group on one of the nitrogen atoms of a urea molecule. In the body, the drug scavenges tyrosyl

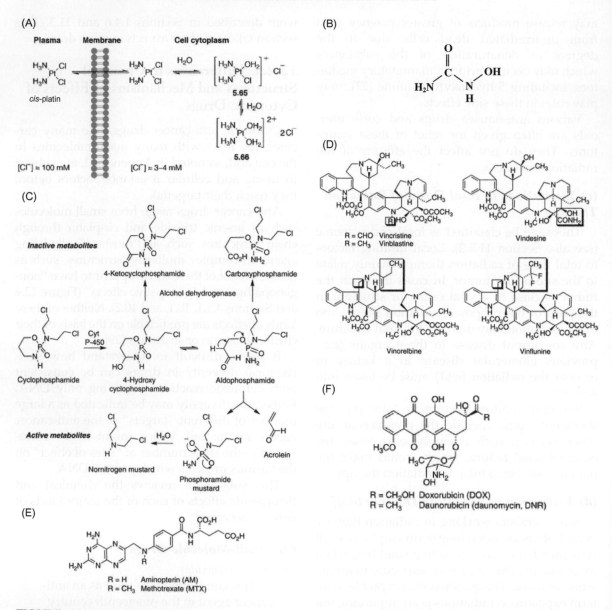

FIGURE 12.4 Structures of some commonly used anti-cancer drugs. (A) cis-platin bioactivation. (B) Hydroxyurea. (C) Cyclophosphamide showing bioactivations and inactivations. (D) Vinca alkaloids. (E) Aminopterin and Methotrexate. (F) Doxorubicin and Daunomycin. (G) Actinomycin. (H) Semisynthesis of taxanes. *Reproduced with permission. Source: Ref [51].*

free radicals, resulting in inhibition of conversion of ribonucleotides to deoxyribonucleotides, and hence the synthesis of DNA [32]. On this basis, it

may be classified as an anti-metabolite (see "Methotrexate," below). Any agent that affects the enzymes responsible for DNA or RNA synthesis can be included as a

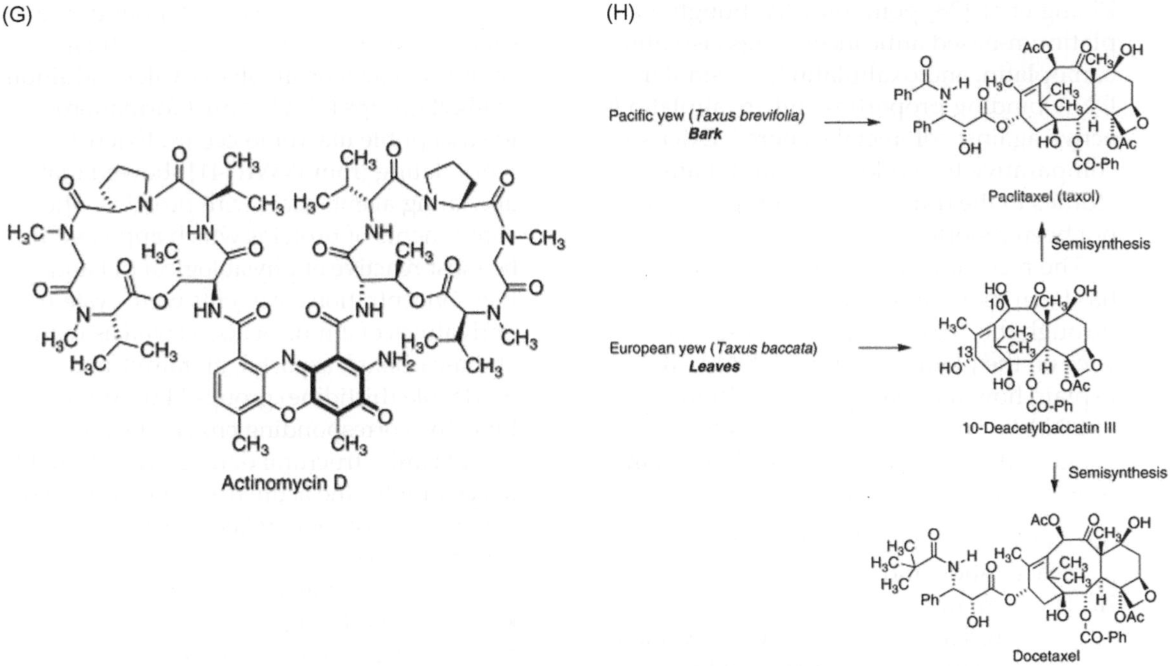

(G)

(H)

FIGURE 12.4 (Continued). Some synthetic modifications in the Taxol group of drugs are included.

"genopathic agent" by the definition used in this book (Section 4.1.1).

Hydroxyurea may well have other actions, because it has beneficial effects in patients with sickle cell anemia [33]. These effects are unexplained, but are unlikely to be caused by inhibiting DNA synthesis.

Nitrosourea drugs have a nitroso group on one nitrogen atom of urea. The various analogues in the group have additional groups on the other nitrogen atom. They react with DNA and are classed as alkylating agents. These drugs pass the blood–brain barrier, and are currently one of the "first-line" therapies for glioblastoma [34].

iii. Cis-*platin*

This is one of the most widely used alkylating drugs in anti-cancer therapy (Figure 12.4A). It was discovered in 1965, and the clinical trials of its usefulness as an anti-cancer drug began in the 1970s. It consists of a platinum ion with two

methyl and two chloride groups attached. The active drug has the chloride groups adjacent to each other (*cis* form), while the inactive isomer (*trans*-platin) has the chloride ions on opposite sides. Both isomers enter the cell where the chloride ions dissociate and both form adducts on DNA (the position of the methyl groups then being the only difference between the active drug and its inactive analogue). Platin drugs do not enter into tumor cells more than they penetrate normal cells. The greater sensitivity of tumor cells compared with normal cells is not explained [35,36].

Clinically, the drug is superior to nitrogen mustard alkylating agents because, although it causes more nausea and vomiting, it causes less suppression of the bone marrow. However, its side effects are numerous and complex [37].

There are tumor type-specific differences in sensitivity to *cis*-platin, for example, as

Zhang et al. [38] point out: "Although the platinum-based anticancer drugs cisplatin, carboplatin, and oxaliplatin have similar DNA-binding properties, only oxaliplatin is active against colorectal tumors." Extensive comparative toxico-kinetic and dynamic studies in the different tumor types have not been reported.

The mechanism of action of *cis*-platin has been thought to involve cross-linking through the "naked" chloride-dissociation sites on the platinum ion with DNA. To explain how the *trans*-platin can bind to DNA but not kill cells, it has been suggested that *trans*-platin inactivity stems from two major factors:

i. The kinetic instability promoting its deactivation and
ii. The formation of DNA adducts characterized by a regioselectivity and a stereochemistry different from those of *cis*-platin [39].

It is known that barely 1% of nuclear *cis*-platin is bound to DNA, the remainder to unspecified nuclear proteins. It is clearly possible that *cis*-platin acts also on proteins associated with processes relevant to the genome, but little information by way of detailed studies of this possibility, especially in comparison to its inactive *trans*-analogue, is available.

(b) Intermediate-Sized Drugs

i. *Nitrogen mustard alkylating agents*

These agents have been studied in considerable detail since the 1940s [40]. The proto-alkylating agent was the WWI poison gas *bis*(2-chloroethyl) sulfide, which smells like mustard but has no other relationship to that plant. The gas has acute irritant effects on exposed tissues, but also causes reversible bone marrow suppression in exposed individuals. Analogues in which a nitrogen atom replaces the sulfur atom are less acutely toxic, but retain cell-killing effects even in small systemic doses. The compounds are called "alkylating agents" because their chemical reactions involve covalent addition of alkyl groups (alkyl adduct formation) to susceptible macromolecules. Extensive work dating from WWII [41] showed that alkylating agents react with proteins. The components of proteins which appeared to be most reactive at physiological pH and low concentrations of agents were cysteinyl derivatives of amino acids, as well as with alpha-carboxyl, aspartyl, glutamyl, and imidazole (histidine) groups [42]. At the time, the corresponding enzymatic and cytoplasmic structural damage was thought to account for the acute toxic effects of these agents, but not for any late effects such as carcinogenesis.

In the 1960s, the mechanisms of cell killing by alkylating agents were often considered to involve direct covalent reactions with DNA, especially the *N7* site on guanine, as well as lesser reactivities with other sites on guanine and on other bases.

While this idea is repeated in many books as the probable mechanism of action of alkylating agents, there is evidence that the important actions of many active anti-cancer drugs are not with DNA. For example, chlorambucil (4-[bis(2chloretyl) amino]benzenebutanoic acid) alkylates DNA but is effective in chronic lymphocytic leukemia, in which cell proliferation is not a marked feature. Chlorambucil may induce apoptosis via actions on non-DNA structures such as proteins, in the cell [43].

(c) Large-Molecule Drugs

i. *Methotrexate*

This drug is classified as an "anti-metabolite" because it interferes specifically with a vital metabolic process in the cell (Figure 12.4C). Methotrexate is an analogue of folate, and is taken up into the cytoplasm of cells through the low pH folate transporter or reduced folate carriers, where it then inhibits the specific enzyme dihydrofolate reductase [44]. As a result, the cell has reduced capacity to synthesize new purine and thymidine nucleotides, as well as proteins. However, the drug is also thought to inhibit purine metabolism, as well as inhibiting methyl transferase activity, with secondary effects on immune functions [45]. The main side effects of the drug occur in labile tissues (those which constantly produce specialized cells from local tissue stem cells—hematologic, gastro-intestinal, see Section A1.3.2). Hepatic and renal toxic effects also are common, presumably reflecting defensive factors (Sections A3.2, A3.3). There appears to be no selective uptake of methotrexate into tumor cells, and the greater susceptibility of tumor cells to methotrexate over normal cells may relate to lower concentrations of the enzyme in tumor cells (Section A4.2.1). Methotrexate does not enter the nucleus. Apart from the inhibition of DNA synthesis, this drug has no genopathic effects such as the induction of chromosomal aberrations [46] (and see Sections 4.1.1 and 10.2).

ii. *Anti-microtubule agents*

These drugs do not react with DNA, but rather affect the genomic process-related proteins, specifically for mitosis, the centrosome and the mitotic spindle. These drugs prevent either microtubule formation or the dissociation of microtubules to individual tubulin molecules [47]. Examples of the first of these groups are the *Vinca* alkaloids, vincristine, and vinblastine (Figure 12.4D). These were discovered in the 1950s and found to have clinically useful anti-tumor activities. These drugs are now known to react with soluble tubulin in the cytoplasm of cells. Vincristine and vinblastine disrupt the microtubules which form part of the spindle of fibers used by the chromosomes to separate in telophase. Mitoses, therefore, are disrupted, and cells die by "mitotic death" (see 10.6.3) or nuclear disruptive events [48]. However, these drugs also affect other

proteins, and have been reported to interfere with DNA, RNA, and lipid synthesis [48].

In contrast to drugs that de-stabilize the tubulin polymers, agents including the taxanes, epothilones, and discodermolides [49] "stabilize" (prevent the physiological dissociations of) the microtubules and arrest cells in the G2-M part of the cell cycle.

iii. *Anti-tumor drugs related to antibiotics*

Actinomycin D (dactomycin) (Figure 12.4G) was isolated from *Actinomyces antibioticus* in the 1940s, during searches for antibacterial substances in microorganisms. Actinomycin D was found to have anti-tumor activities, but is not used for any antibacterial purpose. It is commonly used in anti-cancer therapies against a limited range of relatively rare tumor types, including choriocarcinoma of the placenta, Wilm's tumor, and Ewing's sarcoma [50]. Its main side effects are against labile cell types, such as the hematologic, gastro-intestinal, cutaneous, and immune systems. The drug has a weak potency for causing chromosomal aberrations (i.e., a weak clastogen, see Section 4.1.1) and is not known to be carcinogenic other than perhaps causing leukemia through topoisomerase inhibition (see mention of etoposide, below).

The mechanism(s) of action of actinomycin D are not fully understood [51]. Its effect may be via inhibition of DNA-dependent synthesis of RNA (i.e., RNA polymerases; transcription) through its noncovalent binding to specific DNA motifs near transcriptional complexes. It may bind the minor groove of DNA and/or have an "interfacial" (i.e., intercalatory, noncovalent) action on the DNA–RNA polymerase sites [51].

Other anti-tumor antibiotics that probably have actions on genomic process-related proteins include the anthracyclines (doxorubicin and daunorubicin) (Figure 12.4E), which probably mainly act on the functions of topoisomerases in cells by forming a stable DNA–drug–enzyme complex [52]. Etoposide (an analogue of podophyllotoxin) acts on topoisomerase II, especially when the enzyme is complexed with DNA. The drug reacts only weakly with DNA in the absence of the enzyme. The mechanisms by which etoposide might cause chromosomal aberrations are discussed in Ref. [53].

12.2.2 Activation of Pro-drugs to Active Compounds

Just as for some chemical carcinogens (Section 4.4.1), some drugs are given as pro-drugs, and are activated in the body. The activating enzymes are mainly members of the CPY/P450 family of enzymes, which activate carcinogens, and also deactivate toxins (Sections 11.4.5 and A3.1.3).

Examples include anti-metabolites such as 5-fluoruracil (5-FU) and mercaptopurine; alkylating agents such as melphalan and cis-platin; antibiotics such as mitomycin C; as well as others such as cyclophosphamide (Figure 12.4B), etoposide, and paclitaxel [54] and (Figure 12.4F). The enzymes involved are mainly the P450 cytochromes, and other oxidoreductases and transferases [54].

12.2.3 Unexplained Differences in the Potencies for a Variety of Biological Effects among Different Analogues in the Same Chemical Class

Just as for carcinogens (see Section 4.4.2), wide differences in potencies occur between analogues of anti-cancer drugs, as noted for example in Ref. [40]. The differences in efficacies are often considered to be based on variable effectiveness of defensive factors of the tissues and cells against the analogues (see Appendix 3). However, many biologically inactive analogues are known to reach the genome (evidenced by forming adducts on DNA, for example, cis-platin [55]. Hence, it would seem that exquisite structural sensitivity in some non-DNA target could be involved, as is suggested in Section 4.4.2 for carcinogenic analogues versus non-carcinogenic analogues.

Predicting biological activity from chemical structures is recognized to be complex [56,57].

12.2.4 Multiplicity of Targets: "Polypharmacology"

As in the explanation of the actions of carcinogens (see Section 1.2.3), the facts that many individual drugs have different cell biological effects in different normal cells suggest that these drugs may have effects on multiple different "targets," and hence modes of action, e.g. [58].

Alkylating agents, for example, are simply highly reactive substances capable of binding to a variety of molecules, and thus are liable to have many effects, probably dependent in detail on pharmacokinetic factors.

This phenomenon is now fully recognized in the term "polypharmacology." It represents a challenge in drug development, which may be overcome by better drug designs [59,60].

12.3 "SELECTIVE"/"TARGETED" DRUGS

12.3.1 General

Up until the last few decades, drugs were evaluated by their ability to kill tumor cells with relative sparing of normal cells. The "targets" were thought to be general kinds of chemicals, such as DNA, and it was assumed that the same targets may be present in different kinds of cells. This was especially because toxic side effects, for example, on the cells of normal bone marrow, hair follicles, and nerves occur so regularly.

In recent decades, there has been increasing knowledge of specific molecules that are often overactive in tumor cell populations (Section 6.3). It has been found that different cases of the same tumor type may have different disturbances in these growth factors and pro-growth pathways. Because of this, search for drugs that can selectively inhibit biochemical targets in tumor cells have been undertaken. These drugs also produce relatively fewer side effects, and have been called "non-toxic" anti-cancer drugs [61–63].

12.3.2 Selective Delivery of Cytotoxic Agents

The term "selective targeting" can also refer to chemical modifications of drugs so that they are only taken up into selected cells [64,65].

Antibodies were one of the earliest agents for targeting drugs to specific cells [66,67]. While the principle of antibody-mediated damage to cells is well recognized, the efficacies of these agents are affected by factors related to their targeting and pharmacokinetics. These factors include that tumor-specific antigens may be difficult to identify and may only be variably expressed by the tumor cells. Further, antibody conjugates, being abnormal proteins by definition, may be removed from the bloodstream by the reticuloendothelial system (Section A1.1). In addition, the antibody recognition site may bind to the cancer cell but the conjugate may not enter the cell. This may occur especially in poorly specialized tumors that may lack effective endocytosis mechanisms (see Section 10.1.4).

12.3.3 Antibodies against Specific Cell Surface Receptors

These are mainly antibodies directed against cell type-specific proteins that act as receptors for signaling pathways on the surfaces of cells [68–70] (see Table 12.1). They are all named with the suffix "-mab".

Among the most extensively studied at the present are

i. Trastuzumab. This is an antibody against the HER-2 receptor (see Section 6.3.2), and is used for patients with breast carcinomas which express this surface protein.

ii. Rituximab is an antibody against the surface receptor known as CD20. This antigen is expressed almost exclusively by B lymphocytes, and hence the drug is useful against B-cell tumorous conditions including leukemias and lymphomas, as well as in some autoimmune conditions.

12.3.4 Drugs against Intracellular Signaling Enzymes

The importance of target-selective drugs is that there are a significant number of different targets that are suitable for blocking by drugs [61,71–74]. They mainly are named with the suffix "-ib".

To affect these targets, these drugs must enter cells, must be small, and must be able to access the functional part of the target which can then be inhibited by a drug.

The most established drugs in this group are imatinib for gastrointestinal stromal tumor (GIST), and vemurafenib (a BRAF protein inhibitor) for late-stage melanoma.

Particular proteins can be inhibited, for example, the heat shock protein Hsp90 by geldanmycin, a product of *Streptomyces hygroscopicus*.

12.3.5 Difficulties in Drugging Certain Targets

Many biochemical activities depend on small "pocket" enzymatic sites (see Figure 12.5). These are druggable in the sense that they are susceptible to being blocked by molecules that are able to enter the cells. However, some biochemical activities depend on large areas of interaction of proteins, especially involving

TABLE 12.1   Anti-Cancer Drugs Approved for Common Tumors by the FDA since January 2012

These quotations indicate the circumstances in which the drug is approved for use. Unapproved uses of the drug are often termed "off-label" uses, and costs may not be reimbursed by insurers.

(For the website and for updates, see [70]; for a list of targeted therapies for various cancers, see [70].)

LUNG CARCINOMA

2014.   "Ramucirumab (Cyramza Injection, Eli Lilly and Company) for use in combination with docetaxel for the treatment of patients with metastatic non-small cell lung cancer (NSCLC) with disease progression on or after platinum-based chemotherapy."

2014.   "Ceritinib (ZYKADIA, Novartis Pharmaceuticals Corporation) for the treatment of patients with anaplastic lymphoma kinase (ALK)-positive, metastatic non-small cell lung cancer (NSCLC) with disease progression on or who are intolerant to crizotinib."

2013.   "Crizotinib (Xalkori, Pfizer, Inc.) capsules for the treatment of patients with metastatic non-small cell lung cancer (NSCLC) whose tumors are anaplastic lymphoma kinase (ALK)-positive as detected by an FDA-approved test."

2013.   "Afatinib (Gilotrif tablets, Boehringer Ingelheim Pharmaceuticals, Inc.), for the first-line treatment of patients with metastatic non-small cell lung cancer (NSCLC) whose tumors have epidermal growth factor receptor (EGFR) exon 19 deletions or exon 21 (L858R) substitution mutations as detected by an FDA-approved test. The safety and efficacy of afatinib have not been established in patients whose tumors have other EGFR mutations. Concurrent with this action, FDA approved the therascreen EGFR RGQ PCR Kit (QIAGEN) for detection of EGFR exon 19 deletions or exon 21 (L858R) substitution mutations."

2013.   "Erlotinib (Tarceva, Astellas Pharma Inc.) for the first-line treatment of metastatic non-small cell lung cancer (NSCLC) patients whose tumors have epidermal growth factor receptor (EGFR) exon 19 deletions or exon 21 (L858R) substitution mutations. This indication for erlotinib is being approved concurrently with the cobas EGFR Mutation Test, a companion diagnostic test for patient selection."

BREAST CARCINOMA

2015.   "Palbociclib (IBRANCE, Pfizer, Inc.) for use in combination with letrozole for the treatment of postmenopausal women with estrogen receptor (ER)-positive, human epidermal growth factor receptor 2 (HER2)-negative advanced breast cancer as initial endocrine-based therapy for their metastatic disease."

2013.   "Pertuzumab injection (PERJETA, Genentech, Inc.) for use in combination with trastuzumab and docetaxel for the neoadjuvant treatment of patients with HER2-positive, locally advanced, inflammatory, or early stage breast cancer (either greater than 2 cm in diameter or node positive) as part of a complete treatment regimen for early breast cancer."

2012.   "Everolimus tablets (Afinitor®, Novartis Pharmaceuticals Corporation) for the treatment of postmenopausal women with advanced hormone receptor-positive, HER2-negative breast cancer in combination with exemestane, after failure of treatment with letrozole or anastrozole."

2012.   "Pertuzumab injection (PERJETA, Genentech, Inc.) for use in combination with trastuzumab and docetaxel for the treatment of patients with HER2-positive metastatic breast cancer who have not received prior anti-HER2 therapy or chemotherapy for metastatic disease. Pertuzumab is a recombinant humanized monoclonal antibody that targets the extracellular dimerization domain (Subdomain II) of HER2, and thereby blocks ligand-dependent heterodimerization of HER2 with other HER family members, including EGFR, HER3 and HER4."

COLORECTAL CARCINOMA

2013.   "Bevacizumab (Avastin, Genentech U.S., Inc.) for use in combination with fluoropyrimidine-irinotecan or fluoropyrimidine-oxaliplatin based chemotherapy for the treatment of patients with metastatic colorectal cancer (mCRC) whose disease has progressed on a first-line bevacizumab-containing regimen."

(Continued)

TABLE 12.1 (Continued)

2012.	"Regorafenib (Stivarga tablets, Bayer HealthCare Pharmaceuticals, Inc.), for the treatment of patients with metastatic colorectal cancer (mCRC) who have been previously treated with fluoropyrimidine-, oxaliplatin-, and irinotecan-based chemotherapy, an anti-VEGF therapy, and, if KRASwild type, an anti-EGFR therapy."
2012.	"Ziv-aflibercept injection (ZALTRAP, Sanofi U.S., Inc.) for use in combination with 5-fluorouracil, leucovorin, irinotecan (FOLFIRI) for the treatment of patients with metastatic colorectal cancer (mCRC) that is resistant to or has progressed following an oxaliplatin containing regimen."
2012.	"Cetuximab (Erbitux, ImClone LLC, a wholly owned subsidiary of Eli Lilly and Co) for use in combination with FOLFIRI (irinotecan, 5-fluorouracil, leucovorin) for first-line treatment of patients with K-ras mutation-negative (wild-type), EGFR-expressing metastatic colorectal cancer (mCRC) as determined by FDA-approved tests for this use."

PROSTATE CARCINOMA

2012.	"Abiraterone acetate (Zytiga Tablets, Janssen Biotech, Inc.) in combination with prednisone for the treatment of patients with metastatic castration-resistant prostate cancer."
2012.	"Enzalutamide (XTANDI Capsules, Medivation, Inc. and Astellas Pharma US, Inc.), for the treatment of patients with metastatic castration-resistant prostate cancer who have previously received docetaxel."

Most targets are active ligand–receptor sites (for signaling) or enzymic sites (for metabolism)

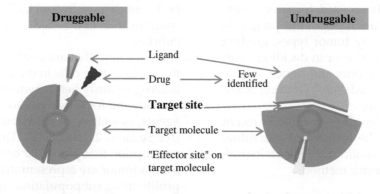

Most of the undruggable sites are relatively large. They are less affected by molecules of sufficiently small size to enter the cell.

FIGURE 12.5 Druggable and undruggable targets.

weak intermolecular bonds. These sites may be too large to be affected by molecules attaching to only a small part of their overall dimensions. For some biochemical actions, this problem may be overcome by a small molecule that binds to a secondary site on the target so that a conformational change is caused [75].

12.4 ASPECTS OF PERSONALIZED MEDICINE

12.4.1 Background

The term "personalized medicine" in relation to tumors is used in three contexts [76–79]:

i. Examining the tumor for characteristics that may assist in choices of the most appropriate treatment for the particular patient.

ii. Assessing the patient's normal biochemical and genomic characteristics to establish which modes of therapy may be most effective in them, and/or from which they may suffer the least side effects.

iii. "Personalized disease prevention." By this is meant calculations in currently healthy individuals, of the risks of future disease, and hence the steps that might be taken to avoid these illnesses. This is described in Section 14.3.5.

12.4.2 Studies of the Patient's Tumor Cells

As noted in Section 1.3.2, grading of histological features has been used for over a century to give an indication of overall prognosis of the patient. For many tumor types, grading is currently used as one factor in deciding treatment protocols for patients and hence, is in a sense, "personalized medicine." Grading is not mentioned further here.

This subsection describes the more recent efforts to quantitate the features of individual cases of malignant tumors by cell biological, biochemical, and genomic methods.

(a) Cell Biological and Pharmacological Assessments of Cells from the Patient's Tumor

One of the general features in the last half century of cancer research has been that no test has been developed for assessing the drug sensitivity of cases of tumor, analogous to the tests in general use in the management of bacterial diseases. In those tests, cultures of bacteria from lesions of patients are tested in vitro with antibiotic drugs for relative bactericidal effects.

When attempted, the growth of tumor cells may be assessed either in in vitro cultures or as grafts into immunodeficient animals [80–82]. The main difficulty has been that no cell lines suitable for the purpose can be grown from most cases of tumors. This is especially because only small amounts of tissue might be available from certain sorts of biopsy such as needle biopsies [82].

In studies in which cultures of cell lines have been successful, there has been little correlation between the sensitivity to anti-cancer agents of cells grown from a tumor and the "responsiveness" (shrinkage) of the tumor in the patient when the same agent(s) is administered.

Reasons may include:

i. The micro-environmental situation of tumor cells—and hence pre-target resistance factors/pharmacodynamics (see Section 11.4.1 and Appendix 3)—in the whole patient or at the relevant site in the body are chemically and biologically more complex than in the media used for in vitro culture.

ii. In tumor lines that are established, the proliferating cells in the tumor are heterogeneously heterogenizing for cell kinetics and for sensitivity to anti-cancer agents, as well as for many other properties (see Sections 2.8 and 8.3.2). Thus the assumption that particular cell cultures from a tumor are representative of all the proliferating subpopulations (present and future) in the tumor mass as a whole is fundamentally unwarranted [83,84]. Hence, even if the factors above are not acting in a particular tumor, the proliferating cells in the tumor masses in a patient may well produce new mutants with different properties by the time the studies of the sensitivities of the cell line cultured earlier become available. This is particularly true of the development of more hyperploid cell types being associated with increasing numbers of copies of defense genes against anti-cancer drugs.

(b) Molecular Assessment of the Patient's Tumor for "Target Profile"

With the discovery of the cell signaling mechanisms (Sections 6.3 and 6.4), it has been possible to assess molecular aspects of individual tumors [85,86].

Current examples of "molecular profiles" being used to determine therapy include:

i. *Lung carcinoma*

The gene expression profiles so far studied have been too variable to be used in clinical applications [87,88].

ii. *Breast carcinoma*

Estrogen, progestogen, and HER2 receptor status of breast cancers remain the principal molecular data used in assessing best therapy. Gene expression profiles, although studied, have not proven sufficiently regular to provide reliable therapeutic guidance or prognostic information [89,90].

iii. *Colorectal carcinoma*

Studies to date have focused on microsatellite instability, chromosomal lesions, and DNA methylation profiles. The status and relative value of these data are unclear [91,92].

iv. *Prostatic carcinoma*

Studies of molecular profiles in cases of this kind of tumor have been carried out [93]. The main treatment options for this malignancy relate to expression of androgen receptors. It is not clear, however, whether these data are more valuable than histological grading (Gleason) in predicting responses to anti-androgen therapies. Further, most cytotoxic drugs and drugs selective for cell signaling (Section 12.3.4) have little effect on this type of malignancy.

v. *Other*

Molecular profiling has been carried out in a variety of other tumors, especially malignant melanoma [94] and squamous cell carcinomas of the head and neck [95], without definite prognostic or therapeutic data emerging.

12.4.3 Patients' Normal Genomes and Therapy (Pharmacogenomics)

Another aspect of personalized medicine is attempting to assess the likelihood of a particular side effect of an agent in the individual person on the basis of their genome [96–98]. As is shown in Section 13.2, potential drugs are tested for unacceptable side effects in "Phase 1" trials. Hence, genomic variations that do not fail the Phase 1 trials by causing these side effects should not be very common in the general population.

Most prescription drugs are metabolized in the body by the cytochrome p450 family of enzymes (see Section 12.2.2). Different enzymes metabolize different drugs. An individual might have excess metabolizing capacity, in which case the drug will be ineffective. On the other hand, another individual might be deficient metabolizing the drug, in which case, effects of overdosing may occur.

With the ability to sequence large amounts of DNA, it has also been suggested that the information about the patient's genome may assist in avoiding the idiosyncratic (rare, individual) side effects of drugs. This has been achieved for some drugs, but few reactions to anti-cancer drugs have been clarified in this way [99].

12.5 OTHER MODES OF NON-SURGICAL THERAPY

12.5.1 Immunostimulation

Specific anti-tumor cell antibody therapy is essentially "passive" immune therapy (see Section 12.3.2) (see also Figure 12.6).

This subsection deals with attempts to stimulate the patients' own immune system to produce an immune response against their tumor cells. This is an "active" form of immune intervention, and is also known as "autologous vaccine therapy" [100].

Essentially, the technique is to prepare extracts or derivatives of the patient's tumor

A. Active methods

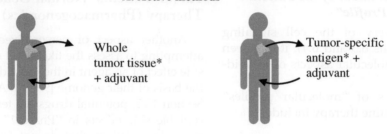

Whole
tumor tissue*
+ adjuvant

Tumor-specific
antigen* +
adjuvant

* In most reports, the antigen preparations are physically or chemically modified in some way.

B. Adoptive T-cell transfer methods

Most suggested therapies of this kind are based on the following assumptions:
(a) The tumor cells express specific antigens.
(b) The T lymphocytes found in tumors are cytotoxic in type (CD8+), and are recognizing the specific antigens and are attempting to kill the tumor cells.
(c) T lymphocytes in the patient's peripheral blood include cytotoxic cells which are
   (i) Anti-tumor memory T cells which are suppressed in some way.
   (ii) Naïve cytotoxic T cells which can be exposed to antigen and then clonally expanded to active anti-tumor cells.

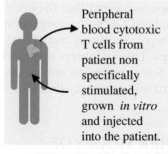

Peripheral blood cytotoxic T cells from patient non specifically stimulated, grown *in vitro* and injected into the patient.

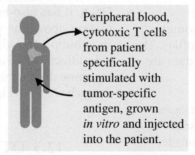

Peripheral blood, cytotoxic T cells from patient specifically stimulated with tumor-specific antigen, grown *in vitro* and injected into the patient.

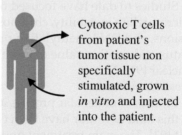

Cytotoxic T cells from patient's tumor tissue non specifically stimulated, grown *in vitro* and injected into the patient.

Antigen-presenting cells (dendritic cells) from patient are specifically stimulated with tumor-specific antigen. These cells are then used to stimulate the patient's peripheral blood T cells, which are then grown *in vitro* and injected into the patient.

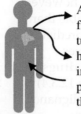

Antigen-presenting cells (dendritic cells) from patient are transfected with gene for tumor-specific antigen. These cells then hyper-express the antigen, causing increased stimulation of the patient's peripheral blood T cells. These T cells are then grown in vitro and injected into the patient (see in [124]).

FIGURE 12.6 Some potential processes in specific immunotherapies for disseminated cancers. Not mentioned here are:
 (i) Passive methods for example involving specific antibodies produced in vitro, or donor T cells,
 (ii) All general immuno-modulatory drugs, e.g. interleukins, inhibitors of 'checkpoint brake' proteins
(iii) Other general, non-specific 'stimulants of the immune system'.

tissue, and increase their antigenicity with immunological adjuvants. Attempts based on this principle have a long history [101]. Both antibody- and cell-mediated immune responses have been targets of stimulation.

A particular form of immunotherapy involves the use of tumor-specific T cells. These are considered to be lymphocytes present in tumors which are assumed to be reacting to the tumor cells but are not necessarily killing the tumor cells. The commonest procedure is to harvest T cells from the patient's tumor tissue, and then develop cell lines from them which have enhanced abilities to kill the patient's tumor cells. This is an example of "adoptive transfer immunotherapy" [102].

For melanomas and ovarian carcinomas, there seems to be evidence that the tumor cells produce immunoinhibiting substances. For these, it has been suggested that immune stimulation might be appropriate therapy [103]. For this, interleukin-2 and interferon-alpha have been used with success in some trials [104].

For renal cell carcinoma, it was once advocated that the tumor should be infarcted, so that massive release of antigens into the patient's bloodstream would occur, and potentially stimulate a successful tumor-toxic reaction. In a study of 30 patients, no clinical benefit was established [105].

12.5.2 Gene Therapy

The principle of gene therapy is to transfect tumor cells with cell death-promoting, or growth-reducing, genes [106,107]. The genes in this context have been mainly tumor suppressors. The procedure can be potentially "personalized" by transfecting a gene that the patient's tumor is known to lack.

The problems of this procedure have included:

i. Developing tumor-specific vectors for the transfection process. Various viruses and viral constructs have been tried.

ii. Infection of normal cells by the viral vector. This has caused severe allergic reactions, and even deaths of patients [106].

iii. Ensuring that all tumor cells in the body are transfected by the vector-gene combination. Here, the phenomenon of "tropism" is relevant, since few normal viruses infect all kinds of cells in the body (see Section 4.5.2).

iv. The theoretical possibility that the viral infection might cause new tumors through viral oncogenic mechanisms (see Section 4.5) [107].

12.5.3 Other

(a) BCG Therapy for Tumors of the Bladder

Bacille Calmette-Guerin (BCG), which is mainly used as a vaccine for tuberculosis, also has adjuvant immunostimulating properties against urothelial tumors of the bladder [108]. Instillation of living BCG into the bladder causes a self-limiting granulomatous inflammatory reaction rather like typical tuberculosis. Superficial urothelial tumors are often destroyed in this process. However, metastases are not affected, indicating no systemic immune reaction against any urothelial tumor antigens. The treatment may be successful because the tumorous urothelium cannot regenerate after damage as effectively as can the normal urothelial cells. Variations in the strains of BCG organisms may contribute to different treatment outcomes between clinics [108].

(b) Other Bacterial Toxins

It has long been known that foci of bacterial infection are associated with formation of toxins which can inhibit tumor growth. The best known example of the clinical application of this observation was Coley's mixed toxins [109]. These mixtures were injected directly into tumor masses. The therapies were found to have some effects against sarcomas in the limbs. Subsequently, however, the bacterial toxin treatments were found to be less effective than radiation therapy, and are currently rarely used.

(c) Endocrine Therapies

These are essentially limited to tumors of the secondary sex organs (breast and endometrium; prostate). The aim is to reduce the proliferation of tumor cells. The cells are not killed. The therapies consist of reductions of the production of natural hormones, or the administration of anti-hormones.

In patients with carcinoma of the breast, estrogen levels in the blood have been reduced by removing the patients' ovaries. The treatment was helpful in a proportion of cases [110], but present therapies for reduction in estrogen stimulation to tumor cells usually involve anti-estrogen drugs, such as tamoxifen. These drugs can also be used in the prevention of carcinomas of the breast and endometrium in high-risk individuals (see Chapter 12).

For carcinoma of the prostate, therapies include removing natural sources of androgens by bilateral orchidectomy. In the past, estrogens, such as stilboestrol, were used, but the side effects of this drug therapy are inconvenient. Presently, anti-androgens drugs are the preferred therapies for this hormonal effect [111].

(d) Electrochemotherapy

These therapies involve brief permeabilizing electric pulses followed by an anticancer drug, often of low toxicity [112]. It can only be used where the location of the tumor allows placement of electrodes. For this reason, it is usually a palliative procedure.

(e) Stem Cell Therapies

These are essentially therapies for the side effects of another therapy, as in bone marrow transplantation (see Section 11.5.3). No stem cells have any actual anti-cancer cell properties.

12.6 ASSESSING EFFECTS OF THERAPIES IN THE INDIVIDUAL CASE

12.6.1 General

When the patient undergoes many forms of therapy, there are the predictable side effects of malaise, vomiting, etc. When the course of anti-cancer therapy has been finished, and these side effects have subsided, it is then appropriate to examine the degree of improvement in the patient's condition which has occurred. In general, these assessments of responses depend on clinical, imaging, and laboratory data as follows.

12.6.2 Clinical Measures

There are many kinds of questionnaires for assessing the functional states of patients with chronic debilitating diseases. The Edmonton Symptom Assessment system [113] assesses severity of a number of common symptoms in cancer patients. The European Organization for Research and Treatment of Cancer (EORTC) questionnaire QLC-30 and modifications is a more complex method of assessing quality of life. Typically, they incorporate multi-item scales relating to functioning, symptoms, and general health, as well as single-item scales [114,115].

12.6.3 Imaging

Imaging (Figure 12.7) is one of the most widely used methods for assessing effects of anti-cancer therapies in cases of un-resectable malignancy. Actual shrinkage of a tumor mass mainly occurs through actual cell killing. It can also occur by suppression of proliferation of the tumor cells with preservation of the various mechanisms of cell losses from the mass (see Section 8.3.2 and Chapter 10). Other phenomena such as reduction in the intensity of

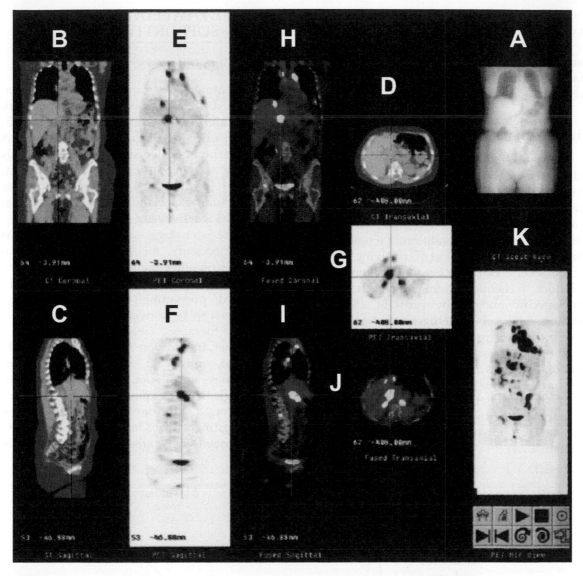

FIGURE 12.7 Images using various techniques of carcinoma of the lung disseminated to the organs of the chest and abdomen. (A) Plain X-ray of chest and abdomen. (B, C, and D) Computerized Tomographic (CT) X-ray images. (E, F, and G) Positron electron (PET) images. (H, I, and J) Fused CT and PET images. (K) Rotating maximum intensity reprojection image (PET-MIP View). *Reproduced with permission from* Abeloff's Clinical Oncology, *5th Edition, John E. Niederhuber, "Imaging", Figure 18-7, Copyright (2014).*

surrounding inflammation may produce a similar radiological change.

There are also difficulties in assessing volumes of tumor masses from two-dimensional images. This has been overcome to certain degrees by three-dimensional scans such as PET-CT. The latter methods are expensive, and may not be sufficiently specific for accurate assessments of changes in tumor volume [116].

The value of these measurements for guidance of further therapy may be affected by the fact that the degrees of shrinkage may be variable from one metastatic deposit to another.

12.6.4 Biomarker Levels

The term "biomarker" in cancer therapy usually refers to chemicals specific for the tumor in the blood of the patient, rather than the wider group of abnormalities observable in cancer patients (see Section 9.1.1). In the context of assessing the condition of a patient who has been diagnosed, and is receiving therapy, "biomarkers" can only be used for tumors that produce a chemical biomarker, such as prostate-specific antigen by prostatic carcinoma. Tumors tend to be heterogeneous for quantitative biomarker production per unit volume of viable tumor tissue. Many common types of tumors, such as non–small cell carcinomas of the lung, have no known reliable biomarker.

It should be noted here that "circulating tumor cells" are sometimes used as a prognostic marker for patients, e.g. [116,117] (see also Section 9.7.4).

12.7 PALLIATIVE AND SUPPORTIVE CARE

12.7.1 General

The treatments for cancer which have been described so far in this book relate to primary medical care—i.e., the physical treatments that are begun once the diagnosis is established. A variety of non-oncologist health professionals contribute to the physical care of the patient during the hospital stays for these treatments. These include especially nurses, physiotherapists, social workers, and spiritual counselors.

At some later stage, further hospitalizations may be necessary, for example, for surgical operations or other medical treatments aimed at improving the quality of life, relieving symptoms, and preventing complications. These treatments can be conveniently referred to as "palliative therapies" (see Figure 12.8).

However, in addition to these major, mainly in-hospital, physical treatments, a cancer patient may need support for other needs. Psychological needs may arise in hospital or later, through the development of depression, adjustment disorders, and anxiety [118]. Physical needs may arise in relation to daily

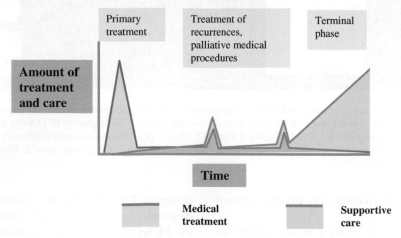

FIGURE 12.8 Supportive care for cancer patients.

living [118]. These can involve fatigue, relief of pain, preservation of mobility, feeding, and general nursing. Financial needs may arise through loss of income and costs of treatment. This is typically an area dealt with by social workers. Social support needs are also common, and may be met through the relatives, or the work of various voluntary charitable organizations [119].

Needs for specific information [120] include the patient's desire for accurate prognosis in his/her case, and in relation to the availability of the support services in his/her particular community.

Spiritual needs vary between persons within and between cultures. These needs may be felt by as many as 75% of persons in Western societies [121]. Attending to these needs is one of the conventional pastoral roles of ministers of religion.

Legal assistance, especially in clarifying the patient's wishes in relation to bequests, may be required. The patient's relatives may need support with grieving or have other needs [118]. Support from religious figures or secular counselors may be appropriate for them. The minimization of distress in the individual patient may depend to a great extent on the relatives and the amount of support they are able to provide.

12.7.2 "Survivorship"

Traditionally, support for patients after discharge from hospital following primary treatment has not been seen as a major issue, because many patients may be cured, or have a long symptom-free interval before recurrences become clinically apparent. When relapses occurred, the patients often had had all possible medical treatment, so that supportive care was all that was available.

In recent years, treatments for the various cancers have become more complicated, and drug and radiation therapies have become commoner in primary care, as well as in the early post-discharge period. Because of this, supportive care has become necessary earlier in the course of disease than previously. The term "survivorship" has been introduced more or less to refer to all supportive care from the time of diagnosis onwards [122].

12.8 "ALTERNATIVE"/ "COMPLEMENTARY" REGIMENS

These kinds of "therapies" have been suggested to be useful in prevention and care of patients with cancers. The regimes include:

i. Broadly passive physical procedures, such as acupuncture, yoga, non-strenuous exercises, and massage of unaffected parts of the body,
ii. Broadly quasi-pharmacological remedies, such as herbal preparations, aroma therapies, and dietary modifications.
iii. Broadly psychological regimes, such as "meditation," "music therapy," "art therapy," and others.

By definition, all these regimes have no known potential biochemical or physiological way of affecting any property, such as excessive growth, of tumor cells.

Historically, they have been associated with charlatanism and quackery. This is because some persons undergoing these regimes have claimed that their methods might be able to prevent and/or slow the growth or spread of tumors.

In the last 20 years, these regimes have enjoyed wider tolerance if not full acceptance in the community. There are probably several factors in this:

i. The persons undergoing the regimes have moderated their claims about their usefulness in relation to disease processes, especially by concentrating on the comforts they can deliver in the "end-of-life" phase of cancer illnesses.

ii. The amount of "regular" anti-cancer therapy has increased massively through improvements in imaging and other tests (leading to more palliative surgery and non-surgical treatment). This is associated with more time in hospital, more uncomfortable investigations, and more side effects of therapies in the cancer patient population as a whole.

iii. The popularity of entering cancer patients into trials has introduced another potential source of anxiety for cancer patients in the traditional medical environment.

iv. The fact that diagnoses are made earlier (through screening methods) means that patients have longer periods of time "with their disease," than ever before.

v. The general decline of formal religion has left many patients without that traditional source of solace.

vi. Smaller families, and the frequency of marital dissolutions, have left many older people with fewer of these traditional supports in times of stress.

All of the supporting professions (see Section 12.7) have become more and more affected by the increasing complexities of the practice of their professions in the medical environment. (One only has to look at the volume of medical and allied health notes in patients' hospital records to appreciate this point.) To a degree, the role of comforter to the sometimes paining and usually anxious, despairing if not distraught, cancer patients may have been neglected in some circumstances. Where this may have occurred, complementary/alternative regimes, with their current, restricted therapeutic claims—and practiced outside the formal institutions of hospitals and clinics—may be filling a gap.

References

[1] Marcu L, Bezak E, Allen BJ. Biomedical physics in radiation therapy for cancer. Collingwood, Victoria, Australia: CSIRO Publishing; 2012. pp. 25–31.

[2] Harrison FL, Anderson SL. Taxonomic and developmental aspects of radiosensitivity. Preprint for Symposium on Ionizing Radiation, Stockholm, Sweden, 1996.

[3] Hendry JH. Survival of cells in mammalian tissues after low doses of irradiation: a short review. Int J Radiat Biol Relat Stud Phys Chem Med 1988;53(1):89–94.

[4] Thoday JM, Read J. Effect of oxygen in the frequency of chromosome aberrations produced by X-Rays. Nature 1947;160:606–8.

[5] Overgaard J. Hypoxic radiosensitization: Adored and ignored. J Clin Oncol 2007;25(26):4066–74.

[6] Hay MP, Hicks KO, Wang J. Hypoxia-directed drug strategies to target the tumor microenvironment. Adv Exp Med Biol 2014;772:111–45.

[7] Alphandéry E. Perspectives of breast cancer thermotherapies. J Cancer 2014;5(6):472–9.

[8] Viglianti BL, Stauffer P, Repasky E, et al. Hyperthermia. In: Perez CA, Brady LW, Halpern EC, Schmidt-Ullrich, editors. Principles and practice of radiation oncology (4th ed.). Philadelphia, PA: LWW/Wolters-Kluwer; 2004. pp. 528–40.

[9] Patel IJ, Pirasteh A, Passalacqua MA, et al. Palliative procedures for the interventional oncologist. AJR Am J Roentgenol 2013;201(4):726–35.

[10] Koller PC. The genetic component of cancerRaven RW, editor. Cancer, vol 1. London: Butterworth & Co; 1957. p. 335–403.

[11] Kronenberg A. Radiation-induced genomic instability. Int J Radiat Biol 1994;66:603–9.

[12] Marder BA, Morgan WF. Delayed chromosomal instability induced by DNA damage. Mol Cell Biol. 1993;13(11):6667–77.

[13] Curtis HJ. Formal discussion of: Somatic mutations and carcinogenesis. Cancer Res 1965;25(8):1305–8.

[14] Morgan WF, Murnane JP. A role for genomic instability in cellular radioresistance? Cancer Metastasis Rev 1995;14(1):49–58.

[15] Nenoi, M. (Ed). Current Topics in Ionising Radiation Research. New Delhi: InTech Publications; 2012.

[16] National Association for Proton Therapy. Proton therapy centers. Available at: http://www.proton-therapy.org/map.htm.

[17] Cancer Research UK. Proton therapy is coming to the UK, but what does it mean for patients? Available at: http://scienceblog.cancerresearchuk.org/2013/09/16/proton-therapy-is-coming-to-the-uk-but-what-does-it-mean-for-patients/.

[18] DeLaney TF, Kooy HM. Proton and charged particle radiation therapy. Philadelphia: LWW/Wolters Kluwer; 2008.

[19] Ma C-MC, Lomax T, editors. Proton and carbon ion therapy (Imaging in Medical Diagnosis and Therapy). Boca Raton, FL: CRC Press; 2013.

[20] Brenner DJ, Hall EJ. Secondary neutrons in clinical proton radiation therapy: A charged issue. Radiother Oncol 2008;86(2):165–70.

[21] Strander H, Turesson I, Cavallin-Ståhl E. A systematic overview of radiation therapy effects in soft tissue sarcomas. Acta Oncol 2003;42(5-6):516–31.

[22] Elgqvist J, Frost S, Pouget JP, et al. The potential and hurdles of targeted alpha therapy—Clinical trials and beyond. Front Oncol 2014;14(3):324.

[23] Karar J, Maity A. Modulating the tumor microenvironment to increase radiation responsiveness. Cancer Biol Ther 2009;8(21):1994–2001.

[24] Wang AZ, Tepper JE. Nanotechnology in radiation oncology. J Clin Oncol 2014;32(26):2879–85.

[25] Purdy JA. Principles of radiologic physics, dosimetry and treatment planning. In: Perez CA, Brady LW, Halpern EC, Schmidt-Ullrich, editors. Principles and practice of radiation oncology (4th ed.). Philadelphia, PA: LWW/Wolters-Kluwer; 2004. pp. 180–218.

[26] Cancer Research UK. General radiotherapy side effects. Available at: http://www.cancerresearchuk.org/about-cancer/cancers-in-general/treatment/radiotherapy/side-effects/general.

[27] Salvo N, Doble B, Khan L, et al. Prophylaxis of radiation-induced nausea and vomiting using 5-hydroxytryptamine-3 serotonin receptor antagonists: a systematic review of randomized trials. Int J Radiat Oncol Biol Phys 2012;82(1):408–17.

[28] Hall EJ. Is there a place for quantitative risk assessment? J Radiol Prot 2009;29(2A):A171–84.

[29] Cheung-Ong K, Giaever G, Nislow C. DNA-damaging agents in cancer chemotherapy: serendipity and chemical biology. Chem Biol 2013;20:648–59.

[30] Kritharis A, Bradley TP, Budman DR. The evolving use of arsenic in pharmacotherapy of malignant disease. Ann Hematol 2013;92(6):719–30.

[31] Davison K, Mann KK, Miller Jr. WH. Arsenic trioxide: Mechanisms of action. Semin Hematol 2002;39(2 Suppl 1):3–7.

[32] Koç A, Wheeler LJ, Mathews CK, Merrill GF. Hydroxyurea arrests DNA replication by a mechanism that preserves basal dNTP pools. J Biol Chem 2004;279:223–30.

[33] Lanzkron S, Strouse JJ, Wilson R, et al. Systematic review: Hydroxyurea for the treatment of adults with sickle cell disease. Ann Intern Med 2008;148(12):939–55.

[34] Olson JJ, Nayak L, Ormond DR, Wen PY, Kalkanis SN. AANS/CNS Joint Guidelines Committee. The role of cytotoxic chemotherapy in the management of progressive glioblastoma: A systematic review and evidence-based clinical practice guideline. J Neurooncol 2014;118(3):501–55.

[35] Trimmer EE, Essigmann JM. Cisplatin. Essays Biochem 1999;34:191–211.

[36] Galea AM, Murray V. The anti-tumor agent, cisplatin, and its clinically ineffective isomer, transplatin, produce unique gene expression profiles in human cells. Cancer Inform 2008;6:315–55.

[37] Dasari S, Bernard Tchounwou P. Cisplatin in cancer therapy: Molecular mechanisms of action. Eur J Pharmacol 2014;740C:364–78.

[38] Zhang S, Lovejoy KS, Shima JE, et al. Organic cation transporters are determinants of oxaliplatin cytotoxicity. Cancer Res 2006;66(17):8847–57.

[39] Coluccia M, Natile G. Trans-platinum complexes in cancer therapy. Anticancer Agents Med Chem 2007;7(1):112–6.

[40] Bignold LP. Alkylating agents and DNA polymerases. Anticancer Res 2006;26(2B):1327–36.

[41] (Various authors). In: Ann NY Acad Sci, vol 68, 1958; vol 163, 1969.

[42] Ross WCJ. Biological alkylating agents. London: Butterworths; 1962.

[43] Begleiter A, Mowat M, Israels LG, Johnston JB. Chlorambucil in chronic lymphocytic leukemia: mechanism of action. Leuk Lymphoma. 1996;23(3-4):187–201.

[44] Visentin M, Zhao R, Goldman ID. The antifolates. Hematol Oncol Clin North Am. 2012;26(3):629–48. ix.

[45] Chan ES, Cronstein BN. Mechanisms of action of methotrexate. Bull Hosp Jt Dis 2013;71(Suppl 1):S5–S8.

[46] Huennekens FM. The methotrexate story: A paradigm for development of cancer chemotherapeutic agents. Adv Enzyme Regul 1994;34:397–419.

[47] Hamel E. An Overview of compounds that interact with tubulin and their effects on microtubule assembly Fojo T, editor. The role of microtubules in cell biology, neurobiology and oncology. New York, NY: Springer; 2009. pp. 1–20.

[48] Kung AL, Zetterberg A, Sherwood SW, Schimke RT. Cytotoxic effects of cell cycle phase specific agents: result of cell cycle perturbation. Cancer Res. 1990;50:7307–17.

[49] Zhao Y, Fang WS, Pors K. Microtubule stabilising agents for cancer chemotherapy. Expert Opin Ther Pat 2009 19:607–22.

[50] National Cancer Institute. Dactinomycin. Available at: http://ncit.nci.nih.gov/ncitbrowser/ConceptReport.jsp?dictionary=NCI%20Thesaurus&code=C412.

[51] Avendaño C, Menéndez JC, editors. Medicinal chemistry of anticancer drugs. Philadelphia: Elesvier; 2008. Chapter 7 – DNA intercalators and topoisomerase inhibitors. pp. 199–228.

[52] Qu X, Wan C, Becker HC, et al. The anticancer drug-DNA complex: femtosecond primary dynamics for anthracycline antibiotics function. Proc Natl Acad Sci U S A 2001;98(25):14212–7.

[53] Bignold LP. Mechanisms of clastogen-induced chromosomal aberrations: A critical review and description

of a model based on failures of tethering of DNA strand ends to strand-breaking enzymes. Mutat Res 2009;681(2-3):271–98.

[54] Rooseboom M, Commandeur JNM, Vermeulen NPE. Enzyme-catalyzed activation of anticancer prodrugs. Pharmacol Rev 2004;56(1):53–102.

[55] Coluccia M, Natile G. Trans-platinum complexes in cancer therapy. Anticancer Agents Med Chem 2007;7(1):111–23.

[56] Dutt R, Madan AK. Classification models for anticancer activity. Curr Top Med Chem 2012;12(24):2705–26.

[57] Maggiora G, Vogt M, Stumpfe D, et al. Molecular similarity in medicinal chemistry. J Med Chem 2014;57(8):3186–204.

[58] Tacar O, Sriamornsak P, Dass CR. Doxorubicin: An update on anticancer molecular action, toxicity and novel drug delivery systems. J Pharm Pharmacol 2013;65(2):157–70.

[59] Boran ADW, Iyengar R. Systems approaches to polypharmacology and drug discovery. Curr Opin Drug Discov Devel 2010;13(3):297–309.

[60] Peters J-W. Polypharmacology in drug discovery. Hoboken, NJ: Wiley; 2012.

[61] National Cancer Institute. Targeted cancer therapies. Available at: http://www.cancer.gov/cancertopics/factsheet/Therapy/targeted.

[62] Atkins JH, Gershell LJ. Selective anticancer drugs. Nat Rev Drug Discov 2002;1(7):491–2.

[63] Avendaño C, Menéndez JC. Drug targeting in anticancer chemotherapy. In: Avendaño C, Menéndez JC, editors. Medicinal chemistry of anticancer drugs. Philadelphia: Elsevier; 2008. pp. 351–85.

[64] Bae YH, Mrsny RJ, Park K, editors. Cancer targeted drug delivery: An elusive dream. New York, NY: Springer; 2013.

[65] Lieu CH, Tan AC, Leong S, et al. From bench to bedside: Lessons learned in translating preclinical studies in cancer drug development. J Natl Cancer Inst 2013;105(19):1441–56.

[66] Stephan JP, Kozak KR, Wong WL. Challenges in developing bioanalytical assays for characterization of antibody-drug conjugates. Bioanalysis 2011;3(6):677–700.

[67] Ducry L, Stump B. Antibody-drug conjugates: Linking cytotoxic payloads to monoclonal antibodies. Bioconjug Chem 2010;21(1):5–13.

[68] Scott AM, Wolchok JD, Old LJ. Antibody therapy of cancer. Nat Rev Cancer 2012;12(4):278–87.

[69] Sherbet GV. Growth factors and their receptors in cell differentiation, cancer and cancer therapy. Philadelphia, PA: Elsevier; 2011.

[70] US Food and Drug Administration. Approved drugs – notifications. Available at: http://www.fda.gov/Drugs/InformationOnDrugs/ApprovedDrugs/ucm381452.htm.

[71] Fabbro D, Cowan-Jacob SW, Möbitz H, et al. Targeting cancer with small-molecular-weight kinase inhibitors. Methods Mol Biol 2012;795:1–34.

[72] Golubovskaya VM, Cance WG. Targeting the p53 pathway. Surg Oncol Clin N Am 2013;22(4):747–64.

[73] Chatterjee M, Kashfi K, editors. Cell signaling & molecular targets in cancer. Dordrecht: Springer; 2011.

[74] Dietel M, editor. Targeted therapies in cancer. Berlin: Springer; 2007.

[75] Epstein RJ. The unpluggable in pursuit of the undruggable: Tackling the dark matter of the cancer therapeutics universe. Front Oncol 2013;12(3):304.

[76] Trent RJ. Molecular medicine: Genomics to personalized healthcare, 4th ed. Philadelphia: Elsevier; 2012.

[77] Coelho J, editor. Drug delivery systems: Advanced technologies potentially applicable in personalised treatment. New York, NY: Springer; 2013.

[78] Yao Y, Bahija Jallal B, Ranade K, editors. Genomic biomarkers for pharmaceutical development: advancing personalized health care. Philadelphia: Elsevier; 2014.

[79] Hedgecoe AR, editor. The politics of personalised medicine: pharmacogenetics in the clinic. West Nyack, NY: Cambridge University Press; 2004.

[80] Troiani T, Schettino C, Martinelli E, et al. The use of xenograft models for the selection of cancer treatments with the EGFR as an example. Crit Rev Oncol Hematol 2008;65(3):200–11.

[81] Céspedes MV, Casanova I, Parreño M, et al. Mouse models in oncogenesis and cancer therapy. Clin Transl Oncol 2006;8(5):318–29.

[82] Kelland LR. Of mice and men: values and liabilities of the athymic nude mouse model in anticancer drug development. Eur J Cancer 2004;40(6):827–36.

[83] Arnedos M, Vielh P, Soria JC, et al. The genetic complexity of common cancers and the promise of personalized medicine: Is there any hope? J Pathol 2014;232(2):274–82.

[84] De Sousa E, Melo F, Vermeulen L, Fessler E, Medema JP. Cancer heterogeneity--a multifaceted view. EMBO Rep 2013;14(8):686–95.

[85] Pfeffer U, editor. Cancer genomics: molecular classification, prognosis and response prediction. New York, NY: Springer; 2013.

[86] Espina V, Liotta LA, editors. Molecular profiling: methods and protocols" (Methods in Molecular Biology). New York, NY: Humana/Springer; 2012.

[87] Früh M. The search for improved systemic therapy of non-small cell lung cancer—What are today's options? Lung Cancer 2011;72(3):265–70.

[88] Raparia K, Villa C, DeCamp MM, et al. Molecular profiling in non-small cell lung cancer: a step toward personalized medicine. Arch Pathol Lab Med 2013;137(4):481–91.

[89] van de Vijver MJ. Molecular tests as prognostic factors in breast cancer. Virchows Arch 2014;464(3):283–91.

[90] Koka R, Ioffe OB. Breast carcinoma: Is molecular evaluation a necessary part of current pathological analysis? Semin Diagn Pathol 2013;30(4):321–8.

[91] Bogaert J, Prenen H. Molecular genetics of colorectal cancer. Ann Gastroenterol 2014;27(1):9–14.

[92] Silvestri A, Pin E, Huijbers A, et al. Individualized therapy for metastatic colorectal cancer. J Intern Med 2013;274(1):1–24.

[93] Schoenborn JR, Nelson P, Fang M. Genomic profiling defines subtypes of prostate cancer with the potential for therapeutic stratification. Clin Cancer Res 2013;19(15):4058–66.

[94] Tremante E, Ginebri A, Lo Monaco E, et al. Melanoma molecular classes and prognosis in the postgenomic era. Lancet Oncol. 2012;13(5):e205–11.

[95] Lallemant B, Evrard A, Chambon G, et al. Gene expression profiling in head and neck squamous cell carcinoma: Clinical perspectives. Head Neck 2010;32(12):1712–9.

[96] Lam Y-WF, Cavallari LH, editors. Pharmacogenomics: Challenges and opportunities in therapeutic implementation. San Diego: Academic Press/Elsevier; 2013.

[97] Federico Innocenti F, van Schaik RHN, editors. Pharmacogenomics" (Methods and Protocols vol 1015). New York, NY: Springer; 2013.

[98] Dickenson J, Freeman F, Mills CL, et al. Pharmacogenomics Molecular pharmacology: from DNA to drug discovery. Chichester, UK: John Wiley & Sons, Ltd; 2012. pp. 201–26.

[99] Blakey JD, Hall IP. Current progress in pharmacogenetics. Br J Clin Pharmacol 2011;71(6):824–31.

[100] Eggermont LJ, Paulis LE, Tel J, et al. Towards efficient cancer immunotherapy: Advances in developing artificial antigen-presenting cells. Trends Biotechnol 2014;32(9):456–65.

[101] Scott OC. Tumor transplantation and tumor immunity: A personal view. Cancer Res. 1991;51(3):757–63.

[102] King J, Waxman J, Stauss H. Advances in tumor immunotherapy. QJM 2008;101(9):675–83.

[103] Pardoll DM. Cancer-specific vaccines. In: Mendelsohn J, Howley PM, Israel MA, editors. The molecular basis of cancer. Philadelphia, PA: Saunders/Elsevier; 2009. pp. 649–70.

[104] Lee S, Margolin K. Cytokines in cancer immunotherapy. Cancers 2011;3:3856–93.

[105] Gottesman JE, Crawford ED, Grossman HB, Scardino P, McCracken JD. Infarction-nephrectomy for metastatic renal carcinoma. Southwest oncology group study. Urology 1985;25(3):248–50.

[106] Lattime EC, Gerson SL, editors. Gene therapy of cancer: translational approaches from preclinical studies to clinical implementation (3rd ed.). San Diego: Academic Press/Elsevier; 2014.

[107] Hatefi A, Canine BF. Perspectives in vector development for systemic cancer gene therapy. Gene Ther Mol Biol 2009;13:15–19.

[108] Gan C, Mostafid H, Khan MS, Lewis DJ. BCG immunotherapy for bladder cancer—The effects of substrain differences. Nat Rev Urol 2013;10(10):580–8.

[109] Wiemann B, Starnes CO. Coley's toxins, tumor necrosis factor and cancer research: a historical perspective. Pharmacol Ther 1994;64(3):529–64.

[110] Veronesi U, Cascinelli N, Greco M. A reappraisal of oophorectomy in carcinoma of the breast. Ann Surg 1987;205(1):18–21.

[111] Palmbos PL, Hussain M. Non-castrate metastatic prostate cancer: Have the treatment options changed? Semin Oncol 2013;40(3):337–46.

[112] Campana LG, Testori A, Mozzillo N, Rossi CR. Treatment of metastatic melanoma with electrochemotherapy. J Surg Oncol 2014;109(4):301–7.

[113] Watanabe SM, Nekolaichuk CL, Beaumont C. The Edmonton Symptom Assessment System, a proposed tool for distress screening in cancer patients: Development and refinement. Psychooncology 2012;21(9):977–85.

[114] Aaronson NK, Ahmedzai S, Bergman B, et al. The European Organization for Research and Treatment of Cancer QLQ-C30: A Quality-of-Life instrument for Use in international clinical trials in oncology. J Natl Cancer Inst 1993;85:365–76.

[115] van der Kloot WA, Kobayashi K, Yamaoka K, et al. Summarizing the fifteen scales of the EORTC QLQ-C30 questionnaire by five aggregate scales with two underlying dimensions: A literature review and an empirical study. J Psychosoc Oncol 2014;32(4):413–30.

[116] Alunni-Fabbroni M, Müller V, Fehm T, Janni W, Rack B. Monitoring in metastatic breast cancer: Is imaging outdated in the era of circulating tumor cells? Breast Care (Basel) 2014;9(1):16–21.

[117] Peach G, Kim C, Zacharakis E, et al. Prognostic significance of circulating tumor cells following surgical resection of colorectal cancers: A systematic review. Br J Cancer 2010;102(9):1327–34.

[118] Taylor K, Currow D. A prospective study of patient identified unmet activity of daily living needs among cancer patients at a comprehensive cancer care centre. Aust Occupational Therap J 2003;50:79–85.

[119] Adler NE, Page AEK, editors. Cancer care for the whole patient: meeting psychosocial health needs. Institute of Medicine (US) Committee on Psychosocial Services to Cancer Patients/Families in a Community Setting. Washington DC: National Academies Press; 2008. pp. 23–50.

[120] (Various authors). In: Handbook of Communication in Oncology and Palliative Care. Oxford University Press, Oxford, 2010, Sections C - D, pp. 101–582.

[121] National Cancer Institute. Spirituality in cancer care. Available at: http://www.cancer.gov/about-cancer/coping/day-to-day/faith-and-spirituality/spirituality-pdq.

[122] National Cancer Institute. Survivorship. Available at: http://www.cancer.gov/cancertopics/coping/survivorship.

[123] Pommier Y, Marchand C. Interfacial inhibitors of protein-nucleic acid interactions. Curr Med Chem Anticancer Agents 2005;5(4):421–9.

[124] Shurin MR, Gregory M, Morris JC, Malyguine AM. Genetically modified dendritic cells in cancer immunotherapy: A better tomorrow? Expert Opin Biol Ther 2010;10(11):1539–53.

L.P. Bignold: Principles of Tumors.
DOI: http://dx.doi.org/10.1016/B978-0-12-801565-0.00013-5

The previous two chapters have dealt with the therapies given to patients with tumors. It is widely felt, however, that the system of care of individuals with tumors—and with malignant tumors in particular—includes aspects of research methods, cost, and ethics which are of concern to the individual patient, as well as the general community.

This chapter gives the principles in these issues.

Sections 13.1 and 13.2. outline steps in the development of new therapeutic agents. Section 13.3 deals with cost factors in non-surgical therapies for malignant tumors, and Section 13.4 with ethical issues in cancer research and practice.

13.1 DEVELOPING NEW THERAPEUTIC AGENTS: CHEMICAL AND LABORATORY PHASES

This is a large, current field, with numerous excellent texts, articles, and websites becoming available in the last few years, e.g., Refs [1–5].

13.1.1 Origins of New Drugs

The first agents to be studied for potential uses as anti-cancer drugs have been among long-established toxins for humans. Arsenic has been known to be toxic to humans for centuries. However, it was only in the 1870s with advances in chemistry that specific arsenic compounds could be synthesized. These were promptly tested for anti-tumor effects (see Section 11.3.1). Another example is the group of "mustard" alkylating agents. These were first investigated when the myelo-suppressive effects of the proto-agent "mustard gas" (see Ref. [68] in Chapter 11) were recognized. All analogues which included the sulfur atoms have been too toxic for medicinal purposes, but compounds in which nitrogen replaces sulfur are less toxic.

In the 1940s, the discovery of penicillin [6]—as a naturally occurring biological compound specifically toxic to certain bacteria but not to mammals—stimulated the search for naturally occurring compounds which might be toxic to cancer cells, but not normal cells. These searches yielded many drugs still in use today, including the "antibiotic" group (see Section 12.2).

Occasionally, a new chemical may be developed by the chemical industry for intended use in the cosmetics, or agricultural chemical industries. However, all chemicals to which humans may be exposed are tested for toxic effects. Cytotoxic effects of possible use in anti-cancer therapy may be discovered in this way.

Another avenue is research into inhibitors of the endogenous growth factors (6.3) and the associated intracellular signaling cascades which may play roles in the excess growth and diminished physiological losses of cells in the tumor cell populations (see Section 11.3.3).

Occasionally, a chemical is found through unrelated studies—as was *cis*-platin [7].

The need to ensure that drugs are effective and safe has led to a complex process of evaluation and testing of new compounds for potential use against cancers (see Figure 13.1). Several "phases" are involved as follows.

13.1.2 Preparation of Analogues

When a new chemical—often called the "lead compound"—is found to have a cytotoxic effect, its analogues (chemical variants) are tested. The details of the chemical processes involved in preparing analogues involve much of the field of organic chemistry.

Analogues must be prepared selectively. A simple calculation shows this. If a proto-molecule has 10 modifiable chemical sites, and there are 10 possible modifications—oxidations, reductions, acylations, additions of rings etc., the number of possible permutations and combinations of these modifications with these

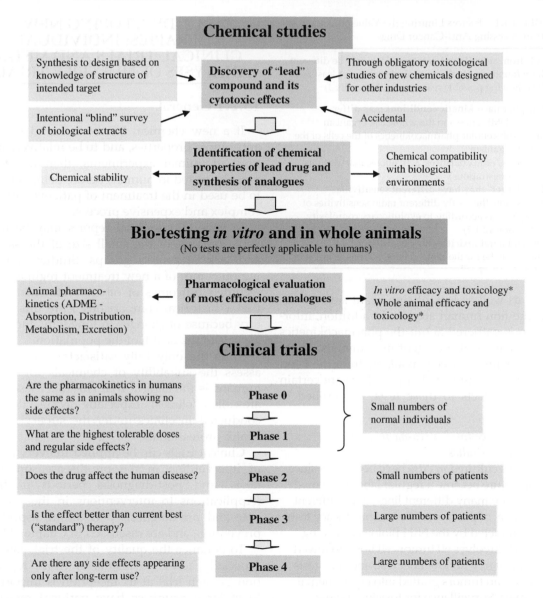

FIGURE 13.1 Steps in the development of new medicinal compounds.

groups is 10^{10}, or ten billion. Thus the search for possibly clinically useful analogues of toxic chemicals is a huge project.

Each potential compound must be chemically characterized for stability, temperature sensitivity, and bio-compatibility.

13.1.3 Toxicity and Pharmacokinetic Evaluation in Laboratory Animals

Before a potentially acceptable drug can be considered for administration to humans, there must be evidence of its possible anti-tumor

TABLE 13.1   Factors Limiting the Value of Laboratory Tests in Assessing Anti-Cancer Drugs

Results from animals and *in vitro* models may be different to those found eventually in clinical trials for at least some of the following possible reasons (see Appendix 3)

1. **The pharmaco-kinetic conditions may differ**
   In the ADME phases in the whole test organism
   In the sub-cellular pharmacokinetics of the cells of the test organism
      (Influx/efflux balances, activations, inactivations, sequestrations).
2. **The "target" may have different sensitivities**
   In this, note the vastly different radio-sensitivities of organisms according to evolutionary complexity (Section 12.1.2).
3. **The collateral toxicities may be different**
   These may be for the same general reasons as in 1 and 2.

---

effect, as well as lack of significant side effects, in some non-human studies. In addition, information is gathered about the pharmacokinetics and dynamics of the agent in mammals. These requirements are freely available from the Food and Drug Administration [8]. There are certain complex aspects to these tests, as outlined in Table 13.1.

a. *Initial "screening" assessments of effect in laboratory studies*

Cells cultured *in vitro* have been used since cell culture became a regular laboratory model. There are many different lines, with different numbers of passages. The *in vitro* efficacy tests as conducted by the NIH Biological Testing Branch involves 60 tumor cell lines, followed by the *in vivo* hollow fiber test, and then tests on human tumors grafted into experimental animals. Surveillance for toxicity of these animals is conducted at the same time [9].

b. *Subsequent assessments*

Chemicals demonstrating positive properties in these tests are then tested further, according to the results in the screening tests. A major part of this is study of toxicity in a wider range of animals.

13.2 DEVELOPING NEW THERAPIES: INDIVIDUAL CLINICAL TRIALS AND META-ANALYSES OF MULTIPLE TRIALS

13.2.1 General

If a new chemical has been found to have anti-cancer properties, and to be relatively non-toxic in animal experiments, then its effects must be tested in humans before being allowed to be used in the treatment of patients. This is a complex and expensive process.

In general, anecdotal reports may be unreliable because of the small size of the sample and lack of control groups. Studies comparing outcomes of a new treatment regimen with recorded outcomes of other treatments in the past ("historical comparisons") are unsatisfactory because of changes in longevity and other aspects of the health of the population.

Thus the only fully satisfactory method to assess the suitability of chemicals for use in humans is the clinical trial [10–12]. These trials are called "clinical" especially because they are conducted in clinics under the supervision of health professionals.

Clinical trials can be applied to every aspect of intervention in the pre-disease, pre-diagnosis, and post-diagnosis periods of time. (Applications to interventions in the pre-disease and pre-diagnosis periods are aspects of prevention, and are discussed in Chapter 14).

To enhance the quality of the trials system, especially access to results of trials, registration of each trial at its inception is important. Most large countries have national registers, either as part of a government department or in a non-government organization. There are currently over 44,000 cancer-related trials registered at the National Institutes of Health site [13]. All of these trials do not relate to new drugs; many are of new regimens or combinations of agents.

The general aims of clinical trials include:

i. How the chemical is absorbed, distributed, metabolized, and excreted
ii. Its potential benefits and mechanisms of action
iii. The optimum dosage
iv. The route of administration (such as by mouth or by injection)
v. Side effects (often referred to as toxicity)
vi. Whether or not the drug affects different groups of people (such as by gender, ethnicity, or culture) differently
vii. How it interacts with other drugs and treatments
viii. Its effectiveness as compared with similar drugs
ix. Its effectiveness in combination with other therapeutic agents

This information is accumulated through consecutive phases of studies.

13.2.2 The Phases of Clinical Trials

(a) Phase 0 Trials

These involve giving small doses of the new drug to normal individuals and measuring pharmacokinetic and other physiological phenomena. These data are then compared with similar studies in experimental animals in which no side effects were discovered in the pre-clinical trials. Phase 0 studies are designed to ensure that humans are likely to deal with the drug in the same way as the unaffected animals.

(b) Phase I Trials

These involve a small number of patients, to establish maximum tolerable doses, in terms of regular side effects.

(c) Phase II Trials

A larger number of patients are involved to test whether or not the agent has any effect on particular tumors.

(d) Phase III Trials

Larger numbers of patients to assess efficacy, especially in relation to current "best available" regimens.

(e) Phase IV Trials

All of the above trials are necessarily short term, and do not identify long-term complications. Phase IV trials refer to the fact that The Food and Drug Administration requires that it be made aware of all information concerning side effects, outcomes etc., which may become apparent through longer term use of the drugs than was possible in the short-term phase II and III trials. Phase IV trials therefore are expected to have durations of years, if not decades.

(f) Subcategories and Variant Notations

Various trials are denoted as subcategories "A" and "B". The usages of these sub-categorizations have not been consistent.

For example, phase 1A and phase 1B may indicate different schedules by which the maximum tolerable dose for regular side effects is administered.

Phase 2A often means a pilot study for the definitive phase 2 study, which is then called "phase 2B."

Phase 3A and phase 3B are terms sometimes used by companies to indicate trials before and after submissions made to drug-regulating authorities (in the United States, the Food and Drug Administration).

13.2.3 Survival Times as Measures of Benefit

Of all new chemicals which have shown potential as drugs in pre-clinical studies and have exhibited no untoward features in phase 0 and phase I trials, the majority fail in phase II and III trials. In principle, this occurs because

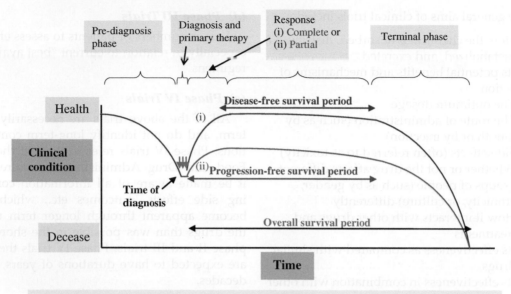

FIGURE 13.2   Terminology of survival periods of patients diagnosed with cancer.

the benefits of the new chemical are not shown to be greater than current standard treatments.

Several measures of survival may be used.

(a) Overall Survival Time

Three pieces of data are critical for assessing the survival of each cancer patient (see Figure 13.2). These are the date of diagnosis, the date of death, and the cause of death [14]. In regard to these, it can be noted that:

i. The date of diagnosis.

This point in time will tend to occur earlier in the patient's life if there is the generally more intense medical activity in the particular community (or social subset), including screening programs for the particular type of tumor. When medical activity is higher in the test group, than in the control group, the earlier diagnosis will tend to increase the post-diagnosis survival time. This is known as "lead-time" bias in relation to survival periods of cancer patients [3,15] (and see Figures 13.3 to 13.5).

ii. The records for establishing date of death (and hence period of survival) may vary from country to country.

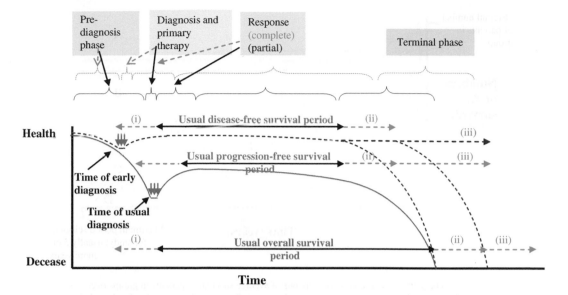

FIGURE 13.3 Potential effects early diagnosis: "lead-time" bias and lengthening of survivals associated with early diagnosis.

In countries where cancer registries exist, the dates of diagnosis are recorded. If the patient remains in the same administrative (for cancer treatment purposes) area, then the death certificate will accurately give the date of death. In this situation, the period of survival is easily established. However, if the patient changes residence—e.g., moves to a state where relatives live, or terminal care is cheaper, or for other reason—then the individual may be lost to follow-up. This will mean that the individual patient's cancer survival outcomes will not be known.

iii. Data concerning the cause of death depend on accuracies of death certificates (see Section 3.3.4). Thus patients may be recorded as dying of other conditions. In these certificates, whether or not tumor is still present at the time of death may not be recorded. Thus the distinction of "cure" versus "died with tumor" versus "died of tumor" may be difficult to establish [16,17].

These may be important factors in determining whether or not differences in survival rates in different decades (e.g., 1960s versus 2000s) for particular types of tumors have occurred.

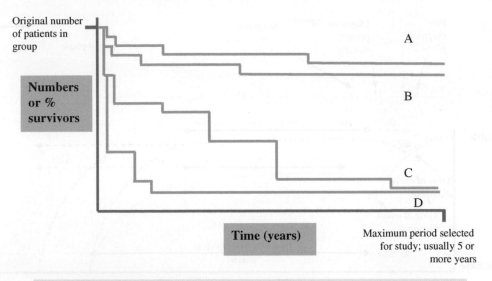

Original number of patients in group

Numbers or % survivors

A

B

C

D

Time (years)

Maximum period selected for study; usually 5 or more years

The plots can be used for comparisons of average survival of patients in groups over time. The survivals are aggregated into intervals, e. g., 1–2 months. This provides the sharp falls at intervals. The groups can be:

- different histological variants of a tumor type,
- different stages of the same tumor type,
- different therapies for the same tumor type.

The lines represent the number or percentage of cases in the original group who are alive after any particular period of time after diagnosis. Usually, some maximum period is selected, for example, 5 years. For some tumors, especially carcinoma of the breast, longer periods of time may be used for more accurate assessments of long-term outcomes for patients.

Some studies show only slight differences ("A" versus "B") and large numbers of cases analysed with statistics may be necessary to establish any beneficial effect.

Some therapies may simply slow the growth of these cells, resulting in a survival plot like "C" in comparison with "D".

FIGURE 13.4   Interpreting the Kaplan-Meier diagram.

(b) Disease-Free Survival

This is the period between complete absence of symptoms to the onset of new symptoms or signs. In leukemias, it is usually the period between complete remission and relapse. For solid tumors, it is usually the period between surgical removal of a tumor when there are no known metastases, to the time at which metastases may appear. This is also called the "latent period" of the disseminated tumor.

(c) Progression-Free Survival

In many cases of tumor, the disease causes no worsening of symptoms or signs over a period of time. This may be called a "plateau" period. However subsequently, more severe, and often accelerated, clinical disease may appear. This is called "progression." The length of the plateau period may be called the "progression-free" survival. A similar measure is "time to treatment failure" (11–17h).

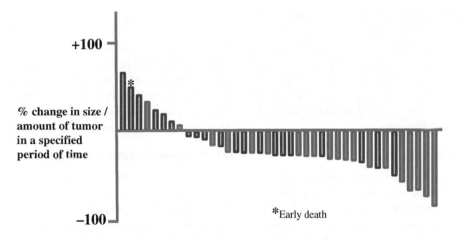

1. This kind of chart can be used for any comparison of groups of cases over time, in which both positive and negative courses may occur.

2. Each bar represents a patient in the clinical trial, arranged from left to right in ascending order of clinical response (greatest response on the right).

3. Two or more regimens can be compared, by indicating cases in each "arm" of the trial with a different color. (In the diagram, the regimen indicated in blue appears to be more effective than the regimen indicated in red.)

4. Additional information on each patient can be included in the chart, e.g., "* early death".

FIGURE 13.5   Interpreting the "waterfall" chart.

(d) Role of Autopsies

This role is to establish whether a patient who has received treatment died of

i. The tumor directly,
ii. The therapy (e.g., pneumonia complicating bone marrow suppression complicating anti-cancer drugs or radiotherapy) or
iii. Of a condition which was unrelated to the tumor or its complications or treatment (e.g., a motor vehicle accident).

With the decline in autopsy rates in most countries in the last half century, this source of information is dwindling.

13.2.4 Other Measures of Benefit

Generally, clinical trials may use the same patient-reported "quality of life" measures are used in assessing the effect of treatment and/or progress of disease, in the individual patient (Section 11.9.1). However, generally, the FDA does not give these data great weight when assessing a chemical for approval as a drug [3,15].

13.2.5 Other Difficulties, Especially Biases, in Clinical Trials

Apart from the difficulty of measuring the benefits of a new agent, clinical trials can be subject to factors which introduce biases into the results. The main general issues are (i) whether or not the "test" and "control" groups in the trial groups are adequately similar, and (ii) whether or not both groups adequately represent the population as a whole. Details of some of these issues, as well as other difficulties, include the following.

(a) Patient Selection Biases

An example of this is the inclusion of African Americans in a trial of a new therapy for carcinoma of the prostate (see Table 13.2). This type of tumor is more aggressive in African Americans than in the remainder of the population. Because of this, either too great a proportion, or too small a proportion of this group in the test and control groups may bias the results of the trial as representative of the population as a whole (see Section 3.2.2).

TABLE 13.2 Reasons for Reluctance on the Part of Patients to Enter Clinical Trials

1. Uncertainty of benefit

If receiving the new therapy

The new therapy may be less effective than "standard treatment" despite the results of the phase II trials[a].

The new therapy may have unsuspected side effects[b].

If receiving the "control" standard therapy

The patient may miss out on the benefit of the new therapy.

2. (Potentially according to medical insurance circumstances) cost factors

These vary enormously from country to country, and individual to individual.

These costs may be unpredictable if the new therapy were to have serious side effects.

[a]Many patients may wish all the results of the pre-clinical trials to be explained to them before making the decision. The attending physician may not be able to provide these explanations in reasonable time.
[b]Like all side effects, these may either regular in humans but not experimental animals, or idiosyncratic to the individual human.

Another example of patient selection bias is by age in relation to side effects. Many trial exclude patients over 65 years of age from toxicity studies. However, for some drugs, the side effects are greater in the elderly than in younger persons. Thus the trial of side effects may not indicate the actual risks of side effects when the drug is given to persons older than 65 years.

(b) Diagnostic Selection Biases

Diagnosis can be a source of bias. As noted in Chapter 3, diagnostic criteria for a disease can change over periods of years. For tumors, diagnostic criteria for benign versus malignant have changed for some types, especially carcinoma of the prostate.

The laboratory study may be histologic, biochemical, or other method. Thus it is possible that the whole study might become unrepresentative of the population as a whole.

(c) Size of Sample/"Power" of Study and Other Statistical Issues

Sometimes the benefits of the new therapeutic agent are only marginally greater than those of the standard therapy (given to the control group). In these situations, large numbers of patients may be required in the two "arms" of the trial to deliver a statistically significant result. There are many complex issues in determining what should be the appropriate sample sizes in proposed trials [18].

(d) In Relation to "Double-Blinding" of Trials

The term "double blind" means that during the trial, neither the patient nor the attending health professionals know whether or not the individual is receiving the "test treatment" of the "control treatment". It is a desirable aspect of trial because it inhibits certain biases occurring during the treatment period of the trial (see later this subsection). Often however, it is not possible to conceal this information from the patients during a treatment period [19].

This may occur because the test treatment is administered in a different way to the control treatment. Another reason is when one treatment has a regular side effect which the other does not: e.g., nausea (see Section 11.5.2).

(e) Compliance Bias

If side effects of the treatment in one "arm" of the trial is more severe than in the other, patients in that "arm" may cease taking the medications without informing the trial staff.

(f) Contamination

When double-blinding is not feasible, the physician and/or the patient may not follow the protocol, and undeclared additional treatments may be given. Even in double-blind trials, patients may cease taking the prescribed agents and/or take additional therapies without informing the organizers. In most countries, almost all drugs, whether officially approved or not, can be bought by the individual via the Internet.

13.2.6. Meta-Analyses

Because there are probably only 20 or so types of tumors which are common enough to mount trials, then for each of these there is likely to be at least 2,000 registered trials. Each of these will be independent, and the degrees of avoidance of the biases mentioned above are likely to differ. Often however, the results of apparently similar trials appear to conflict. Meta-analyses are then often undertaken to provide a summary of the available information and resolve the uncertainties [20]. The authors of meta-analyses may then themselves introduce biases of exclusion of trials not thought to be satisfactory etc. Meta-analyses are also impaired through non-reporting of negative results. This is why it is often advocated that all trials intended for publication should be registered before they begin, so that the negative results can be recorded.

13.2.7 Principles of Meta-Analysis Applied to Comparing Institutions

In different countries, to different extents, for different reasons and in different ways, the apparent outcomes for patients in different cancer treatment facilities are compared. Many websites offer advice directly to prospective patients on the facilities available in most Western countries. The outcomes reported on these sites may range from patient survivals to patients' feelings about the treatment they received and the manners in which the staff provided the treatment [7,21].

Many of the difficulties encountered in meta-analyses of clinical trials apply to these comparisons [22]. The treatment providers at the different clinics may differ, for example, in

i. The types of tumor they select to treat
ii. The degree of spread of the tumor in the particular case which they select to treat
iii. The thoroughness of their follow-up procedures for determining their survival data
iv. The actual modes of therapy which are used.

13.3 THE CANCER BURDEN AND COSTS

13.3.1 Terminology

The term "cancer burden" is used in the general literature for "a measure of the incidence of cancer within the population and an estimate of the financial, emotional, or social impact it creates" [23] (see Table 13.3). However here, it may be used more narrowly for the general reduction in enjoyment of life, social, and psychological impairment, financial loss and loss of life years caused by malignant tumors in a population. A term widely used when attempting to quantitate these burdens

TABLE 13.3 Measures of the Cancer Burden Their Determinants and Limitations

Measure	Definition	Determinants	Limitations
Incidence	Number of new cases, often per 10^5 person-years or absolute number of cases per year.	Burden of exposure to causes of cancer, weighted by the risk imparted by each cause.	Population-based cancer registration limited to a small proportion of global population. Affected by diagnostic intensity, screening, and autopsy rates.
Cumulative incidence	Proportion of people who develop cancer before a defined age.	Incidence.	Requires no loss to follow-up, no competing risks, same period of follow-up time for all study subjects and unchanged exposure status throughout follow-up.
Prevalence	Proportion of population with cancer.	Incidence, prognosis, and mortality from other causes.	Requires population-based registration and follow-up. Cured patients cannot be readily identified.
Survival	Proportion of cancer patients surviving for a specified time after diagnosis.	Natural history of disease. Stage at diagnosis, therapeutic efficacy.	Requires long-term follow-up of large number of patients. Spurious patterns may arise due to lead-time bias Influenced by diagnostic intensity. Sometimes difficult to classify causes of death correctly.
Mortality	Number of cancer deaths, often per 10^5 person-years, or absolute number of deaths per year.	Incidence, prognosis.	Influenced by adequacy of death certification, including autopsy rates.
Life-years lost	Number of years lost between age at death and expected (in the absence of this disease) age at death.	Incidence, age at diagnosis, prognosis.	Requires reliable population life tables.
Disability adjusted life years (DALY)	Combines impact of cancer on both quality of life and survival.	Incidence, age at diagnosis, prognosis, expected longevity, residual disability.	Requires reliable population life tables. Difficult to quantify disability adequately.

Source: "Adami H-O, Hunter D, Trichopoulos D. "Textbook of Cancer Epidemiology", Oxford Scholarship Online, 2nd ed., 2008, p. 35. Taken directly from "Table 2.1. Measures of cancer burden, their determinants and limitations."

is "disability adjusted life years" (DALYs). This yardstick includes the years of morbidity in the course of the illness and the years of life lost (YLLs) by early death [24].

13.3.2 The Financial Costs of Cancer Care

Currently in the United States, treating malignant tumors costs approximately $US 130 billion per year, with expected increases of 2–5% per annum in the foreseeable future [25] (see also Figure 13.6). The total cost of

cancer care is approximately equally divided between

i. Initial diagnostic and treatment costs
ii. Continuing (maintenance) costs and
iii. End-of life (terminal) costs.

Together, cancer costs amount to approximately 5% of total health care costs in the country [25].

There has been considerable attention paid to the prices of some drugs, which can be of the order of $US 10,000 per month [26], while others are only a few dollars.

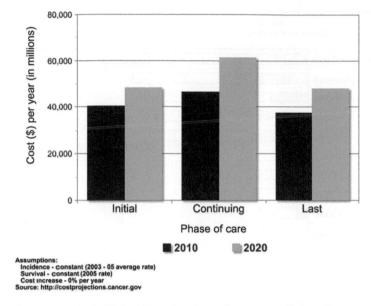

FIGURE 13.6 Costs of cancer care in the United States by phase of treatment: all sites, all ages, male and female, in 2010 dollars. *Source: http://costprojections.cancer.gov/.*

The increases in survival times are usually modest—by months, or a small number of years for most types of tumor. Because cures are rare, the issue becomes, how much should be spent on extending life, considering the ages and social participation of many of the patients? The question may be dealt with differently in different countries, depending especially on competing financial needs of the population (see in Section 13.4 and Figure 13.7).

The costs of drugs are only a proportion of the overall costs of treating cancer patients. Other costs derive from the clinical staff, as well as staff and equipment for radiotherapies, imaging, pathology tests, secondary drugs (antibiotics, analgesics), travel, and accommodation for patient and relatives in some circumstances.

Factors contributing to the rising overall costs to the community are the increasing incidence of tumors as the population ages (see Section 3.2.2), greater prevalence of tumors associated with longer survivals of cases, and developments of further complexities in treatments.

The costs can be divided into those borne by the individual, and those by the community. Their relative proportions vary from country to country according to the extent of government-sponsored health care. The costs include:

i. Direct costs relating to the resources used for prevention, diagnosis, monitoring and treatment, of individual cases. This is discussed in more detail in Section 14.6.

ii. Indirect costs relating to resources lost due to inability to work. These are mainly relevant for diseases that strike in the early years before normal retirement. Indirect costs include costs of lost production due to short-term absence from work, permanent disability, and death before 65 years of age.

Many of these costs can only be estimated in a whole population. An analysis of the methods for assessing costs of cancer is provided in [27].

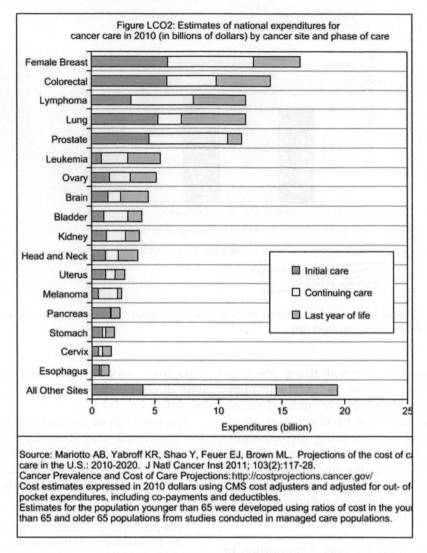

FIGURE 13.7 Costs of cancer care by type and phase. In this cross-sectional snapshot of national expenditures for cancer care in 2010, the proportion of expenditures in each phase of care varies by cancer type. For cancer types with short survival following diagnosis, such as pancreas, stomach, and lung, the majority of expenditures in 2010 are for patients in the initial and last year of life phases, with only a small percentage for patients in the continuing phase. Other cancer types with longer survival, such as female breast, melanoma, and prostate, have a higher percentage of expenditures for patients in the continuing phase of care. *Source: [24].*

13.4 ETHICAL ISSUES

Many aspects of oncology are potentially subject to ethical considerations [28]. Discussion of the ethical management of illnesses in medicine may begin with the Hippocratic dictum, often rendered as "Do no harm" [29]. This clearly applies to the early stages of management of the patient's illness, when therapies are indicated and have prospects of improving the patient's expectation and quality of life.

However in oncology, some aspects of potential treatments are not approachable in this way.

13.4.1 Specific Treatment or Not?

The ethical issues depend on the particular circumstances of the case, which include the following.

a. *The case is of an early stage of a tumor type which is treatable with significant prospects of benefit to the patient.*

There are few ethical issues in these situations. Thus whether or not a resectable carcinoma of the colon should be removed is rarely contentious, except in patients for whom another illness is expected to cause the patient's death before the resection of the carcinoma could have any beneficial effect.

b. *The case is of a later stage of a tumor type for which treatments may have been described, but there is no certainty that the treatment will have beneficial results.*

Patients in this situation are often invited to participate in a clinical trial. This situation is the most ethically difficult [30]. The reported benefits are only an average of a whole group. Some patients in the group may have more severe side effects of the treatment than patients not receiving the treatment. It is also possible that some patients will die of the treatment, and hence sooner than if no treatment had been given. Without clinical trials, however, there would be no way of assessing new therapies [30,31].

c. *The cases in which (i) the tumor is widely spread and is not known to respond to any specific treatment, or (ii) accepted treatment has been given to maximum toxicity, and the tumor is not further specifically treatable.*

Patients in this situation may wish to participate in a clinical trial, and so may the ethical issues in the previous section may

apply. Patients may occasionally approach organizers of trials directly.

This is a different situation to the normal trial, in which the organizers approach known sufferers. The trial budget may not be able to assist in these situations. The ethical issue of allocation of resources then arises. In some countries there are charitable "compassionate funds" available for some patients in this situation.

13.4.2 Hope

Every time a person seeks the advice of a medical attendant, hope is felt that the symptom or illness will not be significant, or if serious, is easily treated. Even faced with a diagnosis of a severe and probably incurable disease, hope persists for many reasons [32]. These include that the patient may be "one of the lucky few" who survives the disease, or that some new treatment will become available before the patient dies.

Historically, medical practitioners sometimes withheld diagnoses of malignancy from patients. Theodor Billroth (1821–1894), the great surgeon in the nineteenth century wrote [32]:

> I consider it the duty of a surgeon, under certain circumstances, to deceive his patients as to the incurability of their disease whenever he considers an operation unadvisable, or when he declines to undertake it. The surgeon, when he cannot remove, ought to relieve the sufferings of his patient, both psychically and physically. Few persons possess that peace of mind, resignation, or strength of character, call it what you will, necessary to enjoy life quietly with the knowledge that they are subjects of a fatal disease. Patients, although outwardly calm, seldom really thank you for too plain a confirmation of what they secretly suspect.

Since the middle of the twentieth century, this position has been largely untenable because the tumors in almost all patients are treatable in some way. Many patients wish to be fully informed about their options concerning therapies, and to contribute to the decision-making process. Nevertheless, although Billroth's era of medical practice has gone, his sentiment concerning the state of mind of the patient after being given the diagnosis of malignant tumor remains valid. A patient in despair suffers severely.

13.4.3 Advance Care Planning/"Directives"

At any stage in a person's life, he or she may be rendered incapacitated to the point of being unable to make decisions concerning his/her own medical care [33]. This occurs particularly following incurable brain injuries and strokes, but also in cases of primary or secondary tumors in the brain. Other causes of incurable incapacity arise during treatments for cancer from coma due to a metabolic condition including septicemia.

Many individuals have written wills before such an event occurs, but may not have indicated their wishes as regards what life-support and other measures the patient wishes to have when they are unconscious with no prospect of recovery.

Generally "Advanced Care Planning" refers to the discussions which may take place of possible actions should the patient become incapable of making decisions for him/herself. "Advanced Care Directives" usually refer to the written document describing what is to be done or not done should the patient suffer that incapacity.

In many countries, Advance Care Directives are legally accepted as guidance to the relatives and/or medical attendances as to how to proceed should a situation arise which is likely to lead to the end of the patient's life. Disputes when they occur in individual cases are settled legally, and there are significant differences in the appropriate laws in different jurisdictions [34].

In oncology practice, the issue of advanced care planning is more complicated. The clauses in the plans drawn up may vary from country to country. In the United States, plans commonly mention specific procedures, such as airway intubation, that the individual does not

wish to undergo. In other countries, the wording describing the situations in which the plan is to be invoked may be less specific.

In Western countries, only approximately 15% of patients with incurable cancer (often stated as "Stage IV") who are offered Advance Care Planning advice, sign a plan.

For health funding agencies, Advance Care Plans may be seen as a way of limiting the costs of treatment of terminal cancer (one third of the total costs—see Section 13.3). For oncologists, however, the plans may be seen as a denial of the ethical responsibility to preserve life, and as jeopardy to the patients' basis of hope (see Section 13.4.2).

13.4.4 Oncology Research

Some ethical issues in clinical trials are mentioned in Sections 13.2 and 13.3. A main issue is whether or not a particular proposed trial should go ahead or not.

First, it is not ethical, and hence rarely undertaken, to conduct a clinical trial of a treatment compared with no treatment, unless no other treatment is known to be beneficial.

To put this another way, the trial is always of the new treatment (A) in comparison with either the "best established treatment" (B), or no treatment (C), if no (B) exists.

Second, in clinical circumstance for which a "standard therapy" is accepted, a new clinical trial should be based on laboratory or clinical evidence of the possible superiority of the new agent or regimen in comparison with "standard" therapy. If one or more previous trials have shown the agent to be ineffective, any proposed new trial should be based on reasons why the new trial may yield better data than the previous trial(s).

Applying these principles may be difficult however, because the range of tumor types against which a new drug may be useful cannot be foretold. Similarly the efficacies of combinations of agents, and of side effects, cannot be predicted in all subgroups of the population.

Other ethical issues relate to phase 0 and phase I clinical trials. In these, normal individuals consent, usually for monetary payment, to being exposed to a potentially harmful agent [35,36]. The subjects should be fully informed, and the payments should reflect the risks.

Phase II and III clinical trials raise additional ethical issues insofar as patients in one or other group in a particular trial may not be treated optimally. However, which group in any particular trial suffers cannot be foreseen (if it were foreseeable, the trial would be redundant and should not be undertaken).

In considering these issues, the benefits to humanity overall may be taken into account. The absence of clinical trials, developments of new treatments would be more difficult (see Section 13.2.1).

Apart from human studies, the main aspects of oncology research relate to use of animals for detecting potential carcinogens and testing potential drugs. Most of the issues relate to welfare of the animals [36].

13.4.5 Allocation of Resources

In communities with limited financial capacities, questions arise as to how best to allocate this resource. This aspect of "Health Economics" is becoming more and more pressing. This is because the costs of delivering all possible medical care to all individuals who may benefit from the many kinds of care continue to rise and exceed many other costs to community. Even the wealthiest communities are beginning to consider this issue.

The decisions required [37–40] can be allocations between:

- Health and non-health expenditure
- Direct patient care and prevention
- Direct patient care and research (if these can be separated, because in many disciplines of medicine, the two activities are interlinked)

- Live-saving and non-life-saving treatments
- The young and the old
- Infrastructure, equipment and salaries of staff.

References

[1] Garrett-Mayer E. Principles of anticancer drug development. New York, NY: Springer; 2011.

[2] Baguley BC, Kerr DJ, editors. Anticancer drug development. Philadelphia, PA: Elsevier; 2002.

[3] US Food and Drug Administration. Patient network newsletter. Available at: http://patientnetwork.fda.gov/learn-how-drugs-devices-get-approved/drug-development-process/step-1-discovery-and-development.

[4] Tan Y, Wu Q, Xia J, et al. Systems biology approaches in identifying the targets of natural compounds for cancer therapy. Curr Drug Discov Technol 2013; 10(2):139–46.

[5] Hait WN. Anticancer drug development: the grand challenges. Nat Rev Drug Discov 2010;9:253–64.

[6] Florey HW. Br Med J 1944;2(4361):169–71.

[7] Macmillan Cancer Support. Macmillan exposes hospitals failing to provide adequate care for cancer patients. Available at: http://www.macmillan.org.uk/Aboutus/News/Latest_News/Macmillanexposeshospitals failingtoprovideadequatecareforcancerpatients.aspx.

[8] American Cancer Society. Finding cancer treatment centers. Available at: http://www.cancer.org/treatment/findingandpayingfortreatment/findingtreatmentcenters/index?sitearea=eto.

[9] US Food and Drug Administration. Guidance for industry: S9 Nonclinical evaluation for anticancer pharmaceuticals. Available at: http://www.fda.gov/downloads/Drugs/.../Guidances/ucm085389.pdf.

[10] Developmental Therapeutics Program. Biological testing branch: available protocols and documents. Available at: http://dtp.nci.nih.gov/branches/btb/available_on_request.html.

[11] ClinicalTrials.gov. Homepage. Available at: http://ClinicalTrials.gov.

[12] Friedman M, Furberg CD, DeMets L, editors. Fundamentals of clinical trials (4th ed.). New York, NY: Springer; 2010.

[13] Piantadosi S. Clinical trials: a methodologic perspective (2nd ed.). Hoboken, NJ: John Wiley & Sons; 2005.

[14] Lagiou P, Adami J, Trichopoulos D. Measures and estimates of cancer burden. In: Adami H-O, Hunter D, Trichopoulos D, editors. Textbook of Cancer Epidemiology. Oxford Scholarship Online, 2008. pp. 34–60.

[15] US Food and Drug Administration. Guidance for industry: Clinical trial endpoints for the approval of cancer drugs and biologics. Available at: http://www.fda.gov/downloads/Drugs/.../Guidances/ucm071590.pdf.

[16] Begg CB, Schrag D. Attribution of deaths following cancer treatment. J Nat Cancer Inst 2002;94:1044–5.

[17] Yin D, Morris CR, Bates JH, et al. Effect of misclassified underlying cause of death on survival estimates of colon and rectal cancer. Source California Cancer Registry, Public Health Institute, Sacramento, CA 95825, USA. J Natl Cancer Inst 2011;103(14):1130–3.

[18] In: Ref. [12]. pp. 251–308.

[19] Cancer Research UK. Randomised trials. Available at: http://www.cancerresearchuk.org/about-cancer/trials/types-of-trials/about-randomised-trials.

[20] Reid K. Interpreting and understanding meta-analysis graphs. Aust Fam Phys 2006;35:635–8.

[21] Begley S, Respaut R. Special Report: Behind a cancer-treatment firm's rosy survival claims. Available at: http://www.reuters.com/article/2013/03/06/us-usa-cancer-ctca-idUSBRE9250L820130306.

[22] http://www.cancer.gov/statistics/glossary#C.

[23] Jönsson B, Wilking N. Cancer burden and costs. Ann Oncol 2007;18(Suppl. 3):iii8–iii22.

[24] Mariotto AB, Yabroff KR, Shao Y, et al. Projections of the cost of cancer care in the United States: 2010–2020. J Natl Cancer Inst 2011;103(2):117–28. Erratum in: J Natl Cancer Inst. 2011, 103(8):699.

[25] Siddiqui M, Rajkumar SV. The high cost of cancer drugs and what we can do about it. Mayo Clin Proc 2012;87(10):935–43.

[26] Barlow WE. Overview of methods to estimate the medical costs of cancer. Med Care 2009;47(7 Suppl. 1):S33–6.

[27] Angelos P (Ed.). Ethical issues in cancer patient care (2nd ed.). New York, NY: Springer; 2007.

[28] Johns Hopkins Sheridan Libraries and University Museum. Hippocratic oath. Available at: http://guides.library.jhu.edu/content.php?pid=23699&sid=190555.

[29] National Institutes of Health. Ethics in clinical research. Available at: http://clinicalcenter.nih.gov/recruit/ethics.html.

[30] Hedlund S. Hope and communication in cancer care: what patients tell us. Cancer Treat Res. 2008;140:65–77.

[31] National Institute on Aging. Advance care planning. Available at: http://www.nia.nih.gov/health/publication/advance-care-planning.

[32] Stein GL, Fineberg IC. Advance care planning in the USA and UK: a comparative analysis of policy, implementation and the social work role. Br J Social Work 2013;43:233–48.

[33] Workman P, Aboagye EO, Balkwill F, et al. Guidelines for the welfare and use of animals in cancer research. Br J Cancer 2010;102(11):1555–77.

[34] Culyer AJ, editor. Encyclopedia of health economics. Philadelphia, PA: Elsevier; 2014.

[35] Glied S, Smith PC, editors. The Oxford handbook of health economics. Oxford: Oxford University Press; 2011.

[36] Williams A. QUALYS and ethics: a health economist's perspective. Soc Sci Med 1996;43:1795–804.

[37] Tenner LL, Helft PR. Ethics in oncology: an annotated bibliography of important literature. J Oncol Pract 2013;9(4):e145–53.

[38] US Food and Drug Administration. Selected FDA GCP/Clinical trial guidance documents. Available at: http://www.fda.gov/ScienceResearch/SpecialTopics/RunningClinicalTrials/ GuidancesInformationSheetsandNotices/ucm113709.htm.

[39] In: Ref. [29]. pp. 65–77.

[40] Billroth T. Lectures on surgical pathology and therapeutics. Translated from the 8th ed. by C Hackley, vol. 2, New Sydenham Society; 1878. p. 494.

Aspects of the Prevention of Tumors

OUTLINE

L.P. Bignold: Principles of Tumors.
DOI: http://dx.doi.org/10.1016/B978-0-12-801565-0.00014-7

Prevention is the major benefit which scientific studies of disease can provide. In general, the methods of prevention are most successful when a definite cause is known. As mentioned in Section 1.5, effective preventive programs for skin cancer involve avoidance of ultraviolet light exposure; prevention of lung cancer is based on non-use of tobacco; and prevention of viral-induced tumors which involves avoidance of infection by the causative viruses (see Sections 4.5 and 5.7).

Along with the search for environmental carcinogens, other methods of prevention have been developed (see Figure 14.1). These have been mainly the screening of sub-groups within the overall populations for tumors at early stages in their development. Screening is aimed at preventing the individual cases of disease causing deaths of the individuals in which they are found. The screening methods may consist of physical examination and imaging of organs, or of measurement of a relevant "biomarker" (see Section 9.7) in a body fluid. For efficiency, screening is usually only applied to sub-groups of the whole population. The sub-groups may be based on age or, to lesser degrees, other risk factors such as relevant family history.

Screening may also be directed to hereditary (germline) genomic events which may

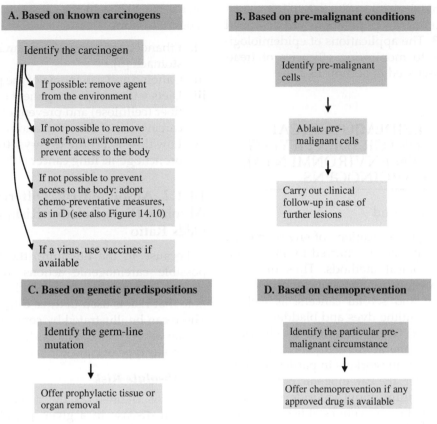

FIGURE 14.1 Overview of the prevention of tumors.

predispose to tumors later in the individual's life (see Chapter 7). In these individuals, there is usually no tumor present at the time that the screening is undertaken. If an individual is found to have the genomic lesion predisposing to the disease, they may or may not accept preventive resection of the organ(s). The main example of this is bilateral mastectomy for persons found to have predisposing germline mutations (Section 7.8.2).

A fourth mode of prevention is the possibility of administering drugs to individuals who are at risk through

i. Known exposure to known carcinogens,
ii. Presence of an unresectable pre-malignant lesion, for example, dysplasia in

bronchial epithelium in a long-term tobacco smoker, and
iii. Germline predisposition in which the relevant organ cannot be removed.

This chapter gives an overview of these aspects of prevention of tumors. Section 14.1 gives a general account of epidemiological methods in the identification of unknown carcinogens. Sections 14.2 deals with laboratory methods for assessing the possible carcinogenic potencies of environmental chemicals to humans. Section 14.3 discusses screening programs for the prevention of deaths from malignant tumors. Section 14.4 is concerned with the roles of cancer preventive drugs in the prevention of tumor formation in high-risk individuals.

Epidemiological issues in determining incidences of tumors in populations are described in Chapter 3. The applications of epidemiological methods to monitoring efficacies of treatments are discussed in Section 13.2.

14.1 EPIDEMIOLOGICAL METHODS FOR IDENTIFYING UNKNOWN ENVIRONMENTAL CARCINOGENS

14.1.1 Background

Most of the identifications of environmental carcinogens have been achieved by fairly simple epidemiological methods. Thus the links between sun exposure and skin cancers, chimney-sweeping and scrotal cancers, as well as working with aniline dyes and bladder cancers (see Section 1.2), were identified by clinicians who observed general markedly increased incidences of tumors in workers in particular occupations [1]. Later, the carcinogenic potencies of uranium, chromium, vinyl chloride monomer, and forms of asbestos were established by rigorous epidemiological studies of workers in particular industries [2].

Formal statistical analyses have also been required in many non-occupational tumor-causation situations. In the 1940s, tobacco smoking in pipes was known to be a cause of carcinomas of the lip and mouth, and was suspected to be a cause of carcinoma of the lung [3]. However, tobacco was only accepted as a cause of lung cancer after careful statistical analyses of data on incidences in relation to tobacco usage [4,5].

Often, it has been difficult to prove by epidemiological methods and statistical analysis that many additional dietary or environmental factors are the causes of any of the common types of malignant tumors. As three examples, there is little agreement as to whether the following

agents have roles in the causation of particular types of tumors:

i. Ethanol consumption and carcinoma of the stomach [6].
ii. Caffeine and carcinoma of the pancreas [7].
iii. Diets with a high component of vegetable fiber (cellulose) and prevention of carcinomas of the colon and rectum [8,9].
iv. Low dose exposure to chrysotile and bronchogenic lung cancer [10].

14.1.2 Absolute Risk, Difference in Absolute Risk, Relative Risk, and Odds Ratio

Because in the literature, the influences of possible carcinogenic actions are often discussed in terms of "risk," it is important to note that risk is used in three different contexts. These can be illustrated by considering the risk of, say, all 40-year-old individuals developing a particular type of cancer in the next 30 years.

(a) Absolute Risk

This is defined as the actual incidence of a given disease in a given population over a given period of time. It is the same as "absolute incidence" (see Section 3.2.1). For example, in the population of 40-year-olds, the absolute incidence/risk of developing a particular cancer in the next three decades of their lives might be 5.25%.

(b) Difference in Absolute Risk

If, say, the group of 40-year-old persons is divided into consumers of a particular dietary additive and non-consumers of the additive. If they are followed for a number of years, it might then be found that the consumers of the additive have an incidence of the tumor of 5.5% and the non-consumers of the additive an incidence of 5.0%.

In this situation, the difference in absolute risk of consumption of additive versus non-consumption of additive for the particular cancer is 0.5%.

(c) Relative Risk

Relative risk is the increase in incidence rate of a disease in a test group as a percentage of the incidence rate in the control group. In the group of 40-year-old persons, the relative risk of additive consumption for development of the particular cancer is $0.5 \div 5.0 \times 100 = 10\%$.

(d) Odds Ratio

This is the ratio between the risk/incidence of an event in the presence of an exposure, and the risk/incidence in the absence of the exposure. In the population of 40-year-old persons, the odds ratio is 5.5:5.0 = 1.1:1.0.

14.1.3 "Risk factors" and Their Identification by Epidemiological Methods

(a) General

A "risk factor is any attribute, characteristic or exposure of an individual that increases the likelihood of developing a disease or injury" [11]. The term includes known and suspected direct carcinogens for man, as well as circumstances which are associated with increased incidences of cancer, through presumably indirect mechanisms.

In general, to identify a causative factor for a disease, the steps in the process followed are

i. To identify the risk factors, and
ii. To investigate on what basis the association might be causative rather than spurious.

The associated factor might not be directly causative of the tumor, but might be associated through some lifestyle, occupational or other activity, with the factor that is the effective cause of the tumor.

To illustrate this: having boxes of matches in one's pockets is associated with increased risk of carcinoma of the lung. However, the boxes of matches do not cause the carcinomas. The association is because the matches are used to light cigarettes, the smoke of which causes the tumors [12]. Logically though there is a second association. It is that the matches in the pockets are associated with lighting cigarettes. This second association would point to the causative factor of carcinoma of the lung.

The two main kinds of studies to establish association are cohort studies and case-controlled studies, as discussed in (b) and (c) following. Both of them rely on accurate diagnosis of cases of tumors in the given populations.

(b) Cohort Studies: Finding Changing Incidences of Disease

These are generally prospective studies (see Figure 14.2). The simplest kind of this study involves identifying particular subsets in a population, and observing whether or not over time a particular disease occurs in the members of the subsets [14] more frequently than in the general population. For etiological factors, the studies always collect data on occupation and lifestyle.

The whole population is usually divided into groups by sex, age, occupational, and lifestyle factors. "Occupation" may mean not just the industry in which the individual works, but the activities of that person in that industry.

Especially when applied to occupational studies, the method depends on

i. Accurately diagnosing the cases of the disease, and hence establishing the incidence in the different periods of time.
ii. Correctly separating out the subgroups within the population in which the tumor type is increased.
iii. Accurately determining the degree of exposure of individuals in that population to the suspected etiological factor(s).

Degree of exposure is important because, in a particular industry, employees in different activities in the industry—clerical,

Cohort studies

Prospective studies,
(usually wait for cases to appear
>10 years)

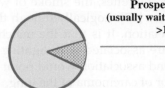

A
At the beginning of the trial, a subpopulation is selected according to the hypothesis of the trial. If the hypothesis is that exposure to a certain agent is associated with the disease, then the subpopulation is of individuals with known exposure to the agent. At this point, it is not known whether the incidence of the disease is different from the incidence in the population as a whole.

B
At the end of the trial, the incidence of the disease in the subpopulation in comparison with the whole population is determined. Hence the existence of an association is either shown or not shown.

Case-controlled studies

Retrospective studies,
(cases are completed before the study is begun)

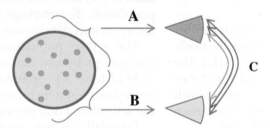

A
From the whole population, the subpopulation of sufferers of the disease is identified.

B
At the same time, a "control" subpopulation is taken from the whole population. The individuals in this subpopulation have the same characteristics (age, gender, occupation, etc.) as individual members in the disease-sufferers group.

C
Identification—by comparison with controls —of occupational, life-style or other factors (colored arrows) which might have been responsible for the disease in the sufferers.

FIGURE 14.2 Cohort and case-controlled studies.

administrative, and ancillary staff as well as employees handling the putative carcinogens— may have quite different degrees of exposure to agents than "factory floor" workers. Thus the incidence of tumor in all the workers may be only slightly above average, but the incidence in the factory floor workers may be greatly in excess of the average for the whole population.

For lifestyle factors, usually only self-reported exposures to putative carcinogenic factors are documented. For example, the data on "pack-years" of exposure to tobacco [13] for each individual are derived from the statements of the individual subjects in the study. These statements are usually not independently assessed.

(c) Case-controlled Studies

These are essentially retrospective studies. The principle [14] involves comparing aspects of the occupations and lifestyles of individuals known to have a particular disease, to the same aspects of members of the general population who are known not to have the disease.

The method depends on accurate diagnoses of cases of the disease in question, and paying attention to changes in terminology (see Section 3.4). Depending on various factors, terminological differences may cause patients to be inappropriately included or excluded from either or both the disease group and the control group. In studies of tumors which depend largely on histopathological diagnosis, it may be necessary to have the histopathological slides of all the cases reviewed—even by multiple pathologists.

Case-controlled studies are cheaper, easier, and carried out more rapidly than prospective studies [15]. There is no need to establish actual incidence rates of the disease being studied.

14.1.4 Assessing Associations for Possible Causative Significance: Bradford Hill's Guidelines

The most widely quoted guidelines are those of Bradford Hill (1965) [16]. They were written in answer to the question "What aspects of an association should we especially consider before deciding that the most likely interpretation of it is causation?"

The guidelines were based on consideration of data according to their

i. "Strength,"
ii. Consistency,
iii. Specificity,
iv. Temporality,
v. Appropriateness of biological gradient (i.e. dose–incidence relationship),
vi. Plausibility,
vii. Coherence,
viii. Consistency with experimental data, and
ix. Consistency with analogous data.

Although changes in detail have been suggested (e.g. [17,18]), Hill's comments on the value of his guidelines remain valid:

> What I do not believe—and this has been suggested—is that we can usefully lay down some hard-and-fast rules of evidence that must be obeyed before we can accept cause and effect. None of my nine viewpoints can bring indisputable evidence for or against the cause-and-effect hypothesis and none can be required as a *sine qua non*. What they can do, with greater or less strength, is to help us to make up our minds on the fundamental question—is there any other way of explaining the set of facts before us: is there any other answer equally, or more, likely than cause and effect?

14.1.5 Aging as a Factor in the Causation of Tumors

Aging has long been accepted as risk factor for most of the common types of tumors [19,20]. This is based on increases in incidence of malignancies with age which have now been clearly documented for over half a century (Table 14.1).

In principle, increasing age could contribute to tumor formation by several possible mechanisms as follows:

i. The individual could have a hereditary predisposition which only manifests late in life. Thus in hereditary non-polyposis colorectal carcinoma, the tumors occur

TABLE 14.1 Death Rates for Malignant Neoplasms, by Age: United States, Selected Years 1950–2010

	1950[1,2]	1970[2]	1990[2]	2010[3]
All persons				
Under 1 year	8.7	4.7	2.3	1.6
1–4 years	11.7	7.5	3.5	2.1
5–14 years	6.7	6.0	3.1	2.2
15–24 years	8.6	8.3	4.9	3.7
25–34 years	20.0	16.5	12.6	8.8
35–44 years	62.7	59.5	43.3	28.8
45–54 years	175.1	182.5	158.9	111.6
55–64 years	390.7	423.0	449.6	300.1
65–74 years	698.8	754.2	872.3	666.1
75–84 years	1,153.3	1,169.2	1,348.5	1,202.2
85 years and over	1,451.0	1,320.7	1,752.9	1,729.5

Abstracted from [21].
Data are based on death certificates. For footnotes, see original.

approximately two decades earlier than in individuals without the syndrome. It is possible that other hereditary predispositions exist which manifest only in the eighth or higher decades in life. The identification of these would be difficult because many of the family members may have died of one or more other causes before the age of the effect of predisposition is reached.

ii. All cells which turnover (labile cells, see Section A1.3.3) may accumulate genomic errors due to inherent defects in replicative fidelity of DNA. These may, in occasional cells, occur in a pro-tumor gene(s) and cause "sporadic" tumors (see Section 4.1.2).

iii. Exposure to a carcinogen, or withdrawal of a carcinogen inhibitor, may occur only in later ages, in relation to lifestyle or other changes in the individual.

14.1.6 Complexities of Geographical/Cultural/Ethnic Factors

It is well known that certain tumor types are more common in certain parts of the world and populations than others [22] (Figure 14.3). In the 1970s, it was widely assumed that all human populations are genetically the same with regard to spontaneous incidences of tumor types. On this assumption, the lowest rate of incidence for any tumor type anywhere in the world was taken as its "spontaneous rate" in all humans. Any excess incidence of a tumor type in a particular part of the world could then be considered due to some environmental influences. Overall, the excess was assessed 80–90% of all tumors [23]. Certain observations supported this. Thus betel nut chewing leads to high incidences of carcinomas of the mouth in India and South-East Asia [24]. Moreover, populations with low incidences of certain types of cancers were reported to show increases in incidences of cancers common in Western communities when those populations abandoned their indigenous cultures and adopted Western lifestyles [25].

However, since then, there have been observations on the differences in incidences in tumors among ethnic groups which appear to share the same geography and culture. For example, African American men suffer prostate cancer more frequently than European Americans [26]. As another example, the high incidence of nasopharyngeal carcinoma in Southern Chinese does not seem to be explained by viral (Epstein-Barr virus) and environmental factors alone [27]. The matter of the possible general role of different genetic make-ups in different incidences of tumors in different ethnic/cultural populations groups has not been clarified.

The most widely accepted view is that the different incidences of tumors in different parts of the world are probably due to largely unrecognized local environmental carcinogens, culturally determined exposures to carcinogens [28], and low-penetrance hereditary factors (Section 7.8).

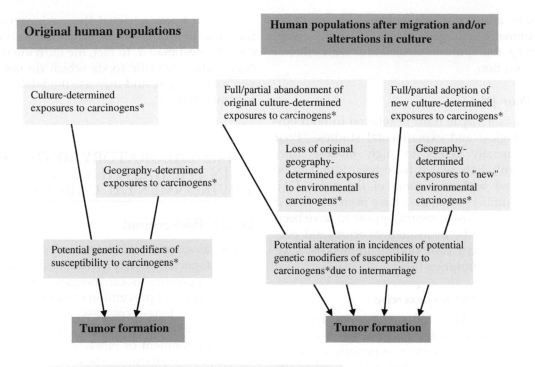

FIGURE 14.3   Some potentially complex aspects of the genetic, cultural, and geographical factors in tumor formation.

14.1.7 Other Risk Factors

Major risk factors [29] for cancer include well-established causative factors, e.g., tobacco usage, ultraviolet B exposure, certain viruses, and certain industrial chemicals. Aging as a risk factor has been discussed in Section 14.1.5.

However, some comments can be made on other factors on the list. This is especially because for most of them, little experimental corroborative evidence has been found.

(a) Affected Family Members

The high impact hereditary factors seem to apply to only a small number of tumor types, while the possibility of multiple low-impact genomic events have not been fully assessed (see Chapter 7).

(b) Alcohol Consumption in Excess

This is shown to be associated with liver cancer, but for other tumors a direct causative relationship is unclear. Of itself, ethanol is a coagulant of all macromolecules, and is not genopathic [30]. Its metabolic product (acetaldehyde) is a Group 1 carcinogen (IARC), and is produced in the liver. Hence a biologically feasible causative mechanism exists for ethanol and liver tumors. It is not metabolized by esophageal cells, so no chemical mechanism is evident for the (disputed) association (see Section 14.1.1). Further, neither ethanol nor acetaldehyde accesses the colon, where increased cancer rates are documented.

To explain the association, it may be remembered that alcohol is consumed in complex mixtures, such as beer, wines, and spirits, which contain additional carcinogens. The putative additional carcinogens might act on the colon, creating the spurious association with ethanol consumption.

(c) Poor Diet

This is a vague term in relation to both epidemiological and experimental studies. "Poor diet" usually refers to high fat–low fiber "Western" diets which have been found to be associated with carcinoma of the colon [31]. Large numbers of studies have been conducted [31], but conclusive results appear to have been made difficult by measurement errors and possible concurrent confounding influences, such as folate consumption [31].

(d) Lack of physical exercise

There is no known direct chemical basis for this. Exercise increases metabolism of energy-producing dietary factors, especially carbohydrates and triglycerides, but as far as is known, does not qualitatively induce any biochemical event. Lack of physical exercise may be associated with certain dietary habits or ill-health due to some other factor or process, which may have carcinogenic potency.

(e) Obesity

Calories in dietary carbohydrates, triglycerides, and amino acids alone are not carcinogenic. A possible suggestion must be that to become obese, lack of exercise and excess calorie intake are necessary [32,33]. In individuals, obesity *per se* may not be a causative factor. For example, the causative factor may be in the food eaten, but may only have its effect if eaten in the large amounts associated with eating the excess calories necessary to become obese. The food which is excessively ingested may contain an incidental excess of carcinogens—or excess inhibitors of inhibitors of naturally occurring carcinogens—which cause the tumors.

Without knowing what the carcinogens are, the value of losing weight by calorie restriction cannot be estimated. In fact, the individual may cease eating calorific foods which do not contain a carcinogen, and increase the intake of the food which does.

14.2 LABORATORY METHODS IN THE IDENTIFICATION OF ENVIRONMENTAL CARCINOGENS

14.2.1 Background

The assessment of suspected carcinogenic agents, especially newly synthesized chemicals, for actual potency as carcinogens in humans is a major aspect of prevention of tumors.

There is a long-recognized need for laboratory tests to identify which chemicals, either in the environment or newly developed by the chemical and pharmaceutical industries, may be carcinogenic in humans. During more than a century of research, hundreds of different tests have been developed. These have used mainly whole animals, eukaryote cells in culture, and bacteria (Table 14.2).

The following three sub-sections outline the principles of some of the laboratory tests. Other uses of laboratory animal experiments are indicated in Figure 14.4. The ways in which these tests are currently used by regulatory authorities is discussed in Section 14.2.5.

14.2.2 Tumors in Animals

Historically, these were the first kinds of test for carcinogenic potency. Successful models began with the discovery of the rabbit and rodent skin model for carcinogens (see Section 1.2.1 and Figure 14.4). These models closely mimic the skin cancers associated with exposures of the skin of workers to chimney soot or mineral oils.

TABLE 14.2   Suitability of Organisms for Tests of Genetic Toxicity and Carcinogenesis

Effect measured	Suitability of organism to test of effect		
	Whole animal	Cells cultured *in vitro*	Cells cultured *in vitro*
Tumor formation	Yes (see Figure 4.4)	Yes: malignant transformation *in vitro*	No
Nuclear structural lesions (esp micronuclei)	Yes	Yes	No
Chromosomal aberrations	Yes	Yes	No
Somatic mutations (sub-cytogenetic damage with phenotypic effect)	Yes, e.g. Big Blue®, Mutamouse®	Yes, e.g., mouse lymphoma test	Yes, e.g., Ames test
DNA damage in extracts of living cells (no phenotypic effect)	Yes, comet assay, SOS test, Unscheduled DNA synthesis, Adducts on DNA	Yes, various tests	Yes, various tests
Germline mutations	Yes[a]	No	No
Chromosomal damage to gametes	Yes[a]	No	No
Lethality to offspring	**Yes[a]**	**No**	**No**

[a]*In various species including in Drosophila and zebra fish.*

Laboratory animals have been used in these tests according to the native susceptibility of the particular species and strain. Smaller animals are preferred to larger ones because of the costs of maintaining them for long periods (usually two years or more) [34–36]. Certain carcinogens cause tumors in a variety of kinds of parent cells ([37], Figure 14.4), while other agents cause tumors in only a few kinds of parent cells. Certain genetically engineered mice (see Section 7.9.3) with particular susceptibilities to classes of carcinogens may be used [38].

However, there are differences in the susceptibilities of different strains of laboratory animals to tumors, and factors in the care, diet, and housing of the animals may influence results [39].

The reasons for the inconclusive results in animals may lie in different defenses against carcinogens in the different species. This is discussed in Section A3.4.1.

These rodent assays nevertheless remain an important test in the "batteries" of tests recommended by regulatory authorities (see 14.2.5).

14.2.3 Enhanced Rates of Malignant Transformation in Cells Cultured *in Vitro*

This was the first *in vitro* model of carcinogenesis to be developed (see Figure 14.5). Although at present it is not favored as a predictor of carcinogenicity, the phenomenon is worth considering in some detail because the most markedly transformed cells are tumorous, and grow as such when transplanted from the culture dish back into the whole animal from which the parent cells were obtained.

(a) Virally versus Non-virally Transformed Cells

It is important at the outset to distinguish transformation caused by viral infections from

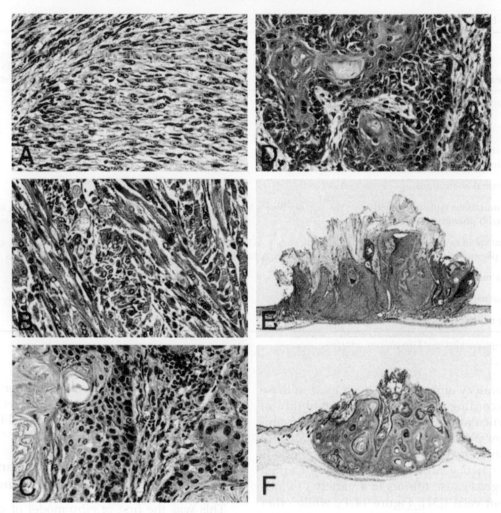

FIGURE 14.4   Micro images of tumors caused by benzo[a]pyrene in mice. (A–C) Histological appearance of a fibrosarcoma (A), a rhabdomyosarcoma (B), and a squamous cell carcinoma (C) induced in AhR(+/+) male mice by subcutaneous injection of B[a]P. (Hematoxylin/eosin staining, ×100.) (D–F) Histological appearance of a squamous cell carcinoma (D), papilloma (E), and keratoacanthoma (F) induced in AhR(+/+) female mice by topical application of B[a]P. (Hematoxylin/eosin staining, ×100 for D, ×5 for E and F.) AhR = aryl hydrogen receptor. *Source: Ref. [37], reproduced with permission.*

transformations induced by non-replicating agents such as chemicals and radiations. Virally transformed cells usually carry replicated viral genes in their genomes (see Sections 4.5 and 5.7) and are thus hybridomas. The properties of these cells may reflect the actions of viral proteins, either directly or through interacting with the host's proteins. However, chemically and radiation-transformed cells have only the assumed altered DNA of the parent cell in their genomes, without any exogenous gene products affecting their properties.

All of the following paragraphs refer to non-viral transformation of cells.

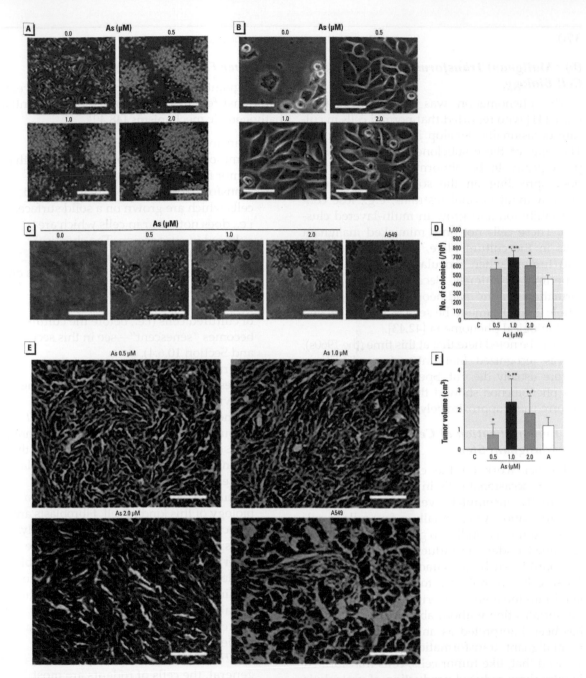

FIGURE 14.5 Transformation of cells *in vitro* and morphology of tumors grown in an experimental animal from the transformed cells. Neoplastic transformation of HELF cells exposed to 0.0, 0.5, 1.0, or 2.0 μM arsenite (As) for about 15 weeks (30 passages). C = control (untreated) HELF cells. A549 carcinoma cells served as the positive control. Morphological images of cells (*A*) after culture with 10% FBS (bars = 500 μm) and (*B*) after culture with 1% FBS (bars = 50 μm). (*C*) Photomicrograph of cell colonies in soft agar; bars = 500 μm. (*D*) The number (mean ± SD) of cell colonies in soft agar (*n* = 3). (*E*) Representative pathological sections of tumors 4 weeks after cells were inoculated into nude/BALB/c mice; tumors induced by arsenite-transformed cells consisted of undifferentiated and spindle cells; bars = 100 μm. (*F*) Volume (mean ± SD) of tumors in nude/BALB/c mice (*n* = 6).

*p < 0.01 compared with control group. **p < 0.05 compared with 0.5 or 2.0 μM arsenite groups. #*p* < 0.05 compared with mice implanted with cells exposed to 0.5 μM arsenite. *Source: Ref. [40], reproduced with permission from* Environmental Health Perspectives.

(b) "Malignant Transformation" in Tumor Cell Biology

The phenomenon was first described by Earle [41] who reported that normal cells in culture occasionally develop immortal subclones. The cells of these subclones were noted morphologically to be abnormally rounded with poor spreading on the solid surface, and to grow without normal restrains, e.g., lose contact inhibition and grow in multi-layered clusters. These cells not only mimicked malignant cells in the culture plate, but they also grew into tumors when inoculated into the source experimental animals. Because of this, "transformation *in vitro*" and neoplastic change in the whole animal came to be seen as closely related cell biological phenomena [42,43].

It can be noted here that at this time (the 1960s), it was recognized that untransformed cells in culture usually die out. Specific investigation of this phenomenon showed that dying of the cells occurred after approximately 60 cell divisions [44].

(c) Immortalization of Cell Cultures as a Pre-Malignant Step

Further work on the changes which may occur in occasional cells in culture revealed an apparently incomplete version of malignant transformation. Occasionally, a few cells in cultures acquire the ability to grow indefinitely but remained under the influence of environmental controls, such as contact inhibition [45–47]. These cells are not able to grow when inoculated into the animal of origin of the cells. This immortalization without ability to form tumors has been interpreted as an essential early step in malignant transformation [47]. Later work showed that, like tumor cells, cells transformed *in vitro* show reduced production of proteolytic enzymes and antigenicity.

Suggested mechanisms of immortalization included loss of apoptosis/senescence/terminal differentiation, all being involved in the basis of tumors *in vivo* [45–47]; see also (Sections 10.3–10.5).

(d) Other Features

Malignant transformation *in vitro* has several additional features which bear on its potential pathogenetic mechanisms [48,49].

i. Transformation *in vitro* is much more common per number of growing cells than tumor formation *in vivo*.

ii. Transformation occurs only in cultured cells which are grown on a solid surface, i.e., does not occur in cells which are cultured in suspension. The chemical nature of the surface seems to be unimportant, as transformation occurs on glass and plastics alike.

iii. It occurs in only a very small proportion of cultured cells (i.e., before the culture becomes "senescent"—see in this section and Section 10.6.4).

iv. The rate of non-viral transformation varies according to the kind of cell being grown. Many kinds of cells, especially epithelial cells, cannot be cultured *in vitro* for sufficient periods of time and so cannot be used for these tests. Because of this, the testing methods usually involve fibroblasts. The ease with which these cells can be cultured *in vitro* accords with the *in vivo* biology of this kind of cell. Fibroblasts are in fact the most durable cells in the body. In scars (Section 10.7.3), fibroblasts replace all other kinds of cells and can remain viable while they lay down enough collagen to seal indigestible noxious agents from the remainder of the body.

v. The cells of different species are more liable to non-viral transformation *in vitro* than cells of other species. In general, the cells of rodents are most susceptible, those of humans are less susceptible, and those of other species rarely susceptible.

vi. The growth of transformed cells occurs in lowered concentrations of serum factors in the medium (see Section 6.2.2).

vii. Chromosomal aberrations occur in the transforming cells at an early stage.

viii. Transformed cells grow in suspension, for example in agar, while the non-transformed cells usually grow only on surfaces.

(e) Non-viral Agents Capable of Inducing/Enhancing Rate of Malignant Transformation in Vitro

Many agents including radiations and chemicals of many different classes are able to cause enhanced rates of malignant transformation *in vitro* [50–52].

Pre-treatment of cell cultures with one type of agent followed by another can increase the rate of transformation with subsequently administered other agents, for example radiations [53] followed by chemicals.

There can be technical difficulties, for example in the choice of cell line for the assay, and culture medium [50]. A recent review has found that a Syrian hamster cell line and the BALBc 3T3 cell lines are more reliable than a C3H10T1/2 line as tests of potential carcinogenicity in humans [54]. However, the same cell line has shown that lead acetate is capable of transforming these cells, while not being an established carcinogen in humans [55].

The general view has been that transformation *in vivo* is a valuable, but not perfect, indicator of carcinogenicity in humans [56,57].

(f) Pathogenesis

Neither the exact nature of the genomic changes required for initiation of transformation, nor how the changes are induced has been established with certainty for all cell lines. These genomic changes may in fact be different in the different cell lines. However, it has become apparent that while the whole population of transformed cells is immortal, individual transformed cells often have reduced life-spans and are also liable to increasing chromosomal aberrations [58]. Moreover, increasing nucleotide errors accumulate in transformed cells [59], suggesting that replicative infidelity of DNA ("mutator phenotype," see Section 5.1) occurs in association with transformation, as well as the long-established karyo-instability [60] (see Section 5.3).

Numerous genes have been proposed as the basis of transformation [61]. A specific suggestion has been that genomic instability begins in cultured cells in two patterns of onset: an "acute" form mediated by deregulation of *c-myc* and a "chronic" (slow) form which requires separate involvement of multiple genes involved in the cell cycle [62]. Another suggestion has been of a specific transformation gene, possibly in association with a *RAS* gene [63].

14.2.4 Other Genopathic Phenomena Used for Testing Potential Carcinogenicity

A list of genopathic effects is given in Section 4.1.1 (see also Table 14.2 and Figure 14.6). This subsection describes the principles of particular tests. The rationale for these tests includes the assumption that if an agent has one or more of these effects in experimental systems, it may also have carcinogenic effects in humans.

(a) In Living Animals

Direct carcinogenicity tests are described in Chapter 1 (Figure 1.4). Whole animals are also used for testing potencies of agents for inducing nuclear structural abnormalities, chromosomal aberrations, developmental defects, and germline mutations. All of these changes may be used as the surrogate event for the beginning of tumor formation [52,64,65].

A commonly recommended assay for nuclear structural abnormalities is the mammalian erythrocyte micronucleus test (OECD Test 474) [66]. Another is the mammalian bone marrow chromosomal aberration test (OECD Test 475) [67,68]. Assays of these kinds are part of the battery of tests for pharmaceuticals recommended by the Food and Drug Administration (FDA).

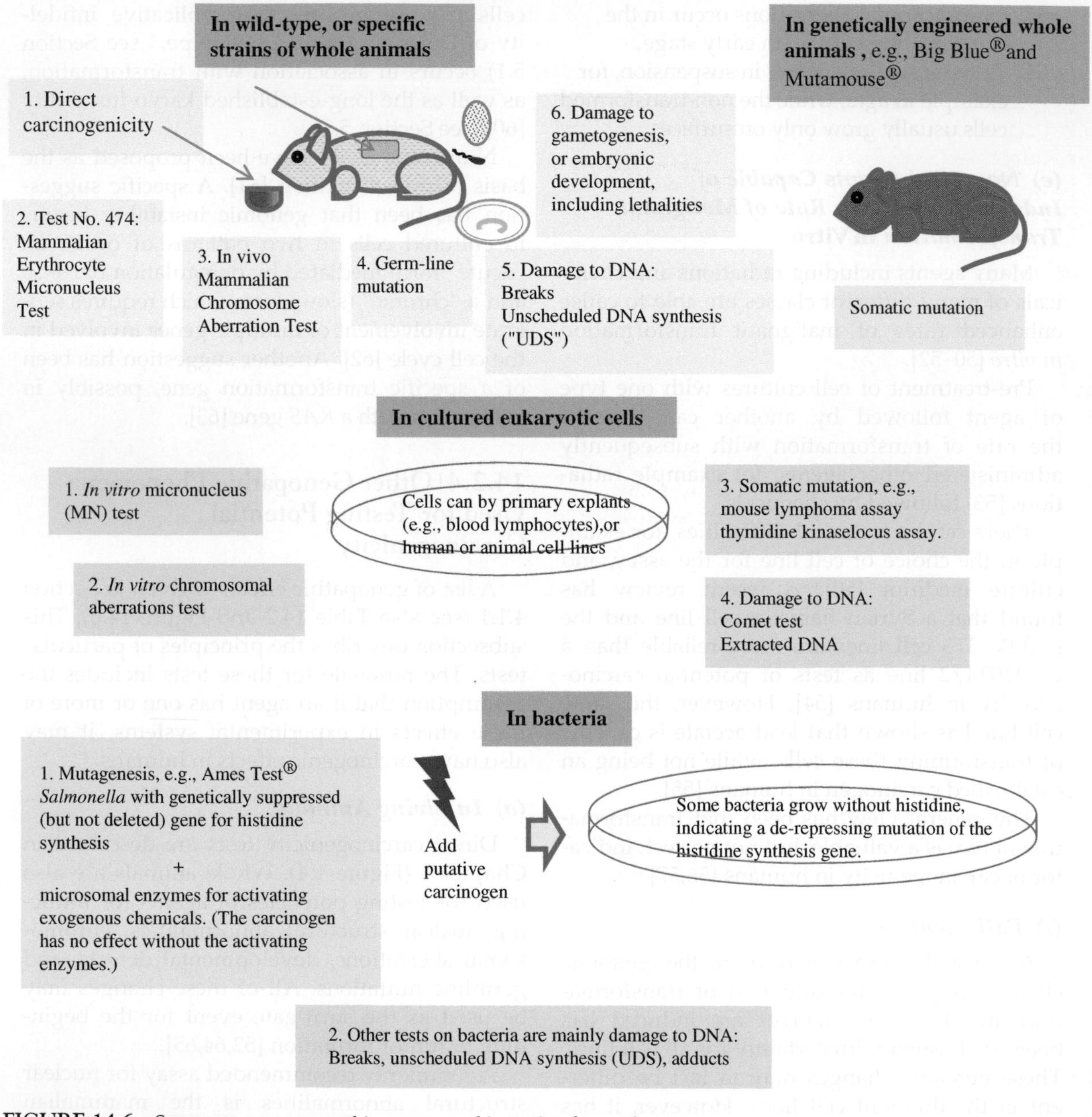

In wild-type, or specific strains of whole animals

1. Direct carcinogenicity

2. Test No. 474: Mammalian Erythrocyte Micronucleus Test

3. In vivo Mammalian Chromosome Aberration Test

4. Germ-line mutation

5. Damage to DNA: Breaks Unscheduled DNA synthesis ("UDS")

6. Damage to gametogenesis, or embryonic development, including lethalities

In genetically engineered whole animals, e.g., Big Blue® and Mutamouse®

Somatic mutation

In cultured eukaryotic cells

1. *In vitro* micronucleus (MN) test

Cells can by primary explants (e.g., blood lymphocytes),or human or animal cell lines

3. Somatic mutation, e.g., mouse lymphoma assay thymidine kinaselocus assay.

2. *In vitro* chromosomal aberrations test

4. Damage to DNA: Comet test Extracted DNA

In bacteria

1. Mutagenesis, e.g., Ames Test® *Salmonella* with genetically suppressed (but not deleted) gene for histidine synthesis

\+

microsomal enzymes for activating exogenous chemicals. (The carcinogen has no effect without the activating enzymes.)

Add putative carcinogen

Some bacteria grow without histidine, indicating a de-repressing mutation of the histidine synthesis gene.

2. Other tests on bacteria are mainly damage to DNA: Breaks, unscheduled DNA synthesis (UDS), adducts

FIGURE 14.6   Some common genopathic assays used in testing for carcinogens.

Examples based on damage appearing during development include embryonic lethality, and the appearance of bodily deformities (teratogenesis) in offspring of exposed parent animals. This has been suggested to imply a genomic event caused in the embryonic or fetal cells of the developing offspring [69]. Other tests involve germline genomic events in rodents, *Drosophila*, or zebra fish [70]. Valid cross-species inferences from these tests may be difficult [71].

(b) In Cultured Cells

In cultures of cells *in vitro*, other kinds of genopathic changes are used. These include tumor-like behavior, such as malignant transformation (see Section 14.2.3). Tests involving induction of micronuclei (see Section 10.2.4) are recommended in the FDA guidelines.

i. In vitro *mammalian cell micronucleus test*
   In this test, cells are exposed to the test agent and then examined for small nuclei, indicating disturbance in the mitotic process, especially in the distribution of chromosomes to daughter cells. A large number of kinds of cells can be used, including human peripheral blood lymphocytes (OECD TG487; [72]).
ii. In vitro *mammalian chromosome aberration test*
   For this test (OECD Test No. 473; [73]), a variety of sources of cells, including primary explants, for example, of peripheral blood leukocytes[74], or of human or mammalian cell lines, can be used. Methods for visualizing aberrations are also various (see Section 9.5).
iii. In vitro *mammalian cell gene mutation assay*
   A widely used example of this kind of test is the mouse lymphoma assay/ thymidine kinase locus assay (OECD Test 476) [75,76]. The cells in the assay are heterozygous for a gene that encodes an enzyme which is capable of incorporating analogues of thymidine into DNA. The test cells are killed when they divide in the presence of a toxic analogue of thymidine (trifluorothymidine) [75,76]. In the presence of a mutagen, the only cells which survive are those that have undergone a loss-of-function mutation in the second allele of the gene.

(c) Tests in Bacteria

In these methods, the carcinogenesis-surrogate genopathic phenomenon is usually a genomic event resulting in a phenotypic change (Ames test) [77,78]. This test in bacteria is widely used as an indicator of potential carcinogenicity in humans.

The basis of the test is as follows. Strains of *Salmonella* are used which have genetically suppressed (but not deleted) gene for histidine synthesis. Histidine is essential for growth of these bacteria.

One sample of the bacteria is then mixed with the chemical being tested, and also microsomal enzymes (often rat liver homogenates). These enzymes are necessary for activating exogenous chemicals (see Section 4.4.1).

Another sample of bacteria are mixed with the microsomal enzymes, but not the test chemical.

The samples are then grown separately on media without histidine. Any cells which grow normally have undergone a de-repressing mutation of the histidine synthesis gene. A higher number of growing cells from the sample containing the test substance compared with the number from the sample without the substance indicates that more of the de-repressing mutations occurred in association with the test chemical, i.e., the test chemical is a mutagen.

The dependency on activation by the liver homogenates raises the issue that such activation may not be appropriate for some chemicals and routes of exposure to carcinogens in humans. Numerous modifications are described [79]. The test is widely considered the best available for many kinds of chemicals [80,81]. Tests of this general nature are recommended in the FDA guidelines [67].

(d) Other Tests

A variety of other tests are used in general genotoxicity (genopathicity, see Section 4.1.1) testing, and are not commonly applied to carcinogenesis (Table 14.2 and [64]). The tests include those for DNA damage, such as the comet assay (for strand breaks), the SOS test, the test for unscheduled DNA synthesis, and the formation of adducts on DNA (see Section 4.4.4). These tests can be applied to a wide variety of experimental systems. Another group of genopathic assays depend on germline mutations and damage to embryos. Almost all of these tests are more expensive than those described above.

14.2.5 Multiplicity of Tests and Methods for Their Use

None of the laboratory organisms are biologically the same as humans, and so the test systems are only surrogates for tumor formation in humans. As such, no single test has been established as a perfect method for indicating carcinogenic potency in humans. The tendency has been to compare the performances of the various tests, and try to establish the best ways of using them.

Because of the large number of possible tests, in the 1970s, government agencies suggested specific algorithms for testing [64]. An algorithmic procedure normally begins with a high-sensitivity, low specificity test. After that, there are successive tests which have progressively higher specificity.

For example in an algorithmic procedure for carcinogenesis the primary screening tests might be (Table 14.2, see in Table II in [64]):

i. A bacterial mutagenesis test (e.g. Ames),
ii. A mammalian cell culture mutagenesis test, and
iii. An *in vivo* chromosomal test.

If the substance produced a positive result in (i) or (ii), it would be tested secondarily by a *Drosophila* embryo lethality assay. If a positive result was found in that test, the substance would be subjected to a tertiary test of cancer bioassay in animals.

If the substance produced a positive result in (iii), it would be tested with cancer bio-assays and rodent embryo lethality assays.

The general outcome of the research is that many tests have various degrees of sensitivity and specificity for carcinogenic effects in humans, and no single test is fully satisfactory in either 100% sensitivity or 100% specificity. Because of this, these algorithms were found to be of restricted value in practice.

As a result, many agencies require only that examples from a list of kinds of test (i.e., a "battery") be performed, without rigid ranking as to precedence. The results of laboratory tests are also often used in conjunction with epidemiological data for assessments of possible carcinogenicity for humans. The IARC publishes monographs [82] on individual substances or groups of substances. Their method for evaluating carcinogenic risks integrates evidence of all relevant kinds:

> An interdisciplinary Working Group of expert scientists meets at IARC in Lyon, France, for eight days. The Working Group reviews the published studies and evaluates the weight of the evidence that an environmental factor can increase the risk of cancer. After performing and discussing a critical review of the published scientific evidence, the Working Group formulates the evaluations. As a result, each agent is classified into one of five categories.

The process does not involve a prescribed set of tests.

The US Environmental Protection Authority has published (2005) its guidelines [83] which are similar to the IARC methods.

14.2.6 Co-carcinogens and Other Multifactorial Circumstances

A difficult aspect of experimental systems for testing individuals is the issue of co-carcinogenesis and multi-factorial etiologies [84,85]; see also Sections 4.1.4 and 4.1.5. In a test system this may mean the necessary sequential or concurrent application of many different substances. To test for these possibilities, large numbers of experiments would be necessary. However, if whole animal models are used, the cost would probably be substantial.

14.2.7 Discordant Relative Potencies for Carcinogenesis in Relation to Other Effects

The fact that different species have different susceptibilities to carcinogens is described in Section 4.2.3. Here it is noted that there are species differences in the susceptibilities to cellular abnormalities which are taken as surrogates for carcinogenesis.

The phenomenon of non-correlating cell biological tests has been extensively studied in relation to alkylating agents (see in [86]). Different species have been used for each test: for example, cells in culture for chromosomal aberrations, bacteria (as in the Ames test (see Section 14.2.4) for mutagenicity and a mammalian species (such as the mouse) for carcinogenicity. The results were overall, that no constant association between carcinogenic and other potencies could be established for these agents [86].

This apparent non-correlation of relative genopathic potencies has been found with many other chemicals. For example, ethlybenzene, which is carcinogenic in rats, has been found to be non-mutagenic in bacteria and yeast, and non-clastogenic (i.e., does not cause chromosomal aberrations, see Section 4.1.1) in Chinese hamster ovary cells [87].

Phenolphthalein induces tumors in rodents and is clastogenic in some experimental test systems, but does not to form adducts on DNA and is not mutagenic in bacterial and mammalian cell models [88]. Acrylonitrile is mutagenic in bacterial assays, but not clastogenic in cultured mammalian cells [89]. Aniline

compounds which are carcinogenic in Fischer 344 rats have been found not to be mutagenic in most test systems, and only clastogenic in some systems in the highest doses [90].

These findings suggest strongly that all mutagens may not be carcinogens [91].

14.2.8 Possible Future Experimental Methods

The previous sections show that perfect methods for predicting carcinogenic potencies of chemicals for humans are still lacking. This may be because, if any agent is found to have an effect in a non-human experimental system, the effect may not be valid for human beings because of toxico-kinetic differences between species (Section A3.1). For human *in vitro* cellular systems, positive results may not be valid for the whole human because the cells in culture are deprived of defensive barriers (Section A3.1).

It is possible that a series of cell-free assay systems might be optimal. These might involve the effects of test substances in cell-free assays of human genome processes for fidelity of replication of DNA, chromosomal aberrations, and cytoskeletal-membrane cohesion among possibly other biochemical phenomena known to occur early in malignant change of cells.

14.3 HUMAN LESION-SCREENING PROGRAMS AND THEIR EFFICACIES IN PREVENTING DEATHS FROM TUMORS

14.3.1 Overview

It was noted at the beginning of this chapter that screening programs of various kinds have become established public health measures. The process of developing a screening program involves [92]:

i. Developing the technology for the test.
ii. Validating the test in terms of specificity for a particular type of tumor (i.e., to avoid

false positive results), and sensitivity (i.e., to avoid false-negative results).

iii. Evidence that a positive result will change the circumstances of the individual. This may mean the availability of a suitable treatment, or indication for the adoption of some life-style change.

iv. Assessment of the proportion of the screened population found to be positive.

v. Related to (iv) are cost–benefit estimations. "Costs" include financial costs of implementing the population testing, and psychological costs for individuals, in particular those provided with inconclusive test results.

This section deals with chemical, pathologic, and genomic bases for screening programs.

14.3.2 Molecular Evidence of Exposures to Known Carcinogens; the Question of Maximum Permissible Exposure Levels

(a) Chemical and Genomic Tests

In Sections 14.1 and 14.2, the complexities of cultural/ethnic as well as occupational factors in exposures to carcinogens were noted. Because self-reported data are not always accurate indications of doses/exposures of individuals in populations to known or putative environmental carcinogens, more precise tests have long been sought. The tests developed have mainly involved chemical residues in blood or tissues of individuals.

In addition, some biological tests have been employed, e.g., the modified lymphocyte micronucleus test [93]. The idea of testing for acquired genomic abnormalities, as for example, in bronchial epithelium in tobacco smokers (e.g., [94]), is rarely undertaken because of cost.

In many surveys, the only samples available are of blood. Members of the general public are sometimes reticent about submitting to more invasive procedures, such as adipose tissue sampling.

These methods are valuable, especially because they may provide accurate dose gradient data [95]. Thus if a study is primarily about the possible carcinogenic effect of exposure to a chemical and the chemical accumulates in the body, then the actual measurements of that accumulation is obviously important (see Section 14.1.3). The data are independent of self- or other-person reports of exposures. In a given work-force, there may be some encouragement to claim exposures if any tumor or other disease develops (see biases in Section 11.11.5). Measuring tissue levels of the putative carcinogen can yield clarifying data [96].

Examples of these methods are as follows:

i. *Presence of carcinogen in body fluids or tissues*
    For arsenic exposure, skin, hair, and nails are the most valuable structures to assess this carcinogen [97]. Levels of tobacco derivatives can be measured in blood samples [98]. Pesticide exposure, for example to organochlorines, can be measured in adipose tissue [99].

ii. *Adducts on particular proteins and/or DNA*
    With improving methods of chemical assays, these can be carried out, not only on serum proteins (e.g., [100]), but also on tissues and proteins from a variety of tissues [101,102].

(b) Maximum Permissible Exposures/"Levels"

As noted in Section 4.1.3, in humans, very few carcinogens are known which can cause tumors in a single dose of exposure. Once a "biological gradient relationship" of exposure to putative carcinogen and incidence of tumor type has been established (see Section 14.1.4), the question in prevention is "what is an acceptable dose and/or exposure rate." For many known chemical carcinogens, low doses with infrequent episodes of exposure may have no effect. As a result, while identifying relationships between dose rate—tumor formation, there may be difficulty nevertheless in establishing how much carcinogenic exposure is "harmless," "acceptable," or "unacceptable" [103–106].

Examples are in workers with radio-active substances. This because human populations are exposed to "background" ionizing radiations, comprising cosmic rays and naturally occurring radioactive elements in minerals, such as thorium in granite [107].

All of the points in the previous section have enormous influence on regulatory bodies dealing with occupational health. The application of this is in establishing "maximum acceptable levels" of exposures.

At the time of writing, the National Institute for Occupational Safety and Health is reviewing its methodology and recommendations concerning human carcinogens [108].

14.3.3 Lesions: Pre-malignant or Early Malignant Tumors

The concept that cancers, if detected early enough, can be treated and cured applies to many types of tumors (see Table 14.3). However, for other types, especially lung cancer, the lesions appear to be almost always incurable before they can be detected by clinical examination or imaging techniques [109]. Each tumor type must be treated separately when considering screening procedures.

There are currently three widely implemented kinds of programs.

(a) Uterine Cervical Cytology

Since its introduction in the 1950s, screening for abnormal cells in this organ has been associated with approximately 70–80% falls in deaths from carcinoma of the cervix in Western countries [111,112] (see Figure 14.7). This has occurred despite an apparent marked rise in the incidence of infection of the cervix with human papilloma virus. The virally infected cells are histologically abnormal, and are destroyed by the treatments. The costs of the screening programs, however, are too great for poorer countries, so that in those regions, no such fall in mortality from cervical cancer has occurred [113].

TABLE 14.3 Sensitivities and Specificities of Screening Methods Based on Detection of Early Pathological Lesions

Tumor	Screening procedure	Sensitivity/specificity
Carcinoma of cervix	Cells from the cervix examined by microscopy	High/high
Carcinoma of breast	Physical examination	Low-moderate/low
	Radiological appearances[a]	Moderate/moderate
Carcinoma of colorectum	Fiber-optic visualization of organ[b]	Very high/very high
	Fecal occult blood	Moderate/moderate
Carcinoma of prostate	Specific antigens in serum	Sensitive for metastatic tumor only
		Low levels in serum have low specificity.
	Digital palpation of prostate	Sensitive for late-stage tumors only
		Moderate–high specificity

N. b. For carcinoma of the lung, screening procedures based on standard radiology or sputum cytology have been tested, but have not been shown to be effective. Current evidence suggests that screening based on low-dose spiral CT scans may reduce mortality in heavy smokers [110].
[a]Costly.
[b]Very costly.

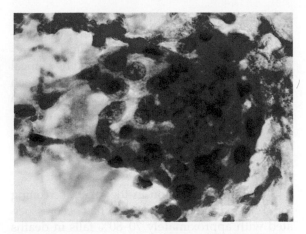

FIGURE 14.7 Abnormal cells in cervical smear. The cluster consists of large, irregular cells, without evidence of specialization.

(b) Breast Cancer Physical Examination and Imaging

Screening for early tumors of the breast began in the 1950s. At first, detection of suspicious masses was by physical examination by a physician or by the woman herself. However, as imaging technology improved, radiology and ultrasound techniques have become the main method of detecting abnormal masses [114] (see Figure 14.8). Improvements of 30–50% in mortality from the disease have been widely reported. However, the precise degrees of the contribution of screening to these improvements are controversial. The difficulties in conducting clinical trials have included

i. The long course of the disease (the disease may be latent for decades before re-appearing in aggressive form—see 8.4),
ii. The changing imaging criteria for lesions to be further assessed,
iii. The changes in histopathological criteria of malignancy, and
iv. The rapid developments in new therapies for the disease (see Section 11.6) [114,115].

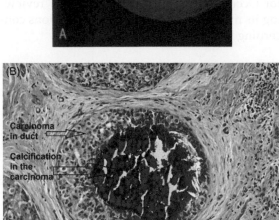

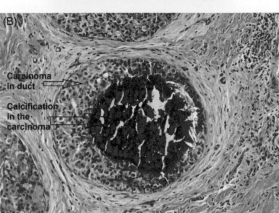

FIGURE 14.8 Screening for carcinoma of breast. (A) Breast carcinoma in mammographic image. (B) Calcification in breast carcinoma.

(c) Occult Blood in Feces and Sigmoidoscopy for Colorectal Cancer

Since the 1970s, screening of individuals older than 60 years has been undertaken based on tests for occult blood in the feces. The sensitivity of the test has been improved by technological improvements [116,117]. The test is said to reduce mortality of colorectal carcinoma by approximately 18% [118].

The other test is by colonoscopy or flexible sigmoidoscopy. This has been used for the detection of carcinomas in patients with polyposis syndromes (see Sections 7.2.2 and 7.4.2). As a random test in asymptomatic individuals over 60 years of age, "once-only" flexible sigmoidoscopy is used as a screening method. This technique is much more expensive than fecal occult blood testing, but has the advantage that it detects small lesions including polys as well as large lesions. The method is said to reduce mortality of colorectal carcinoma by approximately 28% [118], although complications such as perforation of the bowel can occur [119].

14.3.4 Serum Biomarkers

The issue of the use of serum prostate-specific antigen (PSA) as a screening and diagnostic test has been described in Section 3.2.4 (see also [120]). The essential problem is that slightly increased levels of PSA in the serum are not specific for carcinoma of the prostate, and high levels of PSA in the serum indicate disease which has already spread in the body, and hence is not "early" as required by the objectives of a screening program. Improved methods are needed [121].

Serum CA125 levels (see Section 3.3.2) are not currently widely recommended for screening for ovarian carcinoma for similar reasons.

14.3.5 For Germline Genomic Abnormalities (Personalized Disease Prevention through Genomic Studies)

Low-penetrance germline predispositions to tumors, and the possibility of identifying combinations of germline lesions, are discussed in Section 7.9. Such data could be useful in many settings. For example, if there were to be found a "diagnostic combination of genomic lesions" for carcinoma of the colon and rectum, the individual would be able to go onto a colonoscopic surveillance program, similar to that for known hereditary predispositions to the tumor (see Sections 7.2.2 and 7.4.2).

At present, no methods are available which are sufficiently reliable for decision-making [122,123].

14.3.6 Assessing Benefits of Screening

Screening is aimed at detection of asymptomatic tumors. Such tumors are likely to be early in their course, so that they can be treated more effectively than would have been possible if detected later in their course. For screening programs, the participation rate is important. This is the proportion of individuals invited to enter the program who agree to participate. Typically, participation rates vary with the age of the invited person, as well as other factors, such as inconvenience of the test and the expected positive rate (less participation for studies of rarer lesions). Because participants sometimes delay entry into the program, or "drop out" from the program, it is optimal to have data on "participant years."

The most important benefit of a screening program therefore is a reduction in the mortality rate adjusted for participation rate of the disease.

As described in Section 3.5, mortality data for a specific tumor type in a population may depend on various factors. However, if those factors were to remain constant after the inception of the screening program, the following might be expected:

i. For tumor types for which no effective treatment is available regardless of stage of disease (see Section 1.3.3), screening should have no effect on mortality of the tumor type if incidence rates also remain unchanged.
ii. For tumor types for which effective treatment is available, screening leading to earlier treatment should cause a decline in mortality. This appears to be the case with screening for carcinoma of the cervix, see Figure 14.9.

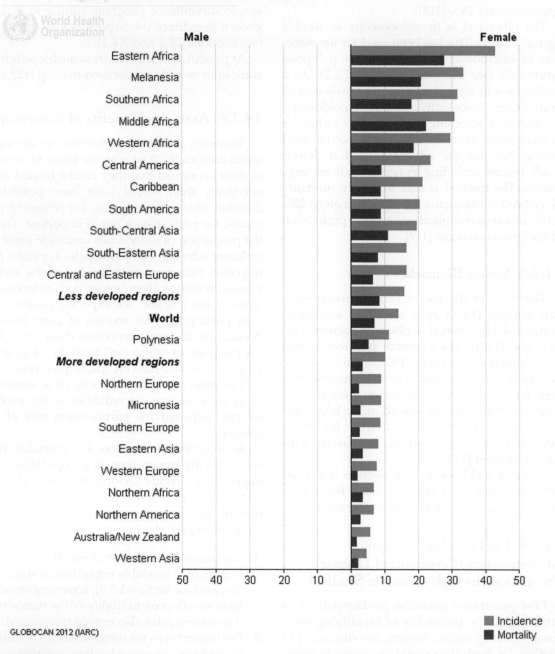

GLOBOCAN 2012 (IARC)

FIGURE 14.9 Aspects of incidence of carcinoma of the cervix. The data from the different countries show marked variations between each other, and in the individual countries, over time. *Reproduced with permission from Ferlay J, Soerjomataram I, Ervik M, Dikshit R, Eser S, Mathers C, et al. GLOBOCAN 2012 v1.0, Cancer Incidence and Mortality Worldwide: IARC CancerBase No. 11 [Internet]. Lyon, France: International Agency for Research on Cancer; 2013. Available from: http://globocan.iarc.fr, accessed on 23 June 2015.*

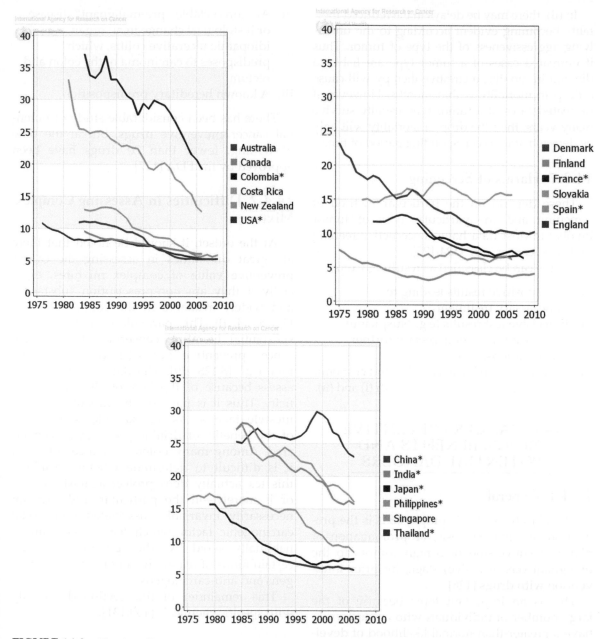

FIGURE 14.9   (Continued).

In (ii), there may be delays in the reduced mortality becoming evident according to the underlying aggressiveness of the type of tumor. Thus if untreated cases of a tumor type are liable to die within months, a curative therapy will cause a drop in mortality within months. However, if untreated cases of a tumor type usually survive many years, then the drop in mortality will only be detected after a corresponding period of time.

14.3.7 Harms of Screening

These depend on the particular method of screening and the particular type of tumor involved. Most data have concerned screening for breast cancer, e.g., [124,125].

Harms can be classified as resulting from:

i. False positive results leading to unnecessary treatment,
ii. Indecisive test results (e.g. "suspicious of...") leading to unnecessary further investigations, and
iii. Anxiety, reduced quality of life, and economic loss which may be associated with (i) and (ii).

14.4 CANCER PREVENTIVE DRUGS: BENEFITS AND POTENTIAL DANGERS

14.4.1 General

The objectives of this prophylaxis is the prevention of carcinogenesis, or prolongation of phase of conversion of benign tumors to the malignant (i.e., invasive) stage, by prior intervention with drugs [126].

This is an important topic because of the large number of individuals who are known to have a greater-than-normal likelihood of developing tumors.

These individuals may have either

i. Prior exposure to a known carcinogen, e.g., tobacco smoker who has not developed tumor,

ii. An unresectable "pre-malignant" disease or histological change in an organ, e.g., idiopathic ulcerative colitis, which predisposes to carcinoma of the colon and rectum.
iii. A known hereditary predisposition.

There has been considerable study of potential cancer preventive drugs, but at the time of writing, fewer than 20 drugs have been approved by the FDA [127].

14.4.2 Difficulties in Assessing Complex Mixtures

At the outset, it should be noted that there are great difficulties in assessing the cancer preventive value of complex mixtures, especially if they are non-prescription substances and widely available in the community (see Figure 14.10). This particularly applies to the substantial literature concerning the possible cancer preventive potencies of dietary factors, e.g., [8,128,129]. The data are difficult to assess because of the lack of chemical specificity. Thus it is particularly difficult with the lifestyle factors. As an example, suggestions of the benefits of drinking tea can be considered. Among many potential sources of error, it is difficult to determine whether drinking this tea actually does protect against tumors, or is merely—in the particular trial, but not necessarily invariably—associated with some carcinogenic factor which tea drinkers might habitually avoid. Another possibility is that certain kinds of tea might contain both carcinogens and anti-carcinogens.

This remainder of this section deals only with specific chemicals [130,131].

14.4.3 Classification of Cancer Preventive Drugs

Cancer preventive drugs have been considered for many years, and many potential modes of action have been identified. Their

A. Agents which prevent activated carcinogens reaching normal cells in the body

Preventing absorption
(a) Through skin
(b) Through the lungs
(c) Through the stomach

Increasing excretion, e.g.,
(d) By the liver
(e) By the kidneys

Preventing activation, e.g.,
(f) By the liver
(g) By the kidneys

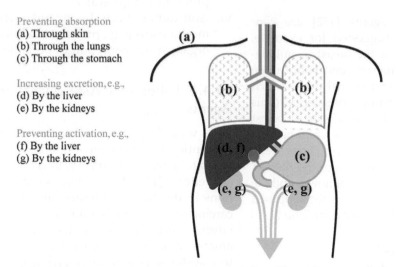

B. Agents that act on tumorous or pre-tumorous lesions. They either prevent benign tumors progressing to malignant ones, or malignant tumors becoming more malignant. They may have the either maturational or suppressive mechanisms.

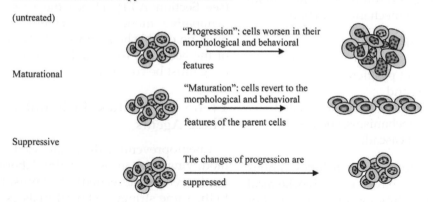

FIGURE 14.10   Potential mechanisms of cancer preventive drugs.

mechanisms of action have been divided into "blocking," "maturing," "suppressive," and "miscellaneous" kinds [132].

(a) Preventing Carcinogens Reaching Susceptible Cells

These have been called "blocking" agents [132]. They may act on any of the "toxico-kinetic/-dynamic factors" (see Section A3.1).

Sunscreen products for the prevention of skin cancers are a major example. A less established example is vegetable fiber in the prevention of colorectal carcinoma. Dietary fiber is said to "trap" carcinogens in the lumen of the colon.

Within the cells, some blocking agents may prevent activation of precarcinogens, and others may enhance detoxification systems of the target cell.

(b) Preventing "Maturation" of Susceptible Cells

These "maturation" agents [132] are controversial. It has been suggested for example, that the terminal ducts of the breast in nulliparous women are "less mature" and hence more liable to carcinogens than terminal ducts which have been "matured" through pregnancy. (This is to account for the higher incidence of breast carcinoma in nulliparous versus child-bearing women). None of these concepts have been proven in a biochemical sense, since the natures of the supposed carcinogens is unknown, and the toxico-kinetic or dynamic aspects of the supposed "immaturity" of the ducts are not established.

(c) "Suppressive agents"

"Suppressing" agents [132] are aimed at preventing benign tumors progressing to malignant ones, or malignant tumors becoming more malignant. Proposed mechanisms include:

i. Induction of "differentiation,"
ii. Preventing further genopathic events,
iii. Selectively inhibit proliferation of malignant cells, and
iv. Inhibition of possible cancer-associated inflammatory mechanisms, such as the arachidonic acid cascade.

The "miscellaneous" group includes various dietary agents which have known biochemical activities, such as inhibition of proteases, but whose pharmacology seems not to be the same as the other groups.

Another classification has a biochemical approach [133]. Drugs are classified into those which cause:

i. Reduction of endogenous production of carcinogens
ii. Reduction in amounts and absorption of exogenous carcinogens
iii. Prevention of activation of pre-carcinogens
iv. Inhibition of binding of carcinogens to DNA

v. Inhibition of tumor promoters and cell proliferation generally
vi. Anti-tumor effects, but by unknown mechanisms (e.g., protease inhibitors, organosulfur compounds, polyphenols).

14.4.4 Laboratory Assessments of These Agents

As with drugs for treating cancers, drugs for prevention of cancer must first be studied with laboratory tests for efficacy, toxico-kinetics and side effects [134]. Many studies examine reductions in the rates of tumors caused by specific carcinogens, or the inhibition of "genopathic effects" induced by known carcinogens, in animal, mammalian cell culture and bacterial test systems (see Section 14.2 and [135,136]). However, these types of study introduce two sets of toxico-kinetic and dynamic variables (see Section A3.1). Hence the validity of the genopathic model for human carcinogenesis must be established, and similarly, the validity of the action of the putative cancer preventive drug must be confirmed.

14.4.5 Difficulties of Clinical Trials of These Agents

Chemopreventive drugs are used as disease-modifying substances, and after laboratory testing (previous subsection), they must be subject to the same stringent clinical trials as are drugs against tumor cells in the body (Section 13.2).

Clinical trials of chemopreventive drugs however, are particularly difficult owing to the long observation periods and large study populations required to measure cancer incidence reduction. Some trials have concentrated on using reductions in "biomarkers" (defined as morphologic and/or molecular alterations in tissue between initiation and tumor invasion) [130]. Nevertheless, valid chemical targets—in respect of the supposed events of this transition from a non-invasive to an invasive state of individual lesions—for the chemopreventive agents have not been clearly identified.

14.4.6 Cancer Preventive Drugs for Particular Tumor Types

In general, the focus of these studies has been on anti-hormonal or anti-inflammatory drugs [137]. However, because each tumor type is a separate disease, and the cells involved have different toxico-kinetic properties, different drugs are probably required for each type. At the time of writing, cancer preventive agents have been investigated in relation to the common malignancies as follows.

a. *Lung carcinoma*

In the 1990s, beta-carotene, retinoic acid, vitamin E, and N-acetyl cysteine were tested in clinical trials. Beta-carotene was found to *increase* the incidence of cancers in tobacco smokers, while the other agents had no effect. Still other agents have been subject to trials, but without definite benefits being observed [138].

b. *Breast carcinoma*

Estrogen-receptor inhibitors, especially tamoxifen, have been shown to be effective in inhibiting the development of carcinoma of the breast, especially in women with positive family histories [139].

c. *Colorectal carcinoma*

Various anti-inflammatory drugs are presently being tested for preventive effects on this disease. The rationale appears to be related to the fact that carcinoma of the colon is a relatively common complication of idiopathic ulcerative colitis. No convincing evidence of the value of anti-inflammatory drugs has been reported [140].

d. *Prostatic carcinoma*

Most studies have focused on possible preventive effects of vitamin E, selenium, and the anti-androgenic drugs 5-alpha-reductase inhibitors. None have been shown to be beneficial in clinical trials [141].

References

[1] Lippman SM, Hawk ET. Cancer prevention: from 1727 to milestones of the past 100 years. Cancer Res 2009;69(13):5269–84.

[2] Siemiatycki J, Richardson L, Boffetta P. Occupation. In: Schottenfeld D, Fraumeni Jr JF, editors. Cancer epidemiology and prevention (3rd ed.). Oxford: Oxford University Press; 2006. pp. 322–54.

[3] Henley SJ, Thun MJ, Chao A, et al. Association between exclusive pipe smoking and mortality from cancer and other diseases. J Natl Cancer Inst 2004;96(11):853–61.

[4] National Library of Medicine. Surgeon General's Report. The health consequences of using smokeless tobacco. 1986. p. xvii–xix. Available at: http://profiles.nlm.nih.gov/ps/access/NNBBFC.pdf.

[5] Doll R, Hill AB. Lung cancer and other causes of death in relation to smoking; a second report on the mortality of British doctors. Br Med J. 1956;ii(5001):1071–81.

[6] Steevens J, Schouten LJ, Goldbohm RA, et al. Alcohol consumption, cigarette smoking and risk of subtypes of oesophageal and gastric cancer: a prospective cohort study. Gut 2010;59(1):39–48.

[7] Anderson LN, Cotterchio M, Gallinger S. Lifestyle, dietary, and medical history factors associated with pancreatic cancer risk in Ontario, Canada. Cancer Causes Control 2009;20(6):825–34.

[8] Asano TK, McLeod RS. Dietary fibre for the prevention of colorectal adenomas and carcinomas. Cochrane Database Syst Rev 2002(2):CD003430.

[9] Burkitt DP. Epidemiology of cancer of the colon and rectum. Cancer 1971;28:3–14.

[10] Bernstein D, Dunnigan J, Hesterberg T, et al. Health risk of chrysotile revisited. Crit Rev Toxicol 2013;43(2):154–83.

[11] World Health Organization. Risk factors. Available at: http://www.who.int/topics/risk_factors/en/.

[12] Kleinsmith LJ. Principles of cancer biology. London: Pearson Higher Education; 2005. p. 63.

[13] National Cancer Institute. Pack year. Available at: http://www.cancer.gov/dictionary?cdrid=306510.

[14] Thun MJ, Jemal A. Cancer epidemiology. In: Hong WK, Bast RC, Hait WN, et al. editors. Holland-frei cancer medicine, 8th ed., People's Medical Publishing House, Shelton CT, pp. 371–85.

[15] Sedgewick P. Case-controlled studies: advantages and disadvantages. BMJ 2014;348:f7707.

[16] Hill AB. The environment and disease: association or causation? Proc R Soc Med 1965;58(5):295–300.

[17] Howick J, Glasziou P, Aronson JK. The evolution of evidence hierarchies: what can Bradford Hill's "guidelines for causation" contribute? J Roy Soc Med 2009; 102(5):186–94.

[18] Goodman SN, Samet JM. Cause and cancer epidemiology. In: Schottenfeld D, Fraumeni Jr JF, editors. Cancer epidemiology and prevention (3rd ed.). Oxford: Oxford University Press; 2006. pp. 3–9.

[19] World Health Organization. Fact sheet: cancer. Available at: http://www.who.int/mediacentre/factsheets/fs297/en/.

[20] American Society of Clinical Oncology. Cancer.net: Prevention and healthy living. Available at: http://www.cancer.net/navigating-cancer-care/prevention-and-healthy-living/understanding-cancer-risk.

[21] US Department of Health and Human Services. Health, United States, 2013. Available at: http://www.cdc.gov/nchs/data/hus/hus13.pdf.

[22] Armstrong B, Doll R. Environmental factors and cancer incidence and mortality in different countries, with special reference to dietary practices. Int J Cancer 1975;15:617–31.

[23] Doll R. Strategy for detection of cancer hazards to man. Nature 1977;265:589–96.

[24] Mendelson RW. Betel nut chewer's cancer. U S Armed Forces Med J 1951;2(9):1371–5.

[25] Friborg JT, Melbye M. Cancer patterns in Inuit populations. Lancet Oncol 2008;9(9):892–900.

[26] Williams H, Powell IJ. Epidemiology, pathology, and genetics of prostate cancer among African Americans compared with other ethnicities. Methods Mol Biol 2009;472:439–53.

[27] Bei JX, Jia WH, Zeng YX. Familial and large-scale case-control studies identify genes associated with nasopharyngeal carcinoma. Semin Cancer Biol 2012;22(2):96–106.

[28] Wild CP. Environmental exposure measurement in cancer epidemiology. Mutagenesis 2009;24(2):117–25.

[29] National Cancer Institute. Cancer prevention overview. Available at: http://www.cancer.gov/about-cancer/causes-prevention/patient-prevention-overview-pdq.

[30] Phillips BJ, Jenkinson P. Is ethanol genotoxic? A review of the published data. Mutagenesis 2001;16(2):91–101.

[31] Aune D, Chan DS, Lau R, et al. Dietary fibre, whole grains, and risk of colorectal cancer: systematic review and dose-response meta-analysis of prospective studies. BMJ 2011;343:d6617.

[32] van Kruijsdijk RC, van der Wall E, Visseren FL. Obesity and cancer: the role of dysfunctional adipose tissue. Cancer Epidemiol Biomarkers Prev 2009;18(10):2569–78.

[33] Irigaray P, Newby JA, Lacomme S, et al. Overweight/obesity and cancer genesis: more than a biological link. Biomed Pharmacother 2007;61(10):665–78.

[34] Alison MR, editor. The cancer handbook. Wiley, Chichester, UK, 2007.

[35] Teicher BA, editor. Tumor models in cancer research. New York, NY: Humana/Springer; 2011.

[36] Abate-Shen C, Politi K, Chodosh LA, editors. Mouse models of cancer: a laboratory manual. Cold Spring Harbor, NY: CSHL Press; 2013.

[37] Shimizu Y, Nakatsuru Y, Ichinose M, et al. Benzo[a]pyrene carcinogenicity is lost in mice lacking the aryl hydrocarbon receptor. Proc Natl Acad Sci U S A 2000;97(2):779–82.

[38] Calvisi DF, Thorgeirsson SS. Molecular mechanisms of hepatocarcinogenesis in transgenic mouse models of liver cancer. Toxicol Pathol 2005;33(1):181–4.

[39] Hardisty JF. Factors influencing laboratory animal spontaneous tumor profiles. Toxicol Pathol 1985;13(2):95–104.

[40] Li Y, Xu Y, Ling M, et al. mot-2–mediated cross talk between nuclear factor-κB and p53 is involved in arsenite-induced tumorigenesis of human embryo lung fibroblast cells. Environ Health Perspect 2010;118:936–42.

[41] Earle WR. Production of malignancy in vitro; mouse fibroblast cultures and changes seen in living cells. J Nat Cancer Inst 1943;4:165–214.

[42] Sanford KK. Malignant transformation of cells in vitro. Int Rev Cytol 1965;18:249–311.

[43] Sanford KK. "Spontaneous" neoplastic transformation of cells in vitro: some facts and theories. Natl Cancer Inst Monogr 1967;26:387–418.

[44] Hayflick L. Current theories of biological aging. Fed Proc 1975;34(1):9–13.

[45] Simons JW, Bols BL, Naaktgeboren JM. Immortalization as an endpoint in studies on malignant transformation. Prog Clin Biol Res 1990;340D:207–17.

[46] Namba M, Mihara K, Fushimi K. Immortalization of human cells and its mechanisms. Crit Rev Oncog 1996;7(1-2):19–31.

[47] Weinberg RA. The biology of cancer, 2nd ed. New York, NY: Garland Science; 2014. pp. 391–437.

[48] Hynes RO, editor. Surfaces of normal and malignant cells. Chichester UK: John Wiley & Sons; 1979.

[49] Borek C. Malignant transformation in vitro: criteria, biological markers, and application in environmental screening of carcinogens. Radiat Res 1979;79(2):209–32.

[50] Mishra NK, Di Mayorca G. In vitro malignant transformation of cells by chemical carcinogens. Biochim Biophys Acta 1974;355(3-4):205–19.

[51] Combes R, Balls M, Curren R, et al. Cell Transformation Assays as Predictors of Human Carcinogenicity. The Report and Recommendations of ECVAM Workshop 39. ATLA 1999;27:745–67.

[52] Barile FA. Principles of toxicology testing, 2nd ed. Boca Raton, FL: CRC Press; 2013. pp 142–156; 229-255.

[53] DiPaolo JA, Casto BC. In vitro carcinogenesis with cells in early passage. Natl Cancer Inst Monogr 1978; 48:245–57.

[54] Vasseur P, Lasne C. OECD Detailed Review Paper (DRP) number 31 on "Cell Transformation Assays for Detection of Chemical Carcinogens": main results and conclusions. Mutat Res 2012;744(1):8–11.

[55] DiPaolo JA, Nelson RL, Casto BC. In vitro neoplastic transformation of Syrian hamster cells by lead acetate and its relevance to environmental carcinogenesis. Br J Cancer 1978;38(3):452–5.

[56] Tennant RW. Relationships between in vitro genetic toxicity and carcinogenicity studies in animals. Ann N Y Acad Sci 1988;534:127–32.

[57] Santella RM. In vitro testing for carcinogens and mutagens. Occup Med 1987;2(1):39–46.

[58] Balogh GA, Russo IH, Balsara BR, et al. Detection of chromosomal aberrations by comparative genomic hybridization during transformation of human breast epithelial cells in vitro. Int J Oncol 2006;29(4):877–81.

[59] Phillips DH, Arlt VM. Genotoxicity: damage to DNA and its consequences. EXS 2009;99:87–110.

[60] Cheung AL, Deng W. Telomere dysfunction, genome instability and cancer. Front Biosci 2008;13:2075–90.

[61] Heeg S, Doebele M, von Werder A, et al. *In vitro* transformation models: modeling human cancer. Cell Cycle 2006;5(6):630–4.

[62] Prochownik EV. c-Myc: linking transformation and genomic instability. Curr Mol Med 2008;8(6):446–58.

[63] Notario V, DiPaolo JA. Molecular aspects of neoplasia of Syrian hamster cells transformed in vitro by chemical carcinogens. Toxicol Lett 1998;96-97:221–30.

[64] Zeiger E. Historical perspective on the development of the genetic toxicity test battery in the United States. Environ Mol Mutagen 2010;51(8-9):781–91.

[65] US Food and Drug Administration. Guidance for Industry: S2(R1) genotoxicity testing and data interpretation for pharmaceuticals intended for human use. Available at: http://www.fda.gov/downloads/Drugs/Guidances/ucm074931.pdf.

[66] Organisation for Economic Co-operation and Development. OECD guideline for the testing of chemicals: Mammalian erythrocyte micronucleus test. Available at: http://www.oecd.org/chemicalsafety/risk-assessment/1948442.pdf.

[67] Organisation for Economic Co-operation and Development. OECD guidelines for the testing of chemicals, Section 4. Test No. 475: Mammalian bone marrow chromosome aberration test. Available at: http://www.oecd-ilibrary.org/environment/test-no-475-mammalian-bone-marrow-chromosome-aberration-test_9789264071308-en.

[68] Norppa H. Cytogenetic biomarkers and genetic polymorphisms. Toxicol Lett 2004;149(1-3):309–34.

[69] Loch-Caruso R, Trosko JE. Inhibited intercellular communication as a mechanistic link between teratogenesis and carcinogenesis. Crit Rev Toxicol 1985;16(2):157–83.

[70] Kari G, Rodeck U, Dicker AP. Zebrafish: an emerging model system for human disease and drug discovery. Clin Pharmacol Ther 2007;82(1):70–80.

[71] Vogel EW, Nivard MJ, Ballering LA, et al. DNA damage and repair in mutagenesis and carcinogenesis: implications of structure-activity relationships for cross-species extrapolation. Mutat Res 1996;353(1-2):177–218.

[72] Organisation for Economic Co-operation and Development. OECD guidelines for the testing of chemicals, Section 4. Test No. 487: In vitro mammalian cell micronucleus test. Available at: http://www.oecd-ilibrary.org/environment/test-no-487-in-vitro-mammalian-cell-micronucleus-test_9789264224438-en.

[73] Organisation for Economic Co-operation and Development. OECD guidelines for the testing of chemicals, Section 4. Test No. 473: In vitro mammalian chromosomal aberration test. Available at: http://www.oecd-ilibrary.org/environment/test-no-473-in-vitro-mammalian-chromosomal-aberration-test_9789264224223-en.

[74] Bonassi S, El-Zein R, Bolognesi C, et al. Micronuclei frequency in peripheral blood lymphocytes and cancer risk: evidence from human studies. Mutagenesis 2011;26(1):93–100.

[75] Organisation for Economic Co-operation and Development. OECD guideline for the testing of chemicals: In vitro mammalian cell gene mutation test. Available at: http://www.oecd.org/chemicalsafety/risk-assessment/1948426.pdf.

[76] Lloyd M, Kidd D. The mouse lymphoma assay. Methods Mol Biol 2012;817:35–54.

[77] Ames BN, Durston WE, Yamasaki E, et al. Carcinogens are mutagens: a simple test system combining liver homogenates for activation and bacteria for detection. Proc Natl Acad Sci U S A 1973;70(8):2281–5.

[78] Gold LS, Slone TH, Ames BN. What do animal cancer tests tell us about human cancer risk? Overview of analyses of the carcinogenic potency database. Drug Metab Rev 1998;30(2):359–404.

[79] Biran A, Yagur-Kroll S, Pedahzur R, et al. Bacterial genotoxicity bioreporters. Microb Biotechnol 2010;3(4):414–27.

[80] Benigni R, Bossa C, Tcheremenskaia O, et al. Alternatives to the carcinogenicity bioassay: *in silico* methods, and the in vitro and in vivo mutagenicity assays. Expert Opin Drug Metab Toxicol 2010;6(7):809–19.

[81] Mortelmans K, Zeiger E. The Ames Salmonella/microsome mutagenicity assay. Mutat Res /Fundamental Molec Mechanisms Mutagenesis 2000;455:29–60.

[82] International Agency for Research on Cancer. IARC monographs homepage. Available at: http://monographs.iarc.fr/index.php.

[83] US Environmental Protection Agency. Basic information about risk assessment guidelines development. Available at: http://www.epa.gov/raf/publications/pdfs/CANCER_GUIDELINES_FINAL_3-25-05.PDF.

[84] Carbone M, Pass HI. Multistep and multifactorial carcinogenesis: when does a contributing factor become a carcinogen? Semin Cancer Biol 2004;14(6):399–405.

[85] Breheny D, Zhang H, Massey ED. Application of a two-stage Syrian hamster embryo cell transformation assay to cigarette smoke particulate matter. Mutat Res 2005;572(1-2):45–57.

[86] Bignold LP. Alkylating agents and DNA polymerases. Anticancer Res 2006;26(2B):1327–36.

[87] Henderson L, Brusick D, Ratpan F, et al. A review of the genotoxicity of ethylbenzene. Mutat Res 2007;635(2-3) 81-13.

[88] Armstrong MJ, Gara JP, Gealy III R, et al. Induction of chromosome aberrations in vitro by phenolphthalein: mechanistic studies. Mutat Res 2000;457(1-2):15–30.

[89] Léonard A, Gerber GB, Stecca C, et al. Mutagenicity, carcinogenicity, and teratogenicity of acrylonitrile. Mutat Res 1999;436(3):263–83.

[90] Bomhard EM, Herbold BA. Genotoxic activities of aniline and its metabolites and their relationship to the carcinogenicity of aniline in the spleen of rats. Crit Rev Toxicol 2005;35(10):783–835.

[91] Cohen SM, Arnold LL. Chemical carcinogenesis. Toxicol Sci 2011;120(Suppl 1):S76–132.

[92] Raffle AE, Muir Gray JA. Screening: evidence and practice. Oxford: Oxford University Press; 2007.

[93] Choy WN. Genetic toxicology and cancer risk assessment. New York, NY: Marcel Dekker; 2001.

[94] Wang X, Chorley BN, Pittman GS, et al. Genetic variation and antioxidant response gene expression in the bronchial airway epithelium of smokers at risk for lung cancer. PLoS One. 2010;5(8):e11934.

[95] Vineis P, Perera F. Molecular epidemiology and biomarkers in etiologic cancer research: the new in light of the old. Cancer Epidemiol Biomarkers Prev 2007;16(10) 1954-165.

[96] Garner RC. Assessment of carcinogen exposure in man. Carcinogenesis 1985;6:1071–8.

[97] Chen CJ, Hsu LI, Wang CH, et al. Biomarkers of exposure, effect, and susceptibility of arsenic-induced health hazards in Taiwan. Toxicol Appl Pharmacol 2005;206(2):198–206.

[98] Maclure M, Katz RB, Bryant MS, et al. Elevated blood levels of carcinogens in passive smokers. Am J Public Health 1989;79(10):1281–4.

[99] Bagga D, Anders KH, Wang HJ, et al. Organochlorine pesticide content of breast adipose tissue from women with breast cancer and control subjects. J Natl Cancer Inst 2000;92(9):750–3.

[100] Gan LS, Skipper PL, Peng XC, et al. Serum albumin adducts in the molecular epidemiology of aflatoxin carcinogenesis: correlation with aflatoxin B1 intake and urinary excretion of aflatoxin M1. Carcinogenesis 1988;9(7):1223–5.

[101] Poirier MC, Santella RM, Weston A. Carcinogen macromolecular adducts and their measurement. Carcinogenesis 2000;21(3):353–9.

[102] Phillips DH, Venitt S. DNA and protein adducts in human tissues resulting from exposure to tobacco smoke. Int J Cancer 2012;131(12):2733–53.

[103] US Department of Labor. Permissible exposure limits. Available at: https://www.osha.gov/dsg/annotated-pels/.

[104] Greim H, Albertini RJ, editors. The cellular response to the genotoxic insult: the question of threshold for genotoxic carcinogens. Cambridge UK: RSC Publishing; 2014.

[105] Huff J. Issues and controversies surrounding qualitative strategies for identifying and forecasting cancer causing agents in the human environment. Pharmacol Toxicol 1993;72(Suppl 1):14–27.

[106] O'Brien J, Renwick AG, Constable A, et al. Approaches to the risk assessment of genotoxic carcinogens in food: a critical appraisal. Food Chem Toxicol 2006;44(10):1614–35.

[107] US Department of Labor. Maximum permissible dose equivalent for occupational exposure. Available at: https://www.osha.gov/SLTC/radiationionizing/introtoionizing/ionizingattachmentsix.html.

[108] Department of Health and Human Services, Centers for Disease Control and Prevention, and National Institute for Occupational Safety and Health. Update of NIOSH carcinogen classification and target risk level policy for chemical hazards in the workplace. Available at: http://www.cdc.gov/niosh/docket/review/docket240A/pdf/EID-CIB-11052013.pdf.

[109] Snell SE, Mancino AT, Edwards MJ. Principles of surgical oncology Pollock RE, Doroshow JH, Khayat D, editors. UICC manual of clinical oncology (8th ed.). Hoboken, NJ: John Wiley and Sons; 2004. p. 215.

[110] National Cancer Institute. Lung cancer. Available at: http://www.cancer.gov/cancertopics/screening/lung.

[111] Hakama M, Auvinen A. Cancer screening. In: Heggenhougen K, Quah S, editors. International encyclopedia of public health. Philadelphia, PA: Elsevier; 2008. pp. 470–1.

[112] Tambouret RH. The evolution of the Papanicolaou smear. Clin Obstet Gynecol. 2013;56(1):3–9.

[113] Denny L. Cytological screening for cervical cancer prevention. Best Pract Res Clin Obstet Gynaecol 2012;26(2):189–96.

[114] Fletcher SW. Breast cancer screening: a 35-year perspective. Epidemiol Rev. 2011;33(1):165–75.

[115] Nelson AL. Controversies regarding mammography, breast self-examination, and clinical breast examination. Obstet Gynecol Clin North Am 2013;40(3): 413–27.

[116] Health Quality Ontario. Fecal occult blood test for colorectal cancer screening: an evidence-based analysis. Ont Health Technol Assess Ser 2009;9(10):1–40.

[117] Lee JK, Liles EG, Bent S, et al. Accuracy of fecal immunochemical tests for colorectal cancer: systematic review and meta-analysis. Ann Intern Med 2014;160(3):171.

[118] Garborg K, Holme Ø, Løberg M, et al. Current status of screening for colorectal cancer. Ann Oncol 2013;24(8):1963–72.

[119] Holme Ø, Bretthauer M, Fretheim A, et al. Flexible sigmoidoscopy versus faecal occult blood testing for colorectal cancer screening in asymptomatic individuals. Cochrane Database Syst Rev 2013;9:CD009259.

[120] Justman S. How did the PSA system arise? J R Soc Med 2010;103(8):309–14.

[121] Bryant RJ, Lilja H. Emerging PSA-based tests to improve screening. Urol Clin North Am 2014;41(2): 267–76.

[122] Azuaje F, editor. Bioinformatics and biomarker discovery: "Omic" data analysis for personalised medicine. Chichester, West Sussex: Wiley-Blackwell; 2010.

[123] Ulahannan D, Kovac MB, Mulholland PJ, et al. Technical and implementation issues in using next-generation sequencing of cancers in clinical practice. Br J Cancer 2013;109(4):827–35.

[124] Pace LE, Keating NL. A systematic assessment of benefits and risks to guide breast cancer screening decisions. JAMA 2014;311(13):1327–35.

[125] Marmot MG, Altman DG, Cameron DA, et al. The benefits and harms of breast cancer screening: an independent review. Br J Cancer 2013;108(11): 2205–40.

[126] Kelloff GJ, Boone CW, Malone WF, et al. Introductory remarks: development of chemopreventive agents for prostate cancer. J Cell Biochem Suppl 1992;16H:1–8.

[127] Patterson SL, Colbert Maresso K, et al. Cancer chemoprevention: successes and failures. Clin Chem 2013;59(1):94–101.

[128] Lambert JD. Does tea prevent cancer? Evidence from laboratory and human intervention studies. Am J Clin Nutr 2013;98(6 Suppl):1667S–75S.

[129] Martin MA, Goya L, Ramos S. Potential for preventive effects of cocoa and cocoa polyphenols in cancer. Food Chem Toxicol 2013;56:336–51.

[130] Kelloff G.J.Hawk ET, Sigman CC, editors. Cancer chemoprevention, 2 vols. New York, NY: Springer; 2004, 2005.

[131] Miller AB, Bartsch H, Boffetta P, et al. Biomarkers in cancer chemoprevention. Lyon: IARC Scientific Publication, No 154; 2001.

[132] Wattenberg LW. What are the critical attributes for cancer chemopreventive agents? Ann N Y Acad Sci 1995;768:73–81.

[133] Rockhill B, Weed D. Increasing the contribution of epidemiology to the primary prevention of cancer. In: [2]. pp. 1292–302.

[134] Steele VE, Lubet RA. The use of animal models for cancer chemoprevention drug development. Semin Oncol 2010;37(4):327–38.

[135] Steele VE, Lubet RA, Moon RC. Preclinical animal models for the development of cancer chemoprevention drugs. In: [133]. vol 2, pp. 39–46.

[136] Crowell JA. The chemopreventive agent development research program in the Division of Cancer Prevention of the US National Cancer Institute: an overview. Eur J Cancer 2005;41(13):1889–910.

[137] Steward WP, Brown K. Cancer chemoprevention: a rapidly evolving field. Br J Cancer 2013;109(1):1–7.

[138] Szabo E, Mao JT, Lam S, et al. Chemoprevention of lung cancer: Diagnosis and management of lung cancer, 3rd ed: American College of Chest Physicians evidence-based clinical practice guidelines. Chest 2013;43(5 Suppl):e40S–60.

[139] Lazzeroni M, DeCensi A. Breast cancer prevention by antihormones and other drugs: where do we stand? Hematol Oncol Clin North Am 2013;27(4):657–72. vii.

[140] Cooper K, Squires H, Carroll C, et al. Chemoprevention of colorectal cancer: systematic review and economic evaluation. Health Technol Assess 2010;14(32):1–206.

[141] Sandhu GS, Nepple KG, Tanagho YS, et al. Prostate cancer chemoprevention. Semin Oncol 2013;40(3): 276–85.

[142] National Library of Medicine. Surgeon-General's Report: Smoking and Health, 1964. Available at: http://profiles.nlm.nih.gov/ps/retrieve/Resource Metadata/NNBBMQ.

Principles of Normal Histology and Related Cell Biology

This appendix gives essential background to understand the diversity in the kinds of cells in the body. Emphasis is given particularly to those features, lives, and environments, which are relevant to causation and therapy of tumors.

Sections A1.1 and A1.2 sketch the structures and functions of the various kinds of normal organs, tissues, and cells, which are relevant to tumors. Section A1.3 describes the aspects of the growth of normal tissues and cells. Section A1.4 gives a sketch of susceptibilities and responses of normal cells to injuries. Section A1.5 is concerned with movements, invasions, and metastases by normal cells.

A1.1 ASPECTS OF NORMAL DEVELOPMENT AND ORGANS AND TISSUES OF THE ADULT

A1.1.1 Meiosis and the Production of Individuality, and Aspects of Early Embryonic Development

The life of the individual begins with fertilization of an ovum by a spermatozoon (Figure A1.1). The individuality of the individual depends on processes in the formation of the genomes of the gametes. The series of cellular events is called *meiosis*, and it begins with a step in which the entire diploid genome in the spermatogonium or oogonium is duplicated without cytokinesis (division of the cell body). The result is a tetraploid cell. The next two steps are "reduction divisions," because the cell undergoes nuclear division and cytokinesis without a prior synthesis of DNA. Hence at the end of the first reduction division of the spermatogonia, there are two diploid cells, and at the end of the second division, there are four spermatozoa. In oogenesis, one of the daughter cells of each reduction division undergoes physiological

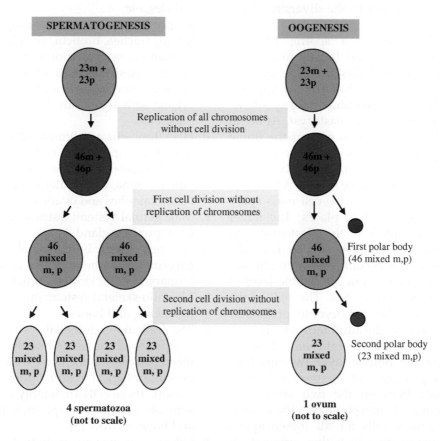

FIGURE A1.1 Meiosis and the production of individuality. The two cell divisions mix whole maternal (m) and paternal (p) chromosomes in resultant cells. As well, parts of parental chromosome may be mixed by the process of 'crossing over'/ exchange of equal lengths of arms between homologous chromosomes (see text). Thus from the same parents, the sons are not identical, and the daughters are not identical.

death (Section 10.3) as "polar bodies." Thus only one viable haploid cell—the ovum—is produced.

This process of doubling of the genome followed by reduction divisions allows for the considerable reassortment of maternal and paternal chromosomes in the gamete. The genetic diversity in individuals created in this way has survival benefits for those species in which consanguineous matings are common.

The phenomenon of "crossing over"/equal exchanges of parts of arms of chromosomes—also called "meiotic recombination" and "homologous meiotic chromosomal exchanges" [1]—contributes further to the diversity of the individual gametes.

After fertilization by a sperm, the egg becomes the "zygote" and begins to develop into the new individual. First there are multiple cell divisions to form the blastula. At approximately day 4, the blastula undergoes spontaneous cavitation to form the blastocyst. At about 5 days a cell mass forms in part of the lining of the cavity of the blastocyst. These cells are pluri-potential for all cells of the embryo, yolk sac, and amniotic membrane (Figure A1.2).

Between days 8 and 10, the cell mass spontaneously cavitates in two places. The larger one is near the middle of the blastocyst, and it becomes the yolk sac. The smaller is on the opposite side, and it becomes the amniotic sac. The embryo develops from the double layer of cells between the two cavities. The cells of the double layer are the "embryonic progenitor cells" of all tissues of the adult individuals.

As the next step in the embryonic development, cells of the double layer (by this time having the appearances of "epithelium") grow into the space between the two layers and develop elongated morphology. These cells are the progenitor cells for all mesenchymal cells. The process by which they originate is called "epithelio-mesenchymal transition." It has a possible role as a normal process which, in deviant or activated form, may underlie tumorous conversion of normal cells (see Section 2.9.1a).

A1.1.2 The Organ Systems of the Adult Body [2–4]

The body consists of approximately a dozen organ systems, each of which comprises variable numbers of individual organs. The systems which most commonly give rise to tumors are as follows.

a. Integumentary system: skin, including epidermis, dermis, sweat glands, hair follicles, etc.
b. Respiratory system: nose, nasopharynx, larynx, trachea, bronchi, alveoli, pleura.
c. Alimentary system: mouth, oropharynx, esophagus, stomach, small and large intestines and anus, as well as accessory organs: salivary glands, liver, and exocrine pancreas.
d. Hematopoietic and lymphoid systems: bone marrow, lymph nodes, and spleen.
e. Female genital system and related organs: vulva, vagina, cervix, uterine corpus, Fallopian tubes and ovaries, as well as breasts.
f. Male genital system: testes and spermatic tract, prostate gland.
g. Nervous system: brain, spinal cord, and coverings; peripheral nerves and ganglia.
h. Urinary system: kidneys, ureter, and bladder.
i. Musculo-skeletal system: muscles, tendons, ligaments, and bones.
j. The "reticulo-endothelial system" comprises all the cells—mainly monocytes/macrophages—in the body which scavenge nonphysiological material from the blood stream. These cells are mainly in the sinusoids of the liver, spleen, lymph nodes, and bone marrow [5].
k. Other organ systems include the endocrine glands—pituitary, thyroid, parathyroid, adrenals, and islets of Langerhans—as well as the organs of special senses.

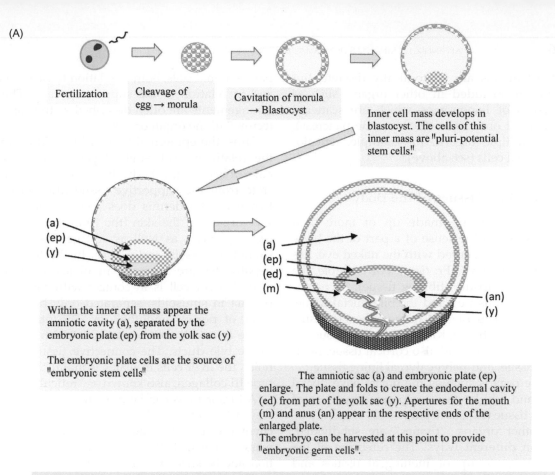

(A)

Fertilization

Cleavage of
egg → morula

Cavitation of morula
→ Blastocyst

Inner cell mass develops in
blastocyst. The cells of this
inner mass are "pluri-potential
stem cells."

(a)
(ep)
(y)

(a)
(ep)
(ed)
(m)

(an)
(y)

Within the inner cell mass appear the
amniotic cavity (a), separated by the
embryonic plate (ep) from the yolk sac (y)

The embryonic plate cells are the source of
"embryonic stem cells"

The amniotic sac (a) and embryonic plate (ep)
enlarge. The plate and folds to create the endodermal cavity
(ed) from part of the yolk sac (y). Apertures for the mouth
(m) and anus (an) appear in the respective ends of the
enlarged plate.
The embryo can be harvested at this point to provide
"embryonic germ cells".

Subsequently, organs (not shown) begin to develop in the folded plate, especially from the covering (ectodermal) and
lining (endodermal) cells.

(B)

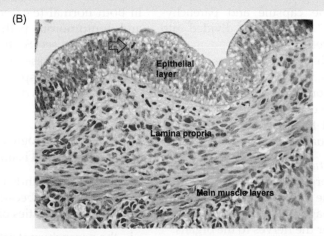

Epithelial
layer

Lamina propria

Main muscle layers

FIGURE A1.2 (A) Early embryonic cells possibly relevant to tumors. (B) *Micrograph of embryonic intestine at approxi-mately 8 weeks gestation (×20).* All of cells are proliferating, characterized by high ratios of nuclear volume: cytoplasmic vol-ume, with occasional mitotic figures (arrowed). In the epithelium, there is little evidence of specialization to vacuolated cells. The lamina propria consists of proliferating blood vessels and connective tissue cells with little specialization. The cells of the main muscle layer can be recognized as smooth muscle cells with relatively large nuclei. These similarities to the appearances of tumor cells led nineteenth century pathologists to suggest that tumor cells are either derived from embry-onic cells, or have reverted from adult cells to embryonic cells (see Chapter 2).

Most organs are anatomically discrete, but a few are included in other organs, such as the islets of Langerhans which are scattered throughout one organ (the exocrine pancreas), or in multiple organs, for example, reticulo-endothelial cells (see above).

A1.1.3 The Tissues of the Body

Most organs are made up of more than one "tissue" in the sense of a part of the body which, when observed with the naked eye, has only one "texture" (Fr: *tissu*). For example, the liver has a capsule (fibrous tissue), its parenchyma (liver tissue), and the bile ductal tissue. Lung comprises pleural tissue, alveolar tissue, and bronchial tissue. Kidney has the fibrous tissue of its capsule, and also cortical tissue, medullary tissue, and pelvi-calyceal lining tissue.

Some organs such as the breast have ductal tissue and lobular tissue separated by areas of fibrous tissue (e.g., as septa).

In other organs, "tissues" are subdivided, albeit in different ways. The cerebrum of the brain is made up of meningeal tissues and "brain tissue." The latter is subdivided into "white matter" and "grey matter." The spleen has a fibrous tissue capsule and parenchymal "splenic tissue." The latter is divided into "white pulp" and "red pulp."

Functionally, each organ comprises first the "parenchymal" cells, which carry out the prime physiological function(s) of the organ, and second the "supportive cells." The supportive cells (sometimes referred to collectively as "connective tissue" cells) usually serve as:

i. the overall *structure* and mechanical strength of the organ,
ii. the avenue of the blood supply and lymphatic drainage,
iii. the pathways for the nerves to the organs.

In the various organs and tissues, there are often quite specific differences in combinations and spatial and functional arrangements of the parenchymal cells, both in relation to each other and in relation to the supporting cells. These arrangements are often described as the "architecture" of the organ or tissue.

Thus, the epidermal tissue of the skin forms the relatively water-tight layer of the external surface of the body. It is completely discrete from the supportive tissue (the dermis). However, the dermis does include the accessory organs of the skin (the "adnexa," such as sweat glands), as well as collagen fibers (for strength), vessels, and nerves. In the liver, the hepatocytes are arranged in plates. One side of each liver cell is in contact with the blood plasma in sinusoids, separated only by a thin layer of porous endothelium. The other side of the liver cell abuts canaliculi which connect to the bile ducts. The connective tissue which holds the liver cells in plates is mainly the fiber type III collagen, also known as "reticulin."

A tumor mass can be thought of in the same way if the tumor cells are considered its parenchymal cells. The supportive cells in tumor masses are usually the same in morphology, function, and behavior as those in the tissue of origin.

The control(s) of the spatial arrangements of parenchymal and supportive cells in tissues in both normal (e.g., embryonic development) and nontumorous pathological situations (e.g., regeneration, see Section A1.4.2) are not well understood. These issues potentially relate to fundamental aspects of the morphology of tumors types, as is discussed in Chapter 8.

A1.1.4 Physiological and Nontumorous Variabilities in Tissues

The different kinds of tissues vary one from the other largely because of the different proportions and qualities of:

i. the parenchymal cells,
ii. the products of parenchymal cells,
iii. the "supporting cells,"
iv. the products of the supporting cells.

In most normal tissues, the ratios of the parenchymal to the supporting cells are constant. However in a few organs, the normal ratios of parenchymal cells to supporting cells vary with physiological circumstances. Thus in the breast, the secretory (lobular secretory) cells increase in proportion to supportive cells during pregnancy and lactation, and return to normal at cessation of lactation.

Also in nontumorous pathological conditions, the cellular proportions vary. In a typical acute inflammatory event, the blood vessels dilate and leukocytes enter the tissue spaces. After that, endothelial cells and fibrocytes change to their respective blastic forms (see Section A1.2.7) and proliferate. In the phase of resolution, all the cellular abnormalities subside, and only a focus of excess collagen may mark the site (i.e., as a scar, see Section 10.7).

These normal and nontumorous variabilities may have relevance to the defensive factors affecting the access of carcinogens and anticancer drugs to cells, as is discussed in Appendix 3. In particular, the susceptibility of any one kind of cell may be affected by the proportions of other kinds of cells present in the same tissue.

A1.2 ASPECTS OF NORMAL CELLS

The general differences between the structures, functions, and regulation of the different kinds of cells are described in most texts of histology [2–4]. Here are mentioned only those aspects which relate to their roles as parent cells of tumors (Figure A1.3).

A1.2.1 The Categories of Cells

(a) Epithelial Cells

These include cells of the epidermis of the skin and the lining cells of the intestine as well as all cells which derive from embryological epidermis or gut lining cells. These derivative cells include the cells of glands or organs which secrete or excrete substances: breast, endometrium, prostate, liver, kidney tubules, exocrine pancreas, thyroid, and anterior lobe of the pituitary gland.

(b) Hematopoietic and Lymphoid Cells

These cells produce the red and white blood cells. Also included are plasma cells, which are only found in tissues, but which derive from circulating lymphocytes.

(c) Cells of the Nervous System

The cells of the central nervous system include the neurons, the specialized supporting cells (glial cells), the ventricular lining, and the plexuses in the ventricles of the brain and spinal cord. The posterior lobe of the pituitary gland is part of the brain.

The cells of the peripheral nervous system are mainly the specialized supporting cells around the nerve axons.

(d) Melanocytes

These are almost entirely in the dermo-epidermal junction of the skin and mucosa of the anal canal.

(e) Cells of "Soft Tissues"

These are mainly those of the supportive (connective tissue) cells, as well as of non-epithelial "soft" tissues such as smooth muscle cells, and skeletal muscle fibers.

(f) Cells of "Hard Tissues"

These are of bone, cartilage, and teeth.

(g) Miscellaneous Cells

These include cells of the gonadal stroma, endocrine cells, e.g., of the pituitary and pancreatic islets, as well as mesothelium (lining of the peritoneal and pleural cavities).

A1.2.2 Cytostructural Regularity of Each Kind of Normal Cell

As mentioned in Section 1.2.1, as far as cytology is concerned, each kind of normal cell is

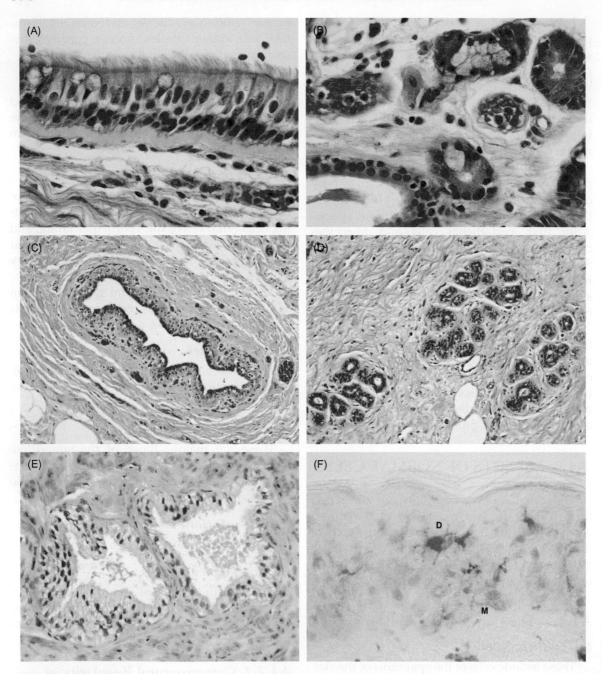

FIGURE A1.3 The cells of the body from which most tumors arise. (A) Normal superficial cells of bronchi, ×40. (B) Normal submucosal glands of bronchi, ×40. The glands include both mucus- and serous (watery-fluid) - producing cells. (C) Normal breast duct ×10. (D) Normal breast lobule ×10. (E) Normal prostate gland ×20. (F) Normal melanocytes (M) and dendritic cell (D) in epidermis, S100 immunoperoxidase stain, ×40. Note: Normal colonic epithelial cells are illustrated in Figure 1.10A.

always the same where ever it is located in the body. However, there are significant differences between the cells of each kind at any one point in time.

Thus in each of the normal kinds of cells which turn over ("labile" cells, see Section A1.3.3), under normal conditions, the destiny of all the specializing cells is completely homogeneous. This means that all dividing cells of the particular kind of cell lead to fully mature descendants, and at the same rate. Moreover, the proportions of local stem cells, transit-amplifying cells, and fully specialized cells are fixed for the whole population of the kind of cell.

In some tissues such as epidermis, the local stem cells and the fully functional cells are in different layers of the epidermis—that is, anatomically separated. However, in the bone marrow, the local stem cells and their specializing descendants are all mixed together. In the liver, many hepatocytes can divide, but the local stem cells (the "oval cells"—see Section A1.3.2) are mainly in the periphery of the lobules. In fractures of bone, new osteocytes come from endosteum and periosteum, not from the preexisting mature osteocytes.

In some kinds of normal cells, the nucleus is at the base of the cell (e.g., goblet mucus producing cells of the gastrointestinal tract) or at the periphery of the cell (as in skeletal muscle fibers). For any given kind of cell, the nucleus is always in the same position and has the same orientation ("polarity") in relation to the other components of the cell.

A1.2.3 Cells Having the Same General Function May Have Different Structural Details in Different Organs

As an example of what is meant here, cells which secrete mucus are not the same in every organ. The mucus cells of the large intestine have the mucus in a large rounded vesicle near the surface of the cell (resembling the bowl of a goblet). This arrangement of the mucus is different to those in mucus-producing cells of the endocervix, salivary glands, and bronchial glands. These differences imply that a mucus-producing cell in one organ has a different pattern of activities of structural genes than a mucus-producing cell in another organ.

A1.2.4 The Cell/Plasma Membrane

This structure separates the external environment of the cell from the cytoplasm [6]. The membrane consists of lipids especially cholesterol, phospholipids, and glycolipids, as well as proteins, many of the latter being glycoproteins. Membrane proteins are classified as:

i. "Integral," being bound to the lipids in the core of the membrane. They are often "transmembrane," which means having extracellular, membrane, and intracellular domains.
ii. "Peripheral," being associated only with the outer and inner surfaces of the membrane.

Many outer surface proteins or parts of proteins are heavily glycosylated and can be demonstrated collectively as the "glycocalyx."

The membrane controls selective movements of many substances into and out of the cell. For some of these, specific pore-like structures are present which pump substances, especially water, as well as sodium and calcium ions, out of the cytoplasm. Depending on the kind of cell, the plasma membrane may also include receptors on its outer surface, e.g., for signaling (Chapter 6). The surface proteins may be specifically bound by virus proteins, contributing to the "tropisms" of infections with these organisms (Section 4.5).

A1.2.5 The Cytoplasmic and Functional Variabilities in Each Kind of Cell

With the low power of the light microscope, it is clear that the specialized features—including the general shapes and sizes—of fully functional cells are quite different, one to another. It can also be seen that for many kinds of cells, these specialized features develop locally in the body.

Thus, epidermal cells begin as rounded "local tissue stem" cells (see Section A1.3.2) in the basal epidermis, and by multiple divisions, produce daughter cells which are flattened with elongated nuclei more or less in their centers. However, with further divisions, the daughter cells accumulate granules of keratin (and are called "granular cells"). Finally at the surface of the epidermis, the daughter cells are in their fully specialized state as flat particles of keratin without any nuclei.

Epithelial cells of the large intestine, on the other hand, develop from inconspicuous local tissue stem cells (see Section A2.6.2) at the bases of the crypts. The daughter cells attain full specialization without clearly distinguishable intermediate forms. The fully functional cells arrange themselves as a single layer, each cell with its nucleus at its base and large amounts of mucus in their cytoplasm. The mucus protects the epithelium and lubricates the movement of the contents along the lumen of the organ.

The term "differentiation" is often applied to this process involving the development of mature, fully functional cells from local precursors. However in this book, "differentiation" is reserved for the changes in cell populations during embryonic development.

The different kinds of cells are also enormously diverse in relation to their exported chemical products. Cells can be categorized on this basis as follows:

i. Cells with no known major exported chemical products. Examples are the cells of all muscle tissues; and all structures which are only conduits, such as the esophagus, ureters, and bladder.
ii. Cells whose products are delivered outside the organ but not via the blood stream.

These include epithelial cells of the sweat glands of the skin, salivary glands, liver, pancreas, gastrointestinal tract, kidney, breast, and prostate.
iii. Cells whose products are released outside the cell into the blood circulation. These are essentially all endocrine glands, including the sex-hormone producing cells in the gonads.
iv. Cells whose products are exported outside the cell, but retained in the organ. These are mainly the cells producing collagen for the supportive tissues, bone and cartilage for the skeleton, or dentine and enamel for the teeth. They also include neurons in the nervous system, including sensory organs, which deliver transmitters directly to other neurons.

The identification of functions in some kinds of cells has depended on higher magnification microscopy including electron microscopy, special stains and ultimately, methods to measure the products of the cells. For example, the functions of all the different kinds of cells, the islet cells of the pancreas, were only established in the 1970s–1990s, when the hormones which they produce had been purified and identified. Functions of some kinds of cells in all probability remain to be discovered.

These differences between the various kinds of cells, especially their specialization products, may affect their susceptibility to carcinogens, as well as their vulnerability to cytotoxic drugs, as is discussed in Section A3.1.

A1.2.6 The Nucleus in General and as the Compartment for the Genome and Genomic Activity

(a) Size

The sizes of nuclei vary greatly according to the kind of cell in which they are located and the metabolic activity of the cells (Figure A1.4). The smallest nuclei include those in inactive cell such as unstimulated "small" lymphocytes,

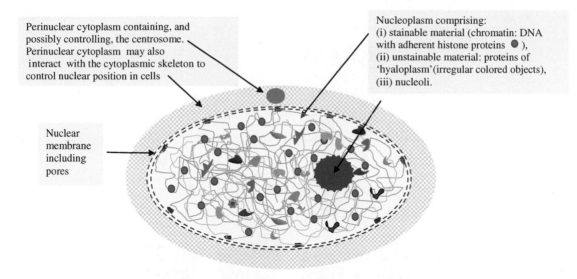

Perinuclear cytoplasm containing, and possibly controlling, the centrosome. Perinuclear cytoplasm may also interact with the cytoplasmic skeleton to control nuclear position in cells

Nucleoplasm comprising:
(i) stainable material (chromatin: DNA with adherent histone proteins ●),
(ii) unstainable material: proteins of 'hyaloplasm'(irregular colored objects),
(iii) nucleoli.

Nuclear membrane including pores

Notes: (i) There is some evidence that in the nondivision period, the centromeres (purple bars in diagram) of at least some of the chromosomes are adherent to the inner surface of the nuclear membrane.
(ii) The concept that the hyaloplasm includes a proteinaceous 'skeletal matrix' is not universally accepted.

FIGURE A1.4   The components of the nucleus.

which have almost no cytoplasm. The largest nuclei are in motor neurons of the cortex of the brain, which—probably not incidentally—have the largest amounts of cytoplasm (axons of these cells are up to 1 m long).

It is generally believed that, with a few exceptions, all normal nuclei in the body have the same amount of DNA. Thus, the differences in nuclear size are probably due entirely to differences in their water, nucleoprotein, and RNA content associated with metabolic activity. This is consistent with the observation that the nuclei of the "blast" versions of many kinds of cells are always larger than in the corresponding inactive "cyte" version (see Section 1.2.7). The appearances of nuclei in cells are markedly affected by the details of histological processing and staining (Figure A1.5).

(b) The Nuclear Membrane and Perinuclear Cytoplasm

The nuclear membrane is not an inert bag, but a physiologically active structure [7,8]. In the interdivision period of the cell cycle (Section A1.3.4), pores in the nuclear membrane control movements of molecules, including RNAs from the nucleus to the cytoplasm. The membrane also controls flows in the reverse direction, for example, transcription factors from the cytoplasm into the nucleus. In prophase, the nuclear membrane dissolves, and some of its components become part of the spindle fibers of the mitotic apparatus. In telophase, the spindle dissolves and the nuclear membrane reforms (see Sections A1.3.4 and A1.3.5).

The perinuclear cytoplasm probably has specific structures and functions. In some cells, this zone contains many cytoskeletal fibers. These may relate to the specific position of the nucleus in certain kinds of cells, such as the basal position in intestinal glandular epithelium.

In nuclear division (see Sections A1.3.4 and A1.3.5), the centrioles move into the perinuclear cytoplasm. The centrioles remain close to the

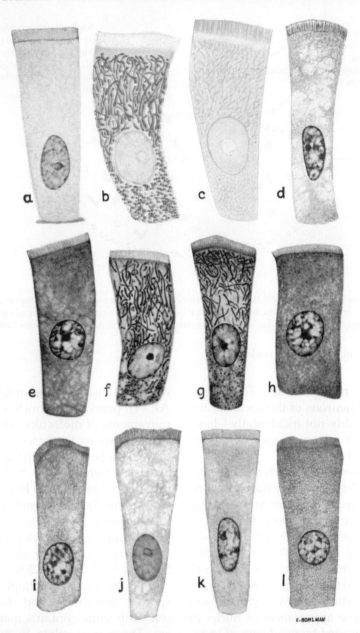

FIGURE A1.5   The appearances of nuclei in different histological stains: Epithelial cells of small intestine of guinea pig. All fixed preparations from same animal. *a*, Zenker-formol and Mallory-azan; *b*, supravital Janus green stain for mitochondria; *c*, supravital, unstained; *d*, absolute alcohol and H and E; *e*, Bouin and iron hematoxylin; *f*, 10 per cent neutral formalin and iron hematoxylin; *g*, Zenker-formol and iron hematoxylin; *h*, Zenker-acetic and iron hematoxylin; *i*, Bouin and H and E; *j*, 10 per cent neutral formalin and H and E; *k*, Zenker-formol and H and E; *l*, Zenker-acetic and H and E.
Source: Maximow AA, Bloom W. A Textbook of Histology, 5th edn, Saunders, Philadelphia, 1948, p 6, reproduced with permission.

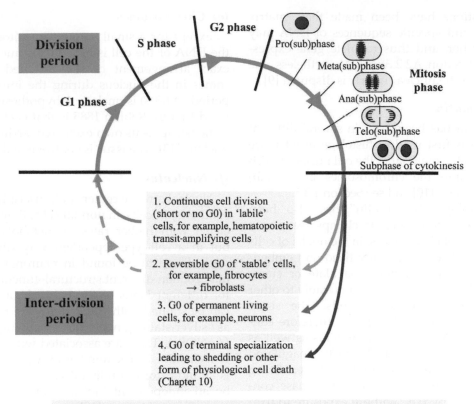

FIGURE A1.6   The parts of the cell cycle according to kind of cell.

surface of the nucleus. They do not move into the general cytoplasm. This restricted movement suggests some kind of mechanical function of the perinuclear cytoplasm, and some specific transnuclear membrane phenomena which coordinate events in the perinuclear cytoplasm and the intranuclear components. The details of this, however, are unclear.

(c) Nonstaining Nuclear Substance, Including "Matrix"/"Scaffold"

As is understood from normal histology, the intranuclear material includes an unstainable component: the "karyoplasm" (syn. "nucleoplasm") which consists of the material which

is neither DNA nor histone protein. The karyoplasm is most obvious (as the "hyaloplasm"), meaning, the background material out of which the chromosomes assemble during the pro(sub) phase of mitosis (see Section 1.3.4).

Conceptually, the "nuclear matrix" is part of the karyoplasm [9]. It is controversial in several ways. Initially, it was defined as all the high m. w. nonhistone protein in the karyoplasm. The molecules are mainly elongate, leading to the suggestion that they form a fibrous mesh of some kind in the nucleus. However, the events of mitosis (A1.3.5 and Figure A1.6) indicate that any such internal structure could not be present in that part of the life of the cell.

Suggestions have been made that "matrix proteins" and specific sequences of DNA may bind together, and thus regulate gene expression (see Section A2.2.3) [9]. Nevertheless, the existence of the nuclear matrix is disputed [9].

(d) Chromatin

This term has been used in several different ways. It was first used in histology in reference to the stainable material in the cell nucleus which assembles into the chromosomes in pro(sub) phase of mitosis [10] and see Section 1.3.5.

In histology, "chromatin" has also been applied to all the stainable clumps seen, on a background of clear spaces, in the nuclei of cells which have been chemically fixed and embedded in paraffin. Fixatives coagulate or cross-link proteins to each other and possibly to other macromolecules of many kinds. The stainable material seen histologically therefore contains the DNA and histone proteins as well as any other molecules which can be coagulated or cross-linked together with it. When living nuclei are examined especially by phase contrast microscopy (i.e., without exposure to fixatives), no such clumping is seen.

In biochemistry, "chromatin" refers to (stainable) DNA and histones in unfixed nuclei in G0. In preparative biochemical studies, certain terms are applied to fractions obtained by ultracentrifugation. "Heterochromatin" refers to the densest chromatin, which sediments at $100g$. "Euchromatin" refers to less dense chromatin (requiring much higher centrifugations—typically $78,000g$ for sedimentation). It appears that the lighter chromatin is more extensively transcribed than the heterochromatin, but this is still disputed [11].

More recently "chromatin" has been used for the first-order condensation structure of DNA in its ultimate assembly into chromosomes. Thus it is applied to the wrapping of DNA duplex strands around histone complexes to form "nucleosomes." During synthesis and transcription of DNA, the nucleosomes are disassembled [12].

(e) Chromosomes

Except for a small amount in mitochondria, the DNA of the cell is held in the nucleus. The exact arrangement of the uncoiled chromosomes in the nucleus during the interdivision period (A1.3.5) is unclear. A hypothesis put forward first by Rabl in 1885 is that each chromosome occupies its own exclusive territory in the nucleus [13]. The issue is controversial [13,14].

(f) Nucleolus

Nucleoli are the centers of transcription (RNA synthesis) in cells. In normal cells, they are associated with higher rates of metabolic activity, but in a cell-type-dependent way (the largest nucleoli are to be found in neurons). Nucleoli may contain different structural–functional units for different kinds of RNAs. Such units which can be stained with silver stains are referred to as "silver-staining nucleolar organizing regions" (AgNORs). These are associated with ribosomal RNA synthesis. The number of AgNORs per cell is an indicator of proliferative activity and may be an independent indicator of prognosis for some types of tumors [15].

A1.2.7 Variability in Activation Status of Cells of the Same Kind—"Cytes" and "Blasts"

This point of terminology is worth explaining because of its profound influence on the nomenclature of tumors (see Section 1.1.2). The term "blast" is used for any cell which gives rise to daughter cells. In embryology, the term refers to almost all cells in embryonic development. Also in the histology of adult tissues, a "blast" usually refers to any cell from which other cells develop. Thus a "fibroblast" is a proliferating version of a "fibrocyte"; an "osteoblast" is the proliferating version of the "osteocyte" and so on. For some tissues however, cells which give rise to other cells do not necessarily have particular names. Hence the basal epidermal cells and early transit-amplifying cells are not usually called "epidermoblasts" except in embryonic skin.

One point of possible significance in the distinction between functionally inactive and active cells is that they may have different defensive properties against harmful agents (see Appendix 3).

A1.2.8 Other Physiological Variabilities Within the One Population of the Same Kind of Normal Cell

This section refers to "labile" cells (A1.3.3)—those which are continuously produced from local stem cells and which proliferate and "mature" to fully specialized forms.

(a) Stage of Specialization

Of potential significance for carcinogenesis is that cells at each stage of specialization contain different relative amounts of various proteins. These proteins may form part of the defensive mechanisms of the cell against noxious agents (A3.1). Because of this, the vulnerability to carcinogens of cells in the different stages of specialization may be different.

In relation to therapy, partly-specialized proliferating cells may be the most sensitive to anticancer agents. Hence the tumors with the smallest proportions of fully specialized cells may have the higher average sensitivity rate of their whole population (11.3.3).

(b) Periods and Phases in the Cell Cycle

The divisions of cells in normal tissues are not synchronized. This applies in all circumstances, that is,

i. in embryonic and fetal life,
ii. in tissues which continuously produce functional kinds from local tissue stem cells,
iii. in which cells are regenerating after loss.

At any particular point in time, different cells will be in interdivision period or one of the phases of the division (G1, S, G2, M, see Section A1.3.4). It is thought that dividing cells in normal tissues complete division at the same rate.

This lack of synchrony in division is potentially relevant to carcinogenesis (Chapter 4). This is because the intracellular defensive conditions (A3.1) in cells *vis à vis* carcinogens may vary. Particularly, in mammalian cells, the nuclear membrane disassembles early in mitosis and reassembles after telophase. Thus if the nuclear membrane were a defensive barrier against a carcinogen, then mitosis may be a particularly susceptible part of the cell cycle in regards to these substances.

Similarly, in anticancer therapy, the drugs may be able to access the relevant targets in tumor cells more easily in some phases of the cell cycle than others (11.4.2).

In tumor cell populations, the rates at which tumors cells may proceed through the phases of the cell cycle are variable. The importance of this is discussed in the section on tumor cytokinetics (see in Section 8.3.2).

A1.3 GROWTH IN NORMAL TISSUES AND CELLS

A1.3.1 General

(a) Overall Controls

Growth in embryonic cell populations is thought to be under the control of generalized growth hormones such as somatotrophin, as well as cell-specific growth factors (GFs), growth-factor oncogenes (GF/Os), and tumor suppressor genes (TSGs) and their regulators (see in Chapter 6). There are almost certainly different GFs, GF/Os, and TSGs and regulators for each kind of cell in the adult.

Moreover, embryonic growth of both the different parts as well as of the different kinds of cells in the body could well be under the control of different particular genes, as evidenced by the multiple independent single-organ growth defects in *Drosophila* (see Section 7.1.2).

The mechanisms of coordination of the growth of multiple kinds of cells—to form the parts (such as limbs) and organs (such as the liver)—are unclear [16,17].

(b) The Phasic Aspect of Cell Production in Embryonic and Normal Cells

Embryonic development depends on populations of cells which (i) proliferate and (ii) cease to proliferate in fixed and coordinated sequences. Thus the melanocytes begin with a proliferation of neural crest cells and, when distributed to the epidermis, cease proliferation.

In adult life, local tissue stem cells divide intermittently. One of the daughter cells does not proliferate and continues like its mother cell as the local tissue stem cell. The other daughter cell proliferates over several generations, as the transit-amplifying cells. These amplifying cells cease proliferation when full specialization is achieved.

In tumors, the ordinary morphological appearances suggest disturbances in these phases. For example, the presence of an abnormally high proportion of cells in one particular phase may suggest that the cells have lost the capacity for final specialization, and correspondingly retain the capacity to proliferate. However, these detailed issues are difficult to study.

(c) Contact Inhibition

Both in the pathology of healing (see Section A1.4.3) and in *in vitro* culture studies, it is confirmed that nontumorous cells stop dividing when surrounded (i.e., in contact with on all sides) by cells of their own kind. This implies some growth control mechanism in the outer surface of the plasma membrane. The phenomenon is not fully understood, but has obvious implications for the phenomena of invasion and metastasis (see Sections 6.5 and 6.6).

(d) Changes with Age of the Individual

In adults, the continued turnover of "labile" cells (see next subsection) of many tissues appears to be independent of any known specific hormones or growth factors. Thus the mucosa of the intestine remains the same from birth to old age and remains unaffected in all the syndromes which are caused by lack of any of the known hormones (e.g., pituitary infantilism and thyroid cretinism).

Certain organs commonly appear to deteriorate with age, but this is probably caused by chronic low-level environmental toxins. Thus the atrophy of the epidermis with age is probably at least partly related to chronic exposure to ultraviolet light.

The main exceptions to this general rule are the gonads, which rely on trophic hormones from the pituitary, and the secondary sex organs, which rely on steroid hormones from the appropriate glands. The growths of cells in these organs are liable to reverse when the hormone is withdrawn. The tissues are said to undergo "involution," also called "atrophy".

A1.3.2 The Different Concepts of "Stem Cells" in Embryology and Adult Histology

The term "stem cells" is used in three distinct situations.

(a) In Embryology

The term is used to refer to any cell which gives rise to any cell which is more differentiated than itself. In this concept, the zygote is the original stem cell of the body. As mentioned in A1.1.1, after that, the inner cell mass (Figure A1.2A) is the "pluripotential" stem cell for embryo, amnion, and yolk sac. Later, individual cells in the embryo-proper (i.e., structures deriving from the embryonic plate) are stem cells for particular lines of cells in the adult (neural cells, muscle cells, epithelial cells, etc.).

The critical biological point is that embryonic stem cells do not undergo asymmetric functional divisions as do local tissue stem cells (see (b) below). No cells in the adult, to our knowledge, are continuously producing cells which are differentiating into adult kinds of cells in different cell groups (A1.2.1). In other words, there are no known normal later embryonic or fetal cells which produce, for example, epithelial and hematopoietic cells, or melanocytes and lymphoid cells.

(b) In Histology, Including Hematology in Adults

The term refers to any cell which, when it divides, produces one daughter cell like itself, and one daughter cell which is "committed" to one specific specialization pathways [18]. This usage refers only the kinds of cells which involve long-lived cells producing short-lived daughter cells which specialize *in situ* (i.e., labile populations, see Section A1.3.3). For example, a hematopoietic stem cell gives rise to a daughter cell which then gives rise to all the kinds of cells in the bone marrow. It does not give rise to mesenchymal, epithelial, or any nonhematopoietic cell. The term is rarely used in relation to cells which produce daughter cells only when specifically stimulated (i.e., "stable" cells, such as those of connective tissues, such as fibrocytes, osteocytes, etc.).

(c) In Studies of Tumor Cell Populations

The usage in these studies is often for any therapy-resistant cell which is still capable of division (see in Section 11.4.3).

A1.3.3 The Different Life Cycles of Different Kinds of Cells in Adults

This is such an important issue that its background is worth noting because it is a prominent aspect of the parent cells of tumors (Section 8.3.2). Understanding of the "cytokinetics" of the different kinds of cells developed

from histological studies in the mid-nineteenth century [19] when it was recognized that the parenchymal components of many tissues comprise:

i. *mutterzellen* ("mother cells," now termed "local tissue stem cells"),
ii. variable numbers of proliferating intermediate cells (now termed "transit amplifying cells"),
iii. the nonproliferative fully functional cells of the tissue.

By the 1890s, it was recognized that tissues of the body can be classified as follows [20]:

a. *"Labile" cells—those in which local tissue stem cells continuously produce functional cells*
    These stem cells, for example, of the epidermis, regularly divide without known stimulus. The division is functionally asymmetric because one daughter stem cell becomes the replacement local stem cells and the other daughter cell proceeds to produce transit-amplifying cells. Examples are the epidermis, the lining of the gastrointestinal tract, and hematopoietic cells.
    These cells have a definable cytokinetic profile, in the sense that there is a balance between production of cells by proliferation and loss of cells through shedding of fully mature cells. Within the overall group of "labile" cells however, there are great differences in the rates of production of cells. The most rapid production of cells occurs in the testis, in which millions of sperm are produced each day. Of other tissues, probably the fastest cell production is that of the bone marrow.
    The periods of time required for completion of the phases of specialization also vary between the different kinds of cells.
    It is also important to note that for some kinds of cells in this group, especially epidermis and bone marrow, each phase of progress from mother cell to fully

specialized cells is marked by different morphological features (see Section A1.2.7).

b. *"Stable" cells—those comprising uniformly long-lived cells which can reproduce their own kind under certain pathological circumstances*

These cells are normally in G0 but can be stimulated into G1 by mechanical and chemical factors. They include mainly connective tissue cells (fibrocytes, osteocytes, endothelial cells) as well as hepatocytes, renal tubular epithelium, and cells of some other organs. These cells characteristically cease proliferating when the stimulus is removed. These cells increase their specialized activity during activation (see Section A1.2.7) but show no other major morphological changes.

c. *"Permanent" cells—those comprising uniformly long-lived cells which cannot reproduce under any circumstances*

These cells are permanently in G0, and include neurons, which do not alter their morphology in adult life.

The proportion of time which each kind of cell spends in each phase of the cell cycle may affect its susceptibility to carcinogens and vulnerability to cytotoxic drugs, as is discussed in Section A1.2.8.

This issue of different cell cycles in different parent cells is also important because the normal cytokinetics are variably disturbed in tumor cell populations as is discussed in Section 8.3.2.

A1.3.4 "Division Period," "Interdivision Period," and the "Cell Cycle"

As was mentioned in the previous subsection, the lives of different kinds of cells comprise different relative amounts of time spent while dividing ("division period") and non-dividing ("interdivision period" or G0 = no growth) (see Figure A1.6). For certain kinds of cells, such as transit-amplifying cells of the hematopoietic system, periods of interdivision

period may be short, so that the cell may be regarded as in a continuous "cycle" of the division phase. However, for other kinds of cells, the interdivision period may be long (as in "stable" cells) or life-long (as in "permanent cells"). In these latter categories of cells, no continuous "cycles" of cell division occur (Figure A1.6).

Cell division/the "division period" comprises four individual phases:

> G1—First growth phase. There is enlargement of the cytoplasm and nucleus of the cell, and also the production of DNA polymerases.
> S—Synthesis of DNA. The centrosome divides and the centrioles move to opposite ends of the nucleus. At the end of this phase, the cell is tetraploid.
> G2—Second growth phase. There is further growth of the cell to approximately double the size of the original cell.
> M—Mitosis. This is the phase in which the chromosomes can be seen.

A1.3.5 Nuclear Division: Mitosis

In mitosis, there is no synthesis of DNA or RNA. The subphases of mitosis are as follows:

a. *Pro(sub)phase*

The nuclear membrane dissolves, and the tetraploid DNA double strands form coils, with new and template strands still joined at their centromeres (i.e., as chromatid pairs).

b. *Meta(sub)phase*

The double chromatid structures become arranged, by attachments of their centromeres, on the fibers of the spindle along the "equator" of the enlarged cell. In meta(sub)phase, the chromosomes are arranged in a ring at right angles to the spindle (Figure A1.7).

c. *Ana(sub)phase*

For all pairs, one chromatid of each pair is drawn to one aster, the other of the pair to the other aster.

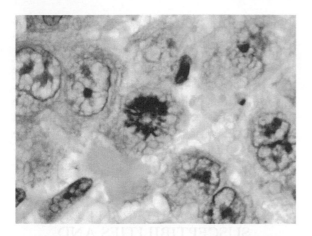

FIGURE A1.7 The ring arrangement of meta(sub)phase chromosomes. Many illustrations show the chromosomes in meta(sub)phase as a 'plate' or column across the long axis of the cell. This may represent oblique sectioning, with or without a coagulative artifact caused by fixation and any further processing. In occasional well-fixed meta(sub) phase cells, the plane of sectioning results in a view of the chromosomes from one end of the cell. These indicate that in each cell, the chromosomes in meta(sub)phase are arranged in a ring. Compared to the 'plate' concept, this ring arrangement would clearly reduce congestion of chromosomes on the spindle fibers so that pair-separations can occur easily. Little is known of how the ring structure is maintained in the spindle fibers.

d. *Telo(sub)phase*

The sets of chromatids are fully clustered at the opposite ends of the cell.

e. *Cytokinesis*

The plasma membrane infolds across the position of the metaphase to divide the double-sized cell into two daughter cells. The chromatids (usually referred to in G0 as chromosomes) at each aster uncoil and nuclear membrane forms around them.

The daughter cells are now in G0.

The acquisition of G0 is permanent if the division is the last division before "terminal specialization" in normally proliferating cells such as epidermis.

There are many aspects of the process which are unclear. The mechanisms of many parts of the process are unknown. Here the following can be pointed to as perhaps the most challenging:

 i. The higher-order arrangements of DNA in G0 and in the chromatids.

 ii. How the individual chromatids in each pair are correctly allocated one to each daughter cell at metaphase–anaphase.

 iii. The "molecular motors" of movements of centrioles and chromosomes.

 iv. The mechanisms of cytokinesis, especially how the cell membrane severs the cytoplasm into two.

Each of these steps can be abnormal in tumor cells (Section 8.1.5).

A1.3.6 Biochemical Aspects of Cell Division

(a) Initiation Through Cyclins

Cell division is probably controlled by proteins known as "cyclins." This term was originally introduced for proteins which are only present in cells in the division phase [21]. It was found that cyclin D1 is present through all the phases of division, while others—cyclins E, A and K—have relatively sharp peaks of concentration at specific division subphases [22,23].

The initiation of cell division is therefore associated with the induction specifically of transcription of gene(s) for cyclin D1. The most-studied cyclin D-inducing factors are nononcogenic and oncogenic growth factors acting through signaling pathways (see Section 6.3 and [22]). These mechanisms may require an additional factor involving plasma membrane–extracellular matrix relationships. This is because the actions of mitogenic factors are often more pronounced on cells grown on surfaces in comparison with those grown in suspension [23], see also Section 14.2.3.

(b) "Check Points" in Cell Division

In normal cell division, the phases and subphases of cell division flow smoothly from one to the next. The time taken for cell division varies according to the kind of cell and the experimental conditions, but can be as short as a few minutes.

In experimental studies, however, it has been shown that at several steps, the process may be halted, or at least, significantly delayed. They are usually noted according to the phases in, or between which, the halt occurs. For example, irradiated fibrocytes may be unable to respond to stimulation (i.e., unable to carry out division), and hence be at a "G0/G1 checkpoint." Excessive damage to DNA may cause failure of the commencement of S phase (a "G1/S checkpoint") or prolongation of S phase, hence an "S phase checkpoint." Cessation of the process of cell division when the chromosomes are fully developed might be considered a "meta(sub) phase checkpoint." Large numbers of ana(sub) phase bridges may be associated with failed completion of ana(sub)phase, and hence "an ana(sub)phase checkpoint" [24,25].

The concept of a "checkpoint" involves an active cellular process, in which there are several parts as follows [25]:

i. an abnormality in the genome of a cell can be "sensed" in some way by mechanisms in the cell,
ii. as consequences of the sensory mechanism detecting the change, two things occur:
  iia. signal transducers inhibit the advance of the whole genome (damaged and undamaged) to the next phase of cell division and
  iib. effector responses are activated which can include repairs to the damage, or alternatively, lead to death of the cell (see Sections 10.4 and 10.5).

An alternative view is that the slowing of cell division by damage to the genome and/or genome processing proteins is essentially a passive phenomenon. That is to say, like all regular sequential processes, each active agent has the dual effects of activating the next phase and inhibiting the previous phase (Section A1.3.3 [26,27]). Cell death is seen to occur because the primary damage may not have been limited to cell division, but also to have affected the integrity of the cell. According to this concept, the individual points are called "arrests." The fact that arrests can occur at one point and not another is because the molecular interactions are different at each point and have different susceptibilities to the agents causing the damage.

A1.4 DIFFERENT SUSCEPTIBILITIES AND RESPONSES OF NORMAL CELLS TO INJURIES

A1.4.1 Metabolic Susceptibilities and Particular Defenses in Cells of Different Kinds

There are striking differences between the different kinds of cells in their capacities to survive adverse metabolic conditions. For example, among all the different kinds of cells, neurons have the lowest tolerance to hypoglycemia and hypoxia. Neurons, however, are less susceptible to radiations than bone marrow cells. Epithelia of the gastrointestinal tract have intermediate sensitivities to hypoxia and radiations. Fibrocytes have probably the highest tolerance of all the different kinds of cells to all adverse conditions, including radiation damage.

These issues are important for anticancer treatments, in which a critical aspect is that the tumor cell populations are more likely to be killed by, and less likely to recover from, cytotoxic agents than normal cells (see Sections 1.4.4, 11.3.3, and A3.3).

The usual explanation for why, for example, bone marrow cells are less resistant than bone cells, has been simply that mitoses are more common in the most sensitive tissues. From this, it has often been assumed that the mitotic apparatus is the target of the slow cell-killing agents (Sections 10.4.3, 10.6.3). Nevertheless, there is no certainty concerning the degrees to which pretarget, target, or posttarget resistance factors are involved (Appendix 3). There have been few studies of whether or not the efficacy of these mechanisms is the same in different kinds of cells or in different species.

A1.4.2 Increased Cell Production After Tissue Loss: Reconstitution, Regeneration, and Repair

"Reconstitution" means replacing whole parts of the body, for example after amputation. It involves coordinated proliferative responses of more than one kind of cell.

Worms such as *Planaria*, certain arthropods, and certain vertebrates such as lizards can regenerate entire parts of the body. The phenomenon strongly indicates that all the cells of these plants and animals contain all the genetic material necessary for the whole organism, but that at any point in time, most of this material is held in an inactive state. By unknown mechanisms, mammals, including humans, have "lost" this capacity during evolution.

"Regeneration" means the replacement of lost tissues from a part of the body [28]. Like reconstitution, this involves degrees of coordination of responses of more than one kind of cell. This occurs in humans, but with enormous variability according to the organ and kind of tissue. Regeneration is particularly seen in the endometrium after menstruation, healing of bones after fractures and in the liver after partial hepatectomy. In all these cases, the residual tissue or organ regrows to precisely its original size before the partial removal, with all the parenchymal and supportive cells properly arranged.

Another well-considered phenomenon related to regeneration is the compensatory growth which occurs in a remaining kidney after the loss of the other kidney. In this, removal of one kidney is followed by enlargement of the other kidney such that the volume of cortex of the remaining kidney is more or less double its normal volume.

"Repair" refers to return to a stable anatomical state through the process of fibrosis [29,30]. In a wound to the skin, the epidermis regenerates. However, the dermis shows a permanent focus of new collagen deposition (i.e., the scar) and is a repair. The two processes are both controlled and coordinated. The epidermis only ceases to proliferate when the edges of the wound join each other (see contact inhibition in Section A1.3.1). The dermal scarring process involves a phase of proliferation of fibroblasts and blood vessel cells, which ceases when the defect between the wound edges is filled in. Both the epidermal and dermal events occur synchronously, suggesting some kind of coordinating control mechanism.

Broadly, however, it is not understood how these controls are exercised. It is also unclear why the reconstitutive and regenerative phenomena are less common and/or effective in "higher" species than in "lower" species in the evolutionary scale.

A1.4.3 Increased Production of Individual Kinds of Cells After Chemical Damage

In relation to toxic injuries, there is also a range of capacities for regeneration. Fibroblasts have an almost limitless capacity to regenerate, as demonstrated in the formation of scars in conditions in which other kinds of cells do not survive. For this reason, it may be no coincidence that fibroblasts are more easily grown in tissue culture conditions than almost any other kind of cell. Also, epidermal cells seem to be able to regenerate to cover wounds to an almost limitless extent.

Some particular situations are well studied as follows:

a. *Ethanol Poisoning of Liver Cells*
   Liver cells can regenerate after poisonings, but in the case of ethanol, fibrosis is also

provoked. The result is that the reticulin framework supporting the liver cell plates collapse and become entwined in the fibrous tissue. The regenerating liver cells then have no normal framework on which to regrow and form poorly-vascularized nodules surrounded by fibrous tissue—i.e., cirrhosis [31].

b. *Cytotoxic Damage to Bone Marrow Cells*

Hematopoietic cells generally regenerate quite well after cytotoxic treatment [32].

c. *Ischemic Damage to Renal Tubular Epithelial Cells*

If the kidney is subjected to complete loss of blood supply, it dies by a process known as "necrosis" (Section 10.4). However, if the blood supply is only partially reduced, the most metabolically active cells—the tubular epithelium—die, while other cells survive. If normal blood supply is reestablished soon enough, the tubular epithelium regenerates, and eventually, normal renal function is restored [33].

d. *Other*

Testicular germ cells appear to have lesser capacity for regeneration than hematopoietic cells. Males are often sterilized by anticancer drug regimes, but females are less affected. Neurons can regenerate their axons, and their dendrites. Neurons which die are not regenerated.

The molecular mechanisms of increases and decreases in cell production in pathological, especially toxic, conditions are little understood.

A1.4.4 Metaplastic Responses

These are peculiar processes, which were identified early in the history of histopathology, especially by Virchow [34]. The essence of metaplasia is that the local proliferating stem cells switch the specialization of their descendant cells from one kind to another. It is a regular feature of the healing of bone fractures, in which connective tissue cells convert to fibroblasts, chondrocytes, and osteocytes. It is also seen in certain epithelia, such as bronchial epithelium near chronic foci of inflammation [35]. However, it is not seen in all chronic inflammations. Thus it never occurs in chronic inflammation of the colon, e.g., in ulcerative colitis, or in the margins of chronic peptic ulcers of the stomach.

"Metaplasia" can be a normal process, as is seen in the uterine cervix. Regularly with age, the mucus cells of the vaginal surface of the endocervix are replaced at the "transformation zone" by squamous ectocervical cells [36].

A1.5 INVASIONS AND METASTASES BY NORMAL INDIVIDUAL CELLS AND POPULATIONS OF CELLS

A1.5.1 Physiological Invasions and Metastases

"Invasion" is widely recognized as a feature of malignant tumor cells (see Section 1.6). However, it is also well known that invasion is carried out by many normal kinds of cells as part of normal physiological processes (Figure A1.8). Examples are as follows.

The invasion of the blastocyst into the secretory endometrium to begin pregnancy is the most obvious example of physiological tissue invasion. In embryonic life, the formation of organs such as the endocrine glands, and the distribution of melanocytic cells from the neural crest to the epidermis, all occur by precursor cell populations relocating through connective tissue to new parts of the body. Again in the embryo, the formation of such organs as the thyroid gland and the central nervous system require some part of the endoderm or ectoderm to grow faster than adjacent cells of the same type, and then to invade the underlying mesodermal connective tissue.

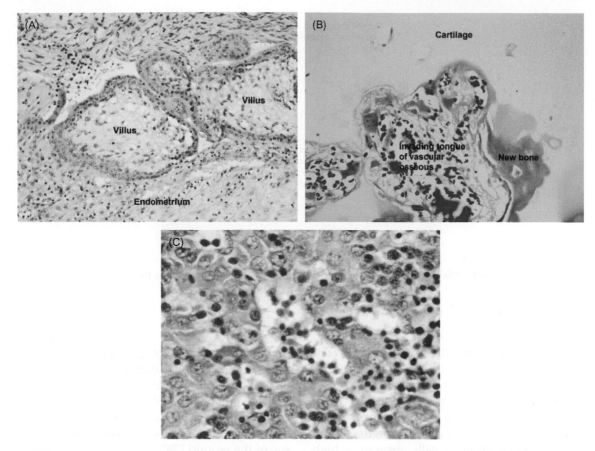

FIGURE A1.8   Physiological invasions and metastases. (A) Invasion of endometrium by placental villi during pregnancy. (B) Invasions of cartilage by bone in normal enchondral ossification In the normal process of development of long bones, precursor cartilaginous tissue is invaded and replaced by bone, ×20. (C) Physiological metastasis of embryonic hematopoietic tissue from the yolk sac to the liver.

In postnatal growth, ossification of the majority of bones occurs by osteoblasts invading cartilaginous precursor structures.

In embryonic life, hematopoietic cells form first in the yolk sac and then "metastasize" to the liver in spleen (Figure A1.8). In adult life, hematopoiesis normally ceases in these organs and continues in the bone marrow.

Also in adult life, the leukocytes, as part of their normal functions, emigrate from blood vessels into connective tissues, epithelia, and, in the case of neutrophils, almost any tissue which has been infected with pyogenic bacteria.

A1.5.2 The Relocalizing of Normal Cell Populations by Differential Localized Growth

In embryonic life, many populations of cells relocate. For example, the thyroid gland begins as an out-pouching from the tongue epithelium. These "movements" of populations of cells may not occur by individual cell movements (next section), but by differential local growth rates [16]. The differential growth rates may occur under the influence of gradients of growth-promoting substances [37]. Blood

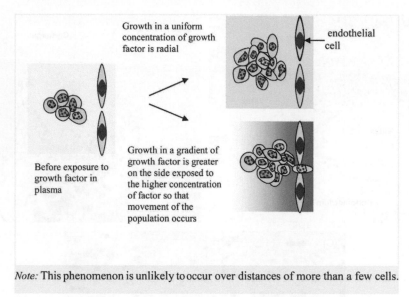

Growth in a uniform concentration of growth factor is radial

endothelial cell

Before exposure to growth factor in plasma

Growth in a gradient of growth factor is greater on the side exposed to the higher concentration of factor so that movement of the population occurs

Note: This phenomenon is unlikely to occur over distances of more than a few cells.

FIGURE A1.9   Population chemotaxis.

certainly contains growth factors (see Section 6.3.1), and it is possible that such factors could promote outward growth of cells from the main mass in one direction (Figure A1.9).

Thus a mass of cells may be considered to have polarity with respect to growth. At one pole of a population of cells, growth is high, and at the other pole, there is no growth. Over time, the center of the population will necessarily "move" in the direction of the first pole (Figure A1.9).

A1.5.3 The Kinds of Active Movements of Individual Normal Cells

There are several different kinds of normal motility of individual cells.

(a) "Sliding Movement"

This is very common, but not widely studied. Sliding movement is exhibited especially by epithelia moving over their basement membranes. For example, the epithelium of the intestine normally slides from the sites of cell division (the bases of crypts) to the surface during normal cell turnover in this organ. In wound healing, a similar sliding movement is exhibited by sheets of new epithelial cells as they are pushed by the mitoses of cells behind the margin to cover the exposed connective tissue of the wound [38].

(b) Ameboid Movement

This is the most widely known kind of individual cell motility. Among mammalian cells, this is most obviously shown by leukocytes. This kind of movement is the subject of various reviews [39–43]. Its mechanism is not fully understood.

(c) The Crawling Movements of Other Cells in Culture (Such as Fibroblasts)

When cells are cultured on solid surfaces such as glass, they may exhibit a crawling-like movement (e.g., [44]). The process may be ameboid movement, modified according to differences in adhesion of the cells to solid substrata compared to cells in tissues.

A1.5.4 Theories of the Induction of Motility in Normally Nonmotile Cells

Overall, the question of how a normal cell acquires motility in nontumorous pathological conditions is similar to the question of how motility is possibly acquired in association with the "tumorous conversion" of cells ([45] and see Section 6.6.3 and Figure 6.9).

The possible mechanisms are the same:

a. *The normal cell always has the motile apparatus present in an active state, but actual movement is prevented by adhesion of the cell to local structures, especially cell–cell junctions (desmosomes), and by cell-to-basement membrane attachments* [46].

These structures have been proposed to perhaps depend on specific carbohydrate moieties in the external coat ("glycocalyx") of the cell (see Section A1.2.4) or CD44, or even a particular monosaccharide species (e.g., [47,48]).

A related notion is that invasion begins with expression (implying gene derepression) by tumor cells of enzymes which lyse basement membrane, thus allowing the inherent motility of cells to become manifest [49].

b. *The normal cell has the motile apparatus present, but in an inactive state. The apparatus can, however, be activated by appropriate stimuli.*

According to the commonest suggestion of this type, cancer cells become motile because the appropriate cytokine (motility factor) is secreted by the tumor cells, as an autocrine phenomenon (e.g., [50]).

c. *The normal cell has no motile apparatus present, but expression of the proteins, etc. for the whole apparatus is induced by derepression of a "motile apparatus master gene."*

Thus "master" (pleiotropic) genomic regulators have been postulated for cell motility and invasion (e.g., [51]). Specific molecules may form complexes and have a general role in the complex process of cell movement (e.g., [52]). The question of how a complex process involving directional morphological changes (pseudopodia formation) and translocation of whole cells is coordinated, is unclear.

References

[1] Baudat F, Imai Y, de Massy B. Meiotic recombination in mammals: localization and regulation. Nat Rev Genet 2013;14(11):794–806.

[2] Meshcer AL, Junqueira LCU. Junqueira's basic histology: text & Atlas, 13th ed. New York, NY: McGraw-Hill; 2013.

[3] Gartner LP, Hiatt JL. Color textbook of histology, 3rd ed. Philadelphia, PA: Saunders/Elsevier; 2007.

[4] Mills SE. Histology for pathologists (4th ed.). Philadelphia, PA: Lippincott, Williams & Wilkins; 2012.

[5] Friedman H, Escobar M, Reichard SM. The reticuloendothelial system: a comprehensive treatise. New York, NY: Plenum Press; 1980–1988.

[6] Alberts B, Johnson A, Lewis J, et al. Molecular biology of the cell (5th ed.). New York, NY: Garland Science; 2008. p. 617–50.

[7] Hetzer MW. The nuclear envelope. Cold Spring Harb Perspect Biol 2010;2(3):a000539.

[8] Doye V, editor. Nuclear pore complexes and nucleocytoplasmic transport—methods. Methods Cell Biol, v 122, San Diego, CA: Academic Press/Elsevier; 2014.

[9] Razin SV. The nuclear matrix and spatial organization of chromosomal DNA domains. New York, NY: Springer/Landes Bioscience; 1997.

[10] The Free Dictionary. Chromatin. Available at: http://www.thefreedictionary.com/chromatin.

[11] Bertaux O, Valencia R, Magnaval R. The nucleus. The biology of Euglena, vol. 4. San Diego, CA: Academic Press; 1989. p. 137–46.

[12] In: Ref. [6]. p. 211–8.

[13] Crèmer T, Cremer C. Rise, fall and resurrection of chromosome territories: a historical perspective. Part I. The rise of chromosome territories. Eur J Histochem 2006;50:161–76.

[14] Bickmore WA. The spatial organization of the human genome. Annu Rev Genomics Hum Genet 2013;14:67–84.

[15] Derenzi X. The AgNORs. Micron 2000;31:117–20.

[16] Conlon I, Raff M. Size control in animal development. Cell 1999;96:235–44.

[17] Behar M, Hoffmann A. Understanding the temporal codes of intra-cellular signals. Curr Opin Genet Dev 2010;20(6):684–93.

[18] Maehle AH. Ambiguous cells: the emergence of the stem cell concept in the nineteenth and twentieth centuries. Notes Rec R Soc Lond 2011;65(4):359–78.

[19] Kölliker A. Manual of human histology, translated and edited by Busk G, Huxley T, 1853-54. London: Sydenham Society.

[20] Bizzozero G. An address on the growth and regeneration of the organism. BMJ 1894;i:728–32.

[21] Evans T, Rosenthal ET, Youngblom J, et al. Cyclin: a protein specified by maternal mRNA in sea urchin eggs that is destroyed at each cleavage division. Cell 1983;33(2):389–96.

[22] Klein EA, Assoian RK. Transcriptional regulation of the cyclin D1 gene at a glance. J Cell Sci 2008;121(Pt 23):3853–7.

[23] Assoian RK, Klein EA. Growth control by intracellular tension and extracellular stiffness. Trends Cell Biol 2008;18(7):347–52.

[24] Weinberg RA. The biology of cancer (2nd ed.). New York, NY: Garland Science; 2014. p. 279–80.

[25] Li WX. Cell cycle checkpoints. New York, NY: Springer; 2011. preface, p. v.

[26] Zetterberg A, Larsson O, Wiman KG. What is the restriction point? Curr Opin Cell Biol 1995;7(6):835–42.

[27] Cooper S. Checkpoints and restriction points in bacteria and eukaryotic cells. Bioessays 2006;28(10):1035–9.

[28] King RS, Newmark PA. The cell biology of regeneration. JCB 2012;196:553–62.

[29] Gurtner GC, Werner S, Barrandon Y, et al. Wound repair and regeneration. Nature 2008;453(7193):314–21.

[30] Tanaka E, Galliot B. Triggering the regeneration and tissue repair programs. Development 2009;136(3):349–53.

[31] Luedde T, Kaplowitz N, Schwabe RF. Cell death and cell death responses in liver disease: mechanisms and clinical relevance. Gastroenterology 2014(14):00914–7.

[32] Nieboer P, de Vries EG, Mulder NH, et al. Long-term haematological recovery following high-dose chemotherapy with autologous bone marrow transplantation or peripheral stem cell transplantation in patients with solid tumours. Bone Marrow Transplant 2001;27(9):959–66.

[33] Goldberg R, Dennen P. Long-term outcomes of acute kidney injury. Adv Chronic Kidney Dis 2008;15(3):297–307.

[34] Bignold LP, Coghlan BL, Jersmann HP. Virchow's "Cellular Pathology" 150 years later. Semin Diagn Pathol 2008;25(3):140–6.

[35] Heard BE, Khatchatourov V, Otto H, et al. The morphology of emphysema, chronic bronchitis, and bronchiectasis: definition, nomenclature, and classification. J Clin Pathol 1979;32(9):882–92.

[36] Robboy SJ, Muttter GL, Prat J, editors. Robboy's pathology of the female genital tract (2nd ed.). Philadelphia, PA: Churchill Livingstone/Elsevier; 2009.

[37] Ettensohn CA. Encoding anatomy: developmental gene regulatory networks and morphogenesis. Genesis 2013;51(6):383–409.

[38] Krawczyk WS. A pattern of epidermal cell migration during wound healing. J Cell Biol 1971;49(2):247–63.

[39] Wells A, editor. Cell motility in Cancer invasion and metastasis. Cancer Metastasis—Biology and Treatment, vol. 8. New York, NY: Springer; 2006.

[40] Yilmaz M, Christofori G. Mechanisms of motility in metastasizing cells. Mol Cancer Res 2010;8(5):629–42.

[41] Bignold LP. Amoeboid movement: a review and proposal of a 'membrane ratchet' model. Experientia 1987;43(8):860–8.

[42] Lämmermann T, Sixt M. Mechanical modes of 'amoeboid' cell migration. Curr Opin Cell Biol 2009; 21(5):636–44.

[43] Friedl P, Borgmann S, Bröcker EB. Amoeboid leukocyte crawling through extracellular matrix: lessons from the Dictyostelium paradigm of cell movement. J Leukoc Biol 2001;70(4):491–509.

[44] Keller HU, Zimmermann A, Cottier H. Crawling-like movements, adhesion to solid substrata and chemokinesis of neutrophil granulocytes. J Cell Sci 1983;64:89–106.

[45] Défamie N, Chepied A, Mesnil M. Connexins, gap junctions and tissue invasion. FEBS Lett 2014;588(8):1331–8.

[46] Jacquemet G, Humphries MJ, Caswell PT. Role of adhesion receptor trafficking in 3D cell migration. Curr Opin Cell Biol 2013;25(5):627–32.

[47] Viguier M, Advedissian T, Delacour D, et al. Galectins in epithelial functions. Tissue Barriers 2014;2:e29103.

[48] Cattaruzza S, Perris R. Proteoglycan control of cell movement during wound healing and cancer spreading. Matrix Biol 2005;24(6):400–17.

[49] Bergman A, Condeelis JS, Gligorijevic B. Invadopodia in context. Cell Adh Migr 2014;8(3).

[50] Fairbank M, St-Pierre P, Nabi IR. The complex biology of autocrine motility factor/phosphoglucose isomerase (AMF/PGI) and its receptor, the gp78/AMFR E3 ubiquitin ligase. Mol Biosyst 2009;5(8):793–801.

[51] Hay ED. An overview of epithelio-mesenchymal transformation. Acta Anat (Basel) 1995;154(1):8–20.

[52] Suozzi KC, Wu X, Fuchs E. Spectraplakins: master orchestrators of cytoskeletal dynamics. J Cell Biol 2012;197(4):465–75.

Aspects of the Normal Genome

It is almost universally accepted that some kind of alterations in the genome of parent cells is involved in the formation of tumors (Chapters 4–7) and that there are large numbers of genomic events in tumor cells (see Chapter 5). However, the nature(s) of the significant genomic events in tumorigenesis of almost all tumor types are not known.

To support the consideration in Chapter 5 of these genomic alterations, this appendix describes relevant aspects of the normal genome.

Section A2.1 mentions some general aspects of the genome. Section A2.2 deals with the production of proteins and RNAs from DNA; and section A2.4 gives an account of the enzymes which act on DNA to achieve processes such as unraveling, synthesis, and transcription.

The material is drawn from main texts [1–5] as well as monographs, selected chapters in books and articles.

A2.1 GENERAL ASPECTS

A2.1.1 Composition

(a) General

The genomes of all eukaryotic organisms are made up of double-stranded DNA. There are enormous variations in the sizes of genomes

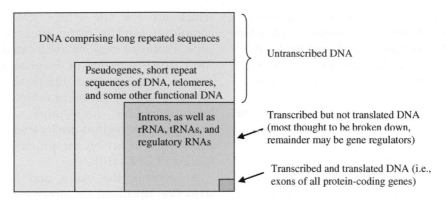

FIGURE A2.1   The components of the DNA in the genome.

between species (Figure A2.1). The haploid human genome comprises approximately 3.1 billion nucleotide pairs divided between 23 chromosomes. In the male, the 23rd chromosome is a "Y" sex chromosome, and in the female, the 23rd is a second "X" sex chromosome. A small piece of DNA is present in mitochondria.

(b) Transcribed and Translated DNA (i.e., Exons of Protein-Coding Genes)

Approximately 1–2% of the DNA codes for messenger RNA (mRNA), which is then translated into a chain of amino acids in proteins used in the cells of the body. Those proteins can be structural for the cell, regulatory for genes (transcription factors), or have other functions. Any particular piece of amino-acid-coding DNA does not necessarily become part of only one protein. By mechanisms including "alternative splicing" of RNA transcript, parts of many different proteins may be derived from the same sequence of DNA.

(c) Transcribed but not Translated DNA

Approximately one quarter of the DNA may be transcribed but not translated. This transcribed RNA includes introns of genes, non-mRNAs, such as ribosomal RNAs (rRNAs) and transfer RNAs (tRNAs). The transcripts of introns are thought to be promptly degraded in the nucleus of the cell to single nucleotides and

reused in new RNAs. However, some intronic RNA may serve as regulators of genes or as RNAs which inhibit ("silence") translation of specific mRNAs (i.e., inhibit the synthesis of specific proteins), see Section A2.2.5(e).

(d) Untranscribed DNA

The remaining three quarters of the entire genome is untranscribed (and hence also untranslated into protein). Its significance and function(s) are unclear. Approximately two-thirds of untranscribed DNA (one half of the total human DNA) comprises repeats of long segments of DNA. Since the 1970s, this has been known as "junk DNA." More recently has been suggested to represent endogenous retrotransposons [6]. These are also described as "decayed" endogenous retroviruses or "human DNA endogenous retroviruses" [7] (see especially Table 1 in this reference).

Most of these retroviral structures and mobile elements are repeated various numbers of times. It is thought that some of the sequences may have gene-regulatory roles in development [8].

The remaining third of the untranscribed DNA includes:

i. "Pseudogenes"—copies of normal human genes which are damaged, and usually without promoter regions. It should be noted that recently, it has been suggested that some of these genes without promoters may in fact be transcribed, and the RNA transcripts may play functional roles *vis à vis* the protein-coding copies of the gene (see Section A2.2.6).

ii. Short (2–10 bases) repeat sequences (8%) of total DNA.

This DNA is also known as "microsatellite DNA." This term dates from the beginnings of the analysis of DNA by using site-specific enzymes. These digests of DNA were separated by ultracentrifugation on density gradients. Most of the digested DNA occurred in one band, but additional "satellite" bands were found. These had different buoyancies because they comprised repeats of nucleotides. The α-satellite consists of repeated units of 171 base pairs and is centromeric DNA. DNA comprising repeated units of 10–50 base pairs are called "mini-satellites" and those comprising repeats of 1–10 base pairs are called "microsatellites." The numbers of repeats in each these sections are highly variable between individuals. Each individual has a unique patterns of these variable lengths of repeats. This fact is used in forensic medicine to identify the sources of blood or tissues ("DNA finger-printing") [9].

iii. Promoter regions, insulators, and other regulatory sequences in flanking regions of genes (see Section A2.2).

iv. Telomeres.

These are the repeat sequences at the ends of chromosomes (see Section 2.9.2).

v. The inactive X chromosome (the Barr body) in human females (see Section A2.3.1).

A2.1.2 Genes

(a) General

The human genome contains approximately 20,000 genes. Each gene has one or sometimes more unique promoter region(s) which are the site where RNA polymerase complexes bind to begin transcription of the gene. Promoter regions are located adjacent to the 5′ (upstream) end of the gene in "flanking" regions. These parts of the gene are not transcribed and may contain multiple short "regulatory sequences," for example, six in the TATA box. Enhancer regions are located distant to the promoter, but act on the promoter regions by folding of DNA [10].

In general, the exons and introns are in sequence upstream to downstream from the promoter region. Between different genes, exons vary in size, number, spacing (introns), and even chromosomal location. On average, there are 10.4 exons per gene. The introns average approximately 20 times the lengths of exons [11].

(b) "Dominance," "Recessivity," and the Functional "Morphisms" of Genes

Mendel introduced the terms "dominant" or "recessive" for parental "factors" (alleles of genes) which have "all-or-nothing" effects (see Section 2.5.1). Dominant and recessive are then applied to genes which are represented by only two active alleles in the genome. If a genomic event in one allele alone is sufficient to produce the phenotypic change, then both the event and the gene can be said to be dominant. If the function of the gene is, for example, to produce an enzyme, and the genomic change results in deficient enzyme production, "dominance" means that the other copy of the gene is not enough (is "haplo-insufficient") to produce physiologically necessary amounts of enzyme. Alternatively, if appropriate genomic events are required in both alleles for loss of gene function, then both the events individually and the genes can be referred to as "recessive" (i.e., each wild-type allele is haplo-sufficient for gene function).

These issues are now understood to be complicated by various possible phenomena:

i. Many individuals may have one of their two copies of alleles permanently shut down (see Section A2.1.3), especially in X-inactivation, and imprinting [12].

ii. There may be more than two alleles of the gene in the genome, of which only some may be functioning (Section A4.2.4).

iii. Some genomic events may result in gain of function—"hypermorphic."

iv. Other genomic events may cause only partial loss of function—the affected gene is then said to be "hypomorphic." This is important in enzymology, because genomic events in genes for enzymes may only cause a partial loss of function. These enzymes are called "leaky," as are the genomic events responsible for them [13].

v. A genomic event may result in gains of one, or possibly more, new and unrelated functions—the gene is "neomorphic."

vi. If the neomorphism results in a gain of function with disease causing effect, the mutant allele is of a "pathogenic neomorphic" type.

vii. A particular version of the pathogenic neomorphism in an allele is the "dominant negative" effect. This refers to the situation in which the product of the altered allele has no physiological function but—as a potentially pathological effect—it inhibits the action of products of the other allele of the gene, or alleles of other genes.

These concepts may well apply to "growth factor/oncogenes" and tumor suppressor genes which can influence the growth of cells (see Sections 6.2 and 6.3). There is no proof that these effects cannot be individually quantitative and, perhaps in some instances, collectively additive.

(c) Other Functional Classifications of Genes

i. "Executive" and "realizator" genes in "cascades"

In theory, if all genes for a complex phenomenon were activated at once,

there would be no sequential phases in any biological phenomenon (e.g., as in Section A1.3.4). Thus it would seem to be essential that genes be activated in sequence in a "cascade" of transcription factors for complex phenomena. This has been extensively studied in embryonic development. Essentially, the ultimate "executive"/"critical"/"master" gene is mRNA in the egg. The protein product (a transcription factor) activates "gap genes," which then activate HOX genes [14]. Each HOX gene is an executive gene for downstream "realizator"/"effector" genes which are necessary for the completion of the corresponding individual processes. A feature of the HOX genes is that they are clustered (colinear) in the genome, in the same order in which their protein products appear in the embryo.

Whether all these phenomena—of gene cascades and of clustered location in the genome—also apply to local tissue specialization—as in keratin production by the epidermal cells. The mechanisms are not clear [14].

ii. "Gate-keeper" and "caretaker" genes

Oncogenes and tumor suppressor genes in any part of cascades which can be activated and enhance growth (see Sections 6.3–6.5) have been called "gate-keeper" genes, in the sense that their wild-type function is to prevent excess growth. Genes which support genomic instability are called "caretaker" genes, because when affected by a germline genomic event, a low-penetrance predisposition to tumor of one or a few kinds of cells (see Section 7.8) is created [15].

iii. "Landscaper" genes

These genes are involved in altering the microenvironment of cells to facilitate growth in cells which have sustained growth-enhancing genomic events in either their "gate-keeper" or their "caretaker" roles [16].

iv. "Executioner" genes

These genes, when activated, kill the cell. Examples include genes for initiating and effector caspases (cysteine-dependent, aspartate-directed proteases) and associated proteins [17]. It may be noted here that almost all cell killing, in which a gene product having lytic effect on some constituent of the cell is involved, is often referred to as 'apoptosis' (see Section 10.5). Conceptually, this is difficult to distinguish from lysosomal activation (see Section 10.1.4).

A2.1.3 Differences Between the Genomes of Individual Humans

The fact that the patterns of repeats sequences of DNA are unique to individual humans has been mentioned above (Section A2.1.3). There are numerous other genomic variations in the human genome [18], including:

i. In chromosomes structure. These are uncommon in the population [19].

ii. In numbers of copies of particular pieces of DNA [20]

The longest of these—in units of hundreds of kilobases to over 1,000 kb—are uncommon and are often called "copy number variants." Shorter repeated pieces—around 10 kb long—occur in over 1% of the population and may be called "copy number polymorphisms." Different numbers of pieces of DNA of around 1 kb long occur in almost every individual in the population, with an average rate of 12 per person. Variations in even shorter pieces (around 50 bases) occur in hundreds per individual.

These differences presumably occur by primarily by amplifications. However, deletions, transpositions, and inversions of pieces in the whole range of sizes may occur.

iii. Single nucleotide "polymorphisms" (substitution changes)

These are now known to be perhaps as common as 0.3–1% of bases, when defined (arbitrarily) as in those occurring in 1% of the population [21]. In this, it should be remembered that citing nucleotides is a difficult and complex matter [22].

A2.1.4 Damage to Nucleotides and the Repairs Which Can Be Made

In the late 1980s and 1990s, it became apparent that the DNA in all living cells is constantly damaged (i) directly by exogenous agents and (ii) indirectly by various endogenous agents to the extent of tens of thousands of DNA lesions per nucleus per day [23] (Figure A2.2).

The kinds of damage are mainly to the purine and pyrimidine bases of the DNA, in the form of depurinations, deaminations, oxidations, and acylations [24].

The types of repairs and their possible consequences for the sequence of nucleotides in DNA are [24]:

a. *Removal by scavenging enzymes, such as methyl-transferases*

There are many enzymes which remove adducts on DNA, including demethylases, deacetylases, and deoxidases. Their functions appear to be "cleaning" the DNA of adducts created during normal or "cell-stress"-related metabolism (see Sections 10.1.5 and 10.1.6). An exception, however, may be cytosine deaminase [25], which has the effect of altering cytosine to uracil. Its role(s) in ordinary metabolism of cells are unclear.

The biochemical modifications in bases are not of themselves alterations in nucleotide sequence since no nonenzymatic molecules (exogenous and endogenous DNA-damaging agents) can directly change

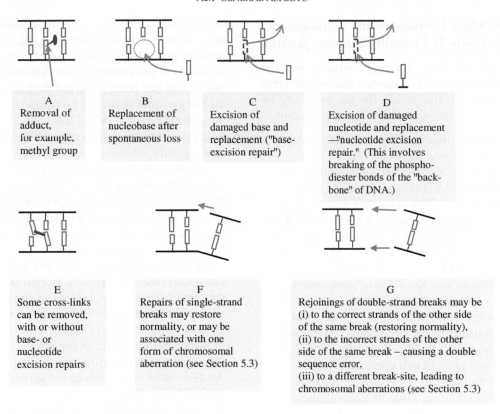

A
Removal of
adduct,
for example,
methyl group

B
Replacement of
nucleobase after
spontaneous loss

C
Excision of
damaged base and
replacement ("base-
excision repair")

D
Excision of damaged
nucleotide and replacement
—"nucleotide excision
repair." (This involves
breaking of the phospho-
diester bonds of the "backbone" of DNA.)

E
Some cross-links
can be removed,
with or without
base- or
nucleotide
excision repairs

F
Repairs of single-strand
breaks may restore
normality, or may be
associated with one
form of chromosomal
aberration (see Section 5.3)

G
Rejoinings of double-strand breaks may be
(i) to the correct strands of the other side
of the same break (restoring normality),
(ii) to the incorrect strands of the other
side of the same break – causing a double
sequence error,
(iii) to a different break-site, leading to
chromosomal aberrations (see Section 5.3)

FIGURE A2.2   The kinds of repairs to DNA. See also Figures 1.6 and 5.3.

the sequence of bases in the DNA chain. Removal/scavenging of the modifications also does not involve the bases *in situ* and therefore do not have a potential effect.

Nevertheless, if nucleotide damage is not repaired, then the modification may cause a mis-pairing in S phase of nucleotides in the new strand of DNA. This latter event is facilitated by "low stringency" polymerases, such as the polymerase Y family. The syntheses which these polymerases carry out are called "translesional" because they can use the damaged nucleotides as templates. If not subsequently corrected by proofreading or mismatch repairases, these mis-pairings persist as genomic events in

descendant cells derived from the new DNA strand [24].

i. Repairs involving excision of either the damaged base or the whole nucleotide [24].

These repairs may cause changes in nucleotide sequence by inserting:

i.   the wrong nucleotide (substitution error),
ii. no nucleotide (deletion) or
iii. two nucleotides in the place of one (insertions)—see Section 5.1.

Nucleotide excision repairs involve single-strand breakage of DNA, and hence may be involved in the mechanism of chromosomal aberrations (see Section 5.3.3).

A2.1.5 Most Damage to the Genomes of Somatic Cells Is Probably Inconsequential

In considering damage which is not repaired, and actually causes a nucleotide error in the genome, the result for the cell may be insignificant. This is for the following reasons.

a. *The genome overall is dilute in respect to genes*

As mentioned in Section A2.1.1 (see Figure A2.1) only perhaps a maximum of 2% of the DNA consists of exons, while approximately 25% is introns and most of the remainder consists of repetitive DNA of uncertain function. Thus damage in at least half of the DNA in the cell may be of little consequence for the affected cell.

b. *In embryonic and adult life, cells presumably use only a small part of their genomes*

Because each of the different kinds of cells in multi-tissue organisms has the whole genome for all the kinds of cells in the body, then each kind of cell presumably uses only small part of the whole genome in its normal existence. Thus in such organisms, the active components of the different kinds of cells are for their viability (genes for essential structures and essential processes, such as respiration) and for their particular specialized activity (lineage genes).

The genomes of local stem cells include – but in the repressed state – the genomic elements for all embryological development from the zygote onward, and for the specializations of all other cell types in the body. Thus if a mutation occurs in one of these repressed parts of the genome of a specialized cell in an adult, it may be of no consequence. For example, it would presumably be of little consequence to a keratinocyte if its globin, myelin, albumin, and pepsin coding regions were all deleted.

This supposition has been confirmed by sequencing whole genomes of species, and especially of humans. Different—but still only small—proportions of the genome seem to be responsible for specialized activity according to the kind of cell. The situation is not entirely clear however, because it has been reported that almost all the genome seems to be transcribed in almost all kinds of cells [26].

c. *Many proteins can sustain alterations of amino acid sequence without altered function*

A uni-nucleotide genomic event in a local stem cell may also be inconsequential because many protein products of genes include large "functionally noncritical" regions. There are reportedly hundreds of thousands of polymorphisms in the exons of the human genome, all of which are all nonlethal (see Section A2.1.3). Part of this robustness of the genome lies in the fact that such polymorphisms—if translated into altered sequence in the amino acid chains of the protein product of the gene—have no significant effect on function of the protein. As discussed in relation to the globin molecule (Section 2.6.2), this is because many of the sites which determine precise functions (e.g., reaction sites on enzymes) are only a small proportion of the whole protein, and amino acid substitutions in the remainder can have no functional effect [27].

d. *Tissue and cell factors in the protection of the normal cell population*

There are two further points to note here. First, significant damage to a structural gene in a cell is most likely to cause its death. However, many kinds of tissues are capable of replacing lost cells (Section A1.4.3), so that loss of cells in these tissues has no long-term consequence.

Second, damage to these genes in transit-amplifying cells (see Section A1.3.3) may be inconsequential because these cells (such as epithelial cells of the colon, keratinocytes of the epidermis, or erythrocytes) are shed or scavenged in the body, along with any genomic events which they may have incurred. The only exception to this may arise if a "senescence" gene is altered along with a pro-growth genomic element. This is one possible mechanism of formation of a tumor (see Section 2.9.1).

A2.2 ASPECTS OF THE PRODUCTION OF PROTEINS AND RNAS FROM GENES

This section gives an account of physiological regulators of genes. Special genomic mechanisms affecting gene expression, such as "dominant negativity," are mentioned in Section A2.1.2b.

A2.2.1 Classification of Gene Regulation into Constant, "One-Off" Physiological Phasic, Recurrent Phasic, and "On Demand" Kinds

From the normal events of embryology and histology (Section A1.1), it can be seen that gene expressions may be "constant," "one-off" physiologically phasic, repetitively physiologically phasic, and "on demand/reactive."

a. *The constant expressions of "structural" genes*
In a mature, nondividing cell, the activities these genes are constant. These genes are for production of the proteins necessary for the components and necessary metabolic processes for life of the cell.

b. *"One-off" physiological phasic expression of genes*
This is the basis for a cell changing its descendants, as in embryonic differentiation and tissue specialization. These have already been mentioned in the discussion of HOX genes (Section A2.1.2c). The relevant genes are never expressed again in the lineage. A particular example in medicine is the

genomic mechanism for achondroplasia. The genomic event is in the gene for fibroblast growth factor receptor 3 (FGR3). The protein produce of this gene is normally expressed in late adolescence, and it acts to terminate growth of bone into cartilage (i.e., normal enchondral ossification) appropriate to final adult stature. The genomic event in the gene causes premature activation of expression of the gene—as early as in fetal life—so that bone production is terminated early in life, and the bones fail to grow to normal lengths. The timing of the activation of the gene—the later the better—determines the severity of the clinical manifestations. Other kinds of cells, however, are not affected [28].

c. *Regularly recurrent phasic expressions of genes*
The main example in the recurrent expressions of cyclins in association with cell division (see Section A1.3.4).

d. *"On demand" expression of genes*
The classic example of this is the *lac* gene, which is inducible in the bacterium *Escherichia coli* by exposure to lactose. Other examples are the many "reactive" cellular responses in nontumorous pathological conditions (see Sections 10.7 and A1.4).

The importance to tumors of these different phasic expressions of genes is that the lack of specialization ("de-differentiation," see Section 8.1.2) apparent in tumor cell populations may be due to changes in the rate of the specialization process, not the presence or absence of genes for parts of the process.

A2.2.2 Regulation Through the Promoter Regions ("Operon" Model)

The operon model remains a major focus of study of the regulation of genes. This model involves promoter regions, which allow RNA polymerases to engage the DNA at sites for the initiation of transcription. "Transcription factors" include both activators and inhibitors of the promoter.

Several kinds of mechanisms can act on promoters [29].

(a) External Agents

In bacteria, the promoter may be acted on by an external agent, such as lactose, but in eukaryotes, intermediary pathways between the external stimulus and the promoter are usually involved.

(b) Regulatory Proteins (Transcription Factors)

These factors are defined as proteins with a binding domain for the DNA of the promoter or other regulatory sequence for another protein-coding gene [30]. Binding of the factor influences transcription of the relevant gene(s). Approximately 2,000 transcription factors are known and at present are thought to amount to approximately 10% of all genes. They can be either enhancers, or inhibitors, and the same transcription factor can have opposite effects on different genes (i.e., are pleiotropic). They may have their effects only in specific combinations. Generally it is thought that there is only one copy of each transcription factor.

Control of transcription of transcription factors themselves appears to be by cascades [31].

A2.2.3 Other Elements or Factors Influencing Transcription

(a) Insulators

Insulator elements are regulators in the sense of permanent total inhibitors of gene activity [32]. They are sequences of DNA which bind specific proteins so that the complexes prevent enhancers of one gene from affecting the promoter region of immediately downstream genes. "Loss of function" mutations of these elements may lead to promoters which are normally inactive becoming active, and hence to inappropriate expression of a DNA coding sequence.

(b) "Scaffold/Matrix-Attachment Regions" ("Matrix-Binding Domains")

The general aspects of this putative structure are discussed in Section A1.2.6. These regions of DNA are thought to bind to intranuclear "matrix" fibers by way of intervening specific proteins ("MAR-binding proteins") such as "Menin" [33]. It has been suggested that the DNA forms loops beginning and ending at matrix fibers, so that the proteins function as "corepressors or coactivators" of gene activity. The protein-binding sites for MAR proteins are thought to average 500bp in length and be spaced approximately 30kb apart. Their sequence is variable, but contains abundant AT-rich regions [34]. Whether they include "origin of replication sites" and are thus components of "DNA factories" (see Section A2.3.2) is controversial [35]. Nevertheless, at the present time, the existence of nuclear matrix is disputed [36].

(c) Local Alternate Structures of DNA

The possible roles of alternate conformations of DNA, such as cruciforms, left-handed DNA, triplexes, and quadruplexes, in gene regulation are controversial [37].

(d) DNA–RNA Complexes

These are discussed in Section A2.2.6.

A2.2.4 "Epigenetic" Regulation

(a) Background

In biology, the term "epigenesis" has been used for two different ideas. To various degrees, the two concepts have been linked together without full proof of the relationship being established. Historically, "epigenesis" was used by Aristotle (fourth century BCE), and also in the seventeenth and eighteenth centuries ACE for the concept that, in embryological development, the parts, organs, and tissues of the adult organism form from the undifferentiated substance in each embryo [38]. In the 1940s, CH Waddington introduced the term "epigenetics" for differential activations and inactivations of genes which he assumed must underlie this classical epigenesis [39,40]. Waddington did not specify any chemical mechanism by which these changes in gene activity might take place.

In the 1950s, DNA was established as the chemical of genes, and in the 1960s, it was shown that untranscribed DNA (as in the inactive X chromosomes in human females) is more frequently methylated than is transcribed DNA. In the 1970s, it was proposed that methylation is the *cause*, rather than an effect, of the nontranscription. This chemical alteration, being "on the genome" rather than "of the genome" (as are transcription factors, see previous subsection), was called "epigenetic gene regulation" [39,40].

Ever since, "epigenetic gene regulation" has referred mainly to chemical alterations of DNA or DNA-related proteins such as histones. It is thought that these modifications to DNA and/or histones cause inhibition of genes [41–43]. Suppressions of genes during both embryological development and specialization of cells in adult life have, in part or entirely, been attributed to these "epigenetic" chemical modifications of DNA. (The alternative view is that the genes are suppressed by negative transcription factors and/or "long RNA regulators," see Section A2.2.6.) However, this definition of epigenomic mechanism is not universally accepted. For example, the definition used by the NIH "Roadmap Epigenomic Project" [44] is much wider, because it includes all mechanisms of alteration of gene expression which do not involve changes in nucleotide sequence in DNA.

(b) Discussion of Chemical Modifications in Nucleobases as a Major Gene-Regulatory Mechanism

In relation to the above, there are difficulties in the completeness of the hypothesis of regulation of transcription through modifications of nucleobases. The main topics include:

i. In the process of cell division, all chemical modifications of DNA are removed before DNA replication. The question then is: how are the modifications faithfully reproduced in both daughter cells to provide the "hereditability" aspect of the regulation? A necessary component of the epigenetic theory would have to include how the "mother cell" produces two copies of instructions—of whatever chemical kind—before DNA synthesis occurs, and then distributes one copy to each daughter cell after DNA synthesis.

ii. Inactive genes in cells may be specifically activated in response to environmental factors. If chemical modification of DNA is a major inactivator of genes, how are the modifications removed specifically from the specified genes? No such specificity for demethlyation of DNA or deacylation of histones has been clearly established.

A2.2.5 RNAs in Cells: General Aspects

(a) Turnover of the Different Types of RNA

Excepting perhaps for the RNA of Xist in Barr bodies (see Section A2.2.6), RNAs are not permanent in cells or even long-lasting. They are synthesized, and when they have served their purpose (if appropriate), they are degraded.

Ninety percent of the RNAs in cells are associated with ribosomes ("rRNAs"). These have half-lives of probably up to 7 days. A few percent are tRNAs which have half-lives of hours to days. Another few percent are mRNAs which have half-lives of minutes to hours [45]. The amount of mRNA depends on the activity status of the particular cell (see Section A1.2.7).

The turnover of the RNAs is achieved by various RNAses. These are present in the nucleus (especially to degrade introns) as well as in the cytoplasm. Some RNAses are released from organisms and remain active in the environment.

The significance of the constant turnover is that at any point in time, cells will contain RNAs in the process of being broken down: that is, small RNAs without necessarily any function other than to return single nucleotides to the "nucleotide pool" [46] in all cells, for synthesis of new nucleic acids.

(b) Measuring mRNAs in Tissues

Many efforts have been made to identify particular mRNAs in tumor cells [47,48]. Methods include identifying content (using microarrays of tissue extracts) and localization of mRNAs in tissues (using *in situ* hybridization). These studies are technically difficult for many reasons, but especially because of the liability of RNAs to be degraded rapidly in tissues removed from the body (by biopsy).

High-throughput RNA sequencing ("RNA-seq") is now available, but is associated with technical difficulties [49,50]. The main issue remains the preservation of cellular RNAs from further degradation by RNAses. The mRNAs can be separated from other RNAs by their characteristic polyA tails. Intronic RNA is particularly difficult to sequence.

(c) Regulation of Editing of RNA Transcripts

In humans, the main examples of RNA editing seem to be the deamination of adenine to inosine and cytosine to uracil. Adenosine-inosine editing may occur in as many as 1,000 human genes [51].

(d) Regulation of Splicing of RNA Transcripts

Alternative splicings of the same mRNA transcript can give rise to many different proteins. Thus the 20,000 or so known protein-coding genes give rise to many times that number of protein products. Clearly, splicing differences may underlie differences between species, as well as differences between cells. Mechanisms of "quality control" of RNA splicing so that the appropriate mRNAs are produced for the particular kind of cell are unclear, but appear to involve actions of sets of regulatory proteins [52].

(e) Regulation of Translation; RNAs in the Cytoplasm

Small RNAs (collectively "microRNAs") have been discovered which can affect gene expression by various mechanisms. These include:

i. inhibition of transcription (in the nucleus),
ii. inhibition of the enzyme specific for RNA synthesis (the ribozyme) in the cytoplasm,
iii. destruction of specific mRNAs in the cytoplasm so that translation of this mRNA does not occur [53].

Generally, these RNAs are up to 100 bases, but with 21–23 bases sequences which determine their specificity. Some are *cis*-acting, in the sense that they are located in the introns of the gene whose function they affect. The others are *trans*-acting, meaning that they are transcribed in one locus and act on another. Some microRNAs affect hundreds of genes. Their actions and significance are a large complex area (see, e.g., in Refs. [54–57]).

One specific point about microRNAs which might be made here is that alteration of a single nucleotide could significantly alter the effect of the molecule. Because the chain is so short, presumably the mutant microRNA could not only fail to inhibit the physiological target of its wild type, but might adversely affect another mRNA (and have a damaging effect)—that is, act as a "pathogenic neomorphic" microRNA (for terminology, see Section 4.1.1).

A2.2.6 RNAs as Direct Regulators of Transcription

This relatively new field relates to the possibility that transcribe RNA may remain in the nucleus, and bind to promoters of genes, preventing, at least temporarily, their being transcribed. It is particularly invoked that "long RNAs" may control normal differentiation of cells in embryonic development [58,59].

A2.2.7 Regulation of Posttranslational Modifications or Protein Products

Posttranslational changes mainly relate to establishment of the functional tertiary or quaternary structure of the peptide chains [60]. Mechanisms include deleting parts of the chain (as in forming insulin from pro-insulin), adding biochemical functional groups, for example, lipids and carbohydrates, and also creating permanent structural disulfide bonds between amino acids in individual peptide chains.

These mechanisms provide many possible ways in which a protein coding gene may be active, but its product may be inactive. These issues are important for interpreting "molecular pathological" studies, as is discussed in Section 9.6.

A2.3 PARTICULAR DISTURBANCES OF REGULATION OF GENE ACTIVITY WHICH ARE POTENTIALLY RELEVANT TO TUMORS

A2.3.1 Differential Chromosomal and Allelic Activities

This phenomenon has been mentioned in Section A2.1.2b. It is potentially of significance for tumor cell biology, because many of the chromosomes in especially hyperploid cancer cells may not be active.

A2.3.2 Position Effect and Reposition Effects

In strict sense, "position effect" refers specifically to experiments with *Drosophila* in which

a gene is inactivated when it is moved to supposed "heterochromatin." Few examples in humans are recorded [61]. The term "reposition effect" is used here for all losses of function or gains of function caused by repositioning of the exons and introns in relation to either their normal promoter or another promoter.

A2.3.3 Other Factors Which Potentially Affect the Production of Functional Protein Products from Genes

As already indicated in Section 9.6.6 (Figure A2.3), there are several steps in the sequence for production of fully functional proteins [62]:

i. activating regulators act on the promoter of a wild-type allele of the gene,
ii. the mRNA is spliced, and possibly edited, appropriately,
iii. the mRNA is permitted (by absence of silencing RNAs, see Section A2.2.6) to translate accurately for a peptide chain,
iv. various subsequent modifications to the structure or the peptide proceed correctly.

A2.3.4 RNAs in Serum ("Circulating"/"Extracellular" RNAs)

These are currently a major area of investigation, especially as diagnostic or "biomarker" (see Section 9.7) indicators for almost every category of disease including cancer [63–65]. The rationale is that each disease may be characterized by a different size-ratio, and/or nucleotide sequence in the dominant microRNAs. It must be pointed out however, that the leukocytes of the blood, as well as the cells lining the blood vessels (the endothelium and reticulo-endothelial cells) turnover, and hence when they die, add their RNAs to the blood stream, in which RNAses digest them. Moreover, the lymph of all the tissues of the body flows into the blood stream, so providing another potential source

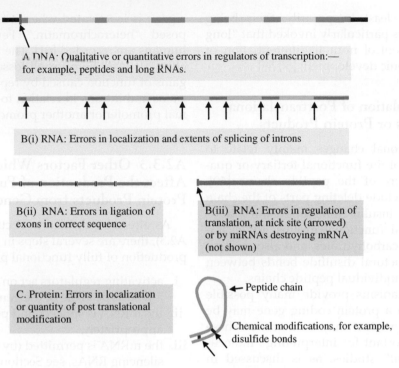

A. DNA: Qualitative or quantitative errors in regulators of transcription:— for example, peptides and long RNAs.

B(i) RNA: Errors in localization and extents of splicing of introns

B(ii) RNA: Errors in ligation of exons in correct sequence

B(iii) RNA: Errors in regulation of translation, at nick site (arrowed) or by miRNAs destroying mRNA (not shown)

C. Protein: Errors in localization or quantity of post translational modification

← Peptide chain

Chemical modifications, for example, disulfide bonds

FIGURE A2.3   Factors and phenomena potentially affecting expression of functional gene product.

of partly digested RNAs to enter the serum. The "background" against which are disease-specific microRNA is to be identified may therefore be substantial.

A2.4 ASPECTS OF ENZYMATIC PROCESSES RELATING TO THE FUNCTIONING OF THE GENOME

DNA itself has no enzymatic functions. All the numerous biochemical events that derive from DNA (replication and repairs to the genome, synthesis of RNA and associated phenomena such as unraveling) are carried out by the protein complexes that carry out processes on it. These proteins may be collectively called the "genomic-processes-associated

nucleoproteins," to distinguish them from the histones and nuclear matrix proteins.

A2.4.1 Structural Biology of the Sites of Molecular Interactions

To carry out their functions, the protein complexes for genomic processes must act at highly specific sites. It is now understood that proteins assemble into specific complexes for specific enzymatic processes [66]. The complexes either localize to, or assemble at, the appropriate site on DNA by noncovalent binding forces between the DNA-binding proteins of the complex and the topology features of the floors and walls of the greater and lesser grooves of the double helix of DNA [67–69] (Figure 4.8). Thus there are proteins in the complex which bind directly

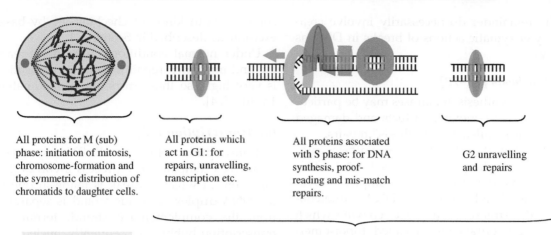

All proteins for M (sub) phase: initiation of mitosis, chromosome-formation and the symmetric distribution of chromatids to daughter cells.

All proteins which act in G1: for repairs, unravelling, transcription etc.

All proteins associated with S phase: for DNA synthesis, proof-reading and mis-match repairs.

G2 unravelling and repairs

Single and double strand breaking do not occur in mitosis (M phase).

All these steps can involve breaking of single or double DNA strands.
Transcription involves single strand breaking.
DNA unravelling (by topoisomerases) involves single strand breaking (Topo-1) and double strand breaking (Topo 2).

FIGURE A2.4 Proteins which carry out the genomic processes.

to DNA and others which bind only to other proteins. So that the complexes do not bind at inappropriate locations (e.g., a DNA polymerase complex binding to a promoter region for transcription), it is believed that when all the proteins "assemble," they affect each other's surface topologies so that the complex when completely assembled can only bind to the appropriate site in the genome (Figure A2.4). The accuracy of site recognition is enhanced by the specific variations in DNA topology (deriving from degrees of bending, twisting, etc. of the strands) which the proteins both induce and can bind to. For an account of allosteric mechanisms in DNA-associated protein complexes and for the complexity of this "pliability-related" DNA topology see Ref. [67].

An extension of these concepts is that chemicals which do not react covalently with proteins or DNA, such as intercalations and "interpositions" (see Section 4.4.1), could affect the functions of these complexes by interfering with "stereological fits." Further, covalent chemical alterations in these proteins themselves (as are induced by radiations or chemicals) could affect the noncovalent binding sites, with similar potential functional disturbances in the protein complexes.

The structures of individual proteins can be established by methods including X-ray crystallography as well as nuclear magnetic resonance spectroscopy, and the probable arrangements of the component proteins in the complex may then be inferred [70].

A2.4.2 The Main Genomic Processes Involve Enzyme-Induced Breaking of DNA Strands

The genomic processes which do not involve breaking DNA strands are those of removing adducts from DNA, and those which replace damaged bases with fresh ones—"base excision repair"—see Section A2.1.4.

The remainder do necessarily involve creation by enzymatic actions of breaks in DNA as follows.

(a) Synthesis [71–73]

Prior to synthesis, repairases may be particularly active in removing adducts and damaged bases. This is called "presynthesis" repairs.

After that step, the next presynthesis process is unraveling of supercoiled DNA. This because, as the polymerase complex moves along the double strand of DNA, supercoiling of the strands occurs. The enzyme which relieves the unraveling is called topoisomerase I. This enzyme breaks single strands of the DNA duplex, allowing the supercoiled duplex to revolve around the other single stand, and hence relieve the supercoil.

The last presynthesis process involves separating the strands by an enzyme called helicase, which does not break the DNA strands.

Synthesis proper is carried out by polymerases of various classes [69] which become attached to the single strands and incorporate nucleotides in a $5'\rightarrow3'$ direction. All eukaryotic DNA polymerases synthesize in this direction. Because the complementary strands of DNA lie in opposite directions, some differences must exist between syntheses of the strand which is in the appropriate direction (the "leading strand") in comparison with the strand which lies in the opposite direction (the "lagging" strand). Lagging strand synthesis is thought to occur via a complex process of synthesis of short pieces ("Okazaki fragments") followed by controlled inversion of these and ligation to the new strand [72].

Nucleotides which are mis-paired by the synthetic site of the DNA polymerase are corrected in two postsynthesis checking mechanisms. The first is "proof-reading," which occurs in the same protein complex as the synthesis site. The other is "mismatch repair," which is carried out by another protein complex. Both kinds of checking are by base-excision, as described in Section 5.1.

Under normal conditions, with proof-reading and mismatch repair, fidelity of replication is very high, so that errors are approximately 1×10^{-9} [74].

(b) Transcription [75]

Transcription refers to all RNA synthesis from DNA template. The process involves first initiation, in which a break is made in one of the DNA duplex, and this strand is separated from the complementary strand, forming a transcription bubble. Subsequently, nucleotides are added to the new chain ("elongation"). Termination is induced by a "stop codon," at which the complex dissociates from the DNA, and the break is ligated (rejoined).

Transcription is not followed by any proof-reading or mismatch repairs to the RNA and is approximately as accurate for RNA synthesis as is DNA polymerase for nucleotide incorporation in DNA synthesis (1 error/10^4 nucleotides).

(c) Unraveling not Related to Synthesis (Chapter 2, in Ref. 54)

"Knots" of DNA duplexes are unraveled by topoisomerase II. During the period of time that one duplex is broken another duplex is passed though the gap. After that, the broken ends of the first duplex are relegated.

(d) Repairs

These are discussed in Sections 1.2.2 and A2.1.4.

A2.4.3 The Support Functions Involved in Actions of Genome-Processing Complexes

In addition to accurate localization (Section A2.4.1) (see also Figure A2.5), efficiency of the particular enzymatic actions described in the

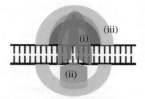
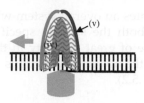

A	B1	B2
All specific processes on DNA probably begin with a locator operation (e. g., for an origin of replication site).	For an immobile process, such as repair of a nucleotide: (i) A specific enzyme is required, (ii) A DNA-topology-modulating protein may be required for the action of the enzymatic site. (iii) For processes involving breaking strands of DNA, some "accessory" or "tether" proteins may be required to hold the broken ends in place until a ligation step re-joins the broken ends.	For a mobile process, such as DNA synthesis and transcription, fixed tether molecules may not be involved. However different necessary functions may include (iv) A molecular motor function to move the complex in one direction, one nucleotide at a time. (v) To support directional movement on the nucleotide chain, an "on-off" binding function by part(s) of the complex to DNA in coordination with nucleotide incorporation.

FIGURE A2.5   Functional components of complexes of proteins for enzymatic actions relevant to DNA.

previous subsection requires the complexes of proteins to have additional functions as follows.

(a) "Tether" Function During the Particular Process

As mentioned in the previous subsection, many enzyme complexes which act on DNA break the strands. The strand breaks can be either single (e.g., for transcription) or both strands of the duplex (e.g., topoisomerase II). Accessory proteins are necessary to hold the broken ends of the DNA strands in place while the break is in existence (see Section 5.3.3).

(b) "Motor Functions" for Polymerase Complexes

The functions of replication of DNA and transcription of RNA involve movement of the relevant complex as a stable structure in one direction only along the DNA chain. The issue of the chemical nature of these particular "molecular motors" is not fully understood [76].

(c) Protein Structures for Integrity of the Complex

The proteins of the some complexes probably bind to each other with sufficient strength to maintain the integrity of the complex. However, other complexes may have proteins which only act as binding between the proteins with the various functions described here.

A2.4.4 Significance of the Complexities of Genome-Process-Associated Proteins to Aspects of Tumors

This complexity in phenomena of the protein complexes has significance for many aspects of tumors. The complexities of the protein assemblies, and the number and susceptibilities of the noncovalent bindings of various kinds which bind them together provide many potential "targets" on which noxins—mutagens, clastogens, carcinogens, and anticancer drugs alike—might act. Such an enormous number of possible chemical targets in these polymerases clearly indicates an overall system which could account for both the striking specificities and the multitude of great diversity in the biological effects of the noxins (see Sections 4.2.2, 4.4.2, 11.3.2, and 11.5.2).

References

[1] Alberts B, Johnson A, Lewis J, Raff M, Roberts K, Walter P. Molecular biology of the cell (5th ed.). New York, NY: Garland Science; 2008.

[2] Watson JD, Baker TA, Bell SP, et al. Molecular biology of the gene (7th ed.). Cold Spring Harbor, NY: CSHL Press; 2014.

[3] Strahan T, Read A. Human molecular genetics (4th ed.). New York, NY: Garland Science; 2011.

[4] Krebs JE, Goldstein ES, Kilpatrick ST. Lewin's genes (11th ed.). Burlington, MA: Jones & Bartlett Learning; 2014.

[5] Lodish H, Berk A, Kaiser CA, et al. Molecular cell biology (7th ed.). New York, NY: WH Freeman; 2013.

[6] Burns KH, Boeke JD. Human transposon tectonics. Cell 2012;149(4):740–52.

[7] Katoh I, Kurata SI. Association of endogenous retroviruses and long terminal repeats with human disorders. Front Oncol 2013;3:234.

[8] Gifford WD, Pfaff SL, Macfarlan TS. Transposable elements as genetic regulatory substrates in early development. Trends Cell Biol 2013;23(5):218–26.

[9] In: Ref. [3]. p. 408.

[10] In: Ref. [1]. pp. 435–9.

[11] In: Ref. [1]. pp. 205–6.

[12] Lee JT, Bartolomei MS. X-inactivation, imprinting, and long noncoding RNAs in health and disease. Cell 2013;152(6):1308–23.

[13] In: Ref. [3]. p. 664.

[14] Mallo M, Alonso CR. The regulation of Hox gene expression during animal development. Development 2013;140(19):3951–63.

[15] van Heemst D, den Reijer PM, Westendorp RG. Ageing or cancer: a review on the role of caretakers and gatekeepers. Eur J Cancer 2007;43(15):2144–52.

[16] Michor F, Iwasa Y, Nowak MA. Dynamics of cancer progression. Nat Rev Cancer 2004;4(3):197–205.

[17] Fulda S, Vucic D. Targeting IAP proteins for therapeutic intervention in cancer. Nat Rev Drug Discov 2012; 11(2):109–24.

[18] In: Ref. [3]. pp. 405–40.

[19] Wyandt HE, Tonk VS. Human chromosome variation: heteromorphism and polymorphism. New York, NY: Springer; 2012.

[20] Abel HJ, Duncavage EJ. Detection of structural DNA variation from next generation sequencing data: a review of informatic approaches. Cancer Genet 2013; 206(12):432–40.

[21] Human Genome Project Information Archive. Available at: http://www.ornl.gov/sci/techresources/Human_Genome.

[22] Human Genome Variation Society. Recommendations for the description of sequence variants. Available at: http://www.hgvs.org/mutnomen/recs.html.

[23] Gupta RC, Lutz WK. Background DNA damage for endogenous and unavoidable exogenous carcinogens: a basis for spontaneous cancer incidence? Mutat Res 1999;424(1-2):1–8.

[24] Friedberg EC, Walker GC, Siede W, editors. DNA repair and mutagenesis (2nd ed.). Washington, DC: ASM Press; 2006.

[25] Nabel CS, Manning SA, Kohli RM. The curious chemical biology of cytosine: deamination, methylation, and oxidation as modulators of genomic potential. ACS Chem Biol 2012;7(1):20–30.

[26] Ponting CP, Nellåker C, Meader S. Rapid turnover of functional sequence in human and other genomes. Annu Rev Genomics Hum Genet 2011;12:275–99.

[27] Kohne E. Hemoglobinopathies: clinical manifestations, diagnosis, and treatment. Dtsch Arztebl Int 2011;108(31-32):532–40.

[28] Krejci P. The paradox of FGFR3 signaling in skeletal dysplasia: why chondrocytes growth arrest while other cells over proliferate. Mutat Res Rev Mutat Res 2014;759:40–8.

[29] In: Ref. [2]. pp. 615–730.

[30] Latchman DS. Eukaryotic transcription factors (5th ed.). Burlington, MA: Academic Press/Elsevier Science; 2007.

[31] Li YQ. Master stem cell transcription factors and signaling regulation. Cell Reprogram 2010;12(1):3–13.

[32] West AG, Gaszner M, Felsenfeld G. Insulators: many functions, many mechanisms. Genes Dev 2002;16:271–88.

[33] Matkar S, Thiel A, Hua X. Menin: a scaffold protein that controls gene expression and cell signaling. Trends Biochem Sci 2013;38(8):394–402.

[34] Boulikas T. Chromatin domains and prediction of MAR sequences. Int Rev Cytol 1995;162A:279–388.

[35] Wilson RH, Coverley D. Relationship between DNA replication and the nuclear matrix. Genes Cells 2013; 18(1):17–31.

[36] Jost KL, Bertulat B, Cardoso MC. Heterochromatin and gene positioning: inside, outside, any side? Chromosoma 2012;121(6):555–63.

[37] Brázda V, Laister RC, Jagelská EB, et al. Cruciform structures are a common DNA feature important for

regulating biological processes. BMC Molec Biol 2011; 12:33.

[38] Stanford Encyclopedia of Philosophy. Epigenesis and preformationism. Available at: http://plato.stanford.edu/entries/epigenesis/.

[39] Holliday R. Epigenetics: a historical overview. Epigenetics 2006;1(2):76–80.

[40] Felsenfeld G. A brief history of epigenetics. Cold Spring Harb Perspect Biol 2014;6(1).

[41] Appasani K, editor. Epigenomics. Cambridge, UK: Cambridge University Press; 2012.

[42] Armstrong L. Epigenetics. New York, NY: Garland Science; 2013.

[43] Hallgrímsson B, Hall BK, editors. Epigenetics: linking genotype and phenotype in development and evolution. Oakland, CA: University of California Press; 2011.

[44] Roadmap Epigenomics Project. Overview of the Roadmap Epigenomics Project. Available at: http://www.roadmapepigenomics.org/overview.

[45] Defoiche J, Zhang Y, Lagneaux L, et al. Measurement of ribosomal RNA turnover *in vivo* by use of deuterium-labeled glucose. Clin Chem 2009;55(10):1824–33.

[46] Mathews CK. Deoxyribonucleotides as genetic and metabolic regulators. FASEB J 2014;28:3832–40.

[47] Darby IA, Bisucci T, Desmoulière A, et al. *In situ* hybridization using cRNA probes: isotopic and nonisotopic detection methods. Methods Mol Biol 2006;326:17–31.

[48] Mahmood R, Mason I. *In-situ* hybridization of radioactive riboprobes to RNA in tissue sections. Methods Mol Biol 2008;461:675–86.

[49] Wang Z, Gerstein M, Snyder M. RNA-Seq: a revolutionary tool for transcriptomics. Nat Rev Genet 2009;10(1):57–63.

[50] Mutz KO, Heilkenbrinker A, Lönne M, et al. Transcriptome analysis using next-generation sequencing. Curr Opin Biotechnol 2013;24(1):22–30.

[51] In: Ref. [1]. pp. 483–5.

[52] In: Ref. [1]. pp. 479–80.

[53] In: Ref. [4]. pp. 872–88.

[54] Appasani K, editor. MicroRNAs. Cambridge, UK: Cambridge University Press; 2008.

[55] Laurie CH, editor. MicroRNAs in medicine. Hoboken, NJ: John Wiley & Sons; 2014.

[56] Sahu SC, editor. MicroRNAs in toxicology and medicine. Chichester, UK: Wiley & Sons; 2013.

[57] Babashah S, editor. MicroRNAs: key regulators of oncogenesis. New York, NY: Springer; 2014.

[58] Kornfeld JW, Brüning JC. Regulation of metabolism by long, non-coding RNAs. Front Genet 2014;25(5):57.

[59] Johnsson P, Lipovich L, Grandér D, et al. Evolutionary conservation of long non-coding RNAs; sequence, structure, function. Biochim Biophys Acta 2014; 1840(3):1063–71.

[60] In: Ref. [1]. pp. 387–99.

[61] In: Ref. [3]. p. 381.

[62] In: Ref. [2]. pp. 423–592.

[63] Sethi S, Ali S, Kong D, et al. Clinical implication of microRNAs in molecular pathology. Clin Lab Med 2013;33(4):773–86.

[64] Anindo MI, Yaqinuddin A. Insights into the potential use of microRNAs as biomarker in cancer. Int J Surg 2012;10(9):443–9.

[65] Etheridge A, Lee I, Hood L, et al. Extracellular microRNA: a new source of biomarkers. Mutat Res 2011;717(1-2):85–90.

[66] In: Ref. [1]. pp. 184–90.

[67] Crothers DM. Sequence-dependent properties of DNA and their role in function. In: Caporale LH, editor. The implicit genome. New York, NY: Oxford University Press; 2006. pp. 3–22.

[68] Kouzine F, Levens D, Baranello L. DNA topology and transcription. Nucleus 2014;5(3):195–202.

[69] Cavalli G, Misteli T. Functional implications of genome topology. Nat Struct Mol Biol 2013;20(3):290–9.

[70] Stratmann SA, van Oijen AM. DNA replication at the single-molecule level. Chem Soc Rev 2014; 43(4):1201–20.

[71] DiPamphilis ML, editor. DNA replication and human disease. Cold Spring Harbor, New York, NY: CHSL Press; 2008.

[72] In: Ref. [2]. pp. 257–312.

[73] In: Ref. [4]. pp. 304–27.

[74] In: Ref. [2]. p. 267.

[75] In: Ref. [2]. pp. 423–64.

[76] Hamdan SM, Richardson CC. Motors, switches, and contacts in the replisome. Annu Rev Biochem 2009;78:205–43.

"Pre-Target," "Target," and "Post-Target" Factors in the Defense of Cells Against Carcinogens and Cytotoxic Agents

OUTLINE

This appendix concerns factors in the resistance of normal cells against carcinogens as well as of tumor cell populations against drugs capable of killing them. In this, it may be remembered that the general environment of most organisms contain many potentially harmful factors which may pass into the skin or otherwise enter the body through the mouth, lungs, or other portals. However, most of these noxious agents do not cause any detectable disease.

There are three groups of defensive factors involved in preserving the tissues and cells of the body from carcinogenic or toxic factors.

These considerations mainly apply to noxious chemicals. In relation to radiations, the only relevant pre-target defenses relate to the availability of endogenous radical-scavenging chemicals (see Sections 4.3 and 11.4). On the other hand, the "target" and "post-target" defenses are probably as relevant to radiation damage as they are to chemical injuries. It may be noted here that radiation-induced damage to cells and mechanisms of the recoveries of cells are not well understood (Sections 4.3 and 12.1.8).

A3.1 "PRE-TARGET" DEFENSES

These are defenses inhibiting access of noxious agents to relevant targets in cells. They are relevant to chemical carcinogens and anticancer drugs. These factors are less relevant to radiations, because these can penetrate most human tissues (see above).

A3.1.1 Barriers and Defenses at the Whole-Body Level

Most discussions of pre-target resistance factors in texts are in terms of whole-body studies. These are often described as "ADME" phenomena: *Absorption, Distribution, Metabolism, Excretion* [1–5]. Broadly, the

resistance mechanisms are considered as "barriers" as follows:

a. *At the portal of entry of the agent*

In this, an obvious example can be mentioned. Fair-skinned people are more susceptible to ultraviolet-induced tumors of the skin than dark-skinned people through absorbing less of the incident radiations in their skin. In this, the lesser amounts of melanin in the skin in fair people increase the access of the etiological agent to the relevant cells.

Among chemicals, if the agent is taken by mouth, it may be absorbed unaltered. Alternatively, some carcinogens might be destroyed by gastric acidity, or they might be retained in the gastrointestinal tract because they bind to vegetable material [6] or retained in the intestinal mucosa [7].

b. *In the circulation and other organs*

In the bloodstream, chemicals may bind to plasma proteins, such as albumin, and may not be available to access the cells (i.e., are present, but not "bioavailable"). Alternatively, carcinogens and drugs may bind to proteins in the blood and hence be "bio-unavailable" in the tissues [5].

c. *In the interstitial spaces of the target tissues*

In different kinds of tissues, there are differences in vascularity; in properties of vascular basement membrane (thickness and composition); in the permeability of vessel walls and the presence of pericapillary cells ("pericytes") (e.g., the "blood–brain barrier" [8]); and the interstitial structures (collagen, other fibrils, glycosaminoglycans, etc.).

A3.1.2 Barriers and Defenses of the Cell and of the Genome Compartment

It has often been assumed that if a substance enters a tissue, then all the cells present—and all their contents—are more or less equally exposed to it (Figure A3.1). However, two

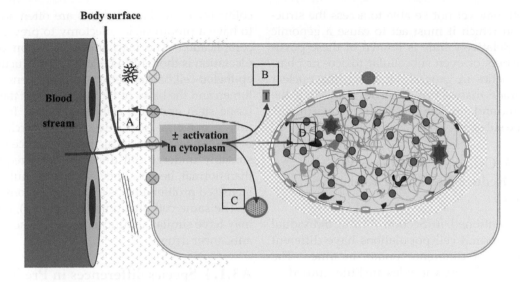

After entering the interstitial space (from the blood or a body surface), a carcinogen or anti-cancer drug may enter cells by passive or active transport. Subsequently, the chemical may or may not be activated in the cytoplasm. After that the agent
A, may be exported from the cell by exporter mechanisms in the cell membrane.
B, may act on a cytoplasmic target.
C, may be neutralized in the cytoplasm by sequestration and/or lysosomal digestion.
D, may enter the nucleus. Here it may act on DNA, on the proteins of genomic processes, or be sequestered by nucleoproteins. Little is known of these last possible phenomena.

FIGURE A3.1 Resistance factors in relation to the access of exogenous or secondary carcinogens, drugs, or other genopathic chemicals to the genome-processing proteins.

additional levels of possible barriers need to be addressed because the target of carcinogens and some anticancer drugs may be located in the genome/genome-processing compartment (Section A1.2.4).

(a) At the Cell Membrane or in the Cytoplasm

The cell membrane and cytoplasm contain various "transport" mechanisms for uptake and for extrusion, as well as for degradation of exogenous agents [1,9] (see Section 11.4.2).

Principal among these are:

i. Transporter proteins such as P-glycoprotein [10].

ii. Enzymes which can deactivate active carcinogens and active drugs (see Sections 4.4.1, 11.4.2, and 12.2.2).

iii. Mechanisms for "trafficking" across the cell membrane and destruction in phagolysosomes (see Section 10.1.4).

In certain specialized cells, intracytoplasmic substances such as mucus or keratin in glandular and keratinizing squamous epithelium, respectively, may possibly play roles as sequestrating agents.

(b) (Potentially) at the Nuclear Membrane or in the Nucleus

These possible levels of barriers are potentially important because factors which enter the cell may yet not be able to access the structures on which it must act to cause a genomic event. Relatively little is known of these possible barriers or even subcellular toxico- or pharmacokinetics in general. However, the nuclear membrane may act as a barrier protecting the "genome and genome-processing compartment" to some substances (see also Section A3.1.4).

A3.1.3 Cellular Phenomena Which Might Reduce Pre-Target Resistance Factors

As mentioned in Section A1.2.6, individual cells in normal cell populations have different properties at different "points in time." The influences of these variables are little studied.

(a) The Events of Cell Division

This may be particularly relevant to mammalian cells in which the nuclear membrane dissolves just before the phase of DNA synthesis (S (synthetic) phase) in nuclear division (see Section A1.2.6). This is because agents—which at other times might be excluded from the interior of the nucleus—may then gain easier access to the chromatin in S phase, and more specifically, to the DNA polymerases, and the uncoiled, single stranded DNA which would be relatively less invested with nucleoproteins (see also Section 11.4.3).

(b) Degree of Specialization

Because potentially any protein in the cell may function as a sequestrator of a toxin, changes in concentrations of proteins according to phase of specialization may potentially cause cells to have greater or lesser cytoplasmic defenses against chemical or other carcinogenic agents (Chapter 4).

(c) Concurrent Pathological Processes

It is well established that the condition of chronic ulcerative colitis predisposes to carcinoma of the colon. In fact, patients with severe colitis for more than 10 years are often advised to have a prophylactic colectomy to prevent this complication. The mechanism of action of the ulceration is thought to be loss of the mucus and epithelial cell barrier between carcinogens in the lumen and the bases of the crypts where the local tissue stem epithelial cells are located. In addition to potential excess exposure to carcinogens in the bowel, these local tissue stem cells presumably undergo cell division more frequently than normal, because they are also stimulated to increased proliferation as a healing response.

The same cell phenomena in (a), (b), and (c) may have similar significances for the actions of anticancer drugs.

A3.1.4 Species differences in Pre-Target Resistance Factors

Differences between species in the relative efficiencies of these mechanisms as possible explanations of species differences in resistance to noxins have been investigated [11]. They are of great importance to assessments of laboratory data relating to the potential potency of chemicals as carcinogens for humans (see Sections 14.2.2 and A3.4). Various models for assisting these assessments have been published [12], including some which adjust data with "scaling factors" for different species [13].

A3.2 "TARGET" DEFENSES

Once a noxious agent has reached the intracellular compartment containing the target, factors relating to the target itself, especially the quantities and qualities of the target itself, may play roles. These factors are relevant to both radiations and chemical carcinogens and anticancer drugs.

A3.2.1 Relative Quantity of Target

This concept applies to situations in which the amount of target present exceeds the maximal dose of noxious agent. For example, an agent may act against a cellular enzyme. If the doses of agent are limited, so that the cell possesses more of the enzyme than the available amount of agent can act on, then the agent may have little effect although qualitatively the target has been reached.

In cytotoxicity, the scheme applies in an "all or nothing" way. Thus agent-induced cell death only occurs if the agent affects enough of the target molecules—say respiratory enzymes—to abolish the functions of the target (for definitions and concepts of cell death, see Section 10.1.1). Clearly if the amount of toxin exceeds the amount of target, the toxin may kill the cell. If, however, there is an excess of target molecules over and above the maximum dose of cytotoxic agent, then this "reserve" may well allow the cell to survive.

In all nondividing cells of the individual, it is assumed that the amount of DNA is the same. In G2, however, the cell is tetraploid. As another consideration, different kinds of cells may have different quantities of possible non-DNA targets of carcinogens and drugs.

A3.2.2 Qualitative Differences in Targets

It is possible that the target of any given noxious agent is different in different species, and that some variants are liable to sustain dysfunction as an effect of the agent, while others are not. This has been little investigated in relation to both carcinogenesis and anticancer drugs.

A3.2.3 Rates of Turnover of Target

The discussions in the previous sections assume that the actual molecules of target are long-lived in the cell. However, carcinogenesis may depend on accumulated damage to a cell (Sections 4.1.2 and 4.3.3). If a target turns over in a cell, and the carcinogen is removed by the same mechanism as removes the target (e.g., by the ubiquitin system [14]), then various possibilities

exist for how these might fail to cause enough damage to cause tumor formation in a cell.

i. The target is short-lived in the cells, and so only very rarely do damaged target molecules persist long enough to cause tumors.
ii. The target is long-existing in the cell but is resistant to fewer than trillions of hits.
iii. The target is long-existing, but is repaired so efficiently that the unrepaired hits accumulate too slowly over time.
iv. The agent kills the cells or renders them incapable of any proliferation (causing atrophy, see Section 10.7) and preventing any formation of tumor from the cell (see also Section 4.3.3).

A3.3 "POST-TARGET" DEFENSES

The capacity of most kinds of cells in the body—and to greater or lesser extents tumor cells—to recover from sublethal doses of cell-damaging agent, is discussed in Section A1.4. For these, the term "post-target" defense is used because these factors operate only after the noxin has had its effect on the target in a cell.

This section applies mainly to anticancer agents. While carcinogens may injure cells, little is known of whether or not the regenerative capacities of cells are affected by this—see Section 4.2.1.

A3.3.1 Capacities for Repairs Within Cells

In principle in relation to toxic injuries, all cells are capable of regenerating cytoplasmic enzymes after mild injury (see Sections 10.1, 10.7, and A1.4). Particular attention has been given to this in experimental studies of cytotoxic agents. Because the regeneration of these constituents is dependent on the synthesis of RNA and proteins, the rate of regeneration may be seen to depend on:

i. The degree of original loss of "target."
ii. The degree of damage to existing enzymes for synthesis of RNA and proteins.

iii. The degree of concurrent damage requiring repair to the DNA of the genes for these enzymes.

iv. (Possibly) the degree of genomic alteration produced in these genes, resulting in hypofunctional enzyme products.

v. Adaptive changes may occur. This means that for periods of time after exposure to an agent, cells may maintain production of larger than normal amounts of defensive substances.

vi. It is possible that the repair mechanisms might themselves be damaged by anticancer agents.

In all these qualities, tumor cells may be relatively deficient, but there is little experimental evidence which allows factors in pre-target, target, and post-target resistance to be distinguished and separately quantitated. This issue can be complicated in practice, because regenerative and adaptational mechanisms themselves may also be targets of anticancer agents [15].

A3.3.2 Capacities for Regenerations with New Cells to Replace Losses

The differences in the capacities of different kinds of cells to regenerate themselves after loss, for example, of liver tissue after partial hepatectomy were noted in Section A1.4.3. These capacities for regeneration are relevant to anticancer agents. For example, the production of hematopoietic cells is commonly depressed by these agents but recovers with cessation of treatment. Ways of increasing regenerative capacities have been suggested to include vitamins, oxygen, and other agents [16].

A3.4 POSSIBLE ROLES OF DEFENSIVE FACTORS IN FALSE-NEGATIVE AND FALSE-POSITIVE RESULTS OF ANIMAL TESTS FOR CARCINOGENS AND ANTICANCER DRUGS

False-negative and false-positive results in tests for carcinogenesis or for cytotoxicity may be obtained in tests involving nonhuman animals, as well as human and animal cells cultured *in vivo*.

Each negative result in a test for carcinogenic or cytotoxic effect may occur because:

i. There is no activator of the pro-agent in the particular model system.

ii. The particular agent does not access its "target" in the cell in sufficient quantities due to efficient defensive factors.

iii. The structures and functions of the "target" in the particular species or tissue of the model are not sensitive to the agent.

iv. The cells of the model do not possess the relevant target (it may possess other enzymes/mechanisms for the relevant function).

v. The agent is toxic for (i.e., kills) the cells in the model, but does not kill the relevant cells in the susceptible species.

vi. The carcinogen is cytostatic for (i.e., prevents proliferation in) the cells in the model, but is not cytostatic for the relevant cells in the susceptible species.

False-positive results can occur when an agent is found to have a genopathic effect in an experimental model, but no such effect occurs in living humans. More generally, the potential causes of false-positives are the reverse of the factors which may cause false-negative results (see above).

These factors may also affect biochemical studies using purified enzymes, substrates, and other factors.

Also, these factors may provide reasons for results indicating that some analogues in a given chemical class have biological effects, but other analogues do not.

Illustrative examples include the following.

In relation to polyaromatic hydrocarbons causing tumors in the skin of rabbits but not dogs (see Section 3.2.5), mechanisms could include differences in metabolism of the carcinogen by different bacterial flora on the skin of two species or of the cells of the particular species. Alternatively, the skin of dogs may prevent absorption due to being thicker, or the sebum of dog skin may bind the carcinogens so that the carcinogens never reach the basal layer of the epidermis of the skin.

In relation to asbestos exposure in man, the causative agent is in the lung tissues and gives rise to tumors of the pleural cells (hyaline plaques and mesotheliomas) but not tumors of the fibroblasts, endothelium, or occasional adipocytes which are present between the alveolar lumen and the pleural surface (see Section 4.2.4). Reasons for the susceptibility of one kind of cell over another may be factors allowing access of the agent, or secondary substance, to the genome/genome compartment of the former cells and not in the latter kinds of cells.

For systemically administered carcinogens such as the D-limonene-induced kidney tumors in male rats (see Section 4.2.3), the differences could depend any of the factors mentioned in above.

In relation to potential anticancer drugs, the differences in resistance of different tumor types (Section 11.3.2) may be due to differences in defensive factors concerning the tumor cells.

In general, these must remain speculative mechanisms, because studies of many of the comparative tests in different species have not been reported.

References

[1] (Various authors). Biochemistry and molecular genetics of drug metabolism. In: Anzenbacher P, Zanger UM, editors. Metabolism of drugs and other xenobiotics. Hoboken, NJ: Wiley-VCH; 2012. Part One. p. 1–284.

[2] Rudek MA, Chau CH, Figg WD, editors. Handbook of anticancer pharmacokinetics and pharmacodynamics (2nd ed.). New York, NY: Springer; 2014.

[3] Kitchen KT, editor. Carcinogenicity: testing: predicting, and interpreting chemical effects. New York, NY: Marcel Dekker; 1999.

[4] Waalkes MP, Ward JM, editors. Carcinogesis (Target organ toxicology series). Boca Raton, FL: CRC Press; 1994.

[5] Neidle S, editor. Cancer drug design and discovery (2nd ed.). Amsterdam: Academic Press, Elsevier; 2014.

[6] Smith-Barbaro P, Hanson D, Reddy BS. Carcinogen binding to various types of dietary fiber. J Natl Cancer Inst 1981;67(2):495–7.

[7] Jibodh RA, Lagas JS, Nuijen B, et al. Taxanes: old drugs, new oral formulations. Eur J Pharmacol 2013;717(1–3):40–6.

[8] Orthmann A, Fichtner I, Zeisig R. Improving the transport of chemotherapeutic drugs across the blood–brain barrier. Expert Rev Clin Pharmacol 2011;4(4):477–90.

[9] Smith DA, Allerton C, Kubinyi H, editors. Pharmacokinetics and metabolism in drug design. Weinheim: Wiley-VCH; 2012.

[10] Glaeser H. Importance of P-glycoprotein for drug–drug interactions. Handb Exp Pharmacol 2011;201:285–97.

[11] Nyman AM, Schirmer K, Ashauer R. Importance of toxicokinetics for interspecies variation in sensitivity to chemicals. Environ Sci Technol 2014;48(10):5946–54.

[12] Dorne JL. Metabolism, variability and risk assessment. Toxicology 2010;268(3):156–64.

[13] Andersen ME. Toxicokinetic modeling and its applications in chemical risk assessment. Toxicol Lett 2003;138(1–2):9–27.

[14] Nath D, Shadan S. The ubiquitin system. Nature 2009;458:421–67.

[15] Cox Jr. LA. A model of cytotoxic dose–response nonlinearities arising from adaptive cell inventory management in tissues. Dose Response 2006;3(4):491–507.

[16] Carlson BM. Principles of regenerative biology. Philadelphia, PA: Elsevier; 2007.

Index

Note: Page numbers followed by "*f*" and "*t*" refer to figures and tables, respectively.

N

Necrobiosis, 273
Necrosis, 412
  electron microscopic appearances
    and biochemical changes, 276
  macroscopic and microscopic
    features, 274–276
  slow death of cells, 276
Necrotic cells, 274
Neo-antigens, 251
Neomorphism, 421
Neoplasia, 38
Neoplasm. See Tumor(s)
Nervous system, 23, 394
  cells, 397
Neurofibromatosis 1 (NF1), 193
Neurofibromatosis 2 (NF2), 191
NF1 gene, 170–171
NF1. See Neurofibromatosis 1 (NF1)
NF2 gene, 170–171
NF2. See Neurofibromatosis 2 (NF2)
Non-genotoxic carcinogen, 92
Non-Hodgkin's lymphoma, 84
  immunostaining and molecular
    pathological techniques, 84
  mixtures of subtypes/variants, 84f
  pure forms, 84
  T-cell/histiocyte-rich large B-cell
    lymphoma, 84–85
Non-Hodgkin's lymphomas, 5
Nongenopathic effects, 91–92, 266
Nonmorphological ectopic
  specializations, 215
Nonstaining nuclear substance,
  403–404
Nonsurgical anticancer therapies, side
  effects of, 28
Nonsurgical treatments, general
  aspects of, 26
Nontumor circumstances, nonimmune
  cells in, 158
Nontumorous human pathological
  conditions, apoptosis in, 279
Nontumorous lesions, 193
  distinction with tumors, 3
  in human hereditary-predisposition
    syndromes, 190
Normal cells
  active movements of individual
    normal cells, 414–415
  body cells from tumors arise, 398f
  categories, 397
  cell division, 409

    check points, 409–410
    initiation through cyclins, 409
  cell function and structural details,
    399
  cell/plasma membrane, 399
  cytoplasmic and functional
    variabilities, 399–400
  cytostructural regularity, 397–399
  growth
    changes with age of individual,
      406
    contact inhibition, 406
    overall controls, 405–406
    phasic aspect of cell production,
      406
  invasions, 412–413, 413f
  life cycles, 407–408
    division period, 408
    interdivision period, 408
  metastases, 412–413, 413f
  nuclear division, 408–409
  nucleus
    appearances in different
      histological stains, 402f
    cell cycle parts, 403f
    chromatin, 404
    chromosomes, 404
    components, 401f
    nonstaining nuclear substance,
      403–404
    nuclear membrane and
      perinuclear cytoplasm, 401–403
    nucleolus, 404
    size, 400–401
  physiological variabilities, 405
  population chemotaxis, 414f
  relocalization by localized growth,
    413–414
  stem cells
    in embryology, 406–407
    in histology, 407
  in studies of tumor cell populations,
    407
  susceptibilities and responses to
    injuries
    cell production after chemical
      damage, 411–412
    cell production after tissue loss,
      411
    metabolic susceptibilities and
      defenses, 410–411
    metaplastic responses, 412
  theories of induction of motility, 415

    variability in activation status,
      404–405
Normal genome, 418–419
  damage to genomes, 424–425
  damage to nucleotides and repairs,
    422–423
  differences between individual
    humans genomes, 422
  disturbances of gene activity
    regulation
    differential chromosomal and
      allelic activities, 429
    factors affecting production of
      functional protein products, 429
    position effect, 429
    reposition effects, 429
    RNAs in serum, 429–430
  DNA components in, 419f
  enzymatic processes, 430
    complexities of genome-process-
      associated proteins, 433–434
    enzyme-induced breaking of
      DNA strands, 431–432
    structural biology of molecular
      interactions sites, 430–431
    supporting functions, 432–433
  genes, 420
    dominance, recessivity, and
      functional morphisms,
      420–421
    functional classifications, 421–422
  proteins, 431f
  repairs to DNA, 423f
  transcribed and not translated DNA,
    419
  transcribed and translated DNA,
    419
  untranscribed DNA, 419–420
Normal tissues
  of body, 396
  growth
    changes with age of individual,
      406
    contact inhibition, 406
    overall controls, 405–406
    physiological and nontumorous
      variabilities, 396–397
Normo-specialization in tumors, 212
Notch protein, 163
Noxins tissue effects, 281–283, 282f
Nuclear abnormalities, 218
  Ag-NORs, 221
  chromatin patterns, 220

Printed and bound by CPI Group (UK) Ltd, Croydon, CR0 4YY

03/10/2024

01040325-0005